Mini-Projects

An Introduction to Statistics

An Introduction to Statistics

George Woodbury
College of the Sequoias

DUXBURY

THOMSON LEARNING

Australia • Canada • Mexico • Singapore • Spain • United Kingdom • United States

DUXBURY

THOMSON LEARNING ™

Sponsoring Editor: *Carolyn Crockett*
Editorial Assistant: *Jennifer Jenkins*
Assistant Editor: *Ann Day*
Marketing: *Tom Ziolkowski/Samantha Cabaluna*
Production Editor: *Tessa Avila*
Production Service: *Helen Walden*
Manuscript Editor: *Helen Walden*

Permissions Editor: *Karyn Morrison*
Interior and Cover Design: *Terri Wright*
Cover Image: *Bill Brooks/Masterfile*
Art Rendering: *Precision Graphics*
Print Buyer: *Jessica Reed*
Typesetting: *TSI Graphics*
Printing and Binding: *Transcontinental Printing*

For more information about this or any other Duxbury product, contact:
DUXBURY
511 Forest Lodge Road
Pacific Grove, CA 93950 USA
www.duxbury.com
1-800-423-0563 (Thomson Learning Academic Resource Center)

For permission to use material from this work, contact us by
www.thomsonrights.com
fax: 1-800-730-2215
phone: 1-800-730-2214

Library of Congress Cataloging-in-Publication Data

Woodbury, George
   An introduction to statistics / George Woodbury.
      p. cm.
   Includes index.
   ISBN 0-534-37755-6
   1. Statistics.   I. Title.

QA276.12.W67 2002
519.5—dc21

To Tina, Dylan, and Alycia

About the Author

George Woodbury is Chair of the Mathematics and Engineering Division at the College of the Sequoias, a public community college in Visalia, California, where he has taught statistics and mathematics since 1994. His teaching has earned him awards and recognition from students and colleagues alike. He earned a B.S. in Mathematics at University of California, Santa Barbara in 1990 and a M.S. in Mathematics at California State University, Northridge in 1994.

George and his wife Tina have two children, Dylan and Alycia. In his free time, George likes to spend time in the garden with his family and on the golf course.

Brief Contents

Contents

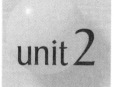

unit 2

• Probability   112

CHAPTER
three   COUNTING AND PROBABILITY   115

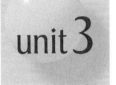

unit 3 • Probability Distributions 166

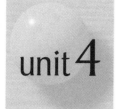

unit 4 • One-Sample Confidence Intervals and Hypothesis Tests   260

unit 5 Two-Sample Inferences, *F*-Tests, Chi-Square Tests 370

UNIT 7 SUMMARY   647

Preface

I believe that students need to understand concepts, rather than mimic steps, to be successful in a math course. This is especially true in an introductory statistics course. One of my main goals in writing this text was to help students understand statistics, to understand the "why" as well as the "how." Toward this end, the many examples in this text are supplemented by clear explanations. Whenever possible, I have tried to show that this is a course of connected material rather than a group of disjointed topics. I try to foreshadow material that will show up in later chapters, while referring back to material covered earlier in the text.

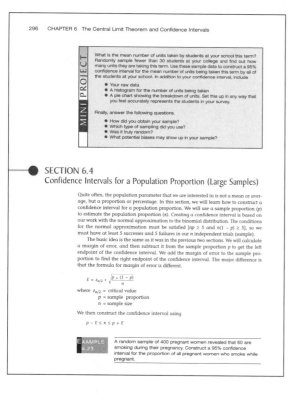

Exercises, Examples, and Mini-Projects

Throughout the book, the exercises and examples describe situations that students can relate to. Some are based on "real data" found in newspapers, magazines, and government reports, and others are based on projects that students have turned in over the years. In class, I try to inspire my students to apply what they are learning to problems found outside of class, and I believe that this text will do the same. The mini-projects that follow many of the sections are written with this in mind. They allow students to apply the techniques learned in a section to data they can collect on their own. One option is to allow students to work in teams in gathering their data.

Focus on Choosing the Appropriate Tool

A common problem seen in introductory statistics courses is that students have a hard time deciding which hypothesis test is appropriate for a given situation. At the end of Chapters 7 through 9 is a section titled "Choosing the Appropriate Tool." Here students find suggestions for deciding which test is appropriate, as well as summaries of the assumptions for each test.

A feature in each section of Chapters 7 through 9 is "What Is Wrong with This Picture?" Each of these presents a flawed statistical project, and students are asked to find the error and explain how to correct it. Some projects do not meet the necessary assumptions, while others use the wrong test.

Extra Material

Topics that are not covered in all statistics courses appear in "Extra" sections. For example, a section on the geometric probability distribution follows Section 4.2 (binomial probability distribution), and a section on determining whether a set of data is normally distributed follows Section 5.3 (finding values for given probabilities).

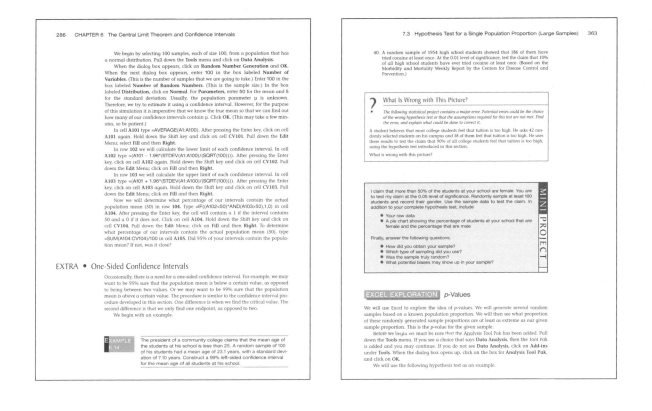

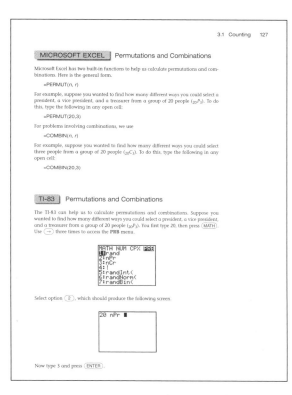

Use of Technology

The availability of technological resources has had a dramatic effect on the way this course is taught and the types of problems that students can solve. I believe that technology should supplement a student's understanding, and have written the technology material in the text with that in mind. I have chosen to incorporate Microsoft Excel as a software package and the Texas Instruments TI-83 as the calculator for this text. Most sections contain directions for using these tools to solve problems found in those sections. Microsoft Excel is used because it is the statistical software most likely to be on a student's own computer. A supplement to the text is available for Minitab.

Microsoft Excel's generation of random numbers makes it easy to run simulations. In several sections you will find "Excel Explorations," which are used to illustrate topics. For example, the Excel Exploration after Section 5.2 examines the normal approximation to the binomial distribution.

To the Student

This book is written with you in mind. You will find plenty of examples to help you learn how to do the exercises, as well as clear explanations to show you why we do what we do. The best course of action for you to take is to read each section before it is discussed in class, so that your instructor can help reinforce what you have read. The homework exercises are extremely important. They are designed to help you gain an understanding of the topics in that section. To assist you, a Student's Solution Manual is available.

Many students dread taking the introductory statistics course (I have heard students refer to it as "sadistics"), but I believe that you will find it to be one of the more rewarding and useful classes you will take in your college career. Good luck!

Learning and Teaching Aids

Duxbury offers a complete set of tools to help students master the course and to assist instructors in teaching.

- *Improving Your Grade for An Introduction to Statistics.* A student solutions manual containing detailed solutions for the textbook's odd-numbered exercises and tips on how to solve these exercises. (ISBN: 0-534-38926-0)
- *Improving Your Grade Version 2 WebTutor on Web CT (and Blackboard) for An Introduction to Statistics.* An electronic resource that offers students real-time access to a full array of study tools, including the student solutions manual (*Improving Your Grade for An Introduction to Statistics*), practice quizzes, flashcards for each chapter with glossary terms, course goals, and objectives, and chapter learning objectives. (WebCT ISBN: 0-534-39621-6 and Blackboard ISBN: 0-534-39622-4)

- *Instructor's Resource Manual.* A complete resource containing suggestions on teaching from the book, ideas for classroom lectures, uses of technology, comments on mini-projects, answers to the "What Is Wrong with This Picture?" sections, answers to all exercises, and discussions of other material in the book. This manual is available through the local sales representative or by submitting a request at info@duxbury.com. (ISBN: 0-534-38921-X)

The publisher also maintains a Web site supporting instruction using *An Introduction to Statistics* at the address http://www.duxbury.com. At this site, following the "Book Companions" and "Data Library" links, are

- Data sets
- Minitab procedures for selected examples from the book
- Lists of known errata (as needed)

Acknowledgments

The publishing of this text follows years of hard work, in which I have received an extraordinary amount of support and help from many people. My editor Carolyn Crockett has been most supportive and encouraging throughout the entire process. Liz Gilley-Poulsen provided the early interest and excitement that got this project off the ground. I have enjoyed my relationship with everyone at Duxbury Press, and I thank all who have helped along the way.

Thanks to the following reviewers:

Michael Allen, Glendale Community College
M. C. Bhattacharjee, New Jersey Institute of Technology
Michael Butler, College of the Redwoods
Guang-Hwa (Andy) Chang, Youngstown State University
Pinyuen Chen, Syracuse University
Carol Curtis, Fresno City College
Gregory J. Davis, University of Wisconsin-Green Bay
Janice Dykacz, Community College of Baltimore County
Charles Johnson, Collin County Community College, Texas
James Lang, Valencia Community College
E. D. McCune, Stephen F. Austin State University
Ron Persky, Christopher Newport University
Geetha Ramachandran, California State University-Sacramento
Glenn Weber, Christopher Newport University
Mark Wilson, University of Charleston

My colleagues at the College of the Sequoias have been most helpful, especially Jared Burch, who I have bounced more ideas off of than I could possibly count. Thanks also to my students, who helped to light the fire in me to be the best possible teacher I can be, for they deserve nothing less than that.

This book is dedicated to the memory of four tremendous people: Joy Bishop, John Schaeffer, Claudia Romero, and Lynn Dwyer. From them I have learned that a positive attitude makes all the difference. All of them left us too early, but left their mark on me and many others as well. They are truly missed.

Thanks to family and friends who have all helped me throughout the years, especially my parents, who were always very supportive. A special thanks to my father for all of the early probability lessons!

Most important, I thank my wife Tina and our children Dylan and Alycia. They make everything possible and worthwhile. I count them as my greatest blessing, and love them all more than words could possibly say.

George Woodbury

An Introduction to Statistics

unit
one

1

Descriptive Statistics

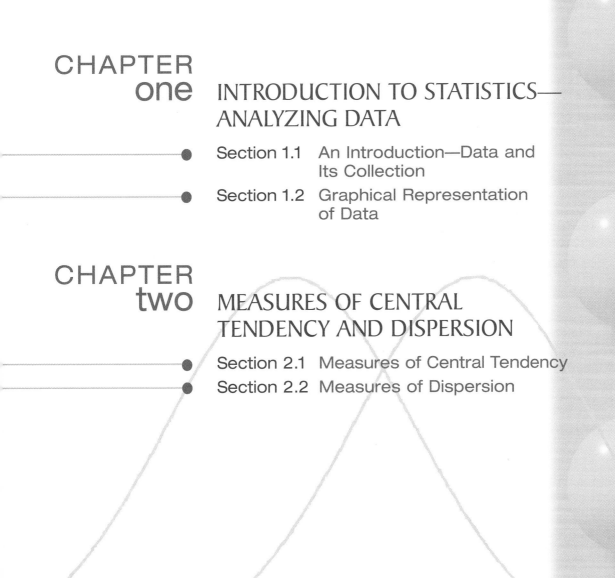

Introduction to Statistics—Analyzing Data

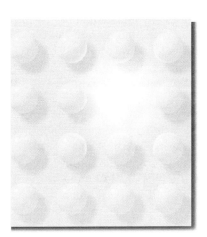

Welcome to statistics! In this course, we will use statistics to formulate answers to questions such as:

- What is the average weight of a newborn baby boy?
- What percentage of people between the ages of 18 and 24 smoke?
- Is the percentage of high school females that plan to attend college greater than the percentage of high school males that plan to attend college?
- Do students who take a prep course for the SAT have a higher average score than students who do not take such a course?
- Is there a relationship between the hours a student works each week and his or her score on a final exam?

To address such questions, we follow three steps:

1. Collect information.
2. Analyze the information.
3. Draw a conclusion or estimate from our analysis.

Clearly, it cannot be as simple as this. However, these are the three main parts of each statistical project. You will learn the finer points as we move along, as well as develop a skeptical eye for what may be going wrong with a project.

The first section presents an introduction to the "big picture" of statistics. You will be introduced to important terms, as well as to the difference between descriptive statistics (the second step above) and inferential statistics (the third step above). After a discussion of what constitutes data, and the different types of data that we will be working with, we will go on to cover the different ways of gathering data. The methods all have advantages and disadvantages, and we shall examine those.

In Section 1.2 we will look at the different ways to represent a set of data graphically. A set of graphs can tell us a great deal about a set of data. They can also be our best tools for presenting our case to someone reading our results.

SECTION 1.1
An Introduction—Data and Its Collection

A senator is considering running for president in the next election and is interested in what percentage of registered voters consider her to be a serious candidate. How can we go about determining this? Contacting all registered voters would be expensive and impractical, if not impossible. Contacting ten registered voters would not produce reliable results. We will take the middle ground, contacting enough registered voters to make the results reliable without contacting too many to make the project impractical for our resources. How many voters should we contact? This question will be answered in a later chapter, where we will see that a survey of approximately 2000 registered voters should give an estimate to within approximately 2% of the actual percentage.

We study two branches of statistics in this course: **descriptive statistics** and **inferential statistics.** Descriptive statistics involves methods that are used to describe data that we have collected. We can describe a set of data using numbers and graphs. For instance, if 280 out of 350 people living in Luserna, Italy have the last name Nicolusi, we can go on to say that 80% of the people in Luserna have the last name Nicolusi. The number 80% describes the set of data that we have. If a car salesman sold four cars this weekend, one car the previous weekend, and one car the weekend before that, we can say that the salesman averaged two cars sold for the last three weekends. A descriptive statement must be verifiable from the data provided.

A **population** is a collection of people, objects, or measurements that we are interested in analyzing. For our senator who is considering running for president, the set of registered voters in the United States is the population, because she is interested in how all registered voters feel. Since it is not practical to contact all registered voters, the senator must contact a **sample** of all registered voters. A sample is a part of the population. If we contacted 2000 registered voters, they would be a sample of all registered voters.

Inferential statistics involves methods that are used to make a generalization, or evaluate a claim, about a population based on sample data. For an inference to be reliable, the sample has to be representative of the population. A sample is **biased** if it does not accurately represent the population. We discuss the selection of a quality sample later in the section, as well as potential biases.

In the last four semesters that an instructor taught intermediate algebra, the following numbers of people passed the class.

17	19	4	20

Which of the following statements are descriptive, and which statements make an inference?

(a) In the last four semesters that the instructor taught intermediate algebra, an average of 15 people passed the class.

This statement is descriptive, because we can verify that the average for these four semesters is 15.

(b) The next time the instructor teaches intermediate algebra, we can expect approximately 15 people to pass the class.

This statement makes an inference, predicting a future event based on the evidence from this sample of four semesters.

(c) This instructor will never pass more than 20 people in an intermediate algebra class.

This statement is an inference; it makes a prediction based on the fact that the instructor did not pass more than 20 students in these four semesters.

(d) In the last four semesters that the instructor taught intermediate algebra, no more than 20 people passed the class.

This statement is descriptive, because we can verify that no more than 20 people passed the course in these four semesters.

(e) Only four people passed one semester because the instructor was in a bad mood the entire semester.

This statement makes an inference, because we are not told that the instructor was in a bad mood that semester.

(f) The last time he taught the class, the instructor passed 20 people to keep the administration from criticizing him for poor results.

Again, this statement makes an inference, because it is not based on any presented fact. ■

A number that describes a characteristic of a population is called a **parameter**, whereas a number that describes a characteristic of a sample is called a **statistic**. In a statistics class of 40 students, 24 had a credit card with them. If we say that 60% of the students in this class had a credit card with them, the 60% is a parameter because the class of 40 students is the population. However, if we make the inference that approximately 60% of all statistics students bring a credit card to class with them, the 60% is a statistic because the class of 40 students is now a sample.

Data

Data are information that we collect, and this information is not necessarily numerical. **Categorical data**, or **qualitative data**, are data that are comprised of names or categories. **Numerical data**, or **quantitative data**, are data that are comprised of numerical values. The colors of M&Ms in a bag are an example of categorical data, whereas the number of M&Ms in a bag is an example of numerical data.

Categorical data can be broken into two levels: **nominal** and **ordinal**. Nominal-level data are categorical data that lack an ordering scheme; ordinal-level data are categorical data that have an ordering scheme. When considering whether a set of data has an ordering scheme, think about whether one value is "better" or "higher" than another value. For example, household income described as low, middle, or high is ordinal because we can tell that a middle-income household earns more than a low-income household. The colors of M&Ms in a bag are an example of nominal data. We cannot say whether a red candy is "higher" or "lower" than a blue candy.

Numerical data can be broken into two levels as well: **interval** and **ratio.** Interval-level data are the next step up from ordinal data, in that we can measure the difference between two values. Interval-level data lack a zero starting point, which means that we cannot construct ratios between two data values. An example of interval-level data would be Celsius temperatures. Suppose we had two Celsius temperatures, 10° and 20°. Can we tell which one is hotter? Yes, 20° is hotter than 10°. Can we tell how much hotter? Yes, it is 10° hotter. Can we say that 20° is twice as hot as 10°? No. This ratio is not valid, because Celsius temperatures do not start at 0°. If we convert these temperatures to Fahrenheit temperatures (50° and 68°), we can see that the higher temperature is not twice the lower temperature. Numerical data that have a zero starting point are ratio-level data. Cash in pocket is an example of ratio-level data. Suppose two people have $10 and $20 in their wallets. Can we say that the person with $20 has twice as much money as the person who has $10? Yes, because the lowest amount of money that a person could have in his or her wallet is $0.

Numerical data can further be classified as either **discrete** or **continuous.** A set of data is discrete if there are only a finite number of values possible or if there is a space on the number line between each two possible values. Discrete data are usually associated with some sort of count. If a five-question quiz is given in a math class, the number of correct answers on a student's quiz is an example of discrete data. The number of correct answers would have to be one of the following: 0, 1, 2, 3, 4, or 5. There are not an infinite number of values, therefore these data are discrete. Also, if we were to draw a number line and place each possible value on it, we would see a space between each pair of values.

For some discrete data there can be an infinite number of possible values, as long as there are gaps between the values. To obtain a taxi license in Las Vegas, a person must pass a written exam regarding different locations in the city. The number of times it would take a person to pass this test is an example of discrete data. A person could take it once, or twice, or three times, or four times, or . . . , so the possible values are 1, 2, 3, ... There are infinitely many possible values, but if we were to put them on a number line, we would see a space between each pair of values.

Continuous data make up the rest of numerical data. This is a type of data that is usually associated with some sort of physical measurement. For any two possible values, there are infinitely many values between them. Think of a number line. For any two points on a number line, there are infinitely many values between them. The same holds true for continuous data. The height of trees at a nursery is an example of continuous data. It is possible for a tree to be 76" tall, and it is possible for a tree to be 77" tall. All decimal heights between 76" and 77" are also possible. One general way to tell whether data are continuous is to ask yourself whether it is possible for the data to take on values that are fractions or decimals. If your answer is yes, these are probably continuous data.

Gathering Data

There are two basic ways to gather data: either through an **experiment** or an **observational study.** In an experiment, we apply a treatment and measure its effect. An example of such an experiment is a test of a new drug for lowering blood pressure.

People in the study would be given the new medication, and after using it for a period of time their blood pressure would be checked to see whether the medication was effective in lowering blood pressure. People who know that they are part of this study may change their daily habits, such as increasing the amount of exercise they get or reducing the amount of fatty foods that they eat. This could make it difficult to determine whether the medication was responsible for the drop in blood pressure. Enter the **control group.** Often an experiment has a group of people that believe they are part of the experiment, but actually are receiving a placebo. If there is a substantial difference between the experimental group and the control group, then this is evidence of the effectiveness of the treatment. Some recent studies suggest that the placebo effect is so strong that perhaps it should be considered as a treatment itself.

Another example of an experiment is a taste test. People are given two different brands of the same product, such as chocolate chip cookies, and asked to tell which of the two they prefer. We may obtain different results if we ask the people in the study to simply name which brand they prefer, without tasting. For instance, if we ask "Do you prefer Burger King french fries or McDonald's french fries?" the results may differ from a taste test of the two types of french fries. A person who has never been to Burger King may answer McDonald's without even knowing what the fries at Burger King taste like. Asking a person to name his or her favorite brand is an example of an observational study.

In an observational study, no treatment is applied to measure a response. Instead, we simply observe the members of our study and record data. Such studies can include opinion surveys. We do not try to influence a person's opinion; the person's opinion is already determined. Other examples include trying to determine the percentage of students at your school that are female or trying to determine the average weight of newborn baby boys.

Any observational study or experiment in which the sample is not representative of the population may produce poor data and unreliable results. The selection of a sample is a very important step in a statistical project, and lays the foundation that all later work will be built upon. We will employ **probability sampling** in choosing our sample. In a probability sample, each member of the population is equally likely to be chosen to be included in the sample.

One method of sampling is **random sampling.** The basic idea is to write down each member of the population on a piece of paper, put all of the pieces in a big hat, mix them all up, and then blindly reach in and select one piece of paper at a time. In practice, we arrange the population in an ordered list. Then we randomly select numbers that correspond to members of the population on the list. Some use a table of random numbers to do this, but technology can help us to do this as well. Microsoft Excel and the TI-83 can randomly generate numbers, as can other computer software packages.

Can a telephone book for a certain city be used to generate a random sample? A problem with using the phone book is that some people have unlisted numbers and are not in the phone book. These people do not have an equal chance of being included in the sample. Would randomly generating phone numbers work? Although this would address the unlisted number problem, there are some people who do not have a telephone at all.

A second sampling technique is **systematic sampling.** A starting point in the list is randomly picked, and then from there we include every kth member on the list. For example, we could begin with person number 19 and then select every 50th person from there. One potential problem with this type of sampling arises with a list that follows a certain pattern. For example, if we had a list of the number of customers at a restaurant every day for an entire year and selected every 21st entry, we would always select the same day of the week.

Stratified sampling is an excellent way to try and ensure that a sample looks just like the population but smaller, although it requires the most resources and information about the population. The population is broken into different groups, or strata, and then we randomly sample from each group. We select a number from each group that is proportional to the breakdown of these groups in the population. Some different strata for an election poll could include gender, race, age, income, political affiliation, or geographical location. If you desire to sample students at your school about some campus-wide issue, and 55% of the students at your school are female students and 45% are male students, then it makes sense to choose your sample in such a way that 55% of your sample is female. We could break the sample down into freshmen, sophomores, juniors, and seniors as well.

Another sampling technique is **cluster sampling.** We divide the population into groups called clusters, usually having a common location. We then randomly select some of these clusters, and include each member of that cluster in the sample. For example, we could divide up a campus by classrooms that are in session at 10 A.M., randomly select some of those classrooms, and include each student in those classrooms at that time. One problem with this example is that it may only be representative of students that take classes at 10 A.M., leaving out evening students, for example.

A **convenience sample** makes use of data that are convenient to gather. For example, a student could stand at the center of campus, or in front of the library or dining hall, and sample people as they pass by. Keep in mind that samples obtained in such a fashion are not necessarily representative of the population. Another convenience sample could use someone else's data, found in a publication or on the Internet. Always be cautious about using data gathered by another party. The data gatherer may have some bias or hidden agenda that could affect the data.

Bias can appear in any set of data, even one that is carefully gathered. One problem is nonresponse. If the people who do not respond have a common characteristic, then that group will be underrepresented. Occasionally, bias is created by poorly worded questions. "Don't you believe that this is another instance of big government interfering with our daily lives?" may lead a participant to agree and say yes. Sometimes respondents do not tell the truth. If you ask strangers whether they wash their hands after using a restroom, very few if any will say no because it is viewed as poor hygiene. However, we know that some people do not wash their hands after using a restroom.

One should never put much faith in the results of a call-in survey, mail-in poll, or on-line poll. The people who make the effort to phone in their opinions or go to a Web site to cast their votes often feel more strongly about the issue. These people can represent the majority in a poll, but the minority in the population.

MICROSOFT EXCEL | Choosing Random Numbers

We will be learning to use Microsoft Excel®, a spreadsheet software package, throughout this course. In this section, we will learn how to use it to randomly generate numbers to help us select a random sample. For this example, we will assume that a college has 10,384 students and we want to select a sample of 100 students.

After opening Microsoft Excel, you will see a rectangular array of boxes that are called cells. Cells are arranged in columns (denoted by the letters A, B, C, . . . across the top of the screen) and rows (denoted by the numbers 1, 2, 3, . . . down the left side of the screen). Cell **A1** is the cell in column **A** and row **1.** In cell **A1**, type the following command and press Enter on your keyboard:

=RANDBETWEEN(1,10384)

RANDBETWEEN is Excel's built-in function to generate a random integer. The first number in the parentheses is the lowest integer that could be generated, and the second number in the parentheses is the highest integer that could be generated. Since there are 10,384 students, we would arrange the list from 1 to 10,384. The number that appears in cell **A1** after you press Enter will tell us the first student to include in the sample.

To generate the remaining 99 numbers for our sample, press the F9 key on your keyboard. This will recalculate the function, giving us a new random number in cell **A1**. Repeat this process until there are 100 different random numbers.

A second way to do this is to begin by typing =RANDBETWEEN(1,10384) in cell **A1**, and pressing Enter on your keyboard. Next, click on cell **A1**. Select the **Edit** menu (click on the word **Edit** at the top of the screen), and choose **Copy.** This copies the formula to the computer's "clipboard" for later use. While holding down the Shift key on your keyboard, press the down arrow on your keyboard until the first column from cell **A1** through **A100** is highlighted. Select the **Edit** menu again, and click on **Paste.** This puts the formula from cell **A1** into all 100 cells. There may be a number that gets repeated, and you will have to select additional random numbers to reach a total of 100 unique random numbers.

▶ ASSIGNMENT

Use Excel to generate the following samples.

1. Select 25 people from a population of 700.
2. Select 40 patients from a group of 295.
3. Select 23 days from a total of 365 days.
4. Select 87 people from a list of 3500 people.

TI-83   Choosing Random Numbers

We will be learning to use the Texas Instruments TI-83 calculator throughout this course. In this section, we will learn how to use it to randomly generate numbers to help us select a random sample. For this example, we will assume that a college has 10,384 students and we want to select a sample of 100 students.

The TI-83 has a built-in function **randInt** that randomly generates a series of integers. Begin by pressing the button labeled (MATH). You should see the screen shown on the right.

Press the right arrow $\boxed{\rightarrow}$ until **PRB** is highlighted, and press $\boxed{5}$ to select the **randInt** function.

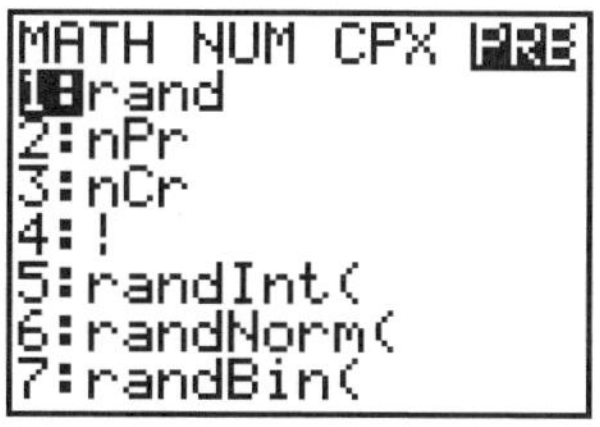

```
MATH NUM CPX PRB
1:rand
2:nPr
3:nCr
4:!
5:randInt(
6:randNorm(
7:randBin(
```

You will be taken back to the main screen. Type

1,10384,100)

after **randInt(** and press the $\boxed{\text{ENTER}}$ key. After a few seconds, you will see the following screen.

```
randInt(1,10384,
100)
{7578 1372 3918…
```

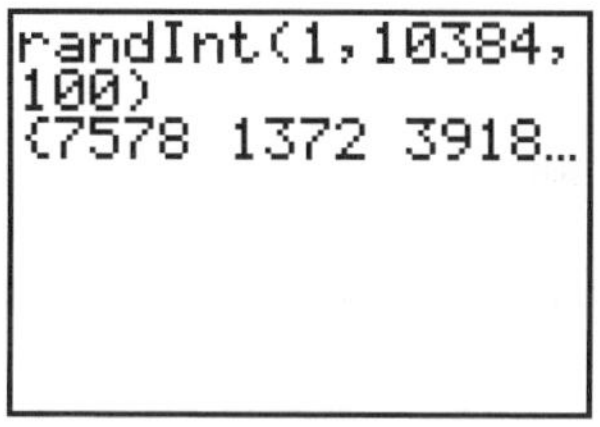

To see all 100 values, we will store these values in a statistical list. Press the key labeled $\boxed{\text{STO}\rightarrow}$, followed by the $\boxed{\text{2nd}}$ key and $\boxed{1}$. This accesses L_1, in orange above $\boxed{1}$ on your calculator, which is the name of a statistical list. After pressing $\boxed{\text{ENTER}}$, you will see the following screen.

```
randInt(1,10384,
100)
{7578 1372 3918…
Ans→L₁
{7578 1372 3918…
```

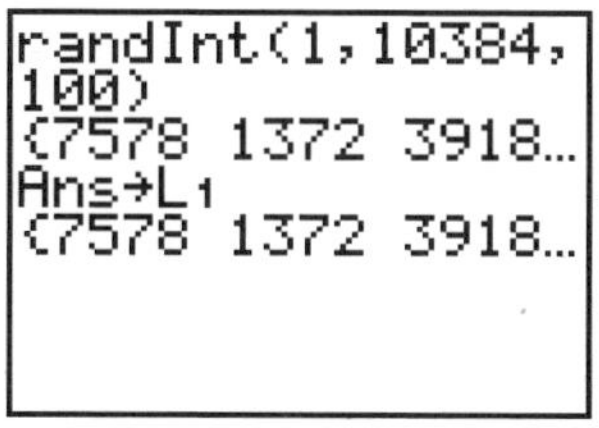

To view this list, press $\boxed{\text{STAT}}$, which produces the following screen.

```
EDIT CALC TESTS
1:Edit…
2:SortA(
3:SortD(
4:ClrList
5:SetUpEditor
```

Press $\boxed{1}$ to select the **Edit** option. In the column labeled L_1, you will find our 100 values. Here is an example of what you should see.

```
L1        L2        L3       1
7578      ------    ------
1372
3918
5889
6058
3511
2314
L1(1)=7578
```

It may be convenient to see these values in ascending order. To do this, press $\boxed{\text{STAT}}$ and select option $\boxed{2}$ under **Edit**. Put L_1 in parentheses after **SortA(** and press $\boxed{\text{ENTER}}$. Your screen should look like this.

```
randInt(1,10384,
100)
{7578 1372 3918…
Ans→L1
{7578 1372 3918…
SortA(L1)
            Done
■
```

Go back to view list L_1 in the $\boxed{\text{STAT}}$ editor, and the values will now be in ascending order.

```
L1        L2        L3       1
59        ------    ------
158
166
519
738
768
926
L1(1)=59
```

Sorting will make it easier to find duplicates in the list. Use the $\boxed{\downarrow}$ key to move your way down the list. If there are duplicates, you can generate more values on the main screen by pressing $\boxed{\text{2ND}}\boxed{\text{MODE}}$, which selects **Quit** and returns you from the $\boxed{\text{STAT}}$ editor to the main screen. Once there, enter randInt(1,10384) as many times as is necessary until you have 100 unique values.

▶ ASSIGNMENT

Use the TI-83 to generate the following samples.

1. Select 25 people from a population of 700.
2. Select 40 patients from a group of 295.
3. Select 23 days from a total of 365 days.
4. Select 87 people from a list of 3500 people.

EXERCISES 1.1

Here are the number of complete games thrown in the American League and the National League for the years 1997 through 1999. (A complete game is a game for which the team's starting pitcher is the only pitcher; he pitches the entire game.)

League	1997	1998	1999
American	123	141	108
National	143	161	127

Determine whether the following statements are descriptive or inferential. Explain the reasoning behind your choice, in your own words.

1. During each of these three seasons, there were at least 120 complete games in the National League.
2. There will always be at least 120 complete games in a season in the National League.
3. In the season of 2005, there will be more complete games in the National League than in the American League.
4. During these three seasons, the American League had fewer complete games than the National League.
5. The National League has more complete games because the hitters in their league are not as good as the hitters in the American League.
6. The American League has fewer complete games because its relief pitchers are superior to the National League's relief pitchers.
7. Based on the data, make a statement of your own that is descriptive.
8. Based on the data, make a statement of your own that is inferential.

Here are the number of homicide arrests of juveniles age 18 and under in the United States for the years 1993 through 1997.

Year	1993	1994	1995	1996	1997
Arrests	3284	3102	2560	2172	1731

(Source: U.S. Justice Department)

Determine whether the following statements are descriptive or inferential. Explain the reasoning behind your choice, in your own words.

9. The number of homicide arrests of juveniles declined each year.
10. A government program is responsible for the decline in arrests.
11. The number of homicide arrests of juveniles in 1992 was higher than 3284.
12. The average number of homicide arrests of juveniles is above 2500.

Here are the test scores for two students in a statistics class.

Student	Test 1	Test 2	Test 3	Test 4	Test 5
Alycia	85	88	83	86	85
Kevin	73	78	81	84	84

Determine whether the following statements are descriptive or inferential. Explain the reasoning behind your choice, in your own words.

13. Alycia's score on the final exam will be in the 80s.
14. Alycia is a better student than Kevin.

15. Alycia's score on the final exam will be higher than Kevin's.
16. Kevin's average for these five tests is 80 points.

Determine whether the following data are nominal-level, ordinal-level, interval-level, or ratio-level data. In addition, if the data are numerical, determine whether they are discrete or continuous.

17. The colors of the marshmallows in a box of cereal
18. A college student's degree (associate, bachelor's, master's, etc.)
19. Temperatures of beers from a tap (°C)
20. Weight of a college student
21. Letter grades on an exam (A, B, C, D, F)
22. Percentage grade on an exam
23. Points scored by a basketball team
24. Ratings of an instructor (superior, average, poor)
25. Elevations of U.S. National Parks, in feet above/below sea level
26. IQ scores
27. Length of commute to school (distance)
28. Length of commute to school (time)
29. 1999 profit/loss of companies on the New York Stock Exchange
30. Number of classes taken by a student this semester
31. Year in school (freshman, sophomore, etc.)
32. Political affiliation (Democrat, Republican, etc.)
33. Average high temperatures of spring break vacation spots (°F)
34. Major in college (mathematics, biology, psychology, etc.)
35. Number of passengers whose baggage was mishandled
36. Height of two-year-old baby boys

For the following scenarios, determine which sampling technique was used. Also list biases that may be present.

37. Average weight of newborn baby boys: Twelve hospitals are selected at random, and the weight of each baby boy born in January is recorded.
38. Percentage of 18- to 25-year-olds who used drugs during the past 30 days: At a shopping mall, people who appear to be in the proper age group are stopped and asked for their age and whether they have used drugs in the past 30 days.
39. Percentage of master's degrees in mathematics earned by females: Several universities are selected at random and asked for the number of master's degrees awarded in mathematics, and how many were awarded to females.
40. Average length (in days) of a sexual harassment trial: The records of a law firm are analyzed, and the lengths of all of their sexual harassment trials are recorded.
41. Effectiveness of a pain reliever against migraine headaches: Patients who have a history of migraines are divided into three groups, using random numbers. The three groups are given a placebo, a half-dose, and a full-dose of the medication. The patients are then asked to rate the effectiveness of the medication on a scale of 1 to 10.
42. Average SAT score for females compared to males: Fifty students are randomly selected from a university's list of students. Their gender and SAT scores are recorded.

43. Explain how you could use cluster sampling to find out how students in the dormitories feel about the quality of food in the dining hall.
44. Explain how you could use convenience sampling to find out how students in the dormitories feel about the quality of food in the dining hall.
45. Explain how you could use stratified sampling to determine how your candidate for governor is doing. List the different strata that you would use.
46. Explain how you could use stratified sampling to find out how students at your school feel about a proposed tuition hike. List the different strata that you would use.

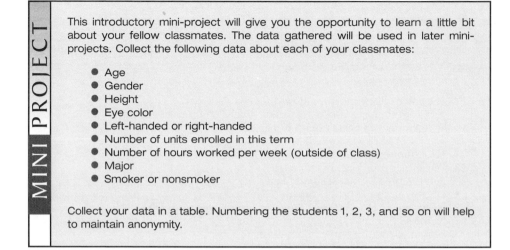

MINI PROJECT

This introductory mini-project will give you the opportunity to learn a little bit about your fellow classmates. The data gathered will be used in later mini-projects. Collect the following data about each of your classmates:

- Age
- Gender
- Height
- Eye color
- Left-handed or right-handed
- Number of units enrolled in this term
- Number of hours worked per week (outside of class)
- Major
- Smoker or nonsmoker

Collect your data in a table. Numbering the students 1, 2, 3, and so on will help to maintain anonymity.

SECTION 1.2
Graphical Representation of Data

Now that we know how to gather data, we can begin to learn how to analyze data. A picture is worth 1000 words, and in this section we will learn some graphical ways to present our data.

There are two ideas that we should keep aware of while analyzing data. First, we need to look for where the data are centered. This will give us an idea of what a typical value is for this set of data. Second, we must note the dispersion of the data. There is a great deal to be learned from how the data are spread out.

Stem-and-Leaf Display

The first graphical tool that we will employ is a stem-and-leaf display. A stem-and-leaf display is a shorthand notation for writing the values in order, from the lowest value to the highest value. From the display we can begin to get a feel for the typical value, or center, of the set of values, as well as how the values are distributed.

To construct a stem-and-leaf display, each value is split into two parts—a "stem" and a "leaf." The stem is the first part of the number, and the leaf is the last part of the number. Although the values may be split in many different ways, depending on the types of values you are working with, we usually let the leaf be the last digit of the value and the stem be all of the preceding digits. For example, the value 46 has a stem of 4 and a leaf of 6. For the value 192, the stem is 19 and the leaf is 2. (Occasionally, we will choose to use a stem of 1 and a leaf of 92 instead. This is a good idea when the values are spread out from 100 to 900, for instance.)

In constructing the actual stem-and-leaf display, we begin by finding the lowest value and the highest value in the set of data. This will give you the first and last stem that is needed in your display. In a vertical column, write down each possible stem from the lowest stem through the highest stem. Then, for each value, write its leaf in the row containing its stem. Once this has been done for each value, each individual row must be sorted from lowest to highest. This step is not necessary if the values have already been sorted. If you are using technology to assist you, such as Microsoft Excel or the TI-83 calculator, it is an excellent idea to sort the values first.

EXAMPLE 1.2   Here are the scores of 32 randomly selected statistics students on an exam on Unit 1. Construct a stem-and-leaf display for these scores.

83	86	65	94	88	51	76	75
86	64	91	47	71	48	68	45
83	76	92	82	96	82	71	56
79	90	92	76	74	98	75	69

The lowest value is 45 (stem: 4, leaf: 5), and the highest value is 98 (stem: 9, leaf: 8). This means that our stems will be 4, 5, 6, 7, 8, and 9.

Stem	Leaf
4	
5	
6	
7	
8	
9	

Begin with the first value, 83. Its stem is 8 and its leaf is 3. We will place a 3 in the row containing the leaf 8.

Stem	Leaf
4	
5	
6	
7	
8	3
9	

After repeating this for all of the values, your display should look like this.

Stem	Leaf
4	7 8 5
5	1 6
6	5 4 8 9
7	6 5 1 6 1 9 6 4 5
8	3 6 8 6 3 2 2
9	4 1 2 6 0 2 8

The values for each stem need to be sorted, which produces the final stem-and-leaf display.

Stem	Leaf
4	5 7 8
5	1 6
6	4 5 8 9
7	1 1 4 5 5 6 6 6 9
8	2 2 3 3 6 6 8
9	0 1 2 2 4 6 8

It is extremely important that you keep the leaves aligned vertically. If you do, you can look at the length of each row to determine which stem has the most values. For these test scores, the stem 7 has the most values (9). Being in the middle of the display indicates that a typical value for this set of data may be somewhere in the 70s.

We should also examine the dispersion of these values. Are they evenly distributed? No, there are more values in the upper three stems (23 values) than in the lower three stems (9 values). ■

Each year, *Forbes* magazine prints a list of the 400 richest people/families in the world. From the 1999 issue, here are the ages of the 54 richest individuals. All of these individuals have a net worth of at least $5 billion. Construct a stem-and-leaf display for this set of data.

43	68	46	43	34	59	55	42
73	71	73	75	71	84	43	36
68	54	77	76	62	59	62	69
31	60	82	88	68	59	81	51
75	79	51	62	41	50	74	82
61	56	66	75	52	59	44	58
51	58	72	62	72	50		

The lowest value in this set is 31 (Pierre Omidyar—eBay, net worth: $7.8 billion) and the highest value is 88 (Walter Haefner—Careal Holdings (Switzerland, computers/cars), net worth: $7.2 billion). Our stems will be 3 through 8. Here is the stem-and-leaf display, unsorted.

Stem	Leaf
3	4 6 1
4	3 6 3 2 3 1 4
5	9 5 4 9 9 1 1 0 6 2 9 8 1 8 0
6	8 8 2 2 9 0 8 2 1 6 2
7	3 1 3 5 1 7 6 5 9 4 5 2 2
8	4 2 8 1 2

Now here is the display after sorting.

Stem	Leaf
3	1 4 6
4	1 2 3 3 3 4 6
5	0 0 1 1 1 2 4 5 6 8 8 9 9 9 9
6	0 1 2 2 2 2 6 8 8 8 9
7	1 1 2 2 3 3 4 5 5 5 6 7 9
8	1 2 2 4 8

The largest number of these individuals is in their 50s, while most are in their 50s, 60s, and 70s. The typical value is probably somewhere in the low to middle 60s. In the next chapter, we will develop methods to come up with an exact value for the typical value.

Sometimes one or more rows will contain too many values. In such a situation, we may break up each row into 2 rows (or more if we wish). To break up a row whose stem is 3, we create two new stems: 3 and 3*. Stem 3 will hold the values from 30 to 34, and 3* will hold the values from 35 to 39. If one row is too long and you want to break it down further, you must break down all of the other rows for consistency. Here is the stem-and-leaf display for the *Forbes* data, with each stem split in 2.

This display gives us more information about these ages. ■

Stem	Leaf
3	1 4
3*	6
4	1 2 3 3 3 4
4*	6
5	0 0 1 1 1 2 4
5*	5 6 8 8 9 9 9 9
6	0 1 2 2 2 2
6*	6 8 8 8 9
7	1 1 2 2 3 3 4
7*	5 5 5 6 7 9
8	1 2 2 4
8*	8

EXAMPLE 1.4

Here are the weights of the 48 players on the 1998 San Francisco Forty-Niners football team, as well as the 25 players on the San Francisco Giants roster on August 11, 1999. Create two stem-and-leaf displays for these two sets of data.

Forty-Niners

290	286	220	260	200	320	315
275	310	262	285	275	275	223
237	250	275	234	260	185	300
195	215	223	235	197	215	195
305	241	278	213	223	240	200
200	270	258	295	185	275	217
250	168	305	220	276	205	

Giants

190	195	220	225	210	215	210
190	205	205	246	190	192	218
210	185	215	187	180	202	185
210	205	202	180			

The stem-and-leaf displays are shown here. As we take a look at these stem-and-leaf displays, we can see that the two sets of weights are distributed in a much different fashion. The weights of the Giants are much closer together than the weights of the Forty-Niners. It should be no surprise that the typical weight of a football player is higher than the typical weight of a baseball player. ■

Stem	Forty-Niners	Stem	Giants
16	8	16	
17		17	
18	5 5	18	0 0 5 5 7
19	5 5 7	19	0 0 0 2 5
20	0 0 0 5	20	2 2 5 5
21	3 5 5 7	21	0 0 0 0 5 5 8
22	0 0 3 3 3	22	0 5
23	4 5 7	23	
24	0 1	24	6
25	0 0 8	25	
26	0 0 2	26	
27	0 5 5 5 5 5 6 8	27	
28	5 6	28	
29	0 5	29	
30	0 5 5	30	
31	0 5	31	
32	0	32	

Frequency Distributions

Score	Frequency
40–50	3
50–60	2
60–70	4
70–80	9
80–90	7
90–100	7

Another graphical tool we can use to analyze data is a frequency distribution. We divide the data into groups called **classes**, and count how many times each class is represented. A frequency distribution for the 32 test scores from the first example in this section is shown on the left.

The first class, 40–50, consists of values from 40 up to, but not including, 50. A value of 50 would be counted in the second class, 50–60. In the first class, 40 is referred to as the **lower class limit**, whereas 50 is the **upper class limit**. The upper class limit is not actually included in the class, but is the boundary point for beginning the next class.

The **class size** for a given class is the distance from its lower limit to its upper limit.

$$\text{Class size} = \text{upper class limit} - \text{lower class limit}$$

The class size for each class in the previous example is 10 ($50 - 40 = 10$, $60 - 50 = 10$, etc.). The **class mark** for a given class is the midpoint of the class.

$$\text{Class mark} = \frac{\text{lower class limit} + \text{upper class limit}}{2}$$

In the previous frequency distribution, the first class mark is 45.

$$\frac{40 + 50}{2} = 45$$

The other class marks are 55, 65, 75, 85, and 95.

Score	Frequency
40–55	
50–65	
60–75	
70–85	
85–100	

Now that we have discussed the terminology for frequency distributions and seen what one looks like, we can focus on the guidelines for creating one. There are two rules that must be followed. First, the classes must be **mutually exhaustive**. This means that each value in the set of data must belong to one of the classes. Consider the classes in the frequency distribution on the left.

There are several values between 40 and 99 that are not accounted for. Suppose you had a value of 50 in your set of data. Which class would that belong in? It does not belong in any of the classes, and this creates a problem. There should be no gaps between your classes.

Score	Frequency
40–48	
52–65	
65–73	
76–89	
89–100	

The second rule is that the classes must be **mutually exclusive**. This means that two classes may not overlap. Consider the classes in the frequency distribution shown on the left. Several of these classes overlap. Suppose you had a value of 52 in your set of data. That value fits into the first class, 40–55, as well as the second class, 50–65. Where should we count it? It would be wrong to arbitrarily place it in one of these classes, and it would be wrong to count it twice. The best way to handle this is to prevent overlapping classes to begin with.

The best way to satisfy both of these rules is to verify that the lowest value fits into the first class, the highest value fits into the last class, and the upper limit of one class is the same as the lower limit of the next class. This prevents overlaps as well as missing values.

There are other suggestions that you may not be able to follow, depending on the set of data. Try to incorporate as many as you can. The first suggestion is about the number of classes. Using too many classes may break down the data too much, and using too few classes may not break down the data enough. A good general rule of thumb is to use somewhere between 5 and 20 classes.

Score	Frequency
40–50	3
50–60	2
60–75	7
75–85	10
85–100	10

Another good idea is to use classes that are of equal size. When we compare the frequency of one class to another, the classes must be of the same size for the comparison to be fair. In our first frequency distribution, each class size was 10. Consider this frequency distribution for the 32 test scores, shown on the right.

These classes do not have equal sizes. The first, second, and fourth classes have a class size of 10, and the third and fifth classes have a class size of 15. This distorts the picture that we got from the first frequency distribution. Although the frequency of the third class seems large, you must keep in mind that it is based on a class size that is 50% bigger than the first, second, and fourth classes.

Another suggestion to consider is to avoid using open-ended classes. An open-ended class is a class that is missing one of its limits. For example, if we were analyzing household incomes, *$100,000 and up* would be an example of an open-ended class. However, *below $20,000* is not an open-ended class because a household's income cannot be lower than $0. Therefore, $0 is really the lower limit for that class. The class should be written as *$0–$20,000*.

Classes that have class sizes of 5 or 10 are often convenient to work with. It is also convenient to use lower class limits that are multiples of the class size. These suggestions will help you at the beginning, but you will find that creating frequency distributions is very intuitive. Often, there is a natural breakdown for classes; the data will dictate what your classes look like.

EXAMPLE 1.5

Let's look again at the *Forbes* magazine list of the 400 richest people/families in the world. From the 1999 issue, here are the ages of the 54 richest individuals. All of these individuals have a net worth of at least $5 billion. Construct a frequency distribution for this set of data.

43	68	46	43	34	59	55	42
73	71	73	75	71	84	43	36
68	54	77	76	62	59	62	69
31	60	82	88	68	59	81	51
75	79	51	62	41	50	74	82
61	56	66	75	52	59	44	58
51	58	72	62	72	50		

I recommend that you begin with the stem-and-leaf display, which was done earlier in this section. Here it is.

Stem	Leaf
3	1 4 6
4	1 2 3 3 3 4 6
5	0 0 1 1 1 2 4 5 6 8 8 9 9 9 9
6	0 1 2 2 2 2 6 8 8 8 9
7	1 1 2 2 3 3 4 5 5 5 6 7 9
8	1 2 2 4 8

The row whose stem is 3 would correspond to a class of 30–40. If it can be done, it is an excellent idea to match your classes to rows from your stem-and-leaf display, because it then becomes a matter of counting how many values are in each row. There are three values in the first row (30–40), seven values in the second row (40–50), and so on. Here is the frequency distribution.

Age	Frequency
30–40	3
40–50	7
50–60	15
60–70	11
70–80	13
80–90	5

We could have used the stem-and-leaf display whose stems were split into two rows. Here is that stem-and-leaf display, along with the corresponding frequency distribution. ∎

Stem	Leaf	Age	Frequency
3	1 4	30–35	2
3*	6	35–40	1
4	1 2 3 3 3 4	40–45	6
4*	6	45–50	1
5	0 0 1 1 1 2 4	50–55	7
5*	5 6 8 8 9 9 9 9	55–60	8
6	0 1 2 2 2 2	60–65	6
6*	6 8 8 8 9	65–70	5
7	1 1 2 2 3 3 4	70–75	7
7*	5 5 5 6 7 9	75–80	6
8	1 2 2 4 4	80–85	4
8*	8	85–90	1

One variation on a frequency distribution that can be made is called a relative frequency distribution. The frequencies are converted to a percentage of the total. This allows a comparison of two sets of data that are similar but are of substantially different sizes. It is impossible in such a case to compare the actual frequencies and gain knowledge about the two sets, but comparing percentages levels the playing field. On the left (top) is an example of a relative frequency distribution for the 32 test scores. Note that the relative frequencies do not total to 100% due to rounding.

If we had another set of similar scores, we could then compare their distributions. On the left (bottom) is another frequency distribution, this time consisting of 76 scores on the first unit examination in a precalculus class.

The relative frequency distribution for both sets of scores follow. Note again that the relative frequencies do not total to 100% due to rounding.

Score	Relative Frequency (%)
40–50	9.4
50–60	6.3
60–70	12.5
70–80	28.1
80–90	21.9
90–100	21.9

Score	Frequency
40–50	4
50–60	3
60–70	15
70–80	26
80–90	20
90–100	8

PRECALCULUS SCORES		STATISTICS SCORES	
Score	Relative Frequency (%)	Score	Relative Frequency (%)
40–50	5.3	40–50	9.4
50–60	3.9	50–60	6.3
60–70	19.7	60–70	12.5
70–80	34.2	70–80	28.1
80–90	26.3	80–90	21.9
90–100	10.5	90–100	21.9

We see that the percentage of scores from 60 to 90 is higher for the precalculus exam than the statistics exam, whereas the percentage of scores from 40 to 60 and 90 to 100 is lower. Based on this evidence, is there a major difference between the scores on these two exams? We can only make an educated guess at this point, but later we will be able to make such an inference.

Another variation of the frequency distribution is the cumulative frequency distribution. In such a distribution we build a cumulative count of the frequencies. In a cumulative less-than frequency distribution we list how many values are in that class or lower. On the right is an example of a cumulative less-than frequency distribution for the 32 test scores.

Score	Frequency	Cumulative Less-Than
40–50	3	3
50–60	2	5
60–70	4	9
70–80	9	18
80–90	7	25
90–100	7	32

To obtain the cumulative frequencies, we add our way down the frequency column. For example, the third cumulative frequency is the sum of the first three frequencies. If we look down the right column containing the cumulative frequencies, the 3 represents the number of scores that are in the first class, or in other words, less than 50. The 5 tells us that there are five values in the first *two* classes, or that there are five values that are less than 60. If a score of 70 is required to pass the test, the value 9 tells us how many students failed to reach that score. If we were looking for the middle of the data set so we could split the data into two equal groups, we would be looking for where the 16th and 17th values are. This cumulative frequency distribution tells us that those values are in the class labeled 70–80. We could then estimate that the typical value is somewhere in the upper 70s.

We can also create a cumulative greater-than frequency distribution in a similar fashion. This time we add the frequencies from the bottom of the distribution up. Each cumulative frequency will tell us how many values are in that class or higher, or how many values are at least as big as the lower class limit of that class. On the right is an example of a cumulative greater-than frequency distribution for the 32 test scores.

Score	Frequency	Cumulative Greater-Than
40–50	3	32
50–60	2	29
60–70	4	27
70–80	9	23
80–90	7	14
90–100	7	7

Taking a look at the cumulative frequency of 27, we see that there are 27 values in the last four classes. In other words, there are 27 test scores that are at least 60.

Cumulative frequency distributions can also be done as relative frequency distributions. This will tell us what percentage of the values fit before or after certain values.

Histograms

A **histogram** is essentially a graph of a frequency distribution. An example of a histogram for the 32 test scores is shown here. We begin by drawing a set of two axes. On the horizontal axis, we label the lower class limits for each class as well as the upper class limit of the last class. These should be scaled precisely, so that none of the bars become too narrow or too wide, which can lead to a deceiving picture. It is not necessary to put the values below the lower class limit of the first class on the horizontal axis, but they have been included here because they are possible test scores. On the vertical axis, we label the frequencies. Since the highest frequency is 9, you must label at least up to 9.

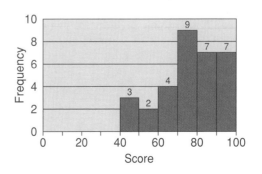

Instead of actual frequencies, you can create a histogram from a relative frequency distribution. Here is the relative frequency distribution (left) for the 32 test scores and the corresponding histogram (right).

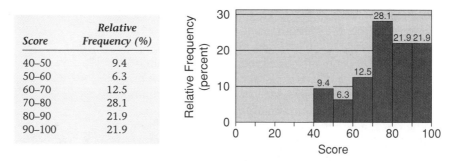

Score	Relative Frequency (%)
40–50	9.4
50–60	6.3
60–70	12.5
70–80	28.1
80–90	21.9
90–100	21.9

This time we had to label the vertical axis past 28.1% instead of 9 as with the previous histogram. Note that the picture looks essentially the same; we are really just looking at a different vertical scale. The picture essentially looks the same as the stem-and-leaf display that was used to create the frequency distribution. Here are the stem-and-leaf display (left) and the first histogram (right) side by side, with the histogram rotated 90° to show the similarity.

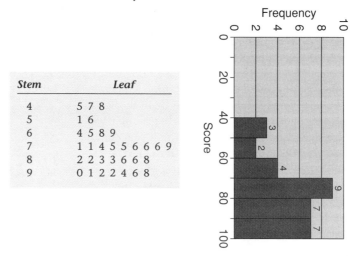

Stem	Leaf
4	5 7 8
5	1 6
6	4 5 8 9
7	1 1 4 5 5 6 6 6 9
8	2 2 3 3 6 6 8
9	0 1 2 2 4 6 8

Now we will take a look at creating a histogram from the beginning.

EXAMPLE 1.6   Once again we use the *Forbes* magazine list of the 400 richest people/families in the world from the 1999 issue. Here is a frequency distribution of the ages of the 54 richest individuals. All of these individuals have a net worth of at least $5 billion. Construct a histogram for this set of data.

Age	Frequency
30–40	3
40–50	7
50–60	15
60–70	11
70–80	13
80–90	5

First we must set up the axes. The horizontal axis should have the values 30, 40, 50, 60, 70, 80, and 90, because they are the class limits. On the vertical axis, we must go to a height of 15, because that is the largest frequency.

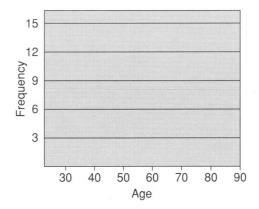

We draw the first bar from 30 to 40, to a height of 3, because its frequency is 3. The next bar, from 40 to 50, is drawn to a height of 7. We draw the next four bars to heights of 15, 11, 13, and 5, respectively. Here is the completed histogram, with the frequencies labeled above each bar. ■

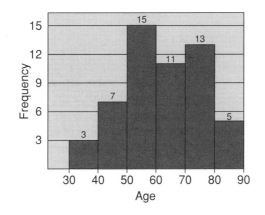

Frequency Polygons

A **frequency polygon** is another way to graph what we see in a frequency distribution. We set up the vertical axis in the same way that we would for a histogram. For the horizontal axis, instead of dealing with the class limits, we begin by finding the class marks. Above each class mark put a point at the height of the frequency for that class. We also put a point on the horizontal axis where the class mark would be if there were one class below the first class, as well as the class mark that would be one class above the last class. At right is an example for the ages of the 54 richest individuals in the world. Note that the graph follows the same shape as the histogram.

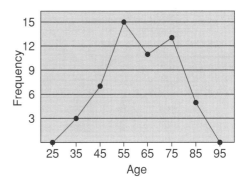

XAMPLE
1.7

Here are the scores of 32 randomly selected statistics students on an exam on Unit 1, arranged in a frequency distribution. Construct a frequency polygon for these data.

Score	Frequency
40–50	3
50–60	2
60–70	4
70–80	9
80–90	7
90–100	7

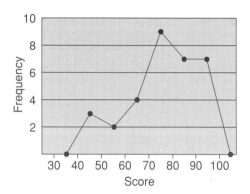

First we make note of the class marks for this frequency distribution: 45, 55, 65, 75, 85, and 95. Then, on the horizontal axis, we place a point at 35 (one class below the first class) and a point at 105 (one class above the last class). We find the first class mark, 45, on the horizontal axis, and go above it to a height of the first class frequency, 3. We place a point there, and then repeat this for the rest of the class marks. We connect each of the points, from left to right, with line segments to produce the frequency polygon shown on the left. ■

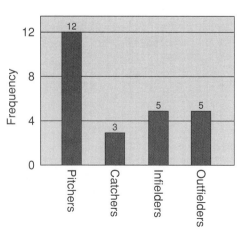

Bar Charts

A **bar chart** is similar to a histogram, but is used for categorical data. There is no scale to worry about on the horizontal axis, but it is important that each bar is of the same width. Also, the bars do not touch each other; there is space between them. On the left is an example of a bar chart for the grades on the exam on Unit 1, created using Microsoft Excel.

XAMPLE
1.8

On the San Francisco Giants current roster there are 12 pitchers, 3 catchers, 5 infielders, and 5 outfielders. Create a bar chart for these data.

There will be four bars in this chart: one for pitchers, a second for catchers, a third for infielders, and a fourth for outfielders. Since the highest total is 12, we must be sure to label the vertical axis to a height of at least 12. At left is the completed chart. ■

Pie Charts

The last graphical way to represent data in this section is the **pie chart.** This type of chart is used for categorical data to show what percentage of the total each category makes up. Here is an example of a pie chart for the letter grade breakdown for 32 students on exam 1 (created using Microsoft Excel).

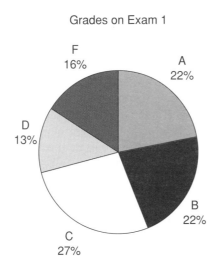

Grades on Exam 1

When creating a pie chart by hand, begin with a circle. The circle represents 100%. Divide the circle into four equal parts, each representing 25%. You can then break down each quarter into five pieces of 5% each. You will end up with a circle that looks like this.

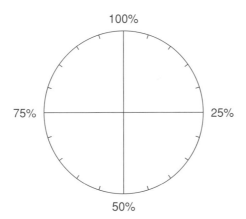

The first wedge of the pie, corresponding to your first category, should begin at 0%. Begin the second wedge where the first wedge ends, and so on.

EXAMPLE 1.9

In 1999, according to *CommerceNet*, 62% of on-line shoppers were men and 38% were women. This is a significant change from 1995, when 81% were men and 19% were women. Create two pie charts to show the breakdown of on-line shoppers by gender.

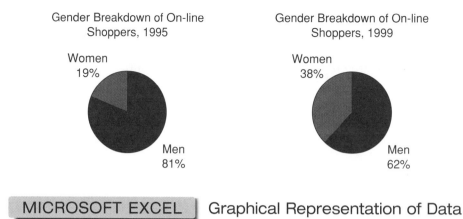

Gender Breakdown of On-line Shoppers, 1995
Women 19%
Men 81%

Gender Breakdown of On-line Shoppers, 1999
Women 38%
Men 62%

MICROSOFT EXCEL   Graphical Representation of Data

We can use Microsoft Excel to assist us with representing a set of data graphically. We will use the *Forbes* magazine richest individuals data to walk through all of the techniques. Here are the data.

43	68	46	43	34	59	55	42
73	71	73	75	71	84	43	36
68	54	77	76	62	59	62	69
31	60	82	88	68	59	81	51
75	79	51	62	41	50	74	82
61	56	66	75	52	59	44	58
51	58	72	62	72	50		

We must put all of the values in a single column to analyze them. Type the first value, 43, in cell **A1**, the second value, 68, in cell **A2**, and so on until the last value, 50, is in cell **A54.**

We will begin with a stem-and-leaf display. Excel will not prepare this display for us, but it can sort the values to make our work easier. Choose the **Data** menu from the toolbar. Then choose **Sort**, and a dialog box will appear on the screen. Excel asks which column you will be sorting by, and whether you want to sort the values in that column in an ascending or descending order. For the stem-and-leaf display, it is best to sort your values in an ascending order. In the dialog box you have the opportunity to tell Excel whether your column has a row at the top of the column that serves as a title for that column (header row). This will prevent you from sorting the title with your values.

Before continuing, note that when sorting with Microsoft Excel, Excel will sort the other columns in your worksheet based on the column you are sorting. This keeps data together for each member of the sample or population that you are analyzing.

Once the values are sorted, it is easy to determine what your highest and lowest stem values are. Once you write the stems in place, you may read down the worksheet to fill in the leaves.

Creating a Histogram from a Set of Data

Before creating a histogram, you must decide what you want to use as upper class limits. Since these values range from 31 through 88, a good choice would be to use 40,

50, 60, 70, 80, and 90 as upper class limits. Excel requires us to put the highest possible value for the class as the upper class limit. So the class *30–40* has an "Excel upper limit" of 39. For this set of data, type these upper class limits in column **B**, putting 39 in cell **B1**, 49 in **B2**, . . . , and 89 in **B6**.

To create the histogram, we will use the Analysis ToolPak, which is an add-in to Microsoft Excel. To see if it has already been added, select the **Tools** menu. If you do not see **Data Analysis** as one of the choices in this menu, then you have to add in the Analysis ToolPak. From the **Tools** menu, select **Add-Ins.** When the dialog box appears, click on the box to the left of **Analysis ToolPak** so that a checkmark appears, then click **OK.**

Select **Data Analysis** from the **Tools** menu. When the dialog box appears, select **Histogram** and click on **OK.** Here is how you need to fill out that dialog box.

- Input Range   **A1:A54**
- Bin Range   **B1:B6**
- Click on **Chart Output** at the bottom of the box.

Once the dialog box is complete, click on **OK.** Your histogram will be created on a new worksheet. Note that Excel leaves gaps between the bars. We can take care of that by clicking on any of the bars with the right mouse button ("right-clicking"), and then selecting **Format Data Series.** When the dialog box appears, click on the **Options** tab. Change the **Gap Width** to 0 and click **OK.** Here is an example of what you should see.

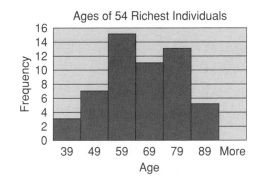

Creating a Histogram from a Frequency Distribution

If you have created a frequency distribution, Excel can assist you in drawing a histogram. We will work with a frequency distribution for the *Forbes* data, shown on the right.

Go back to the original worksheet by clicking on **Worksheet 1** at the bottom left of the screen. Type the classes in column **C**, and the frequencies in column **D**. If you include the header row of Age and Frequency, the last frequency should be in cell **D7**. Click in any open cell, and then choose the **Insert** menu from the toolbar. Then choose **Chart**, and a dialog box will appear for the Chart Wizard. Select **Column** as the chart type, and then click on the button labeled **Next>** to continue to Step 2. For the data range, type **A1:B7**, and be sure that **Columns** is selected under **Series.** Step 3 allows you to put a title on your graph, as well as label the horizontal and vertical axes. When you have completed Step 3, advance to Step 4, where you will be asked whether to include the chart on the same worksheet or

Age	Frequency
30–40	3
40–50	7
50–60	15
60–70	11
70–80	13
80–90	5

somewhere else. Click on **Finish** to see your chart. Readjust the **Gap Width** in the same fashion as for the previous histogram. Here is an example of what you should see.

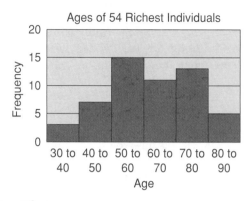

Ages of 54 Richest Individuals

Creating a Pie Chart

To create a pie chart in Microsoft Excel, we begin by typing the categories in one column and the corresponding frequencies in the next column. We then choose the **Insert** menu from the toolbar. Then choose **Chart**, and a dialog box will appear for the Chart Wizard. Select **Pie** as the chart type, and then click on the button labeled **Next>** to continue to Step 2. The range of cells in Step 2 begins at the upper cell in the first column and extends to the lower cell in the second column. While working on Step 3, click on the tab labeled **Legend**, and then unselect **Show legend**. Click on the **Labels** tab, and select **Show value and percent**. Finish Step 4 by deciding where your pie chart should be displayed. Here is what it should look like.

Ages of 54 Richest Individuals

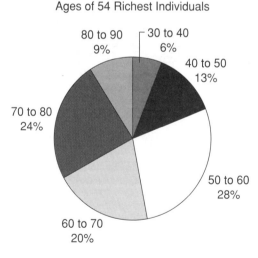

TI-83   Graphical Representation of Data

We can use the TI-83 calculator to assist us with representing a set of data graphically. We will use the *Forbes* magazine richest individuals data to walk through all of the techniques. Here are the data.

43	68	46	43	34	59	55	42
73	71	73	75	71	84	43	36
68	54	77	76	62	59	62	69
31	60	82	88	68	59	81	51
75	79	51	62	41	50	74	82
61	56	66	75	52	59	44	58
51	58	72	62	72	50		

We must put all of the values in a single list to analyze them. We begin by clearing list L_1. Press $\boxed{\text{STAT}}$, and choose option 4 under **Edit**, which is **ClrList**. This will take us to the main screen, where we need to enter the name of the list (L_1) that we want to clear. We do this by pressing $\boxed{\text{2nd}}$ $\boxed{1}$ and then $\boxed{\text{ENTER}}$. Now, to enter the values in the list L_1, press $\boxed{\text{STAT}}$ and select option $\boxed{1}$ under **Edit**. In the column underneath L_1, begin typing the values. Type the first value, 43, and then press the $\boxed{\downarrow}$ key. Type the second value, 68, and press the $\boxed{\downarrow}$ key. Repeat this until you have entered the last value, 50. The TI-83 does not create a stem-and-leaf display, but it can help us by sorting the values. To do this, press $\boxed{\text{STAT}}$ and select option $\boxed{2}$ under **Edit**. Put L_1 in parentheses after **SortA(** and press $\boxed{\text{ENTER}}$. To view L_1, press $\boxed{\text{STAT}}$, and then press $\boxed{1}$ under **Edit**. Use the $\boxed{\downarrow}$ key to move your way down the list and write down the leaves next to the appropriate stems.

To create a histogram for the data, press $\boxed{\text{2nd}}$ $\boxed{\text{Y=}}$ to activate the **STAT PLOT** window. Here is the screen that you will see.

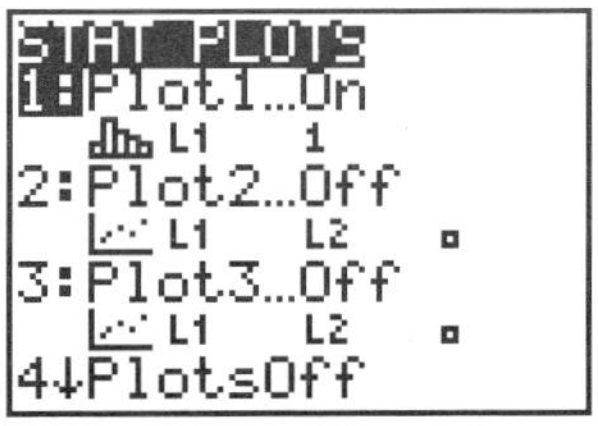

Select option $\boxed{1}$, which leads to the following screen.

On the first line of the display, **Plot1** should be highlighted. On the second line, be sure that the option **On** is highlighted. To move from one line to the next, simply use the $\boxed{\downarrow}$ key. To highlight an option, use the $\boxed{\leftarrow}$ and $\boxed{\rightarrow}$ keys to move back and forth and press $\boxed{\text{ENTER}}$ when you reach the option that you want to highlight. For the type option, highlight the histogram, which is denoted by $\boxed{\text{▦}}$. **Xlist**

should have L_1 next to it, and **Freq** should have 1 next to it. To see the histogram, press ⬭ZOOM⬭ and then option ⬭ 9 ⬭ (**ZoomStat**). Here is what your histogram should look like.

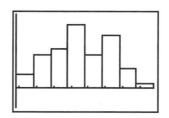

To examine the classes that the TI-83 has selected, and the frequencies of those classes, press ⬭TRACE⬭. The screen should now look like this.

The first class is activated. The lower class limit is 31, the upper class limit is 39.142857, and the frequency is 3. Use the ⬭←⬭ and ⬭→⬭ keys to move from one class to the next.

EXERCISES 1.2

Construct stem-and-leaf displays for the following sets of data.

1. Number of Republicans in the House of Representatives, from the 1949–51 session through the 1997–99 session:

Year	Number of Republicans	Year	Number of Republicans	Year	Number of Republicans
1949–51	171	1967–69	187	1983–85	167
1951–53	199	1969–71	192	1985–87	182
1953–55	221	1971–73	180	1987–89	177
1955–57	203	1973–75	192	1989–91	175
1957–59	201	1975–77	144	1991–93	167
1959–61	153	1977–79	143	1993–95	176
1961–63	175	1979–81	158	1995–97	230
1963–65	176	1981–83	192	1997–99	227
1965–67	140				

(Source: Vital Statistics on Congress 1997–98, reprinted in the New York Times.)

2. First downs gained by the 31 NFL teams in the first two weeks of the 1999 season:

36	35	28	42	47	28	46	38
30	22	22	23	32	31	53	29

(continues)

(continued)

42	38	14	44	43	45	34	36
42	30	42	52	14	35	46	

3. 1999 average SAT verbal scores for the 50 states and the District of Columbia:

State	Verbal Score	State	Verbal Score	State	Verbal Score
AL	561	KY	547	ND	594
AK	516	LA	561	OH	534
AZ	524	ME	507	OK	567
AR	563	MD	507	OR	525
CA	497	MA	511	PA	498
CO	536	MI	557	RI	504
CT	510	MN	586	SC	479
DE	503	MS	563	SD	585
DC	494	MO	572	TN	559
FL	499	MT	545	TX	494
GA	487	NE	568	UT	570
HI	482	NV	512	VT	514
ID	542	NH	520	VA	508
IL	569	NJ	498	WA	525
IN	496	NM	549	WV	527
IA	594	NY	495	WI	584
KS	578	NC	493	WY	546

(Source: The College Board.)

4. 1999 average SAT math scores for the 50 states and the District of Columbia:

State	Math Score	State	Math Score	State	Math Score
AL	555	KY	547	ND	605
AK	514	LA	558	OH	538
AZ	525	ME	503	OK	560
AR	556	MD	507	OR	525
CA	514	MA	511	PA	495
CO	540	MI	565	RI	499
CT	509	MN	598	SC	475
DE	497	MS	548	SD	588
DC	478	MO	572	TN	553
FL	498	MT	546	TX	499
GA	482	NE	571	UT	565
HI	513	NV	517	VT	506
ID	540	NH	518	VA	499
IL	585	NJ	510	WA	526
IN	498	NM	542	WV	512
IA	598	NY	502	WI	595
KS	576	NC	493	WY	551

(Source: The College Board.)

5. Daily volume of the New York Stock Exchange for August 1999 (millions of shares):

649	739	789	859	699	684	836	793
745	692	583	691	683	684	661	682
732	864	719	569	605	860		

6. Percentage of high school graduates that took the 1999 SAT for the 50 states and the District of Columbia:

State	Percentage of Graduates	State	Percentage of Graduates	State	Percentage of Graduates
AL	9	KY	12	ND	5
AK	50	LA	8	OH	25
AZ	34	ME	68	OK	8
AR	6	MD	65	OR	53
CA	49	MA	78	PA	70
CO	32	MI	11	RI	70
CT	80	MN	9	SC	61
DE	67	MS	4	SD	4
DC	77	MO	8	TN	13
FL	53	MT	21	TX	50
GA	63	NE	8	UT	5
HI	52	NV	34	VT	70
ID	16	NH	72	VA	65
IL	12	NJ	80	WA	52
IN	60	NM	12	WV	18
IA	5	NY	76	WI	7
KS	9	NC	61	WY	10

(Source: The College Board.)

7. Time (in minutes) to complete 20 National League baseball games:

166	153	153	141	134	183	167	240	210	165
156	176	172	153	159	196	131	181	188	147

8. Time (in minutes) to complete 16 American League baseball games:

191	161	165	183	176	194	218	162
149	154	154	166	194	200	193	145

Compare this stem-and-leaf display to the previous one for the National League. Can you make any inferences about the length of time that it takes to play a game in the two different leagues? (Compare the "typical" values and the spread of the data.)

For Exercises 9 and 10, explain what is wrong with the stem-and-leaf display. Also explain why the mistake may lead to improper conclusions about the data.

9.

Stem	Leaf
3	1 3 7
6	1 2 5 7 8 8 9
7	0 1 4 5 5 5 6 7 7 9
8	0 1 1 4 8 9
9	0 2

10.

Stem	Leaf
3	0 1 4
3*	5 5 7 8
4	1 1 2 3 4 4
4*	5 5 6 6 6 7 9 9
5	0 0 1 1 2 4 5 6 8 9 9
6	0 1 3 3
6*	5 8 8
7	0 3 8
8	2

Construct frequency distributions for the following sets of data. For Exercises 11–14, you may want to use the stem-and-leaf displays that you created in Exercises 1–4.

11. Number of Republicans in the House of Representatives, from the 1949–51 session through the 1997–99 session:

Year	Number of Republicans	Year	Number of Republicans	Year	Number of Republicans
1949–51	171	1967–69	187	1983–85	167
1951–53	199	1969–71	192	1985–87	182
1953–55	221	1971–73	180	1987–89	177
1955–57	203	1973–75	192	1989–91	175
1957–59	201	1975–77	144	1991–93	167
1959–61	153	1977–79	143	1993–95	176
1961–63	175	1979–81	158	1995–97	230
1963–65	176	1981–83	192	1997–99	227
1965–67	140				

(Source: Vital Statistics on Congress 1997–98, reprinted in the New York Times.*)*

12. First downs gained by the 31 NFL teams in the first two weeks of the 1999 season:

36	35	28	42	47	28	46	38
30	22	22	23	32	31	53	29
42	38	14	44	43	45	34	36
42	30	42	52	14	35	46	

13. 1999 average SAT verbal scores for the 50 states and the District of Columbia:

State	Verbal Score	State	Verbal Score	State	Verbal Score
AL	561	KY	547	ND	594
AK	516	LA	561	OH	534
AZ	524	ME	507	OK	567
AR	563	MD	507	OR	525
CA	497	MA	511	PA	498
CO	536	MI	557	RI	504
CT	510	MN	586	SC	479
DE	503	MS	563	SD	585
DC	494	MO	572	TN	559
FL	499	MT	545	TX	494
GA	487	NE	568	UT	570
HI	482	NV	512	VT	514
ID	542	NH	520	VA	508
IL	569	NJ	498	WA	525
IN	496	NM	549	WV	527
IA	594	NY	495	WI	584
KS	578	NC	493	WY	546

(Source: The College Board.)

14. 1999 average SAT math scores for the 50 states and the District of Columbia:

State	Math Score	State	Math Score	State	Math Score
AL	555	FL	498	LA	558
AK	514	GA	482	ME	503
AZ	525	HI	513	MD	507
AR	556	ID	540	MA	511
CA	514	IL	585	MI	565
CO	540	IN	498	MN	598
CT	509	IA	598	MS	548
DE	497	KS	576	MO	572
DC	478	KY	547	MT	546

(continues)

(continued)

State	Math Score	State	Math Score	State	Math Score
NE	571	OH	538	TX	499
NV	517	OK	560	UT	565
NH	518	OR	525	VT	506
NJ	510	PA	495	VA	499
NM	542	RI	499	WA	526
NY	502	SC	475	WV	512
NC	493	SD	588	WI	595
ND	605	TN	553	WY	551

(Source: The College Board.)

Compare this frequency distribution to the previous one for the SAT verbal scores. Can you make any inferences concerning the scores on these two different tests? (Compare the "typical" values and the spread of the data.)

15. Here are the ages of the presidents of the United States at inauguration (from George Washington through Bill Clinton). Construct a frequency distribution and a relative frequency distribution for these ages.

57	61	57	57	58	57	61	54	68	51	49	64
50	48	65	52	56	46	54	49	51	47	55	55
54	42	51	56	55	51	54	51	60	62	43	55
56	61	52	69	64	46						

16. Here are the numbers of home runs hit by each major league baseball team during the 1999 season. Construct a frequency distribution and a relative frequency distribution for these values.

230	209	151	193	212	203	176	162	145	244
105	212	235	158	223	179	216	161	209	165
188	168	197	187	163	128	194	171	189	153

17. Here are the numbers of earthquakes of magnitude 7.0 or greater for each year from 1900 through 1998. Construct a frequency distribution and a relative frequency distribution for the number of earthquakes per year.

1900	13	1920	8	1940	23	1960	22	1980	18
1901	14	1921	11	1941	24	1961	18	1981	14
1902	8	1922	14	1942	27	1962	15	1982	10
1903	10	1923	23	1943	41	1963	20	1983	15
1904	16	1924	18	1944	31	1964	15	1984	8
1905	26	1925	17	1945	27	1965	22	1985	15
1906	32	1926	19	1946	35	1966	19	1986	6
1907	27	1927	20	1947	26	1967	16	1987	11
1908	18	1928	22	1948	28	1968	30	1988	8
1909	32	1929	19	1949	36	1969	27	1989	7
1910	36	1930	13	1950	39	1970	29	1990	13
1911	24	1931	26	1951	21	1971	23	1991	10
1912	22	1932	13	1952	17	1972	20	1992	23
1913	23	1933	14	1953	22	1973	16	1993	16
1914	22	1934	22	1954	17	1974	21	1994	15
1915	18	1935	24	1955	19	1975	21	1995	25
1916	25	1936	21	1956	15	1976	25	1996	22
1917	21	1937	22	1957	34	1977	16	1997	20
1918	21	1938	26	1958	10	1978	18	1998	16
1919	14	1939	21	1959	15	1979	15		

(Source: U.S. Geological Survey.)

18. Here are the daily volumes (in millions of shares) of the New York Stock Exchange for the first three months of 1999. Construct a frequency distribution and a relative frequency distribution for these volumes.

892	784	987	863	937	818	800	934
797	789	785	905	871	786	728	896
893	852	916	799	845	876	856	871
705	735	721	815	691	653	735	742
700	718	781	782	740	785	669	754
752	771	835	714	804	845	909	826
727	752	752	831	922	662	819	762
784	707	748	729	930			

19. Here are the publisher's prices of 50 books, rounded to the nearest dollar, on *USA Today*'s list of best-selling books (8/26/99). Construct a frequency distribution for these prices, as well as cumulative less-than and greater-than frequency distributions.

17	7	18	9	26	8	8	24	4	8
14	10	8	14	23	5	14	7	28	13
14	20	8	15	23	25	13	8	5	8
7	22	13	5	24	12	5	7	11	6
7	8	22	7	18	22	12	7	4	14

20. Here are the high temperatures, in degrees Fahrenheit, reported in the *New York Times* for 68 foreign cities on August 15, 1999. Construct a frequency distribution for these temperatures, as well as cumulative less-than and greater-than frequency distributions.

93	68	92	60	93	91	71	84	68	81
42	99	87	81	62	87	103	65	65	68
70	76	77	90	65	89	86	89	90	73
73	93	69	88	71	97	89	87	71	99
74	72	70	91	99	85	64	86	69	73
72	64	111	83	91	94	86	85	66	57
89	89	75	89	59	73	70	77		

21. Here are the median incomes, in dollars, for the 50 states and the District of Columbia for the years 1997–98. Construct a frequency distribution for these incomes, as well as cumulative less-than and greater-than frequency distributions.

State	Income ($)	State	Income ($)	State	Income ($)
AL	34,351	KY	35,113	ND	31,229
AK	49,717	LA	32,757	OH	37,811
AZ	35,170	ME	34,461	OK	32,783
AR	27,117	MD	48,714	OR	38,447
CA	40,623	MA	42,511	PA	38,558
CO	45,253	MI	40,583	RI	38,012
CT	45,589	MN	45,576	SC	34,031
DE	42,581	MS	29,031	SD	31,471
DC	32,895	MO	38,662	TN	32,602
FL	33,935	MT	30,622	TX	35,702
GA	37,950	NE	35,823	UT	43,870
HI	41,199	NV	39,608	VT	37,485
ID	35,302	NH	43,297	VA	43,490
IL	42,552	NJ	49,297	WA	46,339
IN	39,613	NM	31,049	WV	27,310
IA	35,664	NY	36,875	WI	40,769
KS	36,875	NC	36,118	WY	34,597

(Source: Census Bureau.)

22. Here are the average number of students per elementary school for the 50 states and the District of Columbia for the years 1996–97. Construct a frequency distribution for these averages, as well as cumulative less-than and greater-than frequency distributions.

State	Number of Students	State	Number of Students	State	Number of Students
AL	489	KY	423	ND	190
AK	357	LA	494	OH	425
AZ	560	ME	262	OK	340
AR	394	MD	551	OR	382
CA	629	MA	437	PA	488
CO	422	MI	433	RI	399
CT	465	MN	434	SC	544
DE	567	MS	522	SD	178
DC	413	MO	392	TN	515
FL	783	MT	187	TX	547
GA	661	NE	172	UT	548
HI	623	NV	595	VT	237
ID	365	NH	329	VA	533
IL	427	NJ	456	WA	452
IN	445	NM	399	WV	296
IA	288	NY	614	WI	371
KS	290	NC	548	WY	201

(Source: U.S. Department of Education National Center for Educational Statistics Common Core of Data Survey.)

23. The following relative frequency distribution is based on a sample of 400 mothers who smoked while pregnant in 1994. Use this to create a standard frequency distribution for the number of cigarettes smoked per day.

Number of Cigarettes	Relative Frequency
1–6	0.237
6–11	0.403
11–16	0.063
16–21	0.250
21–31	0.033
31–41	0.014

(Source: Centers for Disease Control and Prevention—Monthly Vital Statistics Report.)

24. The following relative frequency distribution is based on a total of 195 pro golfers, and shows the number of rounds of golf that they played in 1997. Use this to create a standard frequency distribution for the number of rounds of golf played that year. (Note that relative frequencies do not add to 1 due to rounding.)

Number of Rounds	Relative Frequency
45–55	0.062
55–65	0.108
65–75	0.169
75–85	0.221
85–95	0.205
95–105	0.205
105–115	0.021
115–125	0.010

25. The following is a cumulative less-than frequency distribution of SAT verbal scores of 200 students. Use this to create a standard frequency distribution for the SAT verbal scores of these students.

SAT Verbal Score	Cumulative Frequency
200–300	6
300–400	32
400–500	94
500–600	158
600–700	190
700–800	200

26. The following is a cumulative greater-than frequency distribution of the IQ scores of 368 college students. Use this to create a standard frequency distribution for the IQ scores of these students.

IQ Score	Cumulative Frequency
85–95	368
95–105	298
105–115	207
115–125	97
125–135	30

Construct histograms for the following sets of data. For Exercises 27–30, you may want to use the frequency distributions that you created in Exercises 11–14.

27. Number of Republicans in the House of Representatives, from the 1949–51 session through the 1997–99 session:

Year	Number of Republicans	Year	Number of Republicans	Year	Number of Republicans
1949–51	171	1967–69	187	1983–85	167
1951–53	199	1969–71	192	1985–87	182
1953–55	221	1971–73	180	1987–89	177
1955–57	203	1973–75	192	1989–91	175
1957–59	201	1975–77	144	1991–93	167
1959–61	153	1977–79	143	1993–95	176
1961–63	175	1979–81	158	1995–97	230
1963–65	176	1981–83	192	1997–99	227
1965–67	140				

(Source: Vital Statistics on Congress 1997–98, reprinted in the New York Times.)

28. First downs gained by the 31 NFL teams in the first two weeks of the 1999 season:

36	35	28	42	47	28	46	38
30	22	22	23	32	31	53	29
42	38	14	44	43	45	34	36
42	30	42	52	14	35	46	

29. 1999 average SAT verbal scores for the 50 states and the District of Columbia:

State	Verbal Score	State	Verbal Score	State	Verbal Score
AL	561	KY	547	ND	594
AK	516	LA	561	OH	534
AZ	524	ME	507	OK	567
AR	563	MD	507	OR	525
CA	497	MA	511	PA	498
CO	536	MI	557	RI	504
CT	510	MN	586	SC	479
DE	503	MS	563	SD	585
DC	494	MO	572	TN	559
FL	499	MT	545	TX	494
GA	487	NE	568	UT	570
HI	482	NV	512	VT	514
ID	542	NH	520	VA	508
IL	569	NJ	498	WA	525
IN	496	NM	549	WV	527
IA	594	NY	495	WI	584
KS	578	NC	493	WY	546

(Source: The College Board.)

30. 1999 average SAT math scores for the 50 states and the District of Columbia:

State	Math Score	State	Math Score	State	Math Score
AL	555	KY	547	ND	605
AK	514	LA	558	OH	538
AZ	525	ME	503	OK	560
AR	556	MD	507	OR	525
CA	514	MA	511	PA	495
CO	540	MI	565	RI	499
CT	509	MN	598	SC	475
DE	497	MS	548	SD	588
DC	478	MO	572	TN	553
FL	498	MT	546	TX	499
GA	482	NE	571	UT	565
HI	513	NV	517	VT	506
ID	540	NH	518	VA	499
IL	585	NJ	510	WA	526
IN	498	NM	542	WV	512
IA	598	NY	502	WI	595
KS	576	NC	493	WY	551

(Source: The College Board.)

31. Here are the weights of 40 men before participating in a weight-loss trial. Construct a histogram for these weights.

157	176	179	183	170	184	182	170	160	180
181	169	188	162	155	171	197	170	155	188
174	178	182	175	178	172	178	174	166	180
174	161	175	175	167	180	186	173	168	174

32. Here are the scores of 82 students on the exam for Unit 1. Construct a histogram for these scores.

85	87	40	73	92	87	91	91	77	92
80	96	89	59	79	83	72	41	91	85
89	88	89	86	87	86	96	69	66	82
74	70	92	91	76	57	95	72	81	78
77	92	62	89	81	67	91	60	75	69
93	55	83	82	90	72	55	72	75	82
62	73	87	83	63	75	56	45	89	71
84	71	71	47	77	56	85	78	80	73
91	89								

33. Here are the times, in seconds, of the 66 songs on Bruce Springsteen's "Tracks." Construct a histogram for these times.

267	172	158	118	258	275	213	356	264	505
168	231	197	156	227	269	180	224	195	231
211	223	240	257	291	268	219	164	201	201
266	190	109	231	253	264	286	442	165	185
138	126	280	306	199	213	117	204	174	314
124	211	245	206	167	227	236	158	185	259
277	291	268	179	274	295				

34. Here are the number of complaints, by state, about nursing homes received in 1997. Construct a histogram for the number of complaints that each state received.

State	Number of Complaints	State	Number of Complaints	State	Number of Complaints
AL	998	KY	5,314	ND	546
AK	114	LA	1,992	OH	4,016
AZ	621	ME	499	OK	2,164
AR	1,625	MD	2,260	OR	2,950
CA	17,764	MA	10,438	PA	4,641
CO	7,669	MI	2,674	RI	686
CT	231	MN	2,953	SC	1,896
DE	1,043	MS	661	SD	541
DC	5,199	MO	7,324	TN	1,332
FL	5,029	MT	1,174	TX	9,495
GA	3,274	NE	1,453	UT	1,958
HI	3,952	NV	7,158	VT	330
ID	6,638	NH	593	VA	406
IL	4,731	NJ	2,192	WA	1,670
IN	1,792	NM	9,171	WV	1,043
IA	254	NY	5,483	WI	3,314
KS	4,552	NC	2,457	WY	733

(Source: 1997 preliminary figures, U.S. Department of Health and Human Services, Administration on Aging, USA Today.)

35. Here are the 1999 average SAT verbal scores for the 50 states and the District of Columbia. Use the histogram created in Exercise 29 to create a frequency polygon for these scores.

State	Verbal Score	State	Verbal Score	State	Verbal Score
AL	561	KY	547	ND	594
AK	516	LA	561	OH	534
AZ	524	ME	507	OK	567
AR	563	MD	507	OR	525
CA	497	MA	511	PA	498
CO	536	MI	557	RI	504
CT	510	MN	586	SC	479
DE	503	MS	563	SD	585
DC	494	MO	572	TN	559
FL	499	MT	545	TX	494
GA	487	NE	568	UT	570
HI	482	NV	512	VT	514
ID	542	NH	520	VA	508
IL	569	NJ	498	WA	525
IN	496	NM	549	WV	527
IA	594	NY	495	WI	584
KS	578	NC	493	WY	546

(Source: The College Board.)

36. Here are the 1999 average SAT math scores for the 50 states and the District of Columbia. Use the histogram created in Exercise 30 to create a frequency polygon for these scores.

State	Math Score	State	Math Score	State	Math Score
AL	555	KY	547	ND	605
AK	514	LA	558	OH	538
AZ	525	ME	503	OK	560
AR	556	MD	507	OR	525
CA	514	MA	511	PA	495
CO	540	MI	565	RI	499
CT	509	MN	598	SC	475
DE	497	MS	548	SD	588
DC	478	MO	572	TN	553
FL	498	MT	546	TX	499
GA	482	NE	571	UT	565
HI	513	NV	517	VT	506
ID	540	NH	518	VA	499
IL	585	NJ	510	WA	526
IN	498	NM	542	WV	512
IA	598	NY	502	WI	595
KS	576	NC	493	WY	551

(Source: The College Board.)

37. Here are the number of home runs allowed by each of the 30 major league baseball teams during the 1999 major league season. Construct a frequency polygon for these totals.

AL Team	Home Runs	NL Team	Home Runs
Anaheim	177	Arizona	176
Baltimore	198	Atlanta	142
Boston	160	Chicago	220
Chicago	210	Cincinnati	188
Cleveland	197	Colorado	237
Detroit	209	Florida	171
Kansas City	202	Houston	128
Minnesota	208	Los Angeles	192

(continues)

(continued)

AL Team	Home Runs	NL Team	Home Runs
New York	158	Milwaukee	213
Oakland	160	Montreal	152
Seattle	191	New York	167
Tampa Bay	172	Philadelphia	212
Texas	186	Pittsburgh	160
Toronto	191	San Diego	193
		San Francisco	194
		St. Louis	161

38. Here are amounts spent by each state per welfare case in 1998. Construct a frequency polygon for these amounts.

State	Spending ($)	State	Spending ($)	State	Spending ($)
AL	4,034	KY	3,444	ND	8,061
AK	6,393	LA	3,508	OH	5,189
AZ	5,637	ME	5,096	OK	6,126
AR	4,206	MD	4,817	OR	8,826
CA	5,279	MA	6,917	PA	5,330
CO	6,574	MI	6,268	RI	4,942
CT	5,654	MN	5,530	SC	3,952
DE	4,267	MS	3,764	SD	5,535
DC	4,355	MO	3,613	TN	3,440
FL	5,190	MT	6,415	TX	3,436
GA	4,344	NE	4,339	UT	7,329
HI	5,807	NV	4,322	VT	6,429
ID	17,624	NH	6,129	VA	3,705
IL	3,423	NJ	4,946	WA	5,200
IN	5,212	NM	6,054	WV	6,350
IA	5,226	NY	7,252	WI	21,674
KS	7,326	NC	4,073	WY	17,272

(Source: New York Times *analysis of data from the Department of Health and Human Services.)*

39. The Atlanta Braves made the playoffs eight times in the 1990s, the Cleveland Indians and the New York Yankees each made the playoffs five times, and the Boston Red Sox made the playoffs four times. Draw a bar chart to display these data.

40. Here are the letter grades of 82 students on the exam for Unit 1. Draw a bar chart to display these data.

A	B	C	D	F
15	27	22	8	10

41. A community college offered the following number of sections of math classes this semester. Draw a bar graph to display these data.

Prealgebra	Beginning Algebra	Intermediate Algebra	Statistics	Calculus
28	21	23	7	5

42. Here are the average annual salaries of five occupations in 1998, from the American Federation of Teachers. Draw a bar chart to display these data.

Attorney	Engineer	Systems Analyst	Accountant	Teacher
$72,000	$64,000	$63,000	$46,000	$39,000

43. According to the International Labor Organization, here are the annual number of hours worked per person in five selected countries. Draw a bar chart to display these data.

United States	Japan	Canada	Britain	Norway
1966	1889	1732	1731	1399

44. Here are the market shares of leading personal computer makers for the third quarter of 1998 and 1999, according to Dataquest (*New York Times*). Construct two bar charts to show the distribution of the PC market share for these companies.

Company	Market Share (%) (3rd Quarter 1998)	Market Share (%) (3rd Quarter 1999)
Compaq	13.4	12.8
Dell	8.2	10.8
IBM	8.4	7.6
Hewlett-Packard	5.9	6.2
Gateway	3.8	4.3
Others	60.3	58.3

45. According to a Bureau of Justice statistics study in 1999, 49% of civil trials decided by a jury are won by the plaintiff and 51% are won by the defendant. In contrast, when a judge decides a civil case, 62% are won by the plaintiff and 38% are won by the defendant. Construct a pie chart showing the breakdown for cases decided by a jury. Construct a second pie chart showing the breakdown for cases decided by a judge.

46. According to The College Board, during the 1998–99 school year 70% of high school graduates in Rhode Island took the SAT college admissions test, and 30% did not. For the same school year only 10% of the high school graduates in Wyoming took the test. Construct a pie chart showing the percentage of students in Rhode Island who took the test. Construct a second pie chart showing the breakdown for Wyoming students.

47. A *USA Today*/CNN/Gallup poll asked "What bothers you the most on network TV?" Here are the results.

Response	Percentage
Violence	44
Sexual situations	22
Lewd and profane language	23
Do not watch TV	7
No opinion	4

Construct a pie chart showing the percentages for each response.

48. A *USA Today*/CNN/Gallup poll asked "Have you ever been shocked by something you saw on TV?" Here are the results.

Response	Percentage
Yes	52
No	45
No opinion	3

Construct a pie chart showing the percentages for each response.

49. According to the National Restaurant Association, Saturday is the busiest day for restaurants. For each day of the week, here is a percentage breakdown of business for restaurants.

Day	Percentage
Sunday	14.2
Monday	11.3
Tuesday	11.9
Wednesday	12.6
Thursday	12.6
Friday	18.1
Saturday	19.2

Construct a pie chart showing the percentages for each day of the week.

50. In a study of women age 25–49 by Yankelovich Partners for L'eggs, women were asked how rested they are on waking up. Here are their responses.

Response	Percentage
Extremely refreshed and well rested	8
Fairly well rested	42
Somewhat well rested	32
Not very well rested	13
Not rested at all	5

Construct a pie chart showing the percentages for each response.

51. A 1999 CNN/*Time* poll of 516 potential Democratic primary voters in New Hampshire revealed that 227 preferred Bill Bradley for president, 212 preferred Al Gore, and 77 were undecided. Construct a pie chart showing the percentages for each candidate, as well as the percentage undecided.

52. A survey of 150 fourth- and fifth-graders about their favorite breakfast food produced the following results.

Breakfast Food	Number of Kids
Cereal	63
Eggs	33
Pancakes	21
Toast	12
Doughnuts	9
Toaster pastries	9
Other	3

Construct a pie chart showing the percentages that favor each breakfast food.

MINI PROJECT

In this mini-project, we will use the data gathered in the mini-project of Section 1.1. Construct frequency distributions and histograms for

- Age
- Male height
- Female height
- Number of units enrolled in this term
- Number of hours worked per week (outside of class)

Use these to make an estimate for a typical value for each set of values. Create pie charts showing the percentage breakdown for the following.

- Gender
- Eye color
- Left-handed or right-handed
- Major
- Smoker or nonsmoker

CHAPTER two

Measures of Central Tendency and Dispersion

I n this chapter we will examine two different ways to describe a set of data numerically. First we will describe a "typical value" for a set of data using measures of central tendency. These measures will tell us where the set of data is centered. The most important measure is the mean, which is the arithmetic average.

We will also describe how the data values are spread out using measures of dispersion. This is important because two sets of data may have the same typical value, but are completely different sets due to the way their values are distributed.

SECTION 2.1
Measures of Central Tendency

In this section we will develop ways to find numerical values that we can say are "typical values" for a set of data. These values are called **measures of central tendency**, because they tell us where the data are centered. It makes sense that a typical value for a set of data will be somewhere in the middle of the set.

Mean

The first measure that we examine is the **mean.** The mean is the arithmetic average for a set of data. Recall that to find the arithmetic average for a set of values we add up all the values and then divide that total by how many values there are. We denote the mean of a sample with the symbol $\bar{x}$, which is read "x-bar." We denote the mean of a population with the Greek letter μ, or mu (pronounced "mew," the sound you would get if you crossed a cat with a cow). This is the lower case m in the Greek alphabet (m for mean). We will use our alphabet whenever we refer to a sample statistic, and we will use the Greek alphabet whenever we refer to a population parameter. Recall that the field of inferential statistics makes inferences about population parameters based on sample statistics. For example, we will be using $\bar{x}$ to estimate μ.

The way that we calculate $\bar{x}$ and μ is the same: total the values and divide by how many values there are. Here is the formula for $\bar{x}$.

Sample Mean

$$\bar{x} = \frac{\sum x}{n}$$

The symbol $\sum$ is the capital Greek letter sigma and is used to represent a summation, so $\sum x$ is the sum of all values. The letter n is the number of values in the sample and is referred to as the **sample size.** Here is the formula for μ.

Population Mean

$$\mu = \frac{\sum x}{N}$$

These formulas are the same, with the exception that we use N for the size of a population.

 A manager of a local restaurant is interested in the number of people who eat there on Fridays. Here are the totals for nine randomly selected Fridays:

712	626	600	596	655	682	642	532	526

Find the mean for this set of data.

Does this set of data represent a population or a sample? The manager is interested in the number of patrons on Fridays in general, not only on these nine Fridays, so these data represent a sample. We sum the values and get a total of 5571, and divide by 9 to find the mean.

$$\bar{x} = \frac{\Sigma x}{n}$$

$$= \frac{5571}{9}$$

$$= 619$$

The mean for this set of data is 619 people. ∎

It is a good idea to verify that the mean is a typical value. Does 619 customers seem like a typical value for this set of data? It may help to look at the values after they have been put in order from smallest to largest.

526   532   596   600   626   642   655   682   712

The value 619 seems to fit nicely in the center of these data.

EXAMPLE 2.2   Here are the rushing totals (in yards) for Barry Sanders in the years 1989 through 1998. Find the mean rushing total for these ten seasons.

1470	1304	1548	1352	1115
1883	1500	1553	2053	1491

This time we are looking for a population mean.

$$\mu = \frac{\Sigma x}{n}$$

$$= \frac{15{,}269}{10}$$

$$= 1{,}526.9$$

Barry Sanders' mean rushing total for these ten seasons is 1526.9 yards. ∎

It is important to learn how to use technology, whether it is your calculator or a computer software package, to enter a set of data and use the built-in functions to calculate the mean. Other measures that we will need to calculate later in this chapter and in the next chapter can be easily found once all of the data have been entered. As the sets of data get larger in size, many of the calculations become very tedious, and technology will be your friend.

The mean is the most frequently used measure of central tendency. It is the most reliable measure for inferential purposes. Later in the text, we will be using a sample mean $\bar{x}$ to make an inference about a population mean μ. If many samples are taken from a population, the sample means vary from the population mean less than other measures of central tendency.

Another strength of the mean is that it uses every value in the set of data in its calculation. It should make sense that if we are looking for a value to represent an entire set of data, it should be based on all of the values. We shall see later that other measures of central tendency do not have this property. Other strengths of the mean that other measures do not have include the fact that the mean always exists (we can always calculate the mean of a set of data) and that the mean is unique (a set of data cannot have more than one mean).

The mean does have a weakness in that it is sensitive to outliers. An outlier is a value that is located far away from the rest of the set of data. Here is an example that shows this weakness.

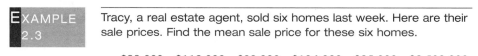

EXAMPLE 2.3   Tracy, a real estate agent, sold six homes last week. Here are their sale prices. Find the mean sale price for these six homes.

$88,000 $112,000 $99,000 $106,000 $95,000 $2,500,000

We are looking for a population mean this time, because there is no suggestion that we are interested in anything other than these six houses.

$$\mu = \frac{\Sigma x}{n}$$
$$= \frac{3,000,000}{6}$$
$$= 500,000$$

The mean sale price is $500,000. ∎

Is the above value typical for the set of sale prices? No, it is off by approximately $400,000 for five of the houses and it is off by $2,000,000 for the other house. If each homeowner was given $500,000 for his or her house there would be five very happy people, and one person would be quite upset. The problem here is the outlier of $2,500,000. The other five prices are grouped closely together, but this one price is substantially higher. The one high value raises the sum of the values and consequently inflates the mean. An outlier that is significantly lower than the other values would make the mean too small to be typical for the set of data. If that value was omitted from the list, the remaining homes would have a mean sale price of $100,000, which would be a typical value for those homes.

We cannot simply omit values that are not convenient. There is a measure that works well in cases where outliers are likely, and it is called the median.

Median

The **median** of a set of data is a value that divides the set of data into two equal groups, after the values have been put in order from lowest to highest. Where do we find the median of a road? In the center of the road.

Two scenarios could occur when finding a median. If there are an odd number of values in the set, there will be one value in the center of the data, and this value is the median. However, if there are an even number of values in the set of data, there is not a single value in the center. In this case, we take the mean of the two values in the center.

EXAMPLE 2.4   A manager of a local restaurant is interested in the number of people who eat there on Fridays. Here are the totals for nine randomly selected Fridays:

712   626   600   596   655   682   642   532   526

Find the median for this set of data.

The first step is to arrange the values in ascending order:

526   532   596   600   626   642   655   682   712

Now we need to divide the set into two equally sized groups. Since there are nine values in the set, we have two groups of four values, with one value left over in the middle.

| 526 | 532 | 596 | 600 | 626 | 642 | 655 | 682 | 712 |

That one value left over in the middle, 626, is the median for this set. Recall that the mean for this set of data was 619. Both values seem typical for this set of data. ∎

EXAMPLE 2.5 Here are the rushing totals (in yards) for Barry Sanders in the years 1989 through 1998. Find the median rushing total for these ten seasons.

1470   1304   1548   1352   1115
1883   1500   1553   2053   1491

Again, we begin by placing the values in ascending order. Since there are ten values, we will have two groups of five values, with no values left over.

| 1115 | 1304 | 1352 | 1470 | 1491 | | 1500 | 1548 | 1553 | 1883 | 2053 |

To find the median, we need to find the mean of the two values in the middle, 1491 and 1500.

$$\text{Median} = \frac{1491 + 1500}{2}$$
$$= 1495.5$$

The median for this set of data is 1495.5 yards. This is fairly close to the mean, which was 1526.9 yards. ∎

Now we will reexamine the example involving home sale prices, which included an outlier.

EXAMPLE 2.6 Tracy, a real estate agent, sold six homes last week. Here are their sale prices. Find the median sale price for these six homes.

$88,000 $112,000 $99,000 $106,000 $95,000 $2,500,000

We begin by putting the values in ascending order. We get two groups of three values, with no values left over in the center.

| $88,000 | $95,000 | $99,000 | | $106,000 | $112,000 | $2,500,000 |

The median is equal to the mean of $99,000 and $106,000.

$$\text{Median} = \frac{99,000 + 106,000}{2}$$

$$= 102,500$$

The median sale price is $102,500. This value is typical of all of the values except for the outlier, and thus describes the set well. ∎

The median handles the problem that outliers create. It also always exists, and it is unique like the mean. It does not use all of the values in its calculation, but instead uses only one or two of the values. The median is used often in situations that produce outliers, such as home prices. Another situation that usually involves the median is household income, which lends itself to high outliers.

Quartiles

The median breaks a set of data into two halves. If we desire to break the set of data into quarters, the appropriate measures are quartiles. The **first quartile**, denoted Q_1, is the value that separates the first quarter of a data set from the rest. The **third quartile**, denoted Q_3, is the value that separates the last quarter of a data set from the rest.

The median divides a set of data into two equal groups of values. The first quartile is the "median" of the first group, while the third quartile is the "median" of the second group.

| EXAMPLE 2.7 | Here are the SAT math scores for 19 randomly selected students. Find the median, as well as the first quartile and third quartile. |

480	370	540	660	650	710	470
490	630	390	430	320	470	400
430	570	450	470	530		

The first step is to put the values in ascending order, and find the median as before. Since there are 19 values, we will have two groups of nine values with one value left in the middle. That middle value is the median.

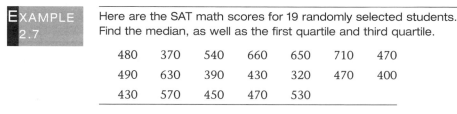

First Group: 320   370   390   400   430   430   450   470   470

Median: 470

Second Group: 480   490   530   540   570   630   650   650   710

The median is 470. Now to find the first quartile, we need to find the "median" of the first group of nine values. This first group will be broken into two groups of four values with one value left in the middle. That middle value is the first quartile, Q_1.

| 320 | 370 | 390 | 400 | 430 | 430 | 450 | 470 | 470 |

The first quartile is 430. This is the score that separates the first quarter of the values from the rest. To find the third quartile, Q_3, we repeat the same procedure with the second group of nine values.

| 480 | 490 | 530 | 540 | 570 | 630 | 650 | 650 | 710 |

The third quartile is 570. This is the score that separates the last quarter of the data from the rest. ∎

EXAMPLE
2.8

Here are the heights in inches of 12 randomly selected college females. Find the median, as well as the first quartile and third quartile.

68	65	63	71	61	66
67	64	66	63	63	60

Here are the heights in ascending order.

| 60 | 61 | 63 | 63 | 63 | 64 | | 65 | 66 | 66 | 67 | 68 | 71 |

We have two groups of six heights, with no value left in the middle. The median will be the mean of the middle two values, 64 and 65.

$$\text{Median} = \frac{64 + 65}{2}$$

$$= 64.5$$

The median is 64.5. To find the first quartile, we work with the first group of six values.

| 60 | 61 | 63 | | 63 | 63 | 64 |

The first quartile is found by taking the mean of the two middle values, which is 63. We repeat this procedure to find the third quartile by using the second group of six values.

| 65 | 66 | 66 | | 67 | 68 | 71 |

The third quartile is found by taking the mean of the two middle values: 66 and 67. The third quartile is 66.5. ∎

EXAMPLE
2.9

Each year, *Forbes* magazine prints a list of the 400 richest people/families in the world. From their 1999 issue, here are the ages of the 54 richest individuals. All of these individuals have a net worth of at least $5 billion. Find the median, the first quartile, and the third quartile.

43	68	46	43	34	59	55	42	73
71	73	75	71	84	43	36	68	54
77	76	62	59	62	69	31	60	82
88	68	59	81	51	75	79	51	62
41	50	74	82	61	56	66	75	52
59	44	58	51	58	72	62	72	50

We begin with the stem-and-leaf display, which was done in the previous chapter:

Stem	Leaf
3	1 4 6
4	1 2 3 3 3 4 6
5	0 0 1 1 1 2 4 5 6 8 8 9 9 9 9
6	0 1 2 2 2 2 6 8 8 8 9
7	1 1 2 2 3 3 4 5 5 5 6 7 9
8	1 2 2 4 8

The advantage of using the stem-and-leaf display is that the values are already in ascending order. Since there are 54 values, we will have two groups of 27 values, with no value left between them. We count until we find the 27th value, which ends the first group. This value is 61. The value that begins the second group is 62. The median for this set is the mean of those two values, which is 61.5.

To find the first quartile, we use the first group of 27 values. This group can be broken into two groups of 13 values, with one value left in the middle, which will be the first quartile. We count until we find the 14th value, which is the second 51 in the list. The first quartile is 51.

To find the third quartile, we use the second group of 27 values, which begins at 62. In a similar fashion, we count until we find the 14th value of that group, which will be the third quartile. The 14th value is the second 73 in the list. The third quartile is 73. ■

Midrange

The **midrange** is a measure of central tendency that is quite easy to calculate. It is found by finding the mean of the lowest value and highest value in the set. The midrange is exactly halfway between the lowest value and the highest value, which often places it in the center of the set of data. However, if there is an outlier the midrange will be pulled away from the center toward that outlier.

EXAMPLE 2.10

A manager of a local restaurant is interested in the number of people who eat there on Fridays. Here are the totals for nine randomly selected Fridays:

| 712 | 626 | 600 | 596 | 655 | 682 | 642 | 532 | 526 |

Find the midrange for this set of data.

The lowest value for this set of data is 526, and the highest value is 712. To find the midrange, we find the mean of these two values.

$$\text{Midrange} = \frac{526 + 712}{2}$$

$$= 619$$

The midrange is 619. ■

For the previous set of data, the midrange happens to be the same as the mean for this set of data as well. Thus this midrange could be considered a good measure of central tendency. This is not always the case. Consider the following.

EXAMPLE 2.11

Tracy, a real estate agent, sold six homes last week. Here are their sale prices. Find the midrange sale price for the six homes.

| $88,000 | $112,000 | $99,000 | $106,000 | $95,000 | $2,500,000 |

The midrange is the mean of the two extreme values: $88,000 and $2,500,000.

$$\text{Midrange} = \frac{88{,}000 + 2{,}500{,}000}{2}$$

$$= 1{,}294{,}000$$

The midrange for this set of data is \$1,294,000. This is far from what we would call a central value. The outlier (\$2,500,000) has pulled the midrange out of the center toward the outlier. ■

The midrange for a set of data is sensitive to outliers, just like the mean. However, unlike the mean it does not use all of the values in its calculation. The use of the midrange should be limited to providing a quick estimate of the center of the data. Further analysis of the data is required to get a better idea of the center of the data.

Mode

The mode of a set of data is the value(s) that occur most frequently in the set.

EXAMPLE 2.12   Here are the ages of the 42 presidents of the United States at inauguration. Find the mode for this set of data.

57	61	57	57	58	57	61	54	68
51	49	64	50	48	65	52	56	46
54	49	51	47	55	55	54	42	51
56	55	51	54	51	60	62	43	55
56	61	52	69	64	46			

The mode for this set of data is 51, because it appears five times in the list. ■

The mode has a few weaknesses for quantitative data. The first problem is that the mode may not be typical for a set of data. Consider the set of data below.

1	2	4	5	6	7	9	10	11
13	14	15	16	17	19	20	100	100

The mode for this set of data is 100 because it is the only value that appears twice, but it is hardly typical. Simply because a value is repeated the most number of times, there is no guarantee that it is in the center of the values. In the preceding example, the value 100 would be considered an outlier.

Another weakness of the mode is that it may not exist. If no value in the set is repeated, then the set of data does not have a mode.

EXAMPLE 2.13   Here are the amounts of vitamin C in a serving of 14 different fruits and vegetables, in milligrams. Find the mode for this set of data.

10	14	15	21	39	44	45
66	70	73	85	93	96	124

Since no value is repeated in this set of data, there is no mode. ■

There is sometimes more than one mode for a set of data. If there are two or more values that are repeated the same number of times, then each value could be a mode.

EXAMPLE 2.14

Each year, *Forbes* magazine prints a list of the 400 richest people/families in the world. From their 1999 issue, here are the ages of the 54 richest individuals. All of these individuals have a net worth of at least $5 billion. Find the mode age for this set of values.

43	68	46	43	34	59	55	42
73	71	73	75	71	84	43	36
68	54	77	76	62	59	62	69
31	60	82	88	68	59	81	51
75	79	51	62	41	50	74	82
61	56	66	75	52	59	44	58
51	58	72	62	72	50		

Again, we begin with the stem-and-leaf display. Since the values are in order, it is easy to see which values are repeating, as well as how many times they repeat.

Stem	Leaf
3	1 4 6
4	1 2 3 3 3 4 6
5	0 0 1 1 1 2 4 5 6 8 8 9 9 9 9
6	0 1 2 2 2 2 6 8 8 8 9
7	1 1 2 2 3 3 4 5 5 5 6 7 9
8	1 2 2 4 8

The values 59 and 62 both appear four times, which is more than any other value. There are two modes for this set of data: 59 and 62. ∎

We should use the mode cautiously with numerical data. For nominal-level data, the mode is a good measure because it shows us which category is represented the greatest number of times. For ordinal-level data, the median is a better measure to use because the values can be put in order. The mode will still be helpful for ordinal-level data, but it will not necessarily show us where the center of the data is.

Weighted Mean

Consider the following example.

EXAMPLE 2.15

A college cross-country team has 12 members. During the off-season, seven of the members run 20 miles per week, four members run 25 miles per week, and one runs 30 miles per week. Find the mean weekly distance for the team during the off-season.

It would be incorrect to simply add 20, 25, and 30, and then divide by 3 to find the mean because there are twelve values, not three. The value 20 needs to contribute to the sum seven times, and the value 25 needs to contribute to the sum four times. Here is the correct calculation.

$$\bar{x} = \frac{\Sigma x}{n}$$

$$= \frac{20 + 20 + 20 + 20 + 20 + 20 + 20 + 25 + 25 + 25 + 25 + 30}{12}$$

$$= \frac{270}{12}$$

$$= 22.5$$

The mean distance for these runners is 22.5 miles per week. ■

In the previous example, the value 20 has a greater relative importance than the other two values when calculating the mean because it shows up seven times. The **weighted mean** takes the relative importance of each value into consideration. The value 20 should count seven times, so we say that the **weight** for the value 20 is 7. The value 25 has a weight of 4, and the value 30 has a weight of 1. Here is the formula for the weighted mean.

$$\bar{x}_w = \frac{\Sigma w \cdot x}{\Sigma w}$$

We use a subscript w to denote the weighted mean. In the formula, w represents the weight for the value x. To calculate the weighted mean, we multiply each value by its weight and total those products. We then divide by the sum of the weights.

 EXAMPLE 2.16   A college cross-country team has 12 members. During the off-season, seven of the members run 20 miles per week, four members run 25 miles per week, and one runs 30 miles per week. Find the mean weekly distance for the team during the off-season.

This time we will use the weighted mean formula.

$$\bar{x}_w = \frac{\Sigma w \cdot x}{\Sigma w}$$

$$= \frac{7(20) + 4(25) + 1(30)}{7 + 4 + 1}$$

$$= \frac{270}{12}$$

$$= 22.5$$

The mean distance for the runners is 22.5 miles per week. ■

The weighted mean is useful when we want to find the combined mean of two or more sets of data that are of different sizes. The size of each set is the weight.

 EXAMPLE 2.17   A college math department gives a common final in its precalculus classes. There are five different versions of the exam. Here are the results from last semester for each version. Find the combined mean for the five versions.

Version	A	B	C	D	E
Number of Students	37	47	40	37	51
Mean	74.0	72.7	72.0	80.1	84.2

The weights for this problem are the number of students that took each version.

$$\bar{x}_w = \frac{\Sigma w \cdot x}{\Sigma w}$$

$$= \frac{37(74.0) + 47(72.7) + 40(72.0) + 37(80.1) + 51(84.2)}{37 + 47 + 40 + 37 + 51}$$

$$= \frac{16{,}292.8}{212}$$

$$= 76.85$$

The combined mean for all these exams is 76.85. ■

The weighted mean can be used to find the mean of two or more percentages as well.

At a community college, 70% of the students who took statistics last year passed the class. At the nearest university, 90% of the students who took statistics last year passed the class. Find the combined percentage of students who passed the class last year at the two schools if 300 students took it at the community college and 700 students took it at the university.

This time the weights are the number of students at the two schools.

$$\bar{x}_w = \frac{\Sigma w \cdot x}{\Sigma w}$$

$$= \frac{300(70) + 700(90)}{300 + 700}$$

$$= \frac{84{,}000}{1{,}000}$$

$$= 84$$

At the two schools, 84% of the students who took statistics passed the class. ■

Note that the percentage in the previous example was not simply the mean of 70% and 90%. The percentage that passed was closer to 90% than it was to 70% because the weight was much higher for the percentage of students who passed at the university.

Estimating a Mean Using a Frequency Distribution

We can use the idea of the weighted mean to estimate the mean of a set of data, even if we do not have the actual values, but instead have a frequency distribution. Consider the frequency distribution for the test scores of 32 statistics students on Exam 1, shown on the next page in the margin.

There are three scores in the first class. What would you estimate the mean of those three values to be? A safe estimate would be the midpoint, or class mark, of that class, which is 45. If the values in a class are evenly distributed throughout the class, then their mean will be close to the class mark. We will estimate the mean of the next two values to be 55, the mean of the next four values to be 65, and so on. We then find the weighted mean for these individual means. Here is a formula.

Score	Frequency
40–50	3
50–60	2
60–70	4
70–80	9
80–90	7
90–100	7

$$\overline{x}_g = \frac{\Sigma \text{ frequency} \cdot \text{class mark}}{\Sigma \text{ frequency}}$$

The frequency for each class will be the weight, and the class mark will be used as the mean of that class. After finding each class mark, multiply the frequency for each class by its class mark. Total those products, and divide that sum by the total of the frequencies. Here are the calculations necessary to estimate the mean using the frequency distribution for the 32 test scores.

Score	Frequency	Class Mark	Frequency · Class Mark
40–50	3	45	135
50–60	2	55	110
60–70	4	65	260
70–80	9	75	675
80–90	7	85	595
90–100	7	95	665
Totals	32		2440

$$\overline{x}_g = \frac{2440}{32}$$

$$= 76.25$$

We estimate the mean to be 76.25. The actual mean is 75.91, so this is an accurate estimate.

EXAMPLE 2.19   Each year, *Forbes* magazine prints a list of the 400 richest people/families in the world. From their 1999 issue, here are the ages of the 54 richest individuals arranged in a frequency distribution. Use the frequency distribution to estimate the mean for this set of data.

Age	Frequency
30–40	3
40–50	7
50–60	15
60–70	11
70–80	13
80–90	5

The class marks are 35, 45, 55, 65, 75, and 85. We multiply each class mark by the frequency for that class, and total those products. To calculate the estimate of the mean, we divide that total by 54, which is the total of the frequencies.

Age	Frequency	Class Mark	Frequency · Class Mark
30–40	3	35	105
40–50	7	45	315
50–60	15	55	825
60–70	11	65	715
70–80	13	75	975
80–90	5	85	425
Totals	54		3360

$$\bar{x}_g = \frac{3360}{54}$$

$$= 62.22$$

Using this frequency distribution, we estimate the mean to be 62.22 years. (The actual mean is 61.35 years.) ∎

MICROSOFT EXCEL   Measures of Central Tendency

We can use Excel to help us calculate quickly measures of central tendency for sets of data. We will walk through an example that demonstrates how to calculate certain measures of central tendency. If you follow the directions carefully, you will reproduce the results shown. You can then adapt these instructions for other sets of data. Following is the set of data that will be used for the example. These are the scores of 57 randomly selected students on a statistics exam.

73	91	97	75	85	99	87	88	63	79
98	66	64	77	75	89	77	94	93	85
92	64	71	99	81	95	92	84	84	97
75	83	86	65	90	46	89	81	56	90
44	50	72	66	66	69	79	71	77	76
78	99	87	90	91	63	94			

To begin, type **Scores** in cell **A1.** Below that, type the values in cells **A2** through **A58.** It is important that all of the values end up in the same column for the upcoming operations. These cells are the "cell range" that contains the data we wish to work with, and to refer to them in Excel we will be typing A2:A58.

MEAN

In cell **B2** type **Mean,** and next to it (**C2**) type

=AVERAGE(A2:A58)

AVERAGE is the built-in Excel function to calculate the mean of a list of numbers. It is important to type the equal sign (=) before the word AVERAGE. Inside the parentheses we type A2:A58 because those cells contain the scores.

MEDIAN

In cell **B3** type **Median,** and next to it (**C3**) type

=MEDIAN(A2:A58)

MEDIAN is the built-in Excel function to calculate the median of a list of numbers.

FIRST QUARTILE

In cell **B4** type First Quartile, and next to it (**C4**) type

=QUARTILE(A2:A58,1)

QUARTILE is the built-in Excel function to calculate a quartile of a list of numbers. We will use it again to calculate the third quartile.

THIRD QUARTILE

In cell **B5** type Third Quartile, and next to it (**C5**) type

=QUARTILE(A2:A58,3).

SMALLEST VALUE

In cell **B6** type Minimum, and next to it (**C6**) type

=MIN(A2:A58)

MIN is the built-in Excel function to find the smallest value in a list of numbers.

LARGEST VALUE

In cell **B7** type Maximum, and next to it (**C7**) type

=MAX(A2:A58)

MAX is the built-in Excel function to find the largest value in a list of numbers.

MIDRANGE

In cell **B8** type Midrange, and next to it (**C8**) type

=(C6+C7)/2

This adds the values in cell **C6** and **C7**, and divides that total by 2.

Here is what the top of your worksheet should look like.

Scores		
73	Mean	79.77
91	Median	81
97	First Quartile	71
75	Third Quartile	90
85	Minimum	44
99	Maximum	99
87	Midrange	71.5
88		
63		
79		
98		
66		

ESTIMATING A MEAN USING A FREQUENCY DISTRIBUTION

Excel can also help us in estimating the mean of a set of data from a frequency distribution. Here is a frequency distribution for these test scores.

Score	Frequency
40–50	2
50–60	2
60–70	9
70–80	13
80–90	14
90–100	17

Using the same worksheet, type Score in cell **D1.** Below it, in cell **D2,** type 40–50. Type the remainder of the classes in cell **D3** through **D7.** In the next column, type Frequency in cell **E1.** Type the respective frequencies in cells **E2** through **E7,** next to their classes.

To estimate the mean using a frequency distribution, we multiply the class mark for each class by the frequency for that class, so we will need a column containing the class marks. In cell **F1,** type Class Marks. Type 45 in cell **F2,** 55 in **F3,** . . . , and 95 in cell **F7.** Now for the multiplication. In cell **G1,** type Product. In cell **G2,** type =E2*F2. This multiplies the frequency of the first class by its class mark. In **G3** type = E3*F3. Repeat these steps to fill column **G** through cell **G7.** To estimate the mean, we need to divide the sum of these products by the sum of the frequencies. In cell **D9** type Estimate of Mean, and in **E9** type =(SUM(G2:G7))/(SUM(E2:E7)). That portion of the worksheet should look like this:

Score	Frequency	Class Mark	Product
40–50	2	45	90
50–60	2	55	110
60–70	9	65	585
70–80	13	75	975
80–90	14	85	1190
90–100	17	95	1615

Estimate of Mean   80.0877193

The actual mean for this data is 79.77, so our estimate of approximately 80.09 is a very good estimate.

TI-83   Measures of Central Tendency

We can use the TI-83 to help us calculate quickly measures of central tendency for sets of data. We will walk through an example that demonstrates how to calculate certain measures of central tendency. The set of data that will be used for the example is the scores of 57 randomly selected students on a statistics exam:

73	91	97	75	85	99	87	88	63	79
98	66	64	77	75	89	77	94	93	85
92	64	71	99	81	95	92	84	84	97
75	83	86	65	90	46	89	81	56	90
44	50	72	66	66	69	79	71	77	76
78	99	87	90	91	63	94			

We must put all of the values in a single list to analyze them. We begin by clearing list L_1. Press (STAT), and choose option 4 under **Edit,** which is **ClrList.** This will take you to the main screen, where we need to enter the name of the list (L_1) that we want to clear. We do this by pressing (2ND)(1) and then (ENTER). Now, to enter the values in the list L_1, press (STAT) and select option (1) under **Edit.** In the column underneath L_1, begin typing the values. Type the first value, 73, and then press the (↓) key. Type the second value, 91, and press the (↓) key. Repeat this until the last value, 94, has been entered. Be sure that you have entered 57 values before proceeding.

We can calculate several measures at the same time. Press (STAT) and use the (→) key to access the **CALC** menu, which looks like this.

```
EDIT CALC TESTS
1:1-Var Stats
2:2-Var Stats
3:Med-Med
4:LinReg(ax+b)
5:QuadReg
6:CubicReg
7↓QuartReg
```

Press (1) to select the **1-Var Stats** option. On the main screen, enter L_1 after **1-Var Stats** by using the (2ND) and (1) keys and press (ENTER). The results take up more than one window. The TI-83 will put an arrow pointing downward in the lower left corner of the window to let us know when this happens. The following two screens show all of the information.

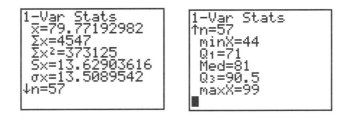

We now see that

- The mean is approximately 79.77.
- The median is 81.
- The first quartile is 71.
- The third quartile is 90.5.
- The smallest value is 44.
- The largest value is 99.

From this information, we can calculate that the midrange is $\dfrac{44 + 99}{2}$ or 71.5.

EXERCISES 2.1

Find the mean of the following sets of values.

1. 32   57   39   85   74   14   45   62   21   19

2. 47   42   46   45   41   49   49   112   42   44

3.	84	93	57	71	94	77	89	78	61	93
	72	87	63	60	55	72	68	80	56	68

4.	19	18	53	19	21	52	52	20	49	50
	50	23	21	17	54	52	53	18	18	51
	49	18	19	22	55	51	20	56	21	53

Find the median and quartiles for the following sets of values.

5.	45	58	50	47	55	60	40	43
	50	55	40	43	48	46	56	46

6.	149	135	128	150	149	144	142	135	127
	147	127	114	138	135	149	142	127	116

7.	60	56	65	19	63	74	63	55	105	23
	30	49	13	68	31	86	101	91	70	76
	15	55	50	98	35	104	57	57	17	107
	98	47	49	84	98	74	33			

8.	204	161	192	109	109	67	145	107	245	182
	111	107	183	176	173	76	146	110	116	243
	181	213	125	105	214	231	239	194	166	185

9. A sample of five college females produced the following heights (in inches). Find the midrange for these heights.

    64   67   65   70   62

10. Here are the scores of six randomly selected professional football teams on a certain Sunday. Find the midrange for these scores.

    31   14   25   17   20   6

11. Here are the math exam scores of eight randomly selected students. Find the midrange for these scores.

    62   75   72   95   82   86   80   71

12. Here are the speeds of seven pitches thrown by a professional baseball pitcher, in miles per hour. Find the midrange of these speeds.

    95   79   88   98   96   94   72

13. Here are the widths of five model year 2000 compact sports utility vehicles (in inches). Find the mode and midrange for these widths.

Make/Model	Width (in.)
Chevy Tracker (4-door)	67.3
Isuzu Amigo	71.4
Jeep Wrangler	66.7
Suzuki Vitara (2-door)	67.3
Toyota RAV4	66.7

(Source: Consumer Guide Car & Truck Test.)

14. Here are the widths of five model year 2000 midsize sports utility vehicles (in inches). Find the mode and midrange for these widths.

Make/Model	Width (in.)
Chevy Blazer (4-door)	67.8
Dodge Durango	71.5
Jeep Cherokee (4-door)	69.4
Nissan Pathfinder	69.7
Toyota 4Runner	66.5

(Source: Consumer Guide Car & Truck Test.)

15. A class for volunteer Braille transcribers has eight students, whose ages are given here. Find the mean and median for these ages.

    55   65   59   30   54   41   60   62

16. A study group has six members, who have the following IQs. Find the mean and median for these IQs.

    108   104   110   104   92   106

17. Here are the math exam scores of eight randomly selected students. Find the median and mean for these scores.

    62   75   72   95   82   86   80   71

18. Here are the speeds of seven pitches thrown by a professional baseball pitcher, in miles per hour. Find the median and mean of these speeds.

    95   79   88   98   96   94   72

19. For the taping of a game show, ten potential contestants are present. Here are the ages of these potential contestants. Find the mean and mode for these ages.

    46   35   37   38   28   42   58   38   32   39

20. The Registrar's Office employs 12 student workers. Here are the number of units each is taking this semester. Find the mean and mode for these amounts.

    9   14   16   12   12   18
    11   10   12   15   12   11

For the sets of data in Exercises 21–34, calculate (a) the mean, (b) the median, (c) the first quartile, (d) the third quartile, (e) the midrange, and (f) the mode.

21. Daily volumes of the New York Stock Exchange for August 1999 (millions of shares):

    649   739   789   859   699   684   836   793
    745   692   583   691   683   684   661   682
    732   864   719   569   605   860

22. Times (in minutes) to complete 20 randomly selected National League baseball games:

    166   153   153   141   134   183   167   240   210   165
    156   176   172   153   159   196   131   181   188   147

23. Times (in minutes) to complete 16 randomly selected American League baseball games:

    191   161   165   183   176   194   218   162
    149   154   154   166   194   200   193   145

24. Number of Republicans in the House of Representatives, from the 1949–51 session through the 1997–99 session:

Year	Number of Republicans	Year	Number of Republicans
1949–51	171	1975–77	144
1951–53	199	1977–79	143
1953–55	221	1979–81	158
1955–57	203	1981–83	192
1957–59	201	1983–85	167
1959–61	153	1985–87	182
1961–63	175	1987–89	177
1963–65	176	1989–91	175
1965–67	140	1991–93	167
1967–69	187	1993–95	176
1969–71	192	1995–97	230
1971–73	180	1997–99	227
1973–75	192		

(Source: Vital Statistics on Congress 1997–98, Reprinted in the New York Times.)

25. Ages of the presidents of the United States at inauguration (George Washington through Bill Clinton):

57	61	57	57	58	57	61	54	68	51	49
64	50	48	65	52	56	46	54	49	51	47
55	55	54	42	51	56	55	51	54	51	60
62	43	55	56	61	52	69	64	46		

26. Market values, in millions of dollars, of the 20 companies making IPOs (initial public offerings) during the week of October 25, 1999:

1538	130	1044	2131	449	202	346	566	283	1883
2275	50	621	408	51	911	1157	129	189	62

(Source: Renaissance Capital in USA Today.)

27. Scores of 20 randomly selected high school sophomores on the PSAT verbal exam:

29	53	36	41	34	52	42	47	57	37
37	52	59	38	47	46	46	47	49	65

28. Scores of eight randomly selected high school juniors on the PSAT math exam:

23	60	42	56	73	52	58	58

29. Scores of 12 randomly selected high school students on the SAT I verbal exam:

300	410	510	500	580	590
390	530	460	590	780	420

30. Scores of 15 randomly selected high school students on the SAT I math exam:

530	700	530	660	670	590	470	560
520	520	550	410	390	740	540	

31. Point totals of 14 algebra students before the final exam (out of 600 points possible):

426	391	495	423	365	393	355
524	483	531	299	533	374	463

32. Number of visitors to ten companies' Web sites in August 1999:

Company	Visitors (in millions)
AOL	53.43
Yahoo	40.24
Microsoft	35.02
Lycos	29.38
Go Network	20.26
Excite@Home	16.06
Amazon	12.54
Time Warner	11.75
Go2Net	11.10
Altavista	10.22

(Source: Media Metrix; Jupiter Communications New York Times.)

33. Number of shares (in millions) being offered by the 19 companies making initial public offerings (IPOs) during the week of October 18, 1999:

6.0	4.0	100.0	2.9	3.5	6.5	9.0	3.4	7.2	5.5
7.8	4.6	42.9	6.0	1.5	6.5	2.0	10.0	3.8	

(Source: Renaissance Capital in USA Today.)

34. Average number of "wet days" in September for 34 cities from around the world:

City	Wet Days	City	Wet Days
Athens	4	Bermuda	10
Beijing	7	Boston	9

(continues)

(continued)

City	Wet Days	City	Wet Days
Budapest	7	Moscow	13
Buenos Aires	8	New York	9
Cairo	0	Paris	13
Chicago	9	Phoenix	3
Delhi	4	Rio de Janeiro	11
Dublin	12	Rome	5
Frankfurt	13	San Francisco	2
Geneva	10	San Juan	18
Hong Kong	12	Stockholm	14
Houston	8	Sydney	12
Jerusalem	0	Tokyo	12
Johannesburg	2	Toronto	12
London	13	Washington	8
Los Angeles	1	Sydney	12
Madrid	6	Tokyo	12

(Source: Times Books World Weather Guide, New York Times.)

35. Given here are the amounts of calcium (in milligrams) in certain foods.

Dairy Products	Amount	Calcium (mg)
Plain fat-free yogurt	1 cup	400
Swiss cheese	$1\frac{1}{2}$ oz.	408
Chocolate milk shake	10 oz.	374
American cheese	2 oz.	348
Milk	8 oz.	300
Chocolate milk	8 oz.	280
Cheddar cheese	$1\frac{1}{2}$ oz.	300
Ricotta cheese	$\frac{1}{2}$ cup	300
Ice cream or frozen yogurt	1 cup	200

Fruits and vegetables		
Collard greens	1 cup cooked	300
Soybeans	$\frac{1}{2}$ cup cooked	130
Broccoli, cooked or raw	1 cup	118
Orange	1	52
Calcium-fortified orange juice	1 cup	300

Fish		
Sardines	3 oz.	370
Oysters, raw	1 cup	220
Canned pink salmon	$3\frac{1}{3}$ oz.	200

Miscellaneous		
Cheese pizza	1 slice	220
Macaroni and cheese	$\frac{1}{2}$ cup	180
Pancakes made with milk	2	72
Taco salad	1	280
Small salad	1	280
Caffe latte	12 oz.	412

(Source: Rockford Register Star, in Visalia Times Delta)

(a) Find the mean amount of calcium in these 23 items.

(b) Find the mean amount of calcium in each of the four subgroups: dairy products, fruits and vegetables, fish, and miscellaneous.

(c) If you add up the four means found in (b) and divide by 4, will this produce the same answer as you found in (a)? Why or why not? If not, explain what needs to be done with these four means to produce the same mean found in (a).

36. Here are the amounts of caffeine (in milligrams) from various sources.

Product	Caffeine (mg)
Coffee	
Brewed (8 oz.)	135
Instant (8 oz.)	95
Decaffeinated	5
Tea	
Tea (8 oz., leaf or bag)	50
Arizona Iced Tea (16 oz.)	30
Snapple Iced Tea (16 oz.)	48
Decaffeinated	5
Soft Drinks (12 oz. can)	
Barq's Root Beer	23
Coca-Cola (classic and diet)	46
Dr. Pepper (regular and diet)	41
Jolt	71
Josta	58
Mellow Yellow	52
Mountain Dew	55
Pepsi (regular and diet)	36
RC Cola	36
Chocolate	
Hot Cocoa (8 oz.)	5
Hershey Bar (1.5 oz.)	10
Hershey's Special Dark (1.5 oz.)	31
Over-the-counter medications	
Anacin (2 tablets)	64
Excedrin (2 tablets)	130
NoDoz (1 tablet, regular strength)	100

(Source: National Coffee Drinking Trends Survey, National Coffee Association of USA Inc.; Nutrition Action Healthletter, Center for Science in the Public Interest—in USA Today *magazine.)*

(a) Find the mean amount of caffeine in these 22 items.
(b) Find the mean amount of caffeine in each of the five subgroups: coffee, tea, soft drinks, chocolate, and over-the-counter medications.
(c) If you add up the five means found in (b) and divide by 5, will this produce the same answer as you found in (a)? Why or why not? If not, explain what needs to be done with these five means to produce the same mean found in (a).

37. Make up a set of data that has at least six distinct values, for which the mean is greater than the median.
38. Make up a set of data that has at least six distinct values, for which the mean is less than the median.
39. Make up a set of data that has at least six distinct values, for which the mean is 75.
40. Make up a set of data that has at least six distinct values, for which the mean is 68.2.
41. Make up a set of data that has at least six distinct values, for which the median is 15.
42. Make up a set of data that has at least six distinct values, for which the median is 17.5.
43. Make up a set of data that has no mode.
44. Make up a set of data that has three modes.

Wilt Chamberlain played center for 14 NBA seasons in Philadelphia, San Francisco, and Los Angeles. Among his notable achievements were scoring 100 points in a single NBA game, averaging over 50 points per game over an entire season, and never fouling out of a basketball game during his entire career. Here are his point totals and rebound totals for those 14 seasons.

Season	Team	Points	Rebounds
1959–60	Philadelphia	2707	1941
1960–61	Philadelphia	3033	2149
1961–62	Philadelphia	4029	2052
1962–63	San Francisco	3586	1946
1963–64	San Francisco	2948	1787
1964–65	San Francisco	2534	1673
1965–66	Philadelphia	2649	1943
1966–67	Philadelphia	1956	1957
1967–68	Philadelphia	1992	1952
1968–69	Los Angeles	1664	1712
1969–70	Los Angeles	328	221
1970–71	Los Angeles	1696	1493
1971–72	Los Angeles	1213	1572
1972–73	Los Angeles	1084	1526

45. (a) Find the mean number of points per season and the median number of points per season.
    (b) An injury during the 1969–70 season cut Wilt's season dramatically short. If we omit that outlier, find the mean number of points per season, and the median number of points per season for the remaining 13 seasons.
    (c) Did the mean number of points change significantly from (a) to (b)? Did the median number of points change significantly from (a) to (b)? Explain these results.

46. (a) Find the mean number of rebounds per season and the median number of rebounds per season.
    (b) An injury during the 1969–70 season cut Wilt's season dramatically short. If we omit that outlier, find the mean number of rebounds per season, and the median number of rebounds per season for the remaining 13 seasons.
    (c) Did the mean number of rebounds change significantly from (a) to (b)? Did the median number of rebounds change significantly from (a) to (b)? Explain these results.

47. During the 46 years from 1950 through 1995, the state of Utah was not struck by a tornado in 19 of those years. Here is a table that shows the frequency of tornadoes in Utah for these 46 years.

Number of Tornadoes	Number of Years
0	19
1	8
2	6
3	4
4	3
5	3
6	3

Use the weighted mean to calculate the mean number of tornadoes per year in Utah.

48. Here are the number of absences for 20 classes at an elementary school.

Absences	Number of Classes
0	11
1	4
2	3
3	0
4	2

Use the weighted mean to calculate the mean number of students absent.

49. A statistics class had four exams. On exam 1, 32 students had a mean score of 81.3. On exam 2, 30 students had a mean score of 68.6. On exam 3, 24 students had a mean score of 74.2. On exam 4, 23 students had a mean score of 83.5. Use the weighted mean to find the mean score of all four exams combined.

50. Here are the number of rushing attempts and average yards per carry for six NFL teams. Use the weighted mean to find the mean number of yards per carry for the six teams combined.

Team	Number of Attempts	Yards per Carry
Kansas City	39	2.5
Washington	24	6.8
Jacksonville	34	5.4
Arizona	19	4.0
Denver	30	4.5
Carolina	18	3.5

51. Here is a frequency distribution showing the scores of 40 students on an algebra midterm exam.

Score	Frequency
30–40	4
40–50	2
50–60	5
60–70	7
70–80	14
80–90	5
90–100	3

Use this frequency distribution to estimate the mean score on this exam.

52. Here is a frequency distribution showing the income level of 98 randomly selected 25-year-old males.

Income	Frequency
$0–$5,000	4
$5,000–$10,000	9
$10,000–$15,000	11
$15,000–$25,000	20
$25,000–$35,000	17
$35,000–$50,000	17
$50,000–$75,000	12
$75,000–$100,000	8

Use this frequency distribution to estimate the mean income of these 25-year-old males.

53. Here is a frequency distribution showing the ages of 121 randomly selected people who have a bachelor's degree or higher.

Ages	Frequency
18–25	8
25–35	32
35–45	35
45–55	23
55–65	11
65–85	9
85–105	3

(Source: Based on the results of a study by the Census Bureau.)

Use this frequency distribution to estimate the mean age of these 121 people.

54. According to the Centers for Disease Control and Prevention, there were 30,903 suicides in the United States in 1996. Here is a frequency distribution for the ages of the 30,879 people whose ages were known. (Twenty-four of the ages were unknown.)

Age	Number
0–10	4
10–20	2115
20–30	5357
30–40	6521
40–50	6050
50–60	3667
60–70	2729
70–80	2667
80–100	1769

(Source: USA Today.)

Use this frequency distribution to estimate the mean age of a person who commits suicide.

55. Here are the SAT I composite scores (verbal + math) of 40 high school students.

580	1060	1030	1130	1230	1240	1030	1170	1120	940
1040	1040	1260	1000	1120	950	1110	1440	910	1010
880	860	920	1240	1160	770	1110	890	1170	1100
930	740	1010	1200	1030	810	1200	680	1000	820

(a) Calculate the mean score for these 40 students.
(b) Create a frequency distribution for these scores.
(c) Use the frequency distribution in (b) to estimate the mean score for these students. Compare your result to (a). Does it seem reasonable?

56. Here are the number of electoral college votes that the 50 states and the District of Columbia had in the 1996 presidential election.

State	Votes	State	Votes	State	Votes
AL	9	DE	3	IN	12
AK	3	DC	3	IA	7
AZ	8	FL	25	KS	6
AR	6	GA	13	KY	8
CA	54	HI	4	LA	9
CO	8	ID	4	ME	4
CT	8	IL	22	MD	10

(continues)

(continued)

State	Votes	State	Votes	State	Votes
MA	12	NM	5	SD	3
MI	18	NY	33	TN	11
MN	10	NC	14	TX	32
MS	7	ND	3	UT	5
MO	11	OH	21	VT	3
MT	3	OK	8	VA	13
NE	5	OR	7	WA	11
NV	4	PA	23	WV	5
NH	4	RI	4	WI	11
NJ	15	SC	8	WY	3

(a) Calculate the mean number of electoral votes for the 50 states and Washington, D.C.

(b) Create a frequency distribution for these scores.

(c) Use the frequency distribution in (b) to estimate the mean score for these students. Compare your result to (a). Does it seem reasonable?

57. Here are the number of earned runs allowed by each of the 30 major league teams during the 1999 baseball season. The teams are divided into American League teams and National League teams.

Al Team	Earned Runs	NL Team	Earned Runs
Anaheim	762	Arizona	615
Baltimore	761	Atlanta	596
Boston	638	Chicago	835
Chicago	787	Cincinnati	643
Cleveland	792	Colorado	957
Detroit	824	Florida	781
Kansas City	845	Houston	622
Minnesota	795	Los Angeles	718
New York	665	Milwaukee	815
Oakland	760	Montreal	749
Seattle	836	New York	691
Tampa Bay	806	Philadelphia	788
Texas	809	Pittsburgh	692
Toronto	788	San Diego	705
		San Francisco	764
		St. Louis	764

(a) Calculate the mean number of earned runs for the American League teams, and then calculate the mean number of earned runs for the National League teams.

(b) Calculate the median number of earned runs for the American League teams, and then calculate the median number of earned runs for the National League teams.

(c) Based on your results from (a) and (b), do you feel that the number of earned runs allowed in the American League and in the National League are substantially different?

58. Here are the number of home runs hit by each of the 30 major league teams during the 1999 baseball season. The teams are divided into American League teams and National League teams.

AL Team	Home Runs	NL Team	Home Runs
Anaheim	158	Arizona	216
Baltimore	203	Atlanta	197
Boston	176	Chicago	189

(continues)

(continued)

AL Team	Home Runs	NL Team	Home Runs
Chicago	162	Cincinnati	209
Cleveland	209	Colorado	223
Detroit	212	Florida	128
Kansas City	151	Houston	168
Minnesota	105	Los Angeles	187
New York	193	Milwaukee	165
Oakland	235	Montreal	163
Seattle	244	New York	179
Tampa Bay	145	Philadelphia	161
Texas	230	Pittsburgh	171
Toronto	212	San Diego	153
		San Francisco	188
		St. Louis	194

(a) Calculate the mean number of home runs for the American League teams, and then calculate the mean number of home runs for the National League teams.

(b) Calculate the median number of home runs for the American League teams, and then calculate the median number of home runs for the National League teams.

(c) Based on your results from (a) and (b), do you feel that the number of home runs hit in the American League and in the National League are substantially different?

Again, use the data collected in the mini-project in Section 1.1. For each of the following, calculate (a) the mean, (b) the median, (c) the first quartile, (d) the third quartile, (e) the midrange, and (f) the mode.

- Age
- Male height
- Female height
- Number of units enrolled in this term
- Number of hours worked per week (outside of class)

MINI PROJECT

SECTION 2.2
Measures of Dispersion

Although measures of central tendency are important when we are trying to describe a set of data, they are not enough. Two sets of data may be centered in the same location, but still be totally different sets of data. Another important characteristic to consider is how the data are dispersed or spread out. In some sets the values are closely grouped together, while in others they are far apart from each other.

Here are the five test scores of two statistics students.

Jack	85	70	55	41	99
Jill	72	68	70	65	75

The mean score for each student is 70, but the two sets of scores are different. Jill's scores are close together; no score is more than 5 points away from her mean score. Jack's scores are more spread out. Two of his scores are 29 points away from his mean score. We can expect Jill to score closer to 70 on the next test than Jack.

Range

The first measure of dispersion that we will discuss is the **range**. To find the range for a set of data, we subtract the lowest value from the highest value.

Range = highest − lowest

Here are the rushing totals (in yards) for Barry Sanders in the years 1989 through 1998. Find the range for these ten seasons.

1470	1304	1548	1352	1115
1883	1500	1553	2053	1491

The highest value is 2053 yards, and the lowest is 1115 yards.

Range = 2053 − 1115

= 938

The range for this set of data is 938 yards. ■

The range tells us how far apart the lowest and highest values are. One problem with the range is that it is sensitive to outliers. If a set of data has an outlier, the range uses it in its calculation. Another problem with the range is that two sets of data can be spread out in totally different fashions, but have the same range.

Here are the scores of two golfers from last month's matches.

Jack	71	72	73	71	73	82	70	72	68
Greg	75	73	77	78	78	81	74	71	85

The range for both golfers is 14 strokes, but the scores are dispersed in a much different fashion. Jack's scores are closely grouped together between 68 and 73, with an outlier at 82. Greg's scores are evenly dispersed from 71 to 85.

The range should be used only as a first step in investigating the dispersion for a set of data. Although it can give us an idea about how spread out the values are, it cannot paint the whole picture for us.

Interquartile Range

Another measure of dispersion is the interquartile range, the distance between the first and third quartiles.

Interquartile range = $Q_3 - Q_1$

Instead of telling us how far apart the two extreme values are, it tells us the range in which we can find the middle 50% of the values. It is not sensitive to outliers.

Here are the SAT math scores for 19 randomly selected students. Find the interquartile range.

480	370	540	660	650	710	470
490	630	390	430	320	470	400
430	570	450	470	530		

Recall that the first step is to put the values in ascending order, and find the median. Then to find the first quartile we find the median of the first group. The third quartile is the median of the second group.

First Group   320   370   390   400   430   430   450   470   470

Median: 470

Second Group   480   490   530   540   570   630   650   650   710

Recall from the last section that the first quartile is 430, and the third quartile is 570.

570 − 430 = 140

The interquartile range for these scores is 140 points. ■

Boxplot

One more way to represent a set of data graphically is with a **boxplot**. A boxplot needs five values for its construction: the lowest value, the first quartile, the median, the third quartile, and the highest value. These five values are often referred to as the **five-number summary** for a set of data. Above a horizontal axis, we draw a box from the first quartile to the third quartile. We put a dashed line in the box at the median. Finally, extend out line segments from the box to the lowest and highest values.

For the previous 19 test scores, here is the five-number summary.

Lowest	Q_1	Median	Q_3	Highest
320	430	470	570	710

Here is a boxplot for this set of data.

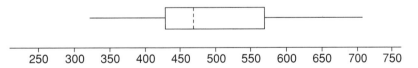

The range is the distance from the far left to the far right, and the interquartile range is the width of the box.

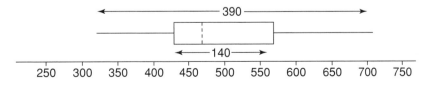

Each year, *Forbes* magazine prints a list of the 400 richest people/families in the world. From their 1999 issue, here are the ages of the 54 richest individuals. All of these individuals have a net worth of at least $5 billion. Construct a boxplot for these data.

43	68	46	43	34	59	55	42
73	71	73	75	71	84	43	36
68	54	77	76	62	59	62	69
31	60	82	88	68	59	81	51
75	79	51	62	41	50	74	82
61	56	66	75	52	59	44	58
51	58	72	62	72	50		

We will begin with the stem-and-leaf display.

Stem	Leaf
3	1 4 6
4	1 2 3 3 3 4 6
5	0 0 1 1 1 2 4 5 6 8 8 9 9 9 9
6	0 1 2 2 2 2 6 8 8 8 9
7	1 1 2 2 3 3 4 5 5 5 6 7 9
8	1 2 2 4 8

Here is the five-number summary; verify it with the work from the previous section.

Lowest	Q$_1$	Median	Q$_3$	Highest
31	51	61.5	73	88

The range is 57 years, and the interquartile range is 22 years. Here is the boxplot. ■

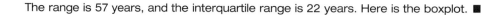
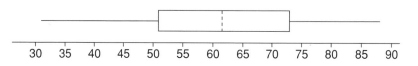

Although the range and interquartile range can give us a good idea of a set of data's dispersion, they do have their drawbacks. They do not use all of the data values in their calculation. They measure dispersion from one end to the other, measuring the width between two values. Another way to look at dispersion is by examining how far the values are from the center of the set. For instance, an important characteristic of a new home is its distance from the center of town. We will continue now by examining three measures that determine dispersion in this fashion.

Mean Deviation

To find the **mean deviation** for a set of data, we begin by finding the distance from each value to the mean. We then find the mean of these distances. This is the mean deviation. Roughly interpreted, it tells us how far from the center of the data the values are on average. A measure of distance is nonnegative, so to calculate each distance we take the absolute value of the difference between the value and the mean. Here is the formula.

$$\text{Mean deviation} = \frac{\Sigma |x - \bar{x}|}{n}$$

This formula uses sample notation, but the procedure is exactly the same for the mean deviation of a population. Simply replace $\bar{x}$ by μ, and n by N. Here are the steps for this calculation.

1. Find the mean.
2. Subtract the mean from each value.
3. Take the absolute value of each difference.
4. Total these distances.
5. Divide the total by the number of values in the set.

 EXAMPLE 2.23

A taxi dispatcher is interested in the number of fares for his drivers on Fridays. He randomly selects seven drivers, and then randomly selects one Friday for each of the drivers. Here are the number of fares for each. Find the mean deviation for these totals.

| 32 | 27 | 30 | 41 | 29 | 38 | 34 |

For this calculation it is best to use a column approach.

x	$x - \bar{x}$	$|x - \bar{x}|$
32	−1	1
27	−6	6
30	−3	3
41	8	8
29	−4	4
38	5	5
34	1	1
$\bar{x} = 33$		28

$$\text{Mean deviation} = \frac{28}{7}$$

$$= 4$$

The mean deviation is 4 fares. We can say that on average, the values are 4 fares away from the mean. ∎

We should note one more thing from the previous example. Look at the middle column, labeled $x - \bar{x}$. The sum of that column is 0. This shows us that the sum of the distances to the right of the mean (positive values) is equal to the sum of the distances to the left of the mean (negative values). This is evidence that the mean truly is in the center of the data.

In practice, that middle column is not necessary. Simply find the distance between each value and the mean, and place that in the right column.

 **EXAMPLE 2.24**

Here is a list of the New York Stock Exchange's daily volumes for one week of trading, in millions of shares. Find the mean deviation for these values.

| 669 | 754 | 752 | 771 | 835 |

Since there is no suggestion that we are interested in anything except these five values, we will treat this set of data as a population.

x	$x - \mu$	$|x - \mu|$
669	–87.2	87.2
754	–2.2	2.2
752	–4.2	4.2
771	14.8	14.8
835	78.8	78.8
$\mu = 756.2$		187.2

$$\text{Mean deviation} = \frac{187.2}{5}$$

$$= 37.44$$

The mean deviation is 37.44 million shares. On the average, the values are 37.44 million shares away from the mean. ∎

Variance

Another measure of dispersion that measures from the inside out is variance. Variance is similar to mean deviation, with two exceptions. The first difference between these two measures is that we square the difference between each value and the mean, rather than taking the absolute value. The other difference is that there are two formulas, depending on whether we are finding the variance of a sample or the variance of a population. Here are the two formulas.

Sample Variance

$$s^2 = \frac{\Sigma(x - \bar{x})^2}{n - 1}$$

Population Variance

$$\sigma^2 = \frac{\Sigma(x - \mu)^2}{N}$$

We use the symbol s^2 to represent sample variance, and the symbol σ^2 to represent population variance. σ is the lowercase Greek letter sigma, which is the Greek letter s.

Note that there is a significant difference in the two formulas. The denominator in the sample variance formula calls for us to subtract 1 from the sample size, whereas the denominator in the population variance uses the population size, without subtracting 1. Why the difference? Although this is beyond the scope of this course, subtracting 1 from the sample size makes the sample variance an *unbiased* estimator of the population variance.

It is crucial that you are able to identify whether a set of data is a sample or a population. Using the sample formula by mistake will produce a variance that is too large. Using the population formula by mistake will produce a variance that is too small.

 EXAMPLE 2.25

A taxi dispatcher is interested in the number of fares for his drivers on Fridays. He randomly selects seven drivers, and then randomly selects one Friday for each of the drivers. Here are the number of fares for each. Find the variance for these totals.

32   27   30   41   29   38   34

Since the dispatcher is interested in all Fridays, these data are a sample. For this calculation, just as with mean deviation, it is best to use a column approach.

x	$x - \bar{x}$	$(x - \bar{x})^2$
32	−1	1
27	−6	36
30	−3	9
41	8	64
29	−4	16
38	5	25
34	1	1
$\bar{x} = 33$		152

$$s^2 = \frac{152}{7 - 1}$$

$$= \frac{152}{6}$$

$$= 25.33$$

The variance is 25.33. ■

A drawback for variance is that it lacks the interpretation that mean deviation has. The idea to keep in mind is that the bigger the variance is, the more spread out the values are. An advantage of variance is that its formula does not involve absolute values, which are difficult to manipulate algebraically.

EXAMPLE 2.26   Here is a list of the New York Stock Exchange's daily volumes for one week of trading, in millions of shares. Find the variance for these values.

669	754	752	771	835

Since there is no suggestion that we are interested in anything except these five values, we will treat this set of data as a population.

x	$x - \mu$	$(x - \mu)^2$
669	−87.2	7,603.84
754	−2.2	4.84
752	−4.2	17.64
771	14.8	219.04
835	78.8	6,209.44
$\mu = 756.2$		14,054.8

$$\sigma^2 = \frac{\Sigma(x - \mu)^2}{N}$$

$$= \frac{14,054.8}{5}$$

$$= 2,810.96$$

The variance is 2,810.96. ■

Standard Deviation

The measure of dispersion that we will use most often in this course is the standard deviation. The **standard deviation** of a set of data is the square root of the variance.

Standard deviation $= \sqrt{\text{variance}}$

We use the letter s to represent sample standard deviation, and σ to represent population standard deviation.

Sample Standard Deviation	Population Standard Deviation
$s = \sqrt{\dfrac{\Sigma(x - \bar{x})^2}{n - 1}}$	$\sigma = \sqrt{\dfrac{\Sigma(x - \mu)^2}{N}}$

 EXAMPLE 2.27

A taxi dispatcher is interested in the number of fares for his drivers on Fridays. He randomly selects seven drivers, and then randomly selects one Friday for each of the drivers. Here are the number of fares for each. Find the standard deviation for these totals.

| 32 | 27 | 30 | 41 | 29 | 38 | 34 |

In a previous example, we calculated the sample variance for these values to be 25.33.

$s = \sqrt{25.33}$

$= 5.03$

The sample standard deviation is 5.03 fares. ■

 **EXAMPLE 2.28**

Here is a list of the New York Stock Exchange's daily volumes for one week of trading, in millions of shares. Find the variance for these values.

| 669 | 754 | 752 | 771 | 835 |

In a previous example, we calculated the population variance for these values to be 2810.96.

$\sigma = \sqrt{2810.96}$

$= 53.02$

The standard deviation is 53.02 million shares. ■

 EXAMPLE 2.29

A student is interested in how old women are when they first get married. To estimate the mean age, she goes into 11 randomly selected chat rooms, and asks randomly selected women how old they were at their first marriage until she gets a response in each room. Here are the 11 ages. Find the standard deviation of these ages.

| 21 | 20 | 19 | 16 | 22 | 21 | 21 | 19 | 18 | 24 | 18 |

Since the student is interested in the mean age for all women, these 11 values represent a sample.

x	$x - \bar{x}$	$(x - \bar{x})^2$
21	1.1	1.21
20	0.1	0.01
19	−0.9	0.81
16	−3.9	15.21
22	2.1	4.41
21	1.1	1.21
21	1.1	1.21
19	−0.9	0.81
18	−1.9	3.61
24	4.1	16.81
18	−1.9	3.61
$\bar{x} = 19.9$		48.91

$$s = \sqrt{\frac{\Sigma(x - \bar{x})^2}{n - 1}}$$

$$= \sqrt{\frac{48.91}{10}}$$

$$= 2.21$$

The standard deviation is 2.21 years. ∎

EXAMPLE 2.30

On July 15, 1999, Bruce Springsteen and the E Street Band began a series of 15 concerts in Bruce's home state, New Jersey. Fans who attended several of the concerts raved about how different the shows were each night. Here is a list of the number of songs played at each of the 15 concerts. Find the standard deviation.

26	26	24	23	23	23	25	23
22	22	23	22	23	25	24	

Since there is no suggestion that we are interested in anything beyond these 15 concerts, we will treat this as a population.

x	$x - \mu$	$(x - \mu)^2$
26	2.4	5.76
26	2.4	5.76
24	0.4	0.16
23	−0.6	0.36
23	−0.6	0.36
23	−0.6	0.36
25	1.4	1.96
23	−0.6	0.36
22	−1.6	2.56
22	−1.6	2.56
23	−0.6	0.36
22	−1.6	2.56
23	−0.6	0.36
25	1.4	1.96
24	0.4	0.16
$\mu = 23.6$		25.6

$$\sigma = \sqrt{\frac{\Sigma(x - \mu)^2}{N}}$$

$$= \sqrt{\frac{25.6}{15}}$$

$$= 1.31$$

The standard deviation is 1.31 songs. ∎

As mentioned before, we should try to take advantage of technology (computer, calculator) when calculating standard deviation. When using the calculator, the process is the reverse of what we have shown by hand. We first calculate the standard deviation, and then if the variance is needed we square the standard deviation.

Skewness

Standard deviation is a measure of dispersion that is used often in inferential statistics. We introduce three of its uses here. The first has to do with **skewness.** A set of data is said to be skewed if it is not symmetrical. Here is a histogram from a set of data that is roughly symmetrical.

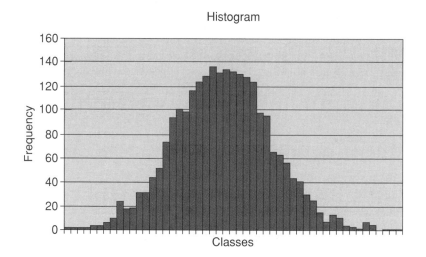

The peak of the previous histogram is located in the center. A set of data is skewed if we notice that the histogram is stretched to the left or stretched to the right, such as in the following two histograms.

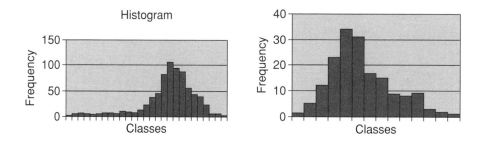

A set of data that is stretched to the left is **negatively skewed.** A set of data is negatively skewed if the values to the left of the median are more spread out than the values on the right side of the median. A low outlier(s) can cause a set of data to be negatively skewed. In such a situation, the mean will be less than the median. (Why?) If a set of data is stretched to the right, we say that it is **positively skewed.**

We have a measure that calculates how skewed a set of data is—the **coefficient of skewness.** Here is the formula.

$$sk = \frac{3(\text{mean} - \text{median})}{\text{standard deviation}}$$

If a set of data is positively skewed, its mean will be greater than its median, causing the coefficient of skewness to be positive. A negative coefficient of skewness indicates that a set of data is negatively skewed. The coefficient of skewness will be a number between –3 and 3. The closer it is to 0, the more symmetric the data are. The farther away from 0 the coefficient of skewness is, the more skewed the data are.

EXAMPLE 2.31

Here are the ages of ten randomly selected women at a college orientation. Find the coefficient of skewness for these data.

| 18 | 25 | 31 | 19 | 22 | 21 | 19 | 25 | 18 | 27 |

The data represent a sample. To calculate the coefficient of skewness we need to find the mean, median, and standard deviation. First the mean.

$$\bar{x} = \frac{\Sigma x}{n}$$
$$= \frac{225}{10}$$
$$= 22.5$$

Next, find the median.

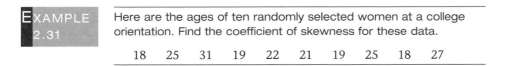

| 18 | 18 | 19 | 19 | 21 | | 22 | 25 | 25 | 27 | 31 |

$$\text{Median} = \frac{21 + 22}{2}$$
$$= 21.5$$

Finally, find the standard deviation. You could use your calculator for this; however, you should also be able to calculate standard deviation through the use of the formula.

x	$x - \bar{x}$	$(x - \bar{x})^2$
18	−4.5	20.25
18	−4.5	20.25
19	−3.5	12.25
19	−3.5	12.25
21	−1.5	2.25
22	−0.5	0.25
25	2.5	6.25
25	2.5	6.25
27	4.5	20.25
31	8.5	72.25
$\bar{x} = 22.5$		172.5

$$s = \sqrt{\frac{\Sigma(x - \bar{x})^2}{n - 1}}$$

$$= \sqrt{\frac{172.5}{9}}$$

$$= 4.38$$

Now we can calculate the coefficient of skewness.

$$sk = \frac{3(\text{mean} - \text{median})}{\text{standard deviation}}$$

$$= \frac{3(22.5 - 21.5)}{4.38}$$

$$= 0.68$$

These data have a coefficient of skewness of 0.68. They are moderately positively skewed. Here is a boxplot to demonstrate the skewness.

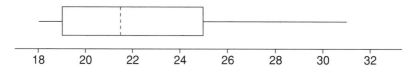

Note that the side to the right of the median is more stretched out than the side to the left of the median. ■

Standard Units

Standard deviation can be used to help us compare two values from two different sets of data. For instance, suppose a man is 6′4″ tall, and his IQ is 130. Both of these values are above the mean, but which stands out the most? The mean height of adult males is 69.0 inches (5′9″), so the man is 7 inches above the mean height. The mean IQ is 100 points, so the man's IQ is 30 points above the mean IQ. We cannot compare the 7 inches to the 30 points, because these are based on two different scales. A fair comparison would be to compare how many standard deviations above the mean these two values are. Finding how many standard deviations away from the mean a value is converts the value to **standard units,** or its **z-score.** To convert a value to standard units, we subtract the mean from the value, and then divide by the standard deviation.

$$z = \frac{\text{value} - \text{mean}}{\text{standard deviation}}$$

Adult male heights have a mean of 69.0 inches, with a standard deviation of 2.8 inches. Here is the z-score that corresponds to 6′4″:

$$z = \frac{\text{value} - \text{mean}}{\text{standard deviation}}$$

$$= \frac{76 - 69.0}{2.8}$$

$$= 2.5$$

This tells us that the man's height is 2.5 standard deviations above the mean. IQ scores have a mean of 100 points and a standard deviation of 15 points. Here is the standard deviation that is associated with an IQ of 130 points.

$$z = \frac{\text{value} - \text{mean}}{\text{standard deviation}}$$

$$= \frac{130 - 100}{15}$$

$$= 2$$

This tells us that the man's IQ is 2 standard deviations above the mean. Relatively speaking, the man's height is greater than his IQ.

Standard units, or z-scores, are positive for values that are to the right of the mean and negative for values that are to the left of the mean. A z-score of 0 indicates that a value is the mean. The bigger the z-score, the farther away from the mean that value is. We will use z-scores a great deal later in the course.

 EXAMPLE 2.32

Here is a list of the New York Stock Exchange's daily volumes for one week of trading, in millions of shares. Convert each volume to its z-score.

669   754   752   771   835

Recall that earlier the mean for these values was 756.2 million shares, and the standard deviation was 53.02 million shares.

669

$$z = \frac{669 - 756.2}{53.02}$$

$$= -1.64$$

754

$$z = \frac{754 - 756.2}{53.02}$$

$$= -0.04$$

752

$$z = \frac{752 - 756.2}{53.02}$$

$$= -0.08$$

771

$$z = \frac{771 - 756.2}{53.02}$$

$$= 0.28$$

835

$$z = \frac{835 - 756.2}{53.02}$$

$$= 1.49 \ \blacksquare$$

In the previous example, the volume of 835 million shares is the farthest from the mean, and therefore has the greatest z-value. Note also that if we total all of the z-values, they add to 0.01. If these values had not been rounded, they would have totaled to 0. This is because the numerators represent the deviations from the mean, which we know total to 0.

Chebyshev's Theorem

Standard deviation can be used to tell us what percentage of the values lie in a certain interval. Our first introduction to this is **Chebyshev's Theorem.** For example, if $k = 2$, then at least $1 - \frac{1}{2^2}$ or $\frac{3}{4}$ of the data must lie within 2 standard deviations of the mean. What percentage of the data must lie within 2.5 standard

> **CHEBYSHEV'S THEOREM** *For any set of data and any constant $k > 1$, at least $1 - \frac{1}{k^2}$ of the data must lie within k standard deviations of the mean on either side.*

deviations of the mean? We apply Chebyshev's Theorem with $k = 2.5$. At least $1 - \frac{1}{2.5^2}$ or 84% of the data must lie within 2.5 standard deviations of the mean.

Chebyshev's Theorem provides us with a minimum guarantee, but we will see later that if a set of data follows a "bell-shaped" distribution, then the percentage of values in the interval is much higher than Chebyshev's Theorem tells us. For example, we saw that Chebyshev's Theorem tells us that at least 75% of the data lie within 2 standard deviations of the mean when "bell-shaped" distributions have approximately 95% of their data within 2 standard deviations of the mean.

 The time required to complete a mathematics placement test has a mean of 40 minutes, with a standard deviation of 8 minutes. At least what percentage of students take from 30 minutes to 50 minutes to finish the exam?

This is a problem requiring Chebyshev's Theorem. We begin by finding the value for k. We must figure out how many standard deviations away from the mean our two values are, so we convert them to their z-scores first.

30 minutes	**50 minutes**
$z = \dfrac{30 - 40}{8}$	$z = \dfrac{50 - 40}{8}$
$= -1.25$	$= 1.25$

Since each value is 1.25 standard deviations away from the mean, we can use 1.25 for k. At least $1 - \frac{1}{1.25^2}$ or 36% of all students take from 30 minutes to 50 minutes to complete the exam. ∎

Estimation of Standard Deviation Using a Frequency Distribution

We conclude this section by learning how to estimate the standard deviation of a set of data if we are given only a frequency distribution. This process is similar to the calculations required to estimate a mean from a frequency distribution. We will estimate the mean first, subtract it from each class mark, square those differences, and then weight these squares according to the class frequencies. We finish by totaling these products, dividing by the appropriate denominator, and taking the square root. Here is the formula for estimating sample standard deviation.

$$s_g = \sqrt{\frac{\Sigma \text{ frequency}(\text{class mark} - \bar{x}_g)^2}{n - 1}}$$

Given here is a frequency distribution for the scores of 32 randomly selected statistics students on an exam on Unit 1.

Score	*Frequency*
40–50	3
50–60	2
60–70	4
70–80	9
80–90	7
90–100	7

We will now estimate the standard deviation for these scores. Keep in mind that this is a sample.

Score	Frequency (f)	Class Mark (x_m)	$f \cdot x_m$
40–50	3	45	135
50–60	2	55	110
60–70	4	65	260
70–80	9	75	675
80–90	7	85	595
90–100	7	95	665
			2440

$$\bar{x}_g = \frac{2440}{32}$$

$$= 76.25$$

Score	Frequency (f)	Class Mark (x_m)	$x_m - \bar{x}_g$	$(x_m - \bar{x}_g)^2$	$f \cdot (x_m - \bar{x}_g)^2$
40–50	3	45	−31.25	976.5625	2929.6875
50–60	2	55	−21.25	451.5625	903.125
60–70	4	65	−11.25	126.5625	506.25
70–80	9	75	−1.25	1.5625	14.0625
80–90	7	85	8.75	76.5625	535.9375
90–100	7	95	18.75	351.5625	2460.9375
					7350

$$s_g = \sqrt{\frac{7350}{31}}$$

$$= 15.40$$

We estimate the standard deviation to be 15.40. The actual standard deviation of those 32 scores is approximately 14.7, so this is not a bad estimate.

MICROSOFT EXCEL   Measures of Dispersion

Excel provides numerous built-in functions for calculating measures of dispersion. In each of the following definitions, **cell range** refers to the range of cells containing data. For example, if you have values in the first 50 cells of column A, type A1:A50 for the cell range.

RANGE

Excel does not have a built-in function to calculate the range, but we can find it using the built-in functions MAX and MIN. MAX returns the largest value in the range of cells, and MIN returns the smallest value. Type the following in any empty cell.

    =MAX(cell range)-MIN(cell range)

If you have already sorted the values in ascending order, the range will be equal to the last value minus the first value.

INTERQUARTILE RANGE

Again, Excel does not have a built-in function for calculating interquartile range, but instead we will use QUARTILE to calculate the third and first quartiles and then subtract them. Type the following in any empty cell.

=QUARTILE(cell range,3)-QUARTILE(cell range,1)

MEAN DEVIATION

Excel does have a built-in function for this one, AVEDEV. Type the following in any empty cell.

=AVEDEV(cell range)

VARIANCE

Excel has two different built-in functions for calculating variance: one for sample variance (VAR) and one for population variance (VARP). To calculate sample variance, type the following in any empty cell.

=VAR(cell range)

To calculate population variance, type the following in any empty cell.

=VARP(cell range)

STANDARD DEVIATION

Excel also has two different built-in functions for calculating standard deviation, one for sample standard deviation (STDEV) and one for population standard deviation (STDEVP). To calculate sample standard deviation, type the following in any empty cell.

=STDEV(cell range)

To calculate population standard deviation, type the following in any empty cell.

=STDEVP(cell range)

STANDARD UNITS (z-SCORE)

To calculate the z-score using Excel, we use the built-in function STANDARDIZE. This function requires three arguments: the value, the mean, and the standard deviation. If you know the mean and standard deviation, you may type the numbers. If there is a set of data, you may use the built-in functions AVERAGE and STDEV. Type one of the following in any empty cell.

=STANDARDIZE(value, mean, standard deviation)

or

=STANDARDIZE(cell, AVERAGE(cell range), STDEV(cell range))

COEFFICIENT OF SKEWNESS

Excel does not have a built-in function to calculate our coefficient of skewness, but we can use the built-in functions AVERAGE, MEDIAN, and STDEV to calculate it. Type the following in any empty cell.

=3*(AVERAGE(cell range)-MEDIAN(cell range))/STDEV(cell range)

There are different coefficients of skewness. Excel's built-in function **SKEW** calculates the coefficient of skewness defined by the formula

$$\frac{n}{(n-1)(n-2)} \Sigma z^3$$

To use this coefficient of skewness, type the following in any empty cell.

=SKEW(cell range)

ESTIMATING STANDARD DEVIATION USING A FREQUENCY DISTRIBUTION

We can also use Excel to estimate the standard deviation of a set of data from a frequency distribution. We will now estimate the standard deviation of a set of test scores based on the following frequency distribution. We will assume that these test scores represent a sample of all students.

Score	Frequency
40–50	2
50–60	2
60–70	9
70–80	13
80–90	14
90–100	17

Before we begin, recall that the formula for estimating sample standard deviation is

$$s_g = \sqrt{\frac{\Sigma \text{ frequency(class mark} - \bar{x}_g)^2}{n-1}}$$

Type **Score** in cell **A1**. Below it, in cell **A2**, type 40 to 50. Type the remainder of the classes in cell **A3** through **A7**. In the next column, type Frequency in cell **B1**. Type the respective frequencies in cells **B2** through **B7**, next to their classes.

To estimate the standard deviation using a frequency distribution, we must first estimate the mean. To do that we multiply the class mark for each class by the frequency for that class. We will need a column containing the class marks. In cell **C1**, type **Class Marks**. Type 45 in cell **C2**, 55 in **C3**, . . . , and 95 in cell **C7**. Now we want to multiply each class mark by its class frequency. In cell **D1**, type **Product**. In cell **D2**, type =B2*C2. This multiplies the frequency of the first class by its class mark. In **D3**, type =B3*C3. Repeat these steps to fill column **D** through cell **D7**. To estimate the mean, we need to divide the sum of these products by the sum of the frequencies. In cell **C9**, type **Estimate of Mean**, and, in **D9**, type =(SUM(D2:D7))/(SUM(B2:B7)).

The next step is to subtract the estimate of the mean from each class mark. In cell **E1**, type **Mark-Mean**. In cell **E2**, type =C2-D9. This subtracts the mean from the class mark of the first class. Continue by typing =C3-D9 in cell **E3** to find the difference between the mark of the second class and the mean. Repeat this in cells **E4** through **E7**.

We follow this by squaring each of these differences. In cell **F2**, type =E2*E2. In cell **F3**, type =E3*E3. Repeat this in cells **F4** through **F7**.

The next step is to multiply each of these squared differences by their respective frequencies. In cell **G2**, type =B2*F2. In cell **G3**, type B3*F3. Repeat this in cells **G4** through **G7**.

Now we will divide the sum of column **G** by one less than the sum of column **B**, and take the square root of that. In cell **C10**, type **Standard Deviation**. In cell **D10**, type =SQRT(SUM(G2:G7)/(SUM(B2:B7)-1)). Here is what the Excel worksheet should look like.

Score	Frequency	Class Marks	Product	Mark - Mean	Squared	Freq* Sq
40 to 50	2	45	90	-35.0877193	1231.148046	2462.296091
50 to 60	2	55	110	-25.0877193	629.3936596	1258.787319
60 to 70	9	65	585	-15.0877193	227.6392736	2048.753463
70 to 80	13	75	975	-5.087719298	25.88488766	336.5035396
80 to 90	14	85	1190	4.912280702	24.13050169	337.8270237
90 to 100	17	95	1615	14.9122807	222.3761157	3780.393967

Estimate of
Mean   80.08772
Standard
Deviation   13.51227

TI-83   Measures of Dispersion

We can use the TI-83 to help us quickly calculate measures of dispersion for sets of data. This is a continuation of the material that began in Section 2.1. We calculate standard deviation in the same way that we calculated the mean, median, and other measures of central tendency. Here is the set of data that will be used for the example. These are the scores of 57 randomly selected students on a statistics exam.

73	91	97	75	85	99	87	88	63	79
98	66	64	77	75	89	77	94	93	85
92	64	71	99	81	95	92	84	84	97
75	83	86	65	90	46	89	81	56	90
44	50	72	66	66	69	79	71	77	76
78	99	87	90	91	63	94			

We must put all of the values in a single list to analyze them. We begin by clearing list L_1. Press (STAT), and choose option 4 under **Edit**, which is **ClrList**. This will take us to the main screen, and we need to enter the name of the list (L_1) that we want to clear. We do this by pressing (2ND)(1) and then (ENTER). Now, to enter the values in the list L_1, press (STAT) and select option (1) under **Edit**. In the column underneath L_1 begin typing the values. Type the first value, 73, and then press the (↓) key. Type the second value, 91, and press the (↓) key. Repeat this until the last value, 94, has been entered. Be sure that you have entered 57 values before proceeding.

We can calculate several measures at the same time. Press (STAT) and use the (→) key to access the **CALC** menu, which looks like this.

Press ⓵ to select the **1-Var Stats** option. On the main screen, enter L$_1$ after **1-Var Stats** by using the (2ND) and ⓵ keys and press (ENTER). The results take up more than one window. The TI-83 will put an arrow pointing downward in the lower left corner of the window to let us know when this happens. The following two screens show all of the information.

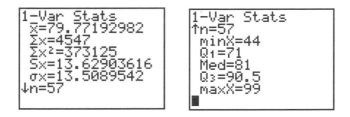

We now see that

- the sample standard deviation is approximately 13.63, and
- the population standard deviation is approximately 13.51.

Also, the interquartile range is 90.5 − 71 or 19.5, and the range is 99 − 44 or 55.

The TI-83 can also create a boxplot for us. Begin by pressing (2ND) (y=) ⓵. For **Type**: select (⊡). Your screen should look like this.

Now press (ZOOM) ⑨ and see the following boxplot.

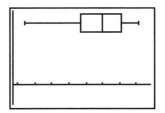

Press (TRACE), and you will be able to see the values of the minimum, Q$_1$, median, Q$_3$, and maximum. The following screen shows the median.

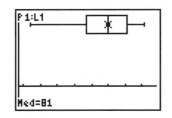

EXERCISES 2.2

Find the range of the following sets of values.

1. 32	57	39	85	74	14	45	62	21	19
2. 47	42	46	45	41	49	49	112	42	44
3. 84	93	57	71	94	77	89	78	61	93
72	87	63	60	55	72	68	80	56	68
4. 19	18	53	19	21	52	52	20	49	50
50	23	21	17	54	52	53	18	18	51
49	18	19	22	55	51	20	56	21	53

Find the interquartile range for the following set of values.

5. 45	58	50	47	55	60	40	43		
50	55	40	43	48	46	56	46		
6. 149	135	128	150	149	144	142	135	127	
147	127	114	138	135	149	142	127	116	
7. 60	56	65	19	63	74	63	55	105	23
30	49	13	68	31	86	101	91	70	76
15	55	50	98	35	104	57	57	17	107
98	47	49	84	98	74	33			
8. 204	161	192	109	109	67	145	107	245	182
111	107	183	176	173	76	146	110	116	243
181	213	125	105	214	231	239	194	166	185

Construct a boxplot for a set of data that has the following five-number summary. Be sure to label both the range and the interquartile range.

9. Lowest	Q_1	Median	Q_3	Highest
15	44	51	60	72

10. Lowest	Q_1	Median	Q_3	Highest
32	37	39	46	58

11. Lowest	Q_1	Median	Q_3	Highest
28	39	50	61	98

12. Lowest	Q_1	Median	Q_3	Highest
44	73	79	87	92

13. A sample of five college females produced the following heights (in inches). Find the mean deviation for these heights.

    64   67   65   70   62

14. Here are the scores of six randomly selected professional football teams on a certain Sunday. Find the mean deviation for these scores.

    31   14   25   17   20   6

15. Here are the scores of eight randomly selected students on a math exam. Find the mean deviation for these scores.

    62   75   72   95   82   86   80   71

16. Here are the speeds of seven pitches thrown by a professional baseball pitcher, in miles per hour. Find the mean deviation of these speeds.

    95   79   88   98   96   94   72

17. A sample of five college females produced the following heights (in inches). Find the variance and standard deviation for these heights.

    64   67   65   70   62

18. Here are the scores of six randomly selected professional football teams on a certain Sunday. Find the variance and standard deviation for these scores.

    31   14   25   17   20   6

19. A class for volunteer Braille transcribers has eight students. Here are their ages. Find the variance and standard deviation for these ages. (Treat the data as a population.)

    55   65   59   30   54   41   60   62

20. A study group has six members, with the following IQs. Find the variance and standard deviation for these IQs. (Treat the data as a population.)

    108   104   110   104   92   106

21. Here are the scores of eight randomly selected students on a math exam. Find the variance and standard deviation for these scores.

    62   75   72   95   82   86   80   71

22. Here are the speeds of seven pitches thrown by a professional baseball pitcher, in miles per hour. Find the variance and standard deviation of these speeds.

    95   79   88   98   96   94   72

23. For the taping of a game show, ten potential contestants are present. Following are the ages of these potential contestants. Find the variance and standard deviation for these ages. (Treat the data as a population.)

    46   35   37   38   28   42   58   38   32   39

24. The Registrar's Office employs 12 student workers. Here are the number of units each is taking this semester. Find the variance and standard deviation for these ages. (Treat the data as a population.)

    9   14   16   12   12   18
    11   10   12   15   12   11

For the data in Exercises 25–36: (a) find the range, (b) find the interquartile range, (c) draw a boxplot, (d) calculate the mean deviation, (e) calculate the standard deviation, and (f) calculate the variance.

25. Amounts of calcium (in milligrams) in nine dairy products:

Dairy Products	Amount	Calcium (mg)
Plain fat-free yogurt	1 cup	400
Swiss cheese	$1\frac{1}{2}$ oz.	408
Chocolate milk shake	10 oz.	374
American cheese	2 oz.	348
Milk	8 oz.	300
Chocolate milk	8 oz.	280
Cheddar cheese	$1\frac{1}{2}$ oz.	300
Ricotta cheese	$\frac{1}{2}$ cup	300
Ice cream or frozen yogurt	1 cup	200

(Source: Rockford Register Star)

26. Amounts of vitamin C (in milligrams) in 16 food sources:

Food	Serving	Vitamin C (mg)
Orange juice, fresh	1 cup	124
Green pepper, raw	$\frac{1}{2}$ cup	96
Grapefruit juice	1 cup	93
Papaya	$\frac{1}{2}$ medium	85
Brussels sprouts	4 medium	73
Broccoli, raw	$\frac{1}{2}$ cup	70
Orange	1 medium	66
Cantaloupe	$\frac{1}{4}$ medium	45
Cauliflower, raw	$\frac{1}{2}$ cup	45
Strawberries	$\frac{1}{2}$ cup	44
Tomato juice	1 cup	39
Cabbage, raw	$\frac{1}{2}$ cup	21
Blackberries	$\frac{1}{2}$ cup	15
Spinach, raw	$\frac{1}{2}$ cup	14
Blueberries	$\frac{1}{2}$ cup	10
Cherries, sweet	$\frac{1}{2}$ cup	8

(*Source:* C for Yourself.)

27. Number of goals scored during each season from 1979–80 through 1998–99 by Wayne Gretzky ("The Great One" played 20 NHL seasons in Edmonton, Los Angeles, St. Louis, and New York):

Season	Team	Goals	Season	Team	Goals
79–80	Edmonton	51	89–90	Los Angeles	40
80–81	Edmonton	55	90–91	Los Angeles	41
81–82	Edmonton	92	91–92	Los Angeles	31
82–83	Edmonton	71	92–93	Los Angeles	16
83–84	Edmonton	87	93–94	Los Angeles	38
84–85	Edmonton	73	94–95	Los Angeles	11
85–86	Edmonton	52	95–96	Los Angeles–St. Louis	23
86–87	Edmonton	62	96–97	New York	25
87–88	Edmonton	40	97–98	New York	23
88–89	Los Angeles	54	98–99	New York	9

28. Number of assists in each season from 1979–80 through 1998–99 by Wayne Gretzky:

Season	Team	Assists	Season	Team	Assists
79–80	Edmonton	86	89–90	Los Angeles	102
80–81	Edmonton	109	90–91	Los Angeles	122
81–82	Edmonton	120	91–92	Los Angeles	90
82–83	Edmonton	125	92–93	Los Angeles	49
83–84	Edmonton	118	93–94	Los Angeles	92
84–85	Edmonton	135	94–95	Los Angeles	37
85–86	Edmonton	163	95–96	Los Angeles–St. Louis	79
86–87	Edmonton	121	96–97	New York	72
87–88	Edmonton	109	97–98	New York	67
88–89	Los Angeles	114	98–99	New York	53

29. Number of pass receptions by Jerry Rice each season from 1985 through 1996 (Jerry Rice began his NFL career as a wide receiver with the San Francisco Forty-Niners in 1985, and currently owns practically every NFL receiving record.)

Year	Receptions
1985	49
1986	86
1987	65
1988	64
1989	82
1990	100
1991	80
1992	84
1993	98
1994	112
1995	122
1996	108

30. Number of yards receiving by Jerry Rice each season from 1985 through 1996

Year	Yards
1985	927
1986	1570
1987	1078
1988	1306
1989	1483
1990	1502
1991	1206
1992	1201
1993	1503
1994	1499
1995	1848
1996	1254

31. Per-site averages for the top ten movies during the first weekend of October 1999:

Film	Average per Site ($)
Double Jeopardy	7,451
Three Kings	5,387
American Beauty	11,599
Blue Streak	2,928
The Sixth Sense	2,490
Drive Me Crazy	3,084
For Love of the Game	1,210
Elmo in Grouchland	2,690
Mystery, Alaska	1,854
Stigmata	1,154

(Source: AC Nielsen EDI Inc. in USA Today.)

32. Gross value of milk produced in Tulare County, California, per year, for 1989 through 1998 (in millions of dollars):

1989	1990	1991	1992	1993	1994	1995	1996	1997	1998
$287	$363	$413	$411	$455	$477	$547	$569	$712	$718

(Source: Visalia Times-Delta.)

33. Winning mutual payouts for 16 races at Santa Anita Park (for a $2 wager):

$6.80	$4.60	$4.00	$40.60	$21.60	$12.80	$6.00	$10.00
$21.20	$3.60	$5.00	$2.60	$5.40	$5.80	$2.40	$6.00

34. Number of horses running in 13 races at Santa Anita Park that were restricted to female horses (fillies and mares):

7	5	8	11	8	6	8
12	12	10	7	9	7	

35. Number of complaints about nursing homes, by state, received in 1997:

State	Complaints	State	Complaints	State	Complaints
AL	998	KY	5,314	ND	546
AK	114	LA	1,992	OH	4,016
AZ	621	ME	499	OK	2,164
AR	1,625	MD	2,260	OR	2,950
CA	17,764	MA	10,438	PA	4,641
CO	7,669	MI	2,674	RI	686
CT	231	MN	2,953	SC	1,896
DE	1,043	MS	661	SD	541
DC	5,199	MO	7,324	TN	1,332
FL	5,029	MT	1,174	TX	9,495
GA	3,274	NE	1,453	UT	1,958
HI	3,952	NV	7,158	VT	330
ID	6,638	NH	593	VA	406
IL	4,731	NJ	2,192	WA	1,670
IN	1,792	NM	9,171	WV	1,043
IA	254	NY	5,483	WI	3,314
KS	4,552	NC	2,457	WY	733

(Source: 1997 preliminary figures, U.S. Department of Health and Human Services, Administration on Aging, in USA Today.)

36. Number of earthquakes per year worldwide that were magnitude 7.0 or higher:

Year	Number	Year	Number	Year	Number	Year	Number	Year	Number
1900	13	1920	8	1940	23	1960	22	1980	18
1901	14	1921	11	1941	24	1961	18	1981	14
1902	8	1922	14	1942	27	1962	15	1982	10
1903	10	1923	23	1943	41	1963	20	1983	15
1904	16	1924	18	1944	31	1964	15	1984	8
1905	26	1925	17	1945	27	1965	22	1985	15
1906	32	1926	19	1946	35	1966	19	1986	6
1907	27	1927	20	1947	26	1967	16	1987	11
1908	18	1928	22	1948	28	1968	30	1988	8
1909	32	1929	19	1949	36	1969	27	1989	7
1910	36	1930	13	1950	39	1970	29	1990	13
1911	24	1931	26	1951	21	1971	23	1991	10
1912	22	1932	13	1952	17	1972	20	1992	23
1913	23	1933	14	1953	22	1973	16	1993	16
1914	22	1934	22	1954	17	1974	21	1994	15
1915	18	1935	24	1955	19	1975	21	1995	25
1916	25	1936	21	1956	15	1976	25	1996	22
1917	21	1937	22	1957	34	1977	16	1997	20
1918	21	1938	26	1958	10	1978	18	1998	16
1919	14	1939	21	1959	15	1979	15		

(Source: U.S. Geological Survey.)

37. Here are the ages of the 16 full-time faculty members in the math department at a community college. Calculate the coefficient of skewness for these ages. Is the skewness positive or negative? Is the skewness moderate or severe?

27	42	40	59	32	28	29	30
32	45	27	44	37	52	54	26

38. Here are the weights of nine 3-year-old girls in a dance class. Calculate the coefficient of skewness for these weights. Is the skewness positive or negative? Is the skewness moderate or severe?

34	33	36	33	32	27	35	33	34

39. Barry Sanders was an NFL running back with the Detroit Lions for 10 NFL seasons. Here are the number of rushing touchdowns that he scored each season for the 1989 through 1998 seasons. Calculate the coefficient of skewness for the number of touchdowns. Is the skewness positive or negative? Is the skewness moderate or severe?

1989	1990	1991	1992	1993	1994	1995	1996	1997	1998
14	13	16	9	3	7	11	11	11	4

40. Here are the number of shares (in millions) being offered by 19 companies through initial public offerings (IPOs) during the week of October 18, 1999. Calculate the coefficient of skewness for the number of shares. Is the skewness positive or negative? Is the skewness moderate or severe?

6.0	4.0	100.0	2.9	3.5	6.5	9.0	3.4	7.2	5.5
7.8	4.6	42.9	6.0	1.5	6.5	2.0	10.0	3.8	

    (Renaissance Capital in USA Today.)

41. Here are the number of electoral college votes that the 50 states and the District of Columbia had in the 1996 presidential election. Calculate the coefficient of skewness for the number of electoral votes. Is the skewness positive or negative? Is the skewness moderate or severe?

State	Votes	State	Votes	State	Votes
AL	9	KY	8	ND	3
AK	3	LA	9	OH	21
AZ	8	ME	4	OK	8
AR	6	MD	10	OR	7
CA	54	MA	12	PA	23
CO	8	MI	18	RI	4
CT	8	MN	10	SC	8
DE	3	MS	7	SD	3
DC	3	MO	11	TN	11
FL	25	MT	3	TX	32
GA	13	NE	5	UT	5
HI	4	NV	4	VT	3
ID	4	NH	4	VA	13
IL	22	NJ	15	WA	11
IN	12	NM	5	WV	5
IA	7	NY	33	WI	11
KS	6	NC	14	WY	3

42. Here are the median household incomes for the 50 states and the District of Columbia for 1997–98. Calculate the coefficient of skewness for these median incomes. Is the skewness positive or negative? Is the skewness moderate or severe?

State	Median Income ($)	State	Median Income ($)	State	Median Income ($)
AL	34,351	KY	35,113	ND	31,229
AK	49,717	LA	32,757	OH	37,811
AZ	35,170	ME	34,461	OK	32,783
AR	27,117	MD	48,714	OR	38,447
CA	40,623	MA	42,511	PA	38,558
CO	45,253	MI	40,583	RI	38,012
CT	45,589	MN	45,576	SC	34,031
DE	42,581	MS	29,031	SD	31,471
DC	32,895	MO	38,662	TN	32,602
FL	33,935	MT	30,622	TX	35,702
GA	37,950	NE	35,823	UT	43,870
HI	41,199	NV	39,608	VT	37,485
ID	35,302	NH	43,297	VA	43,490
IL	42,552	NJ	49,297	WA	46,339
IN	39,613	NM	31,049	WV	27,310
IA	35,664	NY	36,875	WI	40,769
KS	36,875	NC	36,118	WY	34,597

(Source: Census Bureau—USA Today.)

43. Explain, in your own words, why home sale prices would most likely be positively skewed.
44. Explain, in your own words, why household incomes would most likely be positively skewed.
45. If a math instructor gives an easy exam, are the scores more likely to be positively skewed or negatively skewed? Explain your reasoning.
46. If men whose first marriage began 20 years ago were surveyed about the length of their first marriage, are the lengths more likely to be positively skewed or negatively skewed? Explain your reasoning.
47. In fall 1992, a study of part-time instructional faculty in institutions of higher education with regard to the number of students they taught was conducted by the U.S. Department of Education. The following table summarizes the results.

Total Students Taught	Percentage of Part-Time Faculty
Less than 25	36.5
25–49	33.5
50–74	16.2
75–99	6.2
100–149	5.1
150 or more	2.5

Are the numbers of students taught by part-time faculty positively skewed, negatively skewed, or symmetric? Explain your reasoning.

48. A study by the U.S. Department of Education of the ages of college students produced the following table showing the distribution of the ages of full-time students at a two-year institution.

Age	Percentage of Students
Under 18	2.1
18 and 19	35.3
20 and 21	21.4
22–24	13.1
25–29	10.6
30–34	6.4

(continues)

(continued)

Age	Percentage of Students
35–39	4.6
40–49	4.8
50–64	1.1
65 and over	0.2
Age unknown	0.6

Are the ages of full-time students at two-year colleges positively skewed, negatively skewed, or symmetric? Explain your reasoning.

49. Here is the five-number summary for a set of data. Draw a boxplot, and use that boxplot to determine whether the data are positively skewed, negatively skewed, or symmetric. Explain your reasoning.

Lowest	Q_1	Median	Q_3	Highest
37	66	71	74	85

50. Here are the number of public two-year colleges in 1996–97 by state, according to the U.S. Department of Education. Construct a boxplot for the data, and use the boxplot to determine whether the data are positively skewed, negatively skewed, or symmetric. Explain your reasoning.

State	Number of Two-Year Colleges	State	Number of Two-Year Colleges	State	Number of Two-Year Colleges
AL	34	KY	14	ND	9
AK	1	LA	43	OH	39
AZ	20	ME	7	OK	15
AR	24	MD	20	OR	17
CA	108	MA	18	PA	20
CO	15	MI	29	RI	1
CT	12	MN	46	SC	21
DE	3	MS	22	SD	6
DC	0	MO	18	TN	14
FL	28	MT	13	TX	67
GA	34	NE	9	UT	4
HI	7	NV	4	VT	1
ID	3	NH	7	VA	24
IL	49	NJ	19	WA	32
IN	14	NM	20	WV	5
IA	17	NY	47	WI	19
KS	23	NC	58	WY	7

51. Here are the number of colleges and universities in the six New England states. Convert each number to a z-score. Treat the data as a population.

State	Number of Colleges and Universities
Connecticut	42
Maine	34
Massachusetts	124
New Hampshire	29
Rhode Island	12
Vermont	25

(Source: U.S. Department of Education.)

52. Following are the number of home runs hit by Mark McGwire for the 1995–99 seasons. Convert each total to a z-score. Treat the data as a population.

Year	1995	1996	1997	1998	1999
Home Runs	39	52	58	70	65

53. Here are the heights of ten Giant Sequoia trees, in feet. Convert each height to a z-score.

274.9	246.1	267.4	240.9	255.8
243	257.5	268.8	223.8	270.3

54. Here are the number of doctor's degrees awarded to women in 1995–96 by field for certain fields. Convert each total to a z-score.

Field	Doctor's Degrees to Women
Agriculture	336
Business	394
Communications	155
Computer and Information Science	126
Education	4151
English	945
Mathematics	247
Psychology	2452

(Source: U.S. Department of Education.)

55. What percentage of all z-values lie within 3 standard deviations of the mean?
56. What percentage of all z-values lie within 4 standard deviations of the mean?
57. What percentage of all z-values lie within 1.5 standard deviations of the mean?
58. What percentage of all z-values lie within 1.6 standard deviations of the mean?
59. The heights of 12-month-old boys have a mean of 29.8 inches and a standard deviation of 1.2 inches. What percentage of 12-month-old boys have a height that is between 27.7 inches and 31.9 inches?
60. IQ scores have a mean of 100 points, with a standard deviation of 15 points. What percentage of people have an IQ that is between 80 points and 120 points?
61. The heights of adult males have a mean of 69.0 inches and a standard deviation of 2.8 inches. What percentage of adult males are between 5'6" tall and 6' tall?
62. Scores for high school sophomores on the mathematics portion of the PSAT (Preliminary SAT/National Merit Scholarship Qualifying Test) have a mean of 46.1 and a standard deviation of 10.5. (Scores on the PSAT are whole numbers, with a low of 20 and a high of 80.) What percentage of sophomores score at least 30 points on this test?

For Exercises 63 and 64, use the frequency distribution to estimate the standard deviation of the actual values.

63. These are the daily volumes (in millions of shares) of the New York Stock Exchange for the first three months of 1999. (Treat the data as a population.)

Millions of Shares	Frequency
650–700	4
700–750	14
750–800	16
800–850	10
850–900	9
900–950	7
950–1000	1

64. These are the high temperatures, in degrees Fahrenheit, reported in the *New York Times* for 68 foreign cities on August 15. (Treat the data as a sample.)

Temperature (°F)	Frequency
35–45	1
45–55	0
55–65	6
65–75	22
75–85	8
85–95	25
95–105	5
105–115	1

65. These are the number of complaints about nursing homes, by state, received in 1997.

State	Complaints	State	Complaints	State	Complaints
AL	998	KY	5,314	ND	546
AK	114	LA	1,992	OH	4,016
AZ	621	ME	499	OK	2,164
AR	1,625	MD	2,260	OR	2,950
CA	17,764	MA	10,438	PA	4,641
CO	7,669	MI	2,674	RI	686
CT	231	MN	2,953	SC	1,896
DE	1,043	MS	661	SD	541
DC	5,199	MO	7,324	TN	1,332
FL	5,029	MT	1,174	TX	9,495
GA	3,274	NE	1,453	UT	1,958
HI	3,952	NV	7,158	VT	330
ID	6,638	NH	593	VA	406
IL	4,731	NJ	2,192	WA	1,670
IN	1,792	NM	9,171	WV	1,043
IA	254	NY	5,483	WI	3,314
KS	4,552	NC	2,457	WY	733

(Source: 1997 preliminary figures, U.S. Department of Health and Human Services, Administration on Aging, in USA Today.*)*

   (a) Create a frequency distribution for the number of complaints by state.

   (b) Use the frequency distribution to estimate the standard deviation of this population.

   (c) Compare your result in (b) to the standard deviation that you calculated in Exercise 35. Is this estimate a good one?

66. Here are the number of earthquakes worldwide per year that were magnitude 7.0 or higher.

Year	Number	Year	Number	Year	Number	Year	Number	Year	Number
1900	13	1909	32	1918	21	1927	20	1936	21
1901	14	1910	36	1919	14	1928	22	1937	22
1902	8	1911	24	1920	8	1929	19	1938	26
1903	10	1912	22	1921	11	1930	13	1939	21
1904	16	1913	23	1922	14	1931	26	1940	23
1905	26	1914	22	1923	23	1932	13	1941	24
1906	32	1915	18	1924	18	1933	14	1942	27
1907	27	1916	25	1925	17	1934	22	1943	41
1908	18	1917	21	1926	19	1935	24	1944	31

(continues)

(continued)

Year	Number	Year	Number	Year	Number	Year	Number	Year	Number
1945	27	1956	15	1967	16	1978	18	1989	7
1946	35	1957	34	1968	30	1979	15	1990	13
1947	26	1958	10	1969	27	1980	18	1991	10
1948	28	1959	15	1970	29	1981	14	1992	23
1949	36	1960	22	1971	23	1982	10	1993	16
1950	39	1961	18	1972	20	1983	15	1994	15
1951	21	1962	15	1973	16	1984	8	1995	25
1952	17	1963	20	1974	21	1985	15	1996	22
1953	22	1964	15	1975	21	1986	6	1997	20
1954	17	1965	22	1976	25	1987	11	1998	16
1955	19	1966	19	1977	16	1988	8		

(Source: U.S. Geological Survey.)

(a) Create a frequency distribution for the number of earthquakes per year.

(b) Use the frequency distribution to estimate the standard deviation of this population.

(c) Compare your result in (b) to the standard deviation that you calculated in Exercise 36. Is this estimate a good one?

Dan Marino became a quarterback for the Miami Dolphins in 1983. Here are the number of passes he completed each year, as well as the number of touchdown passes that he threw, for the 1983–1998 seasons.

Year	Completions	Touchdowns
1983	173	20
1984	362	48
1985	336	30
1986	378	44
1987	263	26
1988	354	28
1989	308	24
1990	306	21
1991	318	25
1992	330	24
1993	91	8
1994	385	30
1995	309	24
1996	221	17
1997	319	16
1998	310	23

67. (a) Calculate the range, interquartile range, and standard deviation for the number of completions that Dan Marino made in these 16 seasons.

(b) During the 1993 season, Dan Marino missed a majority of the games due to injury. Remove the 91 completions in 1993 from the list, and recalculate the range, interquartile range, and standard deviation for the number of completions that he made in the remaining 15 seasons.

(c) Which of your calculations were greatly affected by deleting that one value, and which ones were not greatly affected? Explain why.

68. (a) Calculate the range, interquartile range, and standard deviation for the number of touchdown passes Dan Marino threw in these 16 seasons.

(b) During the 1993 season, Dan Marino missed a majority of the games due to injury. Remove the 91 completions in 1993 from the list, and recalculate the range, interquartile range, and standard deviation for the number of touchdown passes he threw in the remaining 15 seasons.

(c) Which of your calculations were greatly affected by deleting that one value, and which ones were not greatly affected? Explain why.

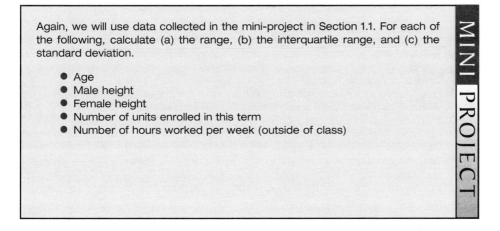

Again, we will use data collected in the mini-project in Section 1.1. For each of the following, calculate (a) the range, (b) the interquartile range, and (c) the standard deviation.

- Age
- Male height
- Female height
- Number of units enrolled in this term
- Number of hours worked per week (outside of class)

MINI PROJECT

Using Samples to Estimate a Population Mean • EXTRA

On the following pages you will find 500 values, numbered from 1 through 500. The mean of these values is 500, and the standard deviation is 46.899. We will draw samples from this population and examine how close our sample means are to the mean of the population.

Here are 12 randomly selected values.

Number	Value
322	503
51	485
460	563
285	491
55	542
441	499
139	432
446	450
244	485
466	505
184	560
77	518

The mean of this sample is 502.75, which is fairly close to the population mean. Will this always be the case? Here are the means of ten samples that were generated in the same fashion.

Sample Number	Mean	Sample Number	Mean
1	502.75	6	500.42
2	482.42	7	485.42
3	494.50	8	522.83
4	485.58	9	498.92
5	499.67	10	522.25

You will notice that some of these means (samples 1, 5, 6, 9) are very close to the population mean, while others (samples 2, 4, 7, 8, 10) are somewhat far away. Thus, an individual sample may vary from the population mean, but look at what happens when we find the mean of these ten sample means. Their mean is 499.476, which is close to 500, and their standard deviation is 14.073.

Repeat this procedure for ten samples of size 30, and answer the following questions.

EXTRA • EXERCISES

1. List the ten sample means.
2. Does this set of means appear to be closer to or farther away from the population mean than the sample means above? Offer an explanation why.
3. Find the mean of these sample means. Does this appear to be as effective for estimating a population mean? Is the mean approximately close to the mean of the ten previous samples or significantly different?
4. Find the standard deviation of these sample means. Is it lower or higher than the standard deviation for the previous ten samples? Does this suggest that as the sample size increases, the sample means as a group get closer to the population mean?

Number	Value	Number	Value	Number	Value	Number	Value	Number	Value
1	400	35	496	69	433	103	595	137	516
2	533	36	549	70	516	104	540	138	450
3	499	37	487	71	464	105	514	139	432
4	451	38	488	72	556	106	525	140	459
5	562	39	550	73	526	107	560	141	471
6	498	40	430	74	449	108	469	142	562
7	589	41	485	75	429	109	563	143	515
8	501	42	554	76	506	110	439	144	426
9	452	43	498	77	518	111	474	145	492
10	407	44	481	78	515	112	481	146	545
11	457	45	478	79	433	113	501	147	530
12	448	46	528	80	484	114	421	148	56
13	477	47	555	81	453	115	572	149	463
14	432	48	470	82	438	116	516	150	543
15	469	49	538	83	586	117	516	151	507
16	548	50	523	84	435	118	458	152	494
17	525	51	492	85	457	119	532	153	400
18	485	52	443	86	418	120	478	154	553
19	469	53	604	87	496	121	503	155	459
20	584	54	499	88	493	122	530	156	521
21	511	55	542	89	520	123	474	157	444
22	591	56	488	90	510	124	565	158	527
23	507	57	486	91	533	125	602	159	426
24	495	58	472	92	446	126	571	160	431
25	528	59	527	93	533	127	524	161	424
26	514	60	513	94	524	128	544	162	509
27	518	61	498	95	537	129	420	163	522
28	523	62	510	96	549	130	473	164	434
29	538	63	474	97	436	131	525	165	482
30	513	64	585	98	425	132	505	166	484
31	412	65	504	99	594	133	444	167	452
32	518	66	517	100	430	134	523	168	622
33	527	67	599	101	439	135	494	169	475
34	522	68	396	102	595	136	490	170	450

(continues)

(continued)

Number	Value	Number	Value	Number	Value	Number	Value	Number	Value
171	449	227	560	283	522	339	547	395	465
172	459	228	587	284	495	340	447	396	541
173	501	229	567	285	491	341	531	397	603
174	428	230	577	286	532	342	503	398	501
175	583	231	490	287	599	343	489	399	467
176	416	232	455	288	576	344	527	400	544
177	518	233	407	289	439	345	473	401	551
178	510	234	519	290	523	346	509	402	544
179	397	235	451	291	489	347	522	403	531
180	442	236	507	292	467	348	538	404	417
181	495	237	601	293	395	349	466	405	545
182	469	238	446	294	416	350	422	406	564
183	518	239	462	295	399	351	517	407	470
184	560	240	448	296	481	352	485	408	487
185	462	241	536	297	544	353	520	409	500
186	407	242	492	298	401	354	565	410	515
187	508	243	449	299	472	355	509	411	634
188	479	244	485	300	459	356	423	412	462
189	523	245	481	301	518	357	493	413	468
190	553	246	514	302	479	358	469	414	588
191	529	247	495	303	373	359	543	415	588
192	500	248	457	304	587	360	595	416	510
193	547	249	449	305	619	361	456	417	524
194	412	250	461	306	476	362	522	418	500
195	527	251	498	307	468	363	485	419	449
196	508	252	572	308	514	364	507	420	589
197	504	253	503	309	526	365	475	421	493
198	437	254	505	310	505	366	457	422	458
199	456	255	570	311	521	367	527	423	438
200	487	256	492	312	554	368	471	424	412
201	465	257	509	313	498	369	570	425	460
202	455	258	523	314	539	370	488	426	496
203	549	259	509	315	508	371	537	427	548
204	469	260	490	316	454	372	501	428	499
205	544	261	473	317	589	373	484	429	476
206	455	262	560	318	516	374	489	430	535
207	633	263	506	319	476	375	460	431	462
208	495	264	506	320	505	376	641	432	488
209	549	265	478	321	503	377	557	433	520
210	494	266	552	322	503	378	530	434	488
211	534	267	494	323	480	379	540	435	515
212	536	268	484	324	574	380	441	436	471
213	477	269	475	325	576	381	581	437	538
214	477	270	470	326	498	382	471	438	427
215	502	271	558	327	436	383	491	437	538
216	480	272	513	328	436	384	466	440	461
217	462	273	461	329	455	385	488	441	499
218	458	274	494	330	472	384	466	442	506
219	536	275	530	331	506	387	510	443	520
220	390	276	526	332	522	388	491	444	469
221	516	277	514	333	479	389	495	445	571
222	511	278	457	334	495	390	558	446	450
223	460	279	508	335	489	391	579	447	472
224	508	280	496	336	408	392	540	448	498
225	428	281	453	337	576	393	512	449	493
226	534	282	553	338	535	394	441	450	514

(continues)

(continued)

Number	Value	Number	Value	Number	Value	Number	Value	Number	Value
451	479	461	542	471	555	481	522	491	499
452	533	462	528	472	490	482	408	492	489
453	484	463	507	473	459	483	497	493	496
454	504	464	498	474	530	484	408	494	506
455	521	465	474	475	488	485	539	495	588
456	553	466	505	476	509	486	508	496	445
457	554	467	516	477	455	487	537	497	541
458	497	468	556	478	481	488	412	498	498
459	449	469	501	479	464	489	489	499	463
460	563	470	490	480	526	490	540	500	518

Overview

We study two types of statistics, **inferential statistics** and **descriptive statistics.** Inferential statistics are techniques and methods that are used to make a generalization, or evaluate a claim, about a **population** based on **sample** data. A population is the complete collection of people, objects, or measurements that we are interested in analyzing. A sample is a part of the population that we are interested in. Descriptive statistics are techniques and methods that are used to describe a set of values, both graphically and numerically.

Data are information that we have collected. **Categorical (qualitative) data** are comprised of values that are names or categories. **Nominal-level data** are a type of categorical data that lack an ordering scheme, while **ordinal-level data** are a type of categorical data that have an ordering scheme to them (such as low, middle, high). **Numerical (quantitative) data** can also be broken into two levels: **interval-level data** and **ratio-level data.** The difference between these two levels is that interval-level data lack a zero starting point, so ratios between two data values are meaningless. Such ratios are possible only with ratio-level data. Numerical data can also be classified as **discrete** or **continuous.** For discrete data, there are gaps between possible data values. Discrete data often involve a count of some type. For continuous data, there are infinitely many possible data values. Continuous data are usually associated with physical measurements.

We gather data through **experiments** and **observational studies.** In an experiment, we apply a treatment and measure its effect. In an observational study, we simply observe and record data. We select our samples using **probability sampling.** Different types of sampling include **random sampling, systematic sampling, stratified sampling, cluster sampling,** and **convenience sampling.**

We can represent a set of data graphically and numerically. One graphic tool that we have is the **stem-and-leaf display.** We break each number into a **stem** (first part) and a **leaf** (last part), and then group the values that have the same stem together. This gives us a rough idea about the "center" and "spread" of the data. This display should be the first step in investigating most sets of numerical data. Another graphical tool is the **frequency distribution.** For a frequency distribution, we divide the data into **classes,** and count the number of times each class is represented. A **histogram** is roughly speaking a graph of a frequency distribution. The classes are placed on the horizontal axis, and bars are drawn above them to the height of each class' frequency. A **pie chart** is of great assistance for picturing categorical data. A circle is divided into wedges for each category, to show the percentage breakdown represented by each category.

For a set of data, we are often interested in the location of the center of the data, as well as how the data are spread out. Measures of central tendency help us investigate the location of the center of the data. The most important measure of central tendency is the **mean.** The mean of a sample is the most reliable measure for predicting or estimating the corresponding population parameter. The mean of a set of values is found by dividing the sum of the values by how many values there are. The **median** is another measure of central tendency. The median is the value for which half of all the values are above it and the other half of the values are below it. The median is especially useful for data that may have **outliers,** values that are far removed from the majority of values. The **first quartile** (Q1) and the **third quartile** (Q3) separate the set of data into four equal parts. Other measures of central tendency include the **midrange** and the **mode.**

Another important measure for a set of data is a measure of dispersion. Such a measure tells us how "spread out" the values are. The measure of dispersion that is used the most in this class is standard deviation. There are two different formulas for **standard deviation,** depending on whether the data represent a sample or a population. Standard deviation is the square root of the **variance.** Other measures that we employ are the **mean deviation, range,** and **interquartile range.**

One simple way to describe a set of data is to use the **five-number summary**. These five numbers are the lowest value, the first quartile, the median, the third quartile, and the highest value. We can use these five values to construct a **boxplot**. A boxplot can show where the data are centered, as well as giving a rough idea of where the data are centered.

Some uses of standard deviation include **z-scores** and **Chebychev's Theorem**. We also use standard deviation in our calculation of the **coefficient of skewness**.

If we are given a frequency distribution, we have the tools to estimate the mean and standard deviation from that frequency distribution.

Unit I Review Exercises

1. In a poll of 125 high school seniors, 25 of them felt that having children was "very important." Identify the following statements as descriptive or inferential.
   (a) Eighty percent of students polled did not feel that having children was very important to them.
   (b) Twenty percent of all high school seniors feel that having children is very important.
   (c) In the future, only 20% of people will have children.
   (d) Many high school seniors will change their feelings about having children before they are 30 years old.

2. You survey the students in each of your classes and find out the amount that each student spent on textbooks this semester.
   (a) Give an example for which these data would be considered a sample.
   (b) Give an example for which these data would be considered a population.

3. Identify the following data as categorical or numerical. If the data are categorical, identify whether they are nominal-level data or ordinal-level data. If the data are numerical, identify whether they are continuous or discrete.
   (a) Number of push-ups that a person can do
   (b) Year in school (freshman, sophomore, junior, senior)
   (c) Race/ethnicity
   (d) Time required to complete an exam

4. You are asked to estimate the number of hours that a student spends studying per week at your school. Explain how you would use the following types of sampling to gather your data.
   (a) Random sampling
   (b) Systematic sampling
   (c) Convenience sampling
   (d) Stratified sampling
   (e) Cluster sampling
   (f) Identify potential biases that may affect the data.

Cal Ripken set a major league record by playing in 2424 consecutive games for the Baltimore Orioles between 1982 and 1998. Here are his at-bats (AB), runs (R), hits (H), home runs (HR), runs batted in (RBI), walks (BB), and strike outs (SO).

Year	AB	R	H	HR	RBI	BB	SO
1982	598	90	158	28	93	46	95
1983	663	121	211	27	102	58	97
1984	641	103	195	27	86	71	89
1985	642	116	181	26	110	67	68
1986	627	98	177	25	81	70	60
1987	624	97	157	27	98	81	77
1988	575	87	152	23	81	102	69
1989	646	80	166	21	93	57	72
1990	600	78	150	21	84	82	66
1991	650	99	210	34	114	53	46
1992	637	73	160	14	72	64	50
1993	641	87	165	24	90	65	58

(continues)

(continued)

Year	AB	R	H	HR	RBI	BB	SO
1994	444	71	140	13	75	32	41
1995	550	71	144	17	88	52	59
1996	640	94	178	26	102	59	78
1997	615	79	166	17	84	56	73
1998	601	65	163	14	61	51	68

5. For the number of hits:
   (a) Construct a stem-and-leaf display.
   (b) Create a frequency distribution.
   (c) Make a histogram.
6. For the number of home runs:
   (a) Calculate the mean.
   (b) Find the median.
   (c) Find the first quartile.
   (d) Find the third quartile.
   (e) Calculate the midrange.
   (f) Find the mode.
   (g) Calculate the range.
   (h) Calculate the interquartile range.
   (i) Construct a boxplot.
   (j) Calculate the standard deviation.
   (k) Calculate the coefficient of skewness.
7. Here are the market values of oranges grown in Tulare County, California, for the years from 1990 through 1998, in millions of dollars.

1990	1991	1992	1993	1994	1995	1996	1997	1998
$387	$93	$359	$356	$375	$472	$396	$454	$478

(Source: Visalia Times Delta.)

   (a) Calculate the mean deviation.
   (b) Calculate the standard deviation.
   (c) Calculate the variance.
   (d) Convert each value to a z-score.
8. SAT I verbal scores have a mean of 505 points and a standard deviation of 111 points. What percentage of students have a score between 330 points and 680 points?
9. The following frequency distribution shows the ages of 480 mathematics instructors at various colleges.

Age	Frequency
25–35	44
35–45	99
45–55	197
55–65	115
65–75	25

   (a) Use this frequency distribution to estimate the mean age of those instructors surveyed.
   (b) Use this frequency distribution to estimate the standard deviation of the ages of those instructors surveyed.

Tying It All Together

A math instructor wrote two versions of the same test. He believed them to be of equal difficulty. He gave the first version to 36 students, and the second version to 41 students. We will consider the two groups to be random samples from the population of all community college statistics students. Your job is to help the instructor decide whether the two tests were of equal difficulty, or whether one of the exams was harder than the other. Here are the scores of the two versions.

Version A

91	79	82	86	88	88	82	88
88	64	98	90	75	60	93	80
86	82	63	77	82	69	79	73
57	92	82	85	94	77	74	90
53	68	62	77				

Version B

69	84	79	94	85	96	94	79
71	94	70	86	82	91	64	86
87	87	92	69	74	95	77	95
94	80	69	98	96	87	76	91
82	89	76	95	95	72	82	82
85							

(a) Construct a stem and leaf display for each set of scores.
(b) Construct a frequency distribution for each set of scores.
(c) Draw a histogram for each set of scores.
(d) Construct a pie chart showing the letter grade breakdown for each set of scores. (90–100: A; 80–89: B; 70–79: C; 60–69: D; 0–59: F)
(e) Construct a pie chart showing the percentage of students who passed and failed the exam for both sets of scores. (0–69: Fail; 70–100: Pass)
(f) Calculate the mean for each set of scores.
(g) Calculate the median for each set of scores.
(h) Calculate the first quartile and third quartile for each set of scores.
(i) Draw a boxplot for each set of scores.
(j) Calculate the mode for each set of scores.
(k) Estimate the mean for each set of scores using the frequency distribution that you created above.
(l) Calculate the range for each set of scores.
(m) Calculate the interquartile range for each set of scores.
(n) Calculate the standard deviation for each set of scores.
(o) Calculate the variance for each set of scores.

Write a brief essay to answer the question "Were the two exams of equal difficulty?" Use as much of the evidence from above as possible to support your case.

Formulas

Sample mean: $\bar{x} = \dfrac{\Sigma x}{n}$

Population mean: $\mu = \dfrac{\Sigma x}{N}$

Midrange: $\text{Midrange} = \dfrac{\text{lowest} + \text{highest}}{2}$

Weighted mean: $\bar{x}_w = \dfrac{\Sigma w \cdot x}{\Sigma w}$

Estimate of a mean using a frequency distribution: $\bar{x}_g = \dfrac{\Sigma \text{ frequency} \cdot \text{class mark}}{\Sigma \text{ frequency}}$

Range: $\text{Range} = \text{highest} - \text{lowest}$

Interquartile range: Interquartile range $= Q_3 - Q_1$

Mean deviation: Mean deviation $= \dfrac{\Sigma |x - \bar{x}|}{n}$

Sample variance: $s^2 = \dfrac{\Sigma(x - \bar{x})^2}{n - 1}$

Population variance: $\sigma^2 = \dfrac{\Sigma(x - \mu)^2}{N}$

Sample standard deviation: $s = \sqrt{\dfrac{\Sigma(x - \bar{x})^2}{n - 1}}$

Population standard deviation: $\sigma = \sqrt{\dfrac{\Sigma(x - \mu)^2}{N}}$

Coefficient of skewness: $sk = \dfrac{3(\text{mean} - \text{median})}{\text{standard deviation}}$

Standard units: $z = \dfrac{\text{value} - \text{mean}}{\text{standard deviation}}$

Chebyshev's Theorem: $1 - \dfrac{1}{k^2}$

Estimate of a sample standard deviation using a frequency distribution:

$$s_g = \sqrt{\dfrac{\Sigma \, \text{frequency}(\text{class mark} - \bar{x}_g)^2}{n - 1}}$$

unit
two
2

Probability

Counting and Probability

Inferential statistics is a very important branch of statistics. It involves using information from a sample to draw conclusions about a population. The principle at work behind the scenes of inferential statistics is probability, and that is the main focus of this chapter. The probability of an event occurring is a measure of how likely it is that the event occurs. We have often heard the weatherman say, "There is a 40% chance of rain tomorrow." That is a probability.

We will begin by studying counting, a tool that is essential to understanding probability.

115

SECTION 3.1
Counting

This section covers the first major area of this chapter—figuring out how many different ways something could happen. Here are some examples of such problems.

- In how many different ways could a statistics instructor choose two students from a class of 38 students to give a presentation on measures of dispersion?
- In how many different ways could a field of 30 figure skaters finish first, second, and third?
- If a pair of dice is rolled, how many different outcomes are possible? How many pairs will have a sum of 9 or more?

We will use four tools to help us: tree diagrams, the Multiplication Principle, permutations, and combinations. Tree diagrams are introduced to develop intuition and lead to the Multiplication Principle.

Tree Diagrams

A **tree diagram** is a systematic, graphical way to list all possible outcomes. It looks similar to a family tree. Each level or step in the problem is similar to each generation in a family tree. Following is a tree diagram showing the possible results of three flips of a coin. Each flip could be either heads (H) or tails (T).

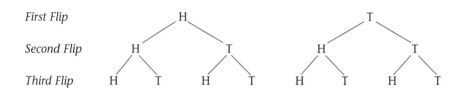

First Flip			H				T		
Second Flip		H		T		H		T	
Third Flip	H	T	H	T	H	T	H	T

The H and T in the top row represent the possible outcomes of the first flip. If the first flip is heads, there are two possible outcomes for the second flip: heads or tails. These are represented by the two branches on the left side that lead to the second row. Scanning across the bottom row, we count eight branches. This tells us that there are eight different possible outcomes. The first possibility is found by tracing the first branch from top to bottom, and it is heads, heads, heads. The second possibility, from the second branch, is heads, tails, heads. Here are the eight different possible outcomes.

Outcome Number	First Flip	Second Flip	Third Flip	Number of Heads
1	H	H	H	3
2	H	H	T	2
3	H	T	H	2
4	H	T	T	1
5	T	H	H	2
6	T	H	T	1
7	T	T	H	1
8	T	T	T	0

We can now answer questions like "How many different ways are there to get exactly two heads?" and "How many different ways are there to get at least one head?"

EXAMPLE 3.1 A tutor can schedule anywhere from zero to two appointments each day. Draw a tree diagram to show in how many different ways she can schedule the next three days.

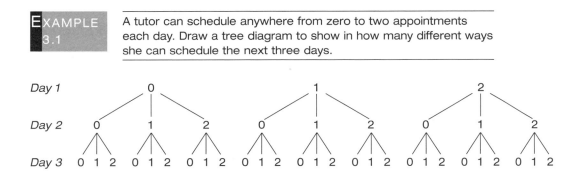

There are 27 different ways that she could schedule her next three days.

0 on Day 1	Total	1 on Day 1	Total	2 on Day 1	Total
0, 0, 0	0	1, 0, 0	1	2, 0, 0	2
0, 0, 1	1	1, 0, 1	2	2, 0, 1	3
0, 0, 2	2	1, 0, 2	3	2, 0, 2	4
0, 1, 0	1	1, 1, 0	2	2, 1, 0	3
0, 1, 1	2	1, 1, 1	3	2, 1, 1	4
0, 1, 2	3	1, 1, 2	4	2, 1, 2	5
0, 2, 0	2	1, 2, 0	3	2, 2, 0	4
0, 2, 1	3	1, 2, 1	4	2, 2, 1	5
0, 2, 2	4	1, 2, 2	5	2, 2, 2	6

We can now answer questions like "In how many different ways can she schedule at least four appointments over the next three days?" The answer would be 10, because there are 10 totals that are 4 or greater. ∎

If a pair of six-sided dice are rolled, how many different outcomes are possible? The key to this problem is to establish a difference between the two dice—for instance, suppose one of the dice is white and the other is yellow. Then a 2 on the white die and a 3 on the yellow die is a different outcome than a 3 on the white die and a 2 on the yellow die. From here on we will simply refer to the two dice as the first die and the second die. We could draw a tree diagram listing the six different possible outcomes of the first die (1–6), and then the six different possible outcomes of the second die for each of those six branches. However, the following chart does the trick. Each roll is represented as an ordered pair (x, y), where x is the result of the first die and y is the result of the second die.

Possible Rolls of a Pair of Dice

		SECOND DIE					
		1	**2**	**3**	**4**	**5**	**6**
	1	(1, 1)	(1, 2)	(1, 3)	(1, 4)	(1, 5)	(1, 6)
FIRST	**2**	(2, 1)	(2, 2)	(2, 3)	(2, 4)	(2, 5)	(2, 6)
DIE	**3**	(3, 1)	(3, 2)	(3, 3)	(3, 4)	(3, 5)	(3, 6)
	4	(4, 1)	(4, 2)	(4, 3)	(4, 4)	(4, 5)	(4, 6)
	5	(5, 1)	(5, 2)	(5, 3)	(5, 4)	(5, 5)	(5, 6)
	6	(6, 1)	(6, 2)	(6, 3)	(6, 4)	(6, 5)	(6, 6)

There are 36 possible rolls. Often we are interested in the sum of the dice. Here is a chart that helps with the sums.

Possible Sums of a Pair of Dice

		SECOND DIE					
		1	**2**	**3**	**4**	**5**	**6**
	1	2	3	4	5	6	7
FIRST	**2**	3	4	5	6	7	8
DIE	**3**	4	5	6	7	8	9
	4	5	6	7	8	9	10
	5	6	7	8	9	10	11
	6	7	8	9	10	11	12

How many different ways are there to obtain a sum that is 9 or higher? By counting, we see that there are 10 different sums that are 9 or higher.

 EXAMPLE 3.2   If a pair of fair six-sided dice are rolled, how many different ways are there to obtain a sum that is 10 or lower?

From the above chart, we can count that there are 33 different ways to obtain a sum of 10 or lower. Note that the only sums that are not 10 or lower are the two rolls of 11 and the one roll of 12. Subtracting these three rolls from the total of 36 leaves us again with 33 different ways to obtain a sum of 10 or lower. ∎

 EXAMPLE 3.3   If a pair of fair six-sided dice are rolled, how many different ways are there to obtain a sum that is between 4 and 10, inclusive?

There are three 4's, four 5's, five 6's, six 7's, five 8's, four 9's, and three 10's. Totaling these, we see that there are 30 different ways to obtain a sum that is between 4 and 10, inclusive. ∎

EXAMPLE 3.4   If a pair of fair six-sided dice are rolled, how many different ways are there to roll "doubles"?

A roll is "doubles" if both dice are the same number. There are six different ways to do this: double 1's, double 2's, double 3's, double 4's, double 5's, and double 6's. ∎

 EXAMPLE 3.5   If a pair of fair six-sided dice are rolled, how many different ways are there to obtain a sum that is 9 or higher or doubles (or both)?

From the first dice example, we know that there are 10 different ways to obtain a sum that is 9 or higher. Two of those 10 rolls are already doubles (double 5's and double 6's), so we need to add four rolls to the total for the four different doubles that have not been counted yet (double 1's, double 2's, double 3's, and double 4's). This gives us a total of 14 different ways to obtain a sum that is 9 or higher or doubles. ∎

 EXAMPLE 3.6   It is rumored that you can get anything you want at Alice's Restaurant. If Alice's has five different types of coffee (regular, decaffeinated, espresso, cappuccino, and latte) and three different types of pie (apple, blueberry, and cherry), in how many different ways could you order coffee and a slice of pie?

We will use A for apple pie, B for blueberry pie, and C for cherry pie. Here is the tree diagram.

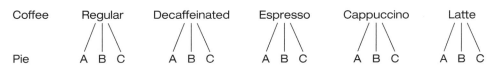

| Coffee | Regular | Decaffeinated | Espresso | Cappuccino | Latte |
| Pie | A B C | A B C | A B C | A B C | A B C |

There are 15 different ways to order coffee and a slice of pie at Alice's. ∎

One nice thing about tree diagrams is that they list each different possible way something could occur. There are times that we do not need all of that detail, but simply need to know the total number of different ways. Is there a quicker, easier way to get the total of 15 different ways out of five types of coffee and three types of pie without using a tree diagram? Here is where the Multiplication Principle comes in.

> **THE MULTIPLICATION PRINCIPLE** *If a choice consists of two steps, the first of which can be made in m different ways and the second can be made in n different ways, then the whole choice can be made in m • n different ways.*

Let's go back to Alice's Restaurant to apply the Multiplication Principle.

EXAMPLE 3.7   It is rumored that you can get anything you want at Alice's Restaurant. If Alice's has five different types of coffee (regular, decaffeinated, espresso, cappuccino, and latte) and three different types of pie (apple, blueberry, and cherry), in how many different ways could you order coffee and a slice of pie?

This is a problem that requires the Multiplication Principle with $m = 5$ and $n = 3$.

$$m \bullet n = 5 \bullet 3$$
$$= 15$$

There are 15 different ways to order coffee and a slice of pie at Alice's Restaurant. ■

If a problem consists of more than two steps, we can extend the Multiplication Principle. Simply multiply the number of ways for the first step by the number of ways for the second step, then continue on for the third step, and so on.

 A community college has 12 math instructors, 14 English instructors, and 5 history instructors. In how many different ways could the president of the college choose a math instructor, an English instructor, and a history instructor to attend an articulation meeting?

There are three steps to this problem. There are 12 different ways to select a math instructor, 14 different ways to select an English instructor, and 5 different ways to select a history instructor. Thus, the president can select the three instructors in $12 \bullet 14 \bullet 5 = 840$ different ways. ■

 A statistics instructor gives a ten-question multiple choice exam. There are five different choices for each problem, of which only one is correct.

1. In how many different ways could a student answer the ten questions?

   How many steps are there to this problem? There are ten steps, one for each question that needs to be answered. In how many different ways could a student answer the first question? Since each question offers five possible answers, there are five different ways to answer the first question. This holds true for the other nine questions as well.

   $$5 \bullet 5 \bullet 5 \bullet 5 \bullet 5 \bullet 5 \bullet 5 \bullet 5 \bullet 5 \bullet 5 = 5^{10}$$
   $$= 9{,}765{,}625$$

   There are 9,765,625 different ways to answer the ten questions.

2. In how many different ways could a student answer all ten questions correctly?

   There are still ten steps to this problem, but now there is only one way to answer each question correctly.

   $$1 \bullet 1 \bullet 1 \bullet 1 \bullet 1 \bullet 1 \bullet 1 \bullet 1 \bullet 1 \bullet 1 = 1^{10}$$
   $$= 1$$

   There is only one way to answer all ten questions correctly (and I'm pretty sure the one way involves a great deal of studying and homework).

3. In how many different ways could a student answer all ten questions incorrectly?

   There are ten steps to this problem, and each question can be answered incorrectly in four different ways.

$$4 \cdot 4 \cdot 4 \cdot 4 \cdot 4 \cdot 4 \cdot 4 \cdot 4 \cdot 4 \cdot 4 = 4^{10}$$
$$= 1,048,576$$

There are 1,048,576 different ways to get all ten questions incorrect. ∎

The next example is different because the number of ways changes from step to step.

If there are 30 skaters competing at the Olympic Games, in how many different ways can the gold, silver, and bronze medals be given?

There are three steps to this problem: determining the number of ways the gold medal could be awarded, determining the number of ways the silver medal could be awarded, and determining the number of ways the bronze medal could be awarded. How many different skaters could win the gold medal? Any of the skaters could win the gold medal, so there are 30. Once we have selected the winner of the gold medal, how many skaters could win the silver medal? Any of the skaters except the gold medal winner could take the silver medal, so that leaves 29 skaters. Finally, now that two medal winners have been determined, how many different ways could the bronze medal be awarded? There are 28 different skaters left, so there are 28 ways.

$$30 \cdot 29 \cdot 28 = 24,360$$

There are 24,360 different ways to give the gold, silver, and bronze medals. ∎

The reason that the number of ways went down by one for each step is that once a person was selected for one of the medals, that person was not eligible for any of the other medals. Here is a similar example.

The I Love Math Club has 24 members. In how many different ways could they elect a president, vice president, secretary, and treasurer?

There are 24 ways to choose a president, which leaves 23 ways to choose a vice president, and so on.

$$24 \cdot 23 \cdot 22 \cdot 21 = 255,024$$

There are 255,024 different ways to elect a president, vice president, secretary, and treasurer. ∎

E XAMPLE
3.12

A recent city council election had nine candidates. In how many different ways could they finish first through ninth?

There are nine candidates that could finish first, eight candidates that could finish second, seven candidates that could finish third, . . . , two candidates that could finish eighth, and one candidate that could finish ninth.

$$9 \cdot 8 \cdot 7 \cdot 6 \cdot 5 \cdot 4 \cdot 3 \cdot 2 \cdot 1 = 362,880$$

There are 362,880 different ways that the nine candidates can finish first through ninth. ■

The product in the previous example is a special product called a **factorial**. The factorial of a nonnegative integer n, denoted by $n!$, is the product of all of the integers between 1 and n. We define 0! to be 1, and 1! to be 1 as well. If n objects are to be arranged in an order from 1 to n, this can be done in $n!$ different ways. Many calculators provide a way to calculate factorials directly.

 EXAMPLE 3.13   A family is planning a trip to six European countries. In how many different ways could they arrange their trip with respect to what order they visit the countries?

Since we are arranging six countries in order from 1 to 6, 6! will do the trick.

$6! = 6 \cdot 5 \cdot 4 \cdot 3 \cdot 2 \cdot 1$

$= 720$

There are 720 different ways that they could arrange their trip. ■

Permutations

A **permutation** is an ordered arrangement of a group of objects. The number of permutations of r objects selected from a set of n distinct objects is given by

$$_nP_r = \frac{n!}{(n-r)!}$$

Let's use some concrete numbers for r and n to help explain exactly what this formula does. Suppose there were ten numbered balls in a bag, and we were going to select three of the balls and record the order that they came out in. So, n would be 10 and r would be 3. This is because we are selecting three out of the ten. The formula would tell us how many different ways there are to select three out of the ten where the order of selection is important. The order of selection is important because it tells us that selecting ball 1, ball 2, and then ball 3 is different from selecting ball 3, ball 2, and then ball 1.

To use this formula, we must be sure that each step involves a choice of what began as n distinct objects. Also, none of the n objects can be selected more than once. Finally, there must be a distinction between being chosen at each different step. In other words, the order of selection matters.

One final note: many calculators are able to calculate permutations for you.

EXAMPLE 3.14   If there are 30 skaters competing at the Olympic Games, in how many different ways can the gold, silver, and bronze medals be given?

Since no skater can win more than one of these medals, and winning the gold medal is different from winning the silver medal (order matters), we can use permutations. This is a permutation problem with $n = 30$ and $r = 3$.

$$_{30}P_3 = \frac{30!}{(30 - 3)!}$$

$$= \frac{30!}{27!}$$

$$= \frac{30 \cdot 29 \cdot 28 \cdot 27!}{27!}$$

$$= 30 \cdot 29 \cdot 28$$

$$= 24{,}360$$

There are 24,360 different ways to award the medals. ∎

If you have to work out permutations by hand instead of using a built-in function on your calculator, working them out in the above fashion is a good idea. We rewrite 30! as $30 \cdot 29 \cdot 28 \cdot 27!$ because 30! is the product of the integers from 30 down to 1, and 27! is the product of the integers from 27 down to 1. 27! is missing only the product of 28, 29, and 30 from 30!. The reason for rewriting it in terms of 27! is to cancel the 27! in the denominator.

Note that we obtained the same result that the Multiplication Principle gave us. Then what is the advantage of doing the problem this way? If your calculator has a built-in function to calculate permutations, then that is easier than multiplying several numbers. This advantage increases as n increases and as the difference between n and r decreases.

EXAMPLE 3.15	The I Love Math Club has 24 members. In how many different ways could they elect a president, vice president, secretary, and treasurer?

This is a permutation problem with $n = 24$ and $r = 4$.

$$_{24}P_4 = \frac{24!}{(24 - 4)!}$$

$$= \frac{24!}{20!}$$

$$= \frac{24 \cdot 23 \cdot 22 \cdot 21 \cdot 20!}{20!}$$

$$= 24 \cdot 23 \cdot 22 \cdot 21$$

$$= 255{,}024$$

There are 255,024 different ways to elect a president, vice president, secretary, and treasurer. ∎

EXAMPLE 3.16	A recent city council election had nine candidates. In how many different ways could they finish first through ninth?

This is a permutation problem with $n = 9$ and $r = 9$.

$$_9P_9 = \frac{9!}{(9 - 9)!}$$

$$= \frac{9!}{0!}$$

$$_9P_9 = \frac{9!}{1}$$

$$= 9!$$

$$= 362{,}880$$

There are 362,880 different ways that the nine candidates can finish first through ninth. ∎

Combinations

There are some problems for which the order of selection does not matter, but in every other way they are like permutation problems. This type of problem involves **combinations**. Suppose that a statistics instructor selected three of his 36 students at random and gave the first a free A on the final exam, gave the second a free C on the final exam, and gave the third a 0 on the final. In how many different ways could this be done? It is a permutation problem with $n = 36$ and $r = 3$. The order of selection matters because getting an A on the final is a different result from getting a C on the final, and both are different from getting a 0 on the final. Now let's change the problem slightly. Suppose that a statistics instructor selected three of his 36 students at random and gave all three a free A on the final exam. Is there any difference between being the first selected, the second selected, or the third selected? No, because the result is the same: a free A on the final. If the instructor picks Ann, Bob and Carl, or if he picks Carl, Bob and Ann, the results are the same. All three get an A.

A combination is similar to a permutation, except that the order in which the objects are selected does not matter. The number of combinations of r objects selected from a set of n distinct objects is given by

$$_nC_r = \frac{n!}{r! \cdot (n - r)!}$$

Let's compare the formulas for permutations and combinations.

Permutations	**Combinations**
$_nP_r = \dfrac{n!}{(n - r)!}$	$_nC_r = \dfrac{n!}{r!(n - r)!}$

The formulas are the same except that the formula for combinations has an extra $r!$ in its denominator. Here is the reason why. Suppose we were selecting three objects in a permutation. The following would be six different results.

A, B, C   A, C, B   B, A, C   B, C, A   C, A, B   C, B, A

But if this were a combination problem, all six would only count as one combination (same letters, but different order). So the number of permutations is six times bigger than the number of combinations. We need to divide the number of permutations by 6 to find the number of combinations. That is where the $r!$ comes in. In this example, $3! = 6$.

EXAMPLE 3.17

A statistics instructor with a class of 36 students has taken leave of his senses, and has decided to select three people to be excused from the final exam with an automatic grade of A. In how many different ways can the instructor select three students?

This is a combinations problem because there is no difference between being selected first, second, or third. We will use 36 for *n* and 3 for *r*.

$$_{36}C_3 = \frac{36!}{3! \cdot (36 - 3)!}$$

$$= \frac{36!}{3! \cdot 33!}$$

$$= \frac{36 \cdot 35 \cdot 34 \cdot 33!}{3! \cdot 33!}$$

$$= \frac{36 \cdot 35 \cdot 34}{3!}$$

$$= 7140$$

The instructor could do this in 7140 different ways. ■

EXAMPLE 3.18

If the I Love Math Club wanted to select four of its 24 members to go to a conference, in how many different ways could this be done?

It does not matter whether a person is selected first, second, third, or fourth because all four people are going to the conference. This is a combinations problem with *n* = 24 and *r* = 4.

$$_{24}C_4 = \frac{24!}{4! \cdot (24 - 4)!}$$

$$= \frac{24!}{4! \cdot 20!}$$

$$= \frac{24 \cdot 23 \cdot 22 \cdot 21 \cdot 20!}{4! \cdot 20!}$$

$$= \frac{24 \cdot 23 \cdot 22 \cdot 21}{4!}$$

$$= 10,626$$

The I Love Math Club could select four of its members to send to a conference in 10,626 different ways. ■

EXAMPLE 3.19

A family dinner special at a local restaurant allows the family to order one entrée for each person from a list of 12 entrées. If no repetition is allowed, how many different ways could a family of five order dinner?

This is a combination problem with *n* = 12 and *r* = 5.

$$_{12}C_5 = \frac{12!}{5! \cdot (12 - 5)!}$$

$$= \frac{12!}{5! \cdot 7!}$$

$$= \frac{12 \cdot 11 \cdot 10 \cdot 9 \cdot 8 \cdot 7!}{5! \cdot 7!}$$

$$_{12}C_5 = \frac{12 \cdot 11 \cdot 10 \cdot 9 \cdot 8}{5!}$$

$$= 792$$

A family of five can order dinner in 792 different ways. ■

A family dinner special at a local restaurant allows the family to order one entrée for each person from a list of 12 entrées. If no repetition is allowed, how many different ways could a family of seven order dinner?

This is a combination problem with $n = 12$ and $r = 7$.

$$_{12}C_7 = \frac{12!}{7! \cdot (12 - 7)!}$$

$$= \frac{12!}{7! \cdot 5!}$$

$$= \frac{12 \cdot 11 \cdot 10 \cdot 9 \cdot 8 \cdot 7!}{7! \cdot 5!}$$

$$= \frac{12 \cdot 11 \cdot 10 \cdot 9 \cdot 8}{5!}$$

$$= 792$$

A family of seven can order dinner in 792 different ways. ■

Are the two previous answers the same by accident, or is there a reason? Think of it this way. A family could order the five entrees that they really like, or they could pick the seven that they really do not like. Either way, the result is the same. The family has decided on five entrees to order. As long as the two numbers add up to *n*, the number of combinations will be the same. For example, a family of eight has the same number of ways to order dinner as a family of four. For further convincing evidence that this is no accident, just look at the denominators of the two calculations. They are identical.

Now here is an example to tie it all together.

If a local pizza parlor has three different crusts and 15 different toppings to choose from, in how many different ways could you order a four-topping pizza? (one crust, four toppings)

There are two steps to this problem. First, we must determine how many different ways there are to select a crust. Then we determine how many different ways there are to select four of the toppings. Finally, we multiply those results.

There are three different ways to select a crust, and $_{15}C_4$ or 1365 different ways to select four of the 15 toppings.

$$3 \cdot {_{15}C_4} = 3 \cdot 1365$$

$$= 4095$$

There are 4095 different ways to order a four-topping pizza. ■

MICROSOFT EXCEL   Permutations and Combinations

Microsoft Excel has two built-in functions to help us calculate permutations and combinations. Here is the general form.

=PERMUT(n, r)

For example, suppose you wanted to find how many different ways you could select a president, a vice president, and a treasurer from a group of 20 people ($_{20}P_3$). To do this, type the following in any open cell:

=PERMUT(20,3)

For problems involving combinations, we use

=COMBIN(n, r)

For example, suppose you wanted to find how many different ways you could select three people from a group of 20 people ($_{20}C_3$). To do this, type the following in any open cell:

=COMBIN(20,3)

TI-83   Permutations and Combinations

The TI-83 can help us to calculate permutations and combinations. Suppose you wanted to find how many different ways you could select a president, a vice president, and a treasurer from a group of 20 people ($_{20}P_3$). You first type 20, then press (MATH). Use (→) three times to access the **PRB** menu.

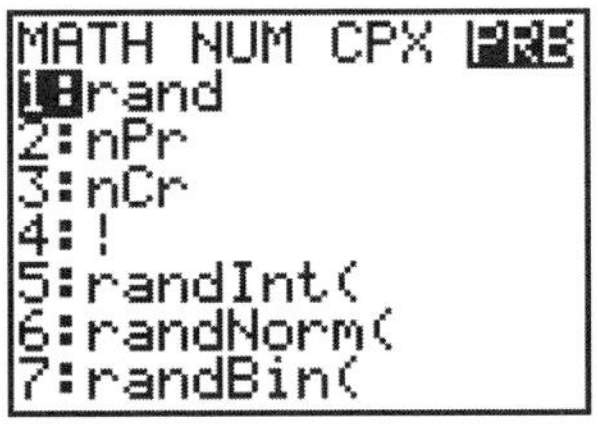

```
MATH NUM CPX PRB
1:rand
2:nPr
3:nCr
4:!
5:randInt(
6:randNorm(
7:randBin(
```

Select option (2), which should produce the following screen.

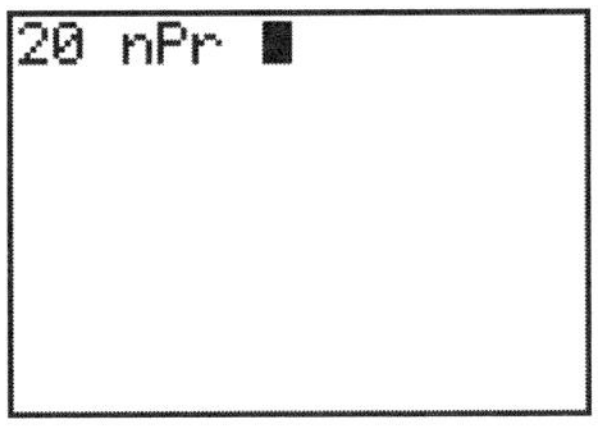

```
20 nPr ▮
```

Now type 3 and press (ENTER).

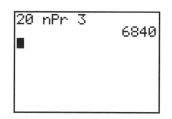

Combinations work exactly the same way, except that we select option (3) ($_nC_r$) from the **PRB** menu.

EXERCISES 3.1

1. Draw a tree diagram showing the possible outcomes of four tosses of a coin. How many different ways are there to get

   (a) No heads?
   (b) Exactly one head?
   (c) Exactly two heads?
   (d) Exactly three heads?
   (e) Four heads?

2. A couple plans to have five children, and they are interested in what the genders will be. To help them out, draw a tree diagram showing all the possible outcomes. How many different ways are there to have

   (a) No girls?
   (b) Exactly one girl?
   (c) Exactly two girls?
   (d) Exactly three girls?
   (e) Exactly four girls?
   (f) Five girls?

3. A pair of dice are rolled and their sum is recorded. How many different ways are there to roll

   (a) An 8?
   (b) A 7 or an 11?
   (c) At least a 6?
   (d) Doubles or a sum that is lower than 8?

4. A pair of dice are rolled and their sum is recorded. How many different ways are there to roll

   (a) A 5?
   (b) A 2, 3, or 12?
   (c) A 9 or less?
   (d) Doubles or a sum that is 8?

5. At a restaurant, you can choose from seven main dishes and 15 side dishes. In how many different ways could you select one main dish and one side dish?

6. An ice cream store has 31 flavors of ice cream and 3 types of cones. In how many different ways could you order a single-scoop ice cream cone (one flavor, one type of cone)?

7. There are 31 professional football teams in the NFL and 10 college football teams in the PAC-10. In how many different ways could a football fan list their favorite NFL team and their favorite PAC-10 team?

8. A pizza parlor has 4 types of crust and 12 different toppings. In how many different ways could you order a one-topping pizza (one type of crust, one topping)?

9. A couple planning their wedding must decide on the menu for the reception. There are ten different appetizer packages, seven main courses, and three dessert packages to choose from. In how many different ways could the couple select an appetizer package, a main course, and a dessert package?

10. A college basketball team has five guards, four forwards, and three centers. The coach must select one guard, one forward, and one center to work at a basketball clinic for kids. In how many different ways could she do this?

11. A math instructor gives a weekly multiple choice quiz. There are five questions, and each question has four possible answers, of which only one is correct.
    (a) In how many different ways could a student answer the five questions?
    (b) In how many different ways could a student answer all five questions correctly?
    (c) In how many different ways could a student answer all five questions incorrectly?

12. An algebra instructor gives an arithmetic quiz the first day of class. There are eight true/false questions.
    (a) In how many different ways could a student answer the eight questions?
    (b) In how many different ways could a student answer all eight questions correctly?
    (c) In how many different ways could a student answer all eight questions incorrectly?

13. A softball league has ten teams in it. In how many different ways could the teams finish first and second?

14. A third-grade class has 25 students in it. In how many different ways could the teacher pick one student to clean the chalkboard and another student to empty the wastebaskets?

15. Bruce Springsteen has released 16 CDs. In how many different ways could a fan select three of his CDs to bring on a road trip?

16. A faculty senate has 23 members. In how many ways could they send four of the senators to a workshop on shared governance?

17. A math instructor has a class with 40 students in it. He suddenly gets hungry in the middle of class. In how many different ways could he select a student to go get him a donut, a second student to go get him a breakfast burrito, and a third student to go get him some coffee?

18. A math instructor has a class with 40 students in it. He suddenly gets hungry in the middle of class. In how many different ways could he select three students to go get breakfast for him?

19. In how many different ways could the Today Show select three jurors from a 12-person jury to interview?

20. Twenty children sign up to test a new medication. In how many different ways could the doctors select eight of the children to receive the medication?

21. In California's Super Lotto game, 6 numbers are drawn from 51 possible numbers (1–51). In how many different ways could 6 numbers be selected from 51?

22. Five cards are drawn from a well-shuffled standard 52-card deck of playing cards. In how many different ways could this be done?

23. Five cards are drawn from a well-shuffled standard 52-card deck of playing cards. In how many different ways could five clubs be drawn?

24. A local Beanie Babies club has 20 members. In how many different ways could they elect a president, a vice president, and a newsletter editor?

25. A pizza parlor has 12 different toppings that you can order for your pizza. In how many different ways could you choose 3 different toppings for a pizza?

26. A pizza parlor has four types of crust, 12 different toppings, and three different cheese blends to choose from. In how many different ways could you order a three-topping pizza (one type of crust, three toppings, and one cheese blend)?

27. A local Beanie Babies club has 20 members. In how many different ways could they elect a president, a vice president, and a three-person newsletter committee?

28. A college basketball team has five guards, four forwards, and three centers. In how many different ways could a coach select a starting lineup of two guards, two forwards, and one center?

SECTION 3.2
Classical Probability

In this section we will calculate how likely it is that something happens, in contrast to the previous section, where we figured out how many different ways something could happen. Something is a rather vague word, so here are some definitions to allow us to proceed.

> **DEFINITION**   An **experiment** is a process that allows us to obtain observations.

This definition agrees with our definition of an experiment in a science class. We perform a certain process, and observe the outcome. In probability, an example of an experiment is flipping a coin and observing whether it is heads or tails. Here are some other experiments.

- Roll a six-sided die and observe the number that is on the top of the die.
- Roll a pair of six-sided dice and observe their sum.
- Draw a card from a 52-card deck and record its suit.
- Have a child and record its gender.
- Randomly select a person and ask whether that person has health insurance.

> **DEFINITION**   An **outcome** of an experiment is a particular result of an experiment.

> **DEFINITION**   The **sample space** (S) for an experiment is the set of all possible outcomes for that experiment.

> **DEFINITION**   An **event** is a subset of the sample space for an experiment.

If we flip a coin and it lands heads-up, heads is the outcome of the experiment. If we roll a single six-sided die, what are the possible outcomes? The possible outcomes are the numbers 1, 2, 3, 4, 5, and 6. When we list all of the possible outcomes of an experiment, we are listing the experiment's **sample space.**

Returning to the experiment of rolling a single six-sided die, we denote the sample space in the following manner: $S = \{1, 2, 3, 4, 5, 6\}$. What is the sample space for tossing a coin? $S = \{$heads, tails$\}$, since those are the only possible results (assuming a coin cannot land on its side).

An event is usually made up of one or more of the possible outcomes of an experiment. We use capital letters to denote an event. If the experiment was rolling a single six-sided die, some possible events are A: the roll is even, and B: the roll is 5 or higher. Event A is then the set $A = \{2, 4, 6\}$ and event B is the set $B = \{5, 6\}$.

Classical Probability

If all of the outcomes in a sample space are equally likely to occur, we denote the probability of event A occurring by $P(A)$ and calculate it using the formula

$$P(A) = \frac{n(A)}{n(S)}$$

In the formula, $n(A)$ represents the number of different ways that event A can occur. We use $n(S)$ to represent the total number of possible outcomes for the experiment.

 EXAMPLE 3.22   A single fair six-sided die is rolled. Find the probability that the roll is even.

The word "fair" implies that each of the possible outcomes is equally likely to occur.

Let A denote the event that the roll is even. Then $A = \{2, 4, 6\}$, and $n(A) = 3$. Since $S = \{1, 2, 3, 4, 5, 6\}$, $n(S) = 6$. Therefore,

$$P(A) = \frac{n(A)}{n(S)}$$

$$= \frac{3}{6}$$

$$= \frac{1}{2}$$

The probability that the roll is even is $\frac{1}{2}$. We could say that there is a 1 in 2 chance that the roll is even, or that there is a 50% chance that the roll is even. We will later see that if this experiment is repeated over and over, approximately $\frac{1}{2}$ of the rolls will be even. ∎

 EXAMPLE
3.23

A single fair six-sided die is rolled. Find the probability that the roll is 5 or higher.

Let B denote the event that the roll is 5 or higher. Then $B = \{5, 6\}$, and $n(B) = 2$. Since $S = \{1, 2, 3, 4, 5, 6\}$, $n(S) = 6$. Therefore,

$$P(B) = \frac{n(B)}{n(S)}$$

$$= \frac{2}{6}$$

$$= \frac{1}{3}$$

The probability that the roll is 5 or higher is $\frac{1}{3}$. ∎

Suppose we rolled a pair of fair six-sided dice and observed the sum of the dice. How large is the sample space? There are 11 possible sums. The possible sums are 2, 3, 4, . . . , 11, and 12. This does not mean that $n(S) = 11$. Recall the following table.

Sums of a Pair of Dice

		SECOND DIE					
		1	**2**	**3**	**4**	**5**	**6**
	1	2	3	4	5	6	7
FIRST	**2**	3	4	5	6	7	8
DIE	**3**	4	5	6	7	8	9
	4	5	6	7	8	9	10
	5	6	7	8	9	10	11
	6	7	8	9	10	11	12

We see that $n(S) = 36$. We will use the following table to calculate the probabilities of events involving the sum of a pair of fair six-sided dice.

Sum	2	3	4	5	6	7	8	9	10	11	12
Number of ways	1	2	3	4	5	6	5	4	3	2	1

EXAMPLE
3.24

A pair of fair six-sided dice are rolled. Find the probability that their sum is 7.

Let event *A* be the event that the sum is 7. Of the 36 different outcomes, how many have a sum of 7? Six of them do, so

$$P(A) = \frac{n(A)}{n(S)}$$

$$= \frac{6}{36}$$

$$= \frac{1}{6}$$

The probability that the sum is 7 is $\frac{1}{6}$. ∎

 A pair of fair six-sided dice are rolled. Find the probability that their sum is 9 or greater.

Let event *B* be the event that the sum is 9 or greater. What sums are 9 or greater? The sums 9, 10, 11, and 12. There are 10 of these sums (four 9's, three 10's, two 11's, and one 12), so

$$P(B) = \frac{n(B)}{n(S)}$$

$$= \frac{10}{36}$$

$$= \frac{5}{18}$$

The probability that the sum is 9 or greater is $\frac{5}{18}$. ∎

Recall the following information about a standard 52-card deck of playing cards.

● There are two colors: red and black. There are 26 cards of each color.
● There are four suits: diamonds (♦), hearts (♥), spades (♠) and clubs (♣). The diamonds and hearts are red, while the spades and clubs are black. There are 13 cards in each suit.
● The 13 cards in each suit are A (ace), 2, 3, 4, 5, 6, 7, 8, 9, 10, J (jack), Q (queen), and K (king). There are 4 of each of these 13 cards in a deck.
● The jack, queen, and king are called face cards, because they have people (with faces) on them. There are 3 face cards in each suit, for a total of 12 face cards in a deck.

EXAMPLE 3.26   A card is drawn from a well-shuffled standard 52-card deck. Find the probability that it is black.

Let *A* be the event that the card is black.

$$P(A) = \frac{n(A)}{n(S)}$$

$$= \frac{26}{52}$$

$$= \frac{1}{2}$$

The probability that the card is black is $\frac{1}{2}$. ∎

A card is drawn from a well-shuffled standard 52-card deck. Find the probability that it is a diamond.

Let *B* be the event that the card is a diamond.

$$P(B) = \frac{n(B)}{n(S)}$$

$$= \frac{13}{52}$$

$$= \frac{1}{4}$$

The probability that the card is a diamond is $\frac{1}{4}$. ∎

 A card is drawn from a well-shuffled standard 52-card deck. Find the probability that it is a face card.

Let *C* be the event that the card is a face card.

$$P(C) = \frac{n(C)}{n(S)}$$

$$= \frac{12}{52}$$

$$= \frac{3}{13}$$

The probability that the card is a face card is $\frac{3}{13}$. ∎

Basic Rules of Probability

- Probabilities are real numbers between 0 and 1.
- If an event is certain to occur, its probability is 1.
- If an event is certain *not* to occur, its probability is 0.
- Special Addition Rule: If events *A* and *B* are mutually exclusive, then *P*(*A* or *B*) = *P*(*A*) + *P*(*B*).
- Complement Rule: *P*(*A*) = 1 − *P*(*A* does not occur).

From our examples, we should believe that the probability of an event is always between 0 and 1. To convince yourself, change 0 and 1 to equivalent percentages: 0% and 100%. Can there ever be more than a 100% chance that an event occurs? Intuitively, the answer is no. Can there ever be less than a 0% chance that an event occurs? Again, no. Let's convince ourselves in a more mathematical fashion. Can the number of different ways that event *A* could occur be less than 0 or more than the size of the sample space? The answer is no, and this gives us $0 \le n(A) \le n(S)$. Take this inequality, and divide each piece by $n(S)$, to verify our original claim.

$$0 \le n(A) \le n(S)$$

$$\frac{0}{n(S)} \le \frac{n(A)}{n(S)} \le \frac{n(S)}{n(S)}$$

$$0 \le P(A) \le 1$$

If an event is guaranteed to happen, then the number of different ways that it could happen must be the same as $n(S)$.

EXAMPLE
3.29 | If a single fair six-sided die is rolled, find the probability that it is lower than a 7.

Let A be the event that the roll is less than 7. In this case, $S = \{1, 2, 3, 4, 5, 6\}$ and $A = \{1, 2, 3, 4, 5, 6\}$ also, so

$$P(A) = \frac{n(A)}{n(S)}$$

$$= \frac{6}{6}$$

$$= 1$$

We know that the roll of a die must produce a number less than 7, so the probability that the roll is less than 7 is 1. ■

If an event is guaranteed not to occur, then the number of different ways that it could occur must be 0. This means that the probability that the event occurs must be 0 as well.

EXAMPLE
3.30 | If a single fair six-sided die is rolled, find the probability that it is a 7.

Let's look at the sample space first. In this case, $S = \{1, 2, 3, 4, 5, 6\}$. Of those six possibilities, how many of them are a 7? None. If we let B be the event that the roll is 7, then B is an empty set and $n(B) = 0$. Therefore,

$$P(B) = \frac{n(B)}{n(S)}$$

$$= \frac{0}{6}$$

$$= 0$$

We know that the roll of a die cannot produce a 7, so the probability that the roll is a 7 is 0. ■

If two events A and B are mutually exclusive, this means that they cannot both occur at the same time. In such a case, we say that the probability of A *and* B occurring is 0. In symbols, we write $P(A \cap B) = 0$. The symbol $\cap$ is the symbol for intersection from set theory. We will use this symbol when we are interested in the probability that both event A occurs and event B occurs.

If we know that the two events A and B are mutually exclusive, then $P(A \text{ or } B) = P(A) + P(B)$. This is known as the **Special Addition Rule.** We will rewrite $P(A \text{ or } B)$ as $P(A \cup B)$. The symbol $\cup$ is the symbol for union from set theory. We will use this symbol when we are interested in the probability that either event A occurs or event B occurs, or maybe both event A and event B occur. The key word for union is *or*. We will be using *or* in the way that we use the phrase *and/or*. This is referred to as the inclusive *or*. For example, if someone offered you coffee or pie, you are not restricted to only one or the other, but could choose both.

To summarize, in symbols we write the *Special Addition Rule* as

$$P(A \cup B) = P(A) + P(B)$$

 EXAMPLE 3.31

If the probability that the San Francisco Forty-Niners win the next Super Bowl is 0.4, and the probability that the Green Bay Packers win the next Super Bowl is 0.25, find the probability that the Forty-Niners or the Packers win the next Super Bowl.

Let *A* be the event that the Forty-Niners win, and let *B* be the event that the Packers win. Are these two events mutually exclusive? Yes, because only one team can win the next Super Bowl.

$$P(A \cup B) = P(A) + P(B)$$
$$= 0.4 + 0.25$$
$$= 0.65$$

The probability that the Forty-Niners or the Packers win the next Super Bowl is 0.65. ∎

 EXAMPLE 3.32

The probability that Joe Student will get an A on the history exam is 0.25, and the probability that he will get a B is 0.35. Find the probability that Joe gets a B or better on the history exam.

Let *A* be the event that Joe gets an A, and let *B* be the event that Joe gets a B. Are these events mutually exclusive? Yes, because Joe cannot get an A and a B on the same exam.

$$P(\text{B or better}) = P(A \cup B)$$
$$= P(A) + P(B)$$
$$= 0.25 + 0.35$$
$$= 0.6$$

The probability that Joe gets a B or better is 0.6. ∎

The *Special Addition Rule* can be extended to situations where there are three or more mutually exclusive events.

EXAMPLE 3.33

A woman needs to buy a new dress to wear to a wedding. If the probability that she buys the dress at Gottschalk's is 0.4, the probability that she buys it at JC Penney's is 0.3, and the probability that she buys it at Mervyn's is 0.1, find the probability that she buys the dress at one of these three stores.

Let *A* be the event that the woman purchases the dress at Gottschalk's, *B* be the event that she buys it at JC Penney's, and *C* be the event that she buys it at Mervyn's.

$$P(A \cup B \cup C) = P(A) + P(B) + P(C)$$
$$= 0.4 + 0.3 + 0.1$$
$$= 0.8$$

The probability that she buys the dress at one of these three stores is 0.8. ∎

Why do we use union in the above situations? When we say that the woman buys the dress at one of these three stores, we are really saying that the woman buys the dress at Gottschalk's *or* JC Penney's *or* Mervyn's. Recall that or is the "magic word" for union.

The **complement** of an event *A*, denoted *A*′, is made up of the possible outcomes of the experiment that are not included in event *A*. Since an event and its complement are mutually exclusive by definition, we can use the Special Addition Rule to show that $P(A \cup A') = P(A) + P(A')$. Since an event either occurs or does not occur, we also know that $P(A \cup A') = 1$. Some easy mathematical manipulation yields the Complement Rule.

$$P(A) + P(A') = 1$$
$$P(A) = 1 - P(A')$$

 EXAMPLE 3.34

Find the probability that it either rains today or does not rain today.

Let *R* be the event that it rains today, and *R*′ be the event that it does not rain today. Since it either has to rain or not rain today, the probability is 1.

$$P(R \cup R') = 1 \quad ∎$$

EXAMPLE 3.35

If the probability that it rains today is 0.35, find the probability that it does not rain.

Let *R* be the event that it rains today, so $P(R) = 0.35$. Since we are interested in the complement of event *R*, we can apply the Complement Rule.

$$P(R') = 1 - P(R)$$
$$= 1 - 0.35$$
$$= 0.65$$

The probability of no rain today is 0.65. ∎

EXAMPLE 3.36

If the probability that the San Francisco Forty-Niners win the next Super Bowl is 0.4, and the probability that the Green Bay Packers win the next Super Bowl is 0.25, find the probability that a team other than the Forty-Niners or the Packers win the next Super Bowl.

Let *A* be the event that the Forty-Niners win the next Super Bowl, and let *B* be the event that the Packers win the next Super Bowl. What we are interested in is the complement of $A \cup B$—that the winner is not the Forty-Niners or the Packers. Use the Complement Rule.

P(A team other than the Forty-Niners or Packers win)

$= 1 - P$(Forty-Niners or Packers win)

$= 1 - P(A \cup B)$

$= 1 - (0.4 + 0.25)$

$= 1 - 0.65$

$= 0.35$

The probability that a team other than the Forty-Niners or Packers win the next Super Bowl is 0.35. ∎

 EXAMPLE 3.37   The probability that Joe Student will get an A on the history exam is 0.25, and the probability that he will get a B is 0.35. Find the probability that Joe gets a C or lower on the history exam.

Let A be the event that Joe gets an A on the exam, and let B be the event that Joe gets a B on the exam. Note that C or lower is the complement of B or better.

P(C or lower) $= 1 - P$(B or better)

$= 1 - P(A \cup B)$

$= 1 - (P(A) + P(B))$

$= 1 - (0.25 + 0.35)$

$= 1 - 0.6$

$= 0.4$

The probability that Joe gets a C or lower on the exam is 0.4. ∎

 EXAMPLE 3.38   The probability that a woman buys a dress to wear at a wedding from Gottschalk's, JC Penney's, or Mervyn's is 0.8. Find the probability that she buys it somewhere other than these three stores.

Let A be the event that the woman buys the dress at Gottschalk's, JC Penney's, or Mervyn's. We are interested in the complement of event A.

$P(A') = 1 - P(A)$

$= 1 - 0.8$

$= 0.2$

The probability that she buys the dress somewhere else is 0.2. ∎

 EXAMPLE 3.39   A pair of fair six-sided dice are rolled. Find the probability that the sum is not 7.

Let A be the event that the sum is 7. (Although S would be a natural choice for the event that the sum is *seven*, recall that S is used for the sample space.) We are interested in $P(A')$, so we will use the Complement Rule. Since 6 of the 36 possible outcomes have a sum of 7, $P(A) = \frac{6}{36}$.

$$P(A') = 1 - P(A)$$

$$= 1 - \frac{6}{36}$$

$$= \frac{30}{36}$$

$$= \frac{5}{6}$$

The probability that the sum is not 7 is $\frac{5}{6}$. ■

 EXAMPLE 3.40   A card is drawn from a well-shuffled standard 52-card deck of playing cards. Find the probability that it is not an ace.

Let A be the event that a card is an ace. We are interested in $P(A')$, so we will use the Complement Rule and the fact that there are four aces in a deck of 52 cards.

$$P(A') = 1 - P(A)$$

$$= 1 - \frac{4}{52}$$

$$= \frac{48}{52}$$

$$= \frac{12}{13}$$

The probability that the card is not an ace is $\frac{12}{13}$. ■

For any two events A and B, we can find the probability that A or B occurs using the **General Addition Rule:**

$$P(A \cup B) = P(A) + P(B) - P(A \cap B)$$

Note that the two events do not need to be mutually exclusive in order to use this rule. However, if the two events are mutually exclusive, this formula will work as well. The only difference between the Special Addition Rule ($P(A \cup B) = P(A) + P(B)$) and the General Addition Rule ($P(A \cup B) = P(A) + P(B) - P(A \cap B)$) is that in the General Addition Rule we subtract $P(A \cap B)$. If events A and B are mutually exclusive, $P(A \cap B) = 0$, and subtracting 0 does not change the result. Let's use the General Addition Rule.

EXAMPLE 3.41   The probability that a foreign visitor to California's Central Valley will visit Sequoia Park is 0.15, and the probability that a foreign visitor will visit Yosemite is 0.22. The probability that the foreign visitor visits both Sequoia Park and Yosemite is 0.08. Find the probability that a foreign visitor to California's Central Valley will visit Sequoia Park or Yosemite.

Let A be the event that the visitor goes to Sequoia Park, and let B be the event that the visitor goes to Yosemite. $P(A) = 0.15$, $P(B) = 0.22$, and $P(A \cap B) = 0.08$. Now use the General Addition Rule.

$$P(A \cup B) = P(A) + P(B) - P(A \cap B)$$
$$= 0.15 + 0.22 - 0.08$$
$$= 0.29$$

The probability that a foreign visitor goes to Sequoia Park or Yosemite is 0.29. ■

   We can use a *Venn diagram* to help us with problems involving the General Addition Rule. When drawing a Venn diagram for probabilities, we begin with a rectangle representing the sample space S. We then draw two circles inside the rectangle for events A and B in such a way that the two circles overlap. We place $P(A \cap B)$ in the region where the circles overlap. We then subtract $P(A \cap B)$ from $P(A)$ to determine what value to place in the rest of circle A. We repeat this for circle B. Finally, the regions inside the box should add up to 1. Here is the Venn diagram that goes along with the previous example.

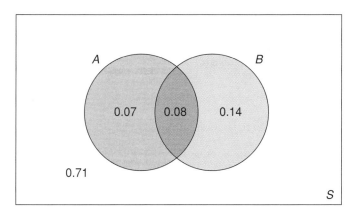

   We put the probability associated with the intersection in the center region. We then subtract the 0.08 from 0.15 (giving us 0.07) and from 0.22 (giving us 0.14). If we add 0.07 + 0.08 + 0.14, we find that the probability of A or B is 0.29. Where does the 0.71 come from? Subtract 0.29 from 1.

EXAMPLE 3.42

A card is drawn from a well-shuffled standard 52-card deck of playing cards. Find the probability that the card is a diamond or a face card.

Let A be the event that the card is a diamond, and let B be the event that the card is a face card. $P(A) = \frac{13}{52}$, $P(B) = \frac{12}{52}$, and $P(A \cap B) = \frac{3}{52}$ because there are three cards in the deck that are diamonds *and* face cards.

$$P(A \cup B) = P(A) + P(B) - P(A \cap B)$$
$$= \frac{13}{52} + \frac{12}{52} - \frac{3}{52}$$
$$= \frac{22}{52}$$
$$= \frac{11}{26}$$

The probability that a card is a diamond or a face card is $\frac{11}{26}$. ■

EXERCISES 3.2

For Exercises 1–4, a fair six-sided die is rolled.

1. Find the probability that the roll is a 4.
2. Find the probability that the roll is at least 4.
3. Find the probability that the roll is less than 4.
4. Find the probability that the roll is a prime number.

For Exercises 5–10, a pair of fair six-sided dice are rolled.

5. Find the probability that the sum is 4.
6. Find the probability that the sum is 4 or higher.
7. Find the probability that the sum is 7 or 11.
8. Find the probability that the sum is 2, 3, or 12.
9. Find the probability that the sum is odd.
10. Find the probability that the sum is even.

For Exercises 11–14, a card is drawn from a well-shuffled standard 52-card deck of playing cards.

11. Find the probability that the card is red.
12. Find the probability that the card is a club.
13. Find the probability that the card is a jack.
14. Find the probability that the card is a queen or an ace.

A child has 12 blocks: 4 red, 4 blue, and 4 green. All of the blocks are put inside a bucket.

15. The child reaches into the bucket and randomly picks a block. Find the probability that the block is red.
16. The child reaches into the bucket and randomly picks a block. Find the probability that the block is not red.
17. The child throws two of the green blocks away. The child reaches into the bucket and randomly picks a block. Find the probability that the block is green.
18. The child throws two of the green blocks away. The child reaches into the bucket and randomly picks a block. Find the probability that the block is red.

For Exercises 19–21, a fair six-sided die is rolled.

19. Find the probability that the roll is at least a 1.
20. Find the probability that the roll is at least a 10.
21. Find the probability that the roll is a $3\frac{1}{2}$.

For Exercises 22–24, a pair of fair six-sided dice are rolled.

22. Find the probability that the sum is 13.
23. Find the probability that the sum is 13 or less.
24. Find the probability that the sum is $3\frac{1}{2}$.

For Exercises 25–28, a card is drawn from a well-shuffled standard 52-card deck of playing cards.

25. Find the probability that the card is a king or a queen.
26. Find the probability that the card is a face card or an ace.
27. Find the probability that the card is a spade or a heart.
28. Find the probability that the card is black or red.

29. The probability that Bob Dylan wins the Grammy award for album of the year is 0.4, and the probability that Paul McCartney wins it is 0.15. Find the probability that Bob Dylan or Paul McCartney wins the award.
30. The probability that Jeff Gordon wins the next Daytona 500 is 0.35, and the probability that Dale Earnhart Jr. wins the race is 0.17. Find the probability that Jeff Gordon or Dale Earnhart Jr. wins the next Daytona 500.
31. Joe Star is a prized baseball free agent. Joe's agent states that there is a 75% chance that his client will sign with the Yankees, but there is only a 10% chance that he

will sign with the Braves. Find the probability that Joe Star will sign with a team other than the Yankees or Braves.

32. The probability that an algebra student earns a final grade of A is 0.1, the probability of earning a final grade of B is 0.12, and the probability of earning a final grade of C is 0.17. Find the probability that an algebra student passes with a grade of C or higher.

33. Find the probability that it either snows or does not snow today.

34. Find the probability that a customer at a coffee shop either buys a donut or does not buy a donut.

35. A pair of fair six-sided dice are rolled. Find the probability that the sum is even or not even.

36. A card is drawn from a well-shuffled standard 52-card deck of playing cards. Find the probability that the card is red or not red.

For Exercises 37–40, a card is drawn from a well-shuffled standard 52-card deck of playing cards.

37. Find the probability that the card is not a queen.

38. Find the probability that the card is not a face card.

39. Find the probability that the card is not red.

40. Find the probability that the card is not a diamond.

41. The probability that it snows today is 0.25. Find the probability that it does not snow today.

42. The probability that Bob Dylan wins the Grammy award for album of the year is 0.4. Find the probability that Bob Dylan does not win the award.

43. The probability that Jeff Gordon wins the next Daytona 500 is 0.35. Find the probability that someone else wins the next Daytona 500.

44. The probability that an algebra student earns a final grade of A is 0.1, the probability of earning a final grade of B is 0.12, and the probability of earning a final grade of C is 0.17. Find the probability that a student does not pass with a grade of C or better.

45. A card is drawn from a well-shuffled standard 52-card deck of playing cards. Find the probability that the card is an ace or a diamond.

46. A card is drawn from a well-shuffled standard 52-card deck of playing cards. Find the probability that the card is red or a face card.

47. A pair of fair six-sided dice are rolled. Find the probability that the sum is 5 or that at least one of the dice is a 3.

48. A pair of fair six-sided dice are rolled. Find the probability that the sum is 8 or that doubles are rolled.

49. The probability that a student passes his or her math class is 0.8, the probability that a student makes the Dean's List is 0.4, and the probability that a student passes his or her math class and makes the Dean's List is 0.25. Find the probability that a student passes his or her math class or makes the Dean's List.

50. The probability that a college freshman takes a math class the first semester is 0.75. The probability that a college freshman takes an English class the first semester is 0.7. The probability that a college freshman takes both a math class and an English class the first semester is 0.6. Find the probability that a college freshman takes a math class or an English class the first semester.

SECTION 3.3
Table Probabilities, Conditional Probability

One hundred fifty students at the College of the Sequoias were asked their preference of burgers. Their responses were categorized as Burger King, McDonald's, In-N-Out, and Other. Here are the results.

	Burger King	McDonald's	In-N-Out	Other	Total
Male	15	18	35	15	83
Female	35	12	10	10	67
Total	50	30	45	25	150

 EXAMPLE 3.43   If a student from this survey is selected at random, find the probability that the student is male.

Let M be the event that the student is male.

$$P(M) = \frac{n(M)}{n(S)}$$

$$= \frac{83}{150}$$

The probability that the student is male is $\frac{83}{150}$. ■

 EXAMPLE 3.44   If a student from this survey is selected at random, find the probability that the student is female.

Let F be the event that the student is female.

$$P(F) = \frac{n(F)}{n(S)}$$

$$= \frac{67}{150}$$

The probability that the student is female is $\frac{67}{150}$. ■

EXAMPLE 3.45   If a student from this survey is selected at random, find the probability that the student prefers Burger King.

Let B be the event that the student prefers Burger King.

$$P(B) = \frac{n(B)}{n(S)}$$

$$= \frac{50}{150}$$

$$= \frac{1}{3}$$

The probability that the student prefers Burger King is $\frac{1}{3}$. ■

EXAMPLE 3.46   If a student from this survey is selected at random, find the probability that the student is male and prefers Burger King.

Let M be the event that the student is male, and let B be the event that the student prefers Burger King. Of the 150 students, 15 are male and prefer Burger King. The 15 students are highlighted in the following table.

	Burger King	McDonald's	In-N-Out	Other	Total
Male	15	18	35	15	83
Female	35	12	10	10	67
Total	50	30	45	25	150

$$P(M \cap B) = \frac{15}{150}$$

$$= \frac{1}{10}$$

The probability that the student is male and prefers Burger King is $\frac{1}{10}$. ∎

EXAMPLE 3.47   If a student from this survey is selected at random, find the probability that the student is male or prefers Burger King.

Let M be the event that the student is male, and let B be the event that the student prefers Burger King. We could use the General Addition Rule.

$$P(M \cup B) = P(M) + P(B) - P(M \cap B)$$

$$= \frac{83}{150} + \frac{50}{150} - \frac{15}{150}$$

$$= \frac{59}{75}$$

The probability that the student is male or prefers Burger King is $\frac{59}{75}$. ∎

We could have also used the table. We need to include all the students in the top row (male) as well as the students in the left column (Burger King). The 118 students are highlighted in the following table.

	Burger King	McDonald's	In-N-Out	Other	Total
Male	15	18	35	15	83
Female	35	12	10	10	67
Total	50	30	45	25	150

Note that 15 + 18 + 35 + 15 + 35 gives us 118, the same total as we had from the General Addition Rule.

$$P(M \cup B) = \frac{118}{150}$$

$$= \frac{59}{75}$$

EXAMPLE 3.48   If a male from the survey is selected at random, find the probability that he prefers Burger King.

Without using any notation, let's use our intuition to find the probability. In this problem we are selecting one of the males, and there are 83 to choose from.

How many of those 83 prefer Burger King? Fifteen of the 83 prefer Burger King. It stands to reason that the probability is $\frac{15}{83}$. This problem is an example of conditional probability. ◼

Conditional Probability

A problem is known as a **conditional probability** problem if there is a given condition that restricts the sample space. In the previous problem, we knew that the student being selected was male, not just a student. The size of the sample space was reduced from the 150 students to the 83 male students.

The probability of event A occurring, given that event B has occurred (B is the condition) is denoted $P(A|B)$. This is read "The probability of A given B." The known condition is put after the vertical line, and the event whose probability we are interested in is put before the vertical line.

 EXAMPLE 3.49   A child has a set of 12 plastic blocks in a bucket. Four of the blocks are red, 4 of the blocks are blue, and 4 of the blocks are green. If a child draws a green block from the bucket, find the probability that the next block drawn from the bucket will be red.

Let R be the event that a block is red, B be the event that a block is blue, and G be the event that a block is green. The condition that has changed our sample space of 12 blocks is that a green block has already been drawn from the bucket. This leaves us with only 11 blocks, 4 of which are red.

$$P(R|G) = \frac{n(R)}{n(S^*)}$$

$$= \frac{4}{11}$$

The notation S^* is used to identify the fact that the sample space has been restricted. The probability that a red block is drawn if a green block has already been drawn is $\frac{4}{11}$. ◼

 EXAMPLE 3.50   A child has a set of 12 plastic blocks in a bucket. Four of the blocks are red, 4 of the blocks are blue, and 4 of the blocks are green. If a child draws a green block and a red block from the bucket, find the probability that the third block drawn from the bucket will be red.

Let R be the event that a block is red, B be the event that a block is blue, and G be the event that a block is green. We are interested in the probability of drawing a red block given that a green block and a red block have already been drawn. In symbols, we write this as $P(R|GR)$. Since one red block has already been removed, there are 3 red blocks remaining in the bucket. Also, since 2 blocks have been removed from the bucket, there are 10 blocks remaining in the bucket.

$$P(R|GR) = \frac{n(R)}{n(S^*)}$$

$$= \frac{3}{10}$$

The probability of drawing a red block after drawing a green block and a red block is $\frac{3}{10}$. ■

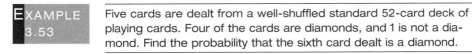

EXAMPLE
3.51

If a club is drawn from a well-shuffled standard 52-card deck of playing cards, find the probability that the next card drawn is a diamond.

Let C be the event that a club is drawn, and let D be the event that a diamond is drawn. We are interested in $P(D|C)$. After the club is drawn, how many diamonds remain in the deck? There are still 13 diamonds (none have been removed). After the club is drawn, how many cards remain in the deck? There are 51 cards in the deck, because 1 card has been removed.

$$P(D|C) = \frac{13}{51}$$

The probability that a diamond is drawn after a club has been drawn is $\frac{13}{51}$. ■

EXAMPLE
3.52

If a diamond is drawn from a well-shuffled standard 52-card deck of playing cards, find the probability that the next card drawn is a diamond.

Let D be the event that a diamond is drawn. We are interested in $P(D|D)$. After the diamond is drawn, how many diamonds remain in the deck? There are now 12 diamonds because 1 has been removed. After the diamond is drawn, how many cards remain in the deck? There are 51 cards in the deck, because 1 card has been removed.

$$P(D|D) = \frac{12}{51}$$

$$= \frac{4}{17}$$

The probability that a diamond is drawn after a diamond has been drawn is $\frac{4}{17}$. ■

EXAMPLE
3.53

Five cards are dealt from a well-shuffled standard 52-card deck of playing cards. Four of the cards are diamonds, and 1 is not a diamond. Find the probability that the sixth card dealt is a diamond.

Let D be the event that a diamond is dealt, and let N be the event that a "non-diamond" is dealt. We are interested in $P(D|DDDDN)$. After the first 5 cards are dealt, how many diamonds remain in the deck? There are now 9 diamonds because 4 have been removed. After the first 5 cards are dealt, how many cards remain in the deck? There are 47 cards in the deck, because 5 cards have been removed.

$$P(D|DDDDN) = \frac{9}{47}$$

The probability that the sixth card dealt is a diamond is $\frac{9}{47}$. ■

Now let's revisit the burger preference example.

	Burger King	*McDonald's*	*In-N-Out*	*Other*	*Total*
Male	15	18	35	15	83
Female	35	12	10	10	67
Total	50	30	45	25	150

EXAMPLE 3.54   If a male from the survey is selected at random, find the probability that he prefers Burger King.

Let M be the event that a male is selected, and B be the event that the person prefers Burger King. Since we know that a male is selected, we are looking for $P(B|M)$.

$$P(B|M) = \frac{n(B)}{n(S^*)}$$

$$= \frac{15}{83}$$

If a male from the survey is selected at random, the probability that he prefers Burger King is $\frac{15}{83}$. ∎

EXAMPLE 3.55   If a person from the survey who prefers Burger King is selected at random, find the probability that the person is male.

Let B be the event that the person prefers Burger King, and M be the event that a male is selected. Since we know that a person who prefers Burger King is selected, we are looking for $P(M|B)$.

$$P(M|B) = \frac{n(M)}{n(S^*)}$$

$$= \frac{15}{50}$$

$$= \frac{3}{10}$$

If a person from the survey who prefers Burger King is selected at random, the probability that the person is male is $\frac{3}{10}$. ∎

EXAMPLE 3.56   If a female from the survey is selected at random, find the probability that she prefers McDonald's.

Let F be the event that a female is selected, and D be the event that the person prefers McDonald's. Since we know that a female is selected, we are looking for $P(D|F)$.

$$P(D|F) = \frac{n(D)}{n(S^*)}$$

$$= \frac{12}{67}$$

Twelve of the 67 females in the survey prefer McDonald's. If a female from the survey is selected at random, the probability that she prefers McDonald's is $\frac{12}{67}$. ■

The probability of *A* given *B* can be calculated using the **conditional probability formula:**

$$P(A|B) = \frac{P(A \cap B)}{P(B)}$$

The best way to remember this formula is to think of the numerator as the probability of "both," and the denominator as the probability of "the given one." Now we use the formula on the previous example.

 EXAMPLE 3.57 If a female from the survey is selected at random, find the probability that she prefers McDonald's.

Let *F* be the event that a female is selected, and *D* be the event that the person prefers McDonald's. Since we know that a female is selected, we are looking for $P(D|F)$.

$$P(D|F) = \frac{P(D \cap F)}{P(F)}$$

$$= \frac{12/150}{67/150}$$

$$= \frac{12}{67}$$

If a female from the survey is selected at random, the probability that she prefers McDonald's is $\frac{12}{67}$. This agrees with our answer found using the tables. ■

Reasoning out the probability from the tables may seem easier than applying the formula, but there will be times when we must use the formula (for example, when we do not have a table).

EXAMPLE 3.58 Sixty percent of the households in Visalia have the *Visalia Times Delta* delivered to their house. Fifteen percent of the households in Visalia have the *Visalia Times Delta* and the *Fresno Bee* delivered to their house. If we know that a Visalia household has the *Visalia Times Delta* delivered to their house, find the probability that they also have the *Fresno Bee* delivered to their house.

Let *D* be the event that a household has the *Times Delta* delivered to their house, and let *B* be the event that a household has the *Fresno Bee* delivered to their house. We are looking for $P(B|D)$, the probability that a house gets the *Bee* if it is known that the house gets the *Times Delta*.

$$P(B|D) = \frac{P(B \cap D)}{P(D)}$$

$$= \frac{0.15}{0.6}$$

$$= 0.25$$

If a Visalia household has the *Visalia Times Delta* delivered to their house, the probability that they also have the *Fresno Bee* delivered to their house is 0.25. Twenty-five percent of households that have the *Times Delta* (not all Visalia households) delivered also have the *Fresno Bee* delivered. ■

 EXAMPLE 3.59   The probability that a student passes their math class is 0.8, and the probability that a student passes their math class and makes the Dean's List is 0.25. If a student passes their math class, find the probability that the student makes the Dean's List.

Let M be the event that the student passes their math class, and let D be the event that the student makes the Dean's List. We are looking for $P(D|M)$.

$$P(D|M) = \frac{P(D \cap M)}{P(M)}$$

$$= \frac{0.25}{0.8}$$

$$= 0.3125$$

If we know that a student has passed their math class, the probability that the student makes the Dean's List is 0.3125. ■

EXERCISES 3.3

Two hundred basketball fans were asked who they felt was responsible for the 1998–1999 NBA lockout—the players or the owners. Here are the responses, divided into "casual" fans and "hardcore" fans.

	Players Are Responsible	*Owners Are Responsible*	*Both/Unsure*
"Casual" Fans	90	22	8
"Hardcore" Fans	20	53	7

For Exercises 1–12, a fan in this survey is selected at random.
1. Find the probability that the fan is a "casual" fan.
2. Find the probability that the fan is a "hardcore" fan.
3. Find the probability that the fan feels the players are responsible for the lockout.
4. Find the probability that the fan feels the owners are responsible for the lockout.
5. Find the probability that the fan is a "casual" fan and feels that the players are responsible for the lockout.
6. Find the probability that the fan is a "hardcore" fan and feels that the players are responsible for the lockout.
7. Find the probability that the fan is a "casual" fan and feels that the owners are responsible for the lockout.
8. Find the probability that the fan is a "hardcore" fan and feels that the owners are responsible for the lockout.
9. Find the probability that the fan is a "casual" fan or feels that the players are responsible for the lockout.
10. Find the probability that the fan is a "hardcore" fan or feels that the players are responsible for the lockout.

11. Find the probability that the fan is a "casual" fan or feels that the owners are responsible for the lockout.
12. Find the probability that the fan is a "hardcore" fan or feels that the owners are responsible for the lockout.
13. If a "casual" fan in this survey is selected at random, find the probability that the fan feels that the players are responsible for the lockout.
14. If a "casual" fan in this survey is selected at random, find the probability that the fan feels that the owners are responsible for the lockout.
15. If a "hardcore" fan in this survey is selected at random, find the probability that the fan feels that the players are responsible for the lockout.
16. If a "hardcore" fan in this survey is selected at random, find the probability that the fan feels that the owners are responsible for the lockout.
17. If a fan in this survey who feels that the players are responsible for the lockout is selected at random, find the probability that the fan is a "casual" fan.
18. If a fan in this survey who feels that the players are responsible for the lockout is selected at random, find the probability that the fan is a "hardcore" fan.
19. If a fan in this survey who feels that the owners are responsible for the lockout is selected at random, find the probability that the fan is a "casual" fan.
20. If a fan in this survey who feels that the owners are responsible for the lockout is selected at random, find the probability that the fan is a "hardcore" fan.

One hundred voters were surveyed in November 1998 for an exit poll. They were asked the following questions:

- On most political matters, do you consider yourself liberal, moderate, or conservative?
- In today's House of Representatives election, did you vote for the Democratic Party candidate or the Republican Party candidate?

Here are the results.

	Liberal	Moderate	Conservative
Democrat	16	27	5
Republican	3	23	26

For Exercises 21–24, a voter from this survey is selected at random.

21. Find the probability that the voter voted for the Democratic Party candidate.
22. Find the probability that the voter considers himself liberal.
23. Find the probability that the voter voted for the Democratic Party candidate or considers himself liberal.
24. Find the probability that the voter voted for the Democratic Party candidate and considers himself liberal.
25. If a voter from this survey who voted for the Democratic Party candidate is selected at random, find the probability that the voter considers himself liberal.
26. If a voter from this survey who considers himself liberal is selected at random, find the probability that the voter voted for the Democratic Party candidate.
27. If a voter from this survey who voted for the Republican Party candidate is selected at random, find the probability that the voter considers himself conservative.
28. If a voter from this survey who considers himself conservative is selected at random, find the probability that the voter voted for the Republican Party candidate.
29. If a voter from this survey who voted for the Republican Party candidate is selected at random, find the probability that the voter considers himself moderate or conservative.
30. If a voter from this survey who considers himself moderate or conservative is selected at random, find the probability that the voter voted for the Republican Party candidate.

A child has a set of 12 plastic blocks in a bucket. Four of the blocks are red, 4 of the blocks are blue, and 4 of the blocks are green.

31. If a child draws a red block from the bucket, find the probability that the next block drawn from the bucket will be red.

32. If a child draws a red block from the bucket, find the probability that the next block drawn from the bucket will be blue.

33. If a child draws three red blocks from the bucket, find the probability that the next block drawn from the bucket will be red.

34. If a child draws three red blocks from the bucket, find the probability that the next block drawn from the bucket will be green.

35. If a child draws a block from the bucket and does not reveal its color, find the probability that the next block drawn from the bucket will be red.

36. A face card is drawn from a well-shuffled standard 52-card deck of playing cards. Find the probability that the next card drawn is also a face card.

37. Three cards are drawn from a well-shuffled standard 52-card deck of playing cards. All three cards are hearts. Find the probability that the next card drawn is also a heart.

38. Five cards are drawn from a well-shuffled standard 52-card deck of playing cards. Three cards are kings, and the other two are not. Find the probability that the next card drawn is a king.

39. A card is drawn from a well-shuffled standard 52-card deck of playing cards, its suit is observed, and then the card is put back into the deck. After the deck is reshuffled, another card is drawn. If the first card selected is a spade, find the probability that the second card is a spade.

40. How would your answer to Exercise 39 differ if the first card were not put back into the deck before the second card is drawn? Explain your answer in your own words.

41. On par-three golf holes, the probability that a PGA professional golfer hits his first shot on the green is 0.8. The probability that a PGA professional golfer hits his first shot on the green and requires only one putt to put the ball in the hole is 0.2. If a PGA professional golfer hits his first shot on the green, find the probability that he only requires one putt to put the ball in the hole.

42. Eighty percent of the flights from Fresno to Las Vegas depart on time; in other words, the probability that a flight from Fresno to Las Vegas departs on time is 0.8. Seventy-two percent of all flights from Fresno to Las Vegas depart on time and arrive on time. If a flight from Fresno to Las Vegas departs on time, find the probability that it also arrives on time.

43. At a middle school, 48% of the students ride the bus to and from school. Thirty percent of the students ride the bus to and from school and buy their lunch at school. If a student rides the bus to and from school, find the probability that the student buys his or her lunch at school.

44. In a recent exit poll, 60% of the respondents felt that the country was generally going in the right direction. Twenty-four percent of the respondents were Republicans who felt that the country was generally going in the right direction. If a person in this poll felt that the country was generally going in the right direction, find the probability that the person was Republican.

45. In a recent exit poll, 40% of the respondents claimed to be Democrats. If a person was a Democrat, the probability that he or she voted for the Democratic candidate was 0.9. Find the probability that a person is a Democrat and voted for the Democratic candidate.

SECTION 3.4
Independent Events, Multiplication Rule, Law of Large Numbers

Independent Events

Two events A and B are said to be **independent** if the probability of event A is unaffected by event B occurring or not occurring. In other words, A and B are independent if $P(A|B) = P(A)$.

EXAMPLE 3.60

Sixty percent of the households in Visalia have the *Visalia Times Delta* delivered to their house. Twenty-five percent of the households in Visalia have the *Fresno Bee* delivered to their house. Fifteen percent of the households in Visalia have the *Visalia Times Delta* and the *Fresno Bee* delivered to their house. Are the events "A household gets the *Times Delta* delivered to their house" and "A household gets the *Fresno Bee* delivered to their house" independent?

Let *D* be the event that a household has the *Times Delta* delivered to their house, and let *B* be the event that a household has the *Fresno Bee* delivered to their house. We know that $P(D) = 0.6$, $P(B) = 0.25$, and $P(D \cap B) = 0.15$. To show that the two events *B* and *D* are independent, we must show that $P(B) = P(B|D)$.

$$P(B|D) = \frac{P(B \cap D)}{P(D)}$$

$$= \frac{0.15}{0.6}$$

$$= 0.25$$

$$= P(B)$$

Therefore, $P(B) = P(B|D)$, and events *B* and *D* are independent. We could have chosen to show that $P(B) = P(D|B)$ to show that events *B* and *D* are independent, and that is left to you as an exercise. The events "A household gets the *Times Delta* delivered to their house" and "A household gets the *Fresno Bee* delivered to their house" are independent, since the probability of a household getting the *Fresno Bee* delivered is the same whether we know the house gets the *Times Delta* or not. ∎

There is another way that we can check for independence. If $P(A) \cdot P(B) = P(A \cap B)$, then events *A* and *B* are independent. Where does this come from? It comes from the formula for conditional probability as shown next.

$$P(A) = P(A|B)$$

$$P(A) = \frac{P(A \cap B)}{P(B)}$$

$$P(A) \cdot P(B) = P(A \cap B)$$

Let's revisit the previous example.

EXAMPLE 3.61

Sixty percent of the households in Visalia have the *Visalia Times Delta* delivered to their house. Twenty-five percent of the households in Visalia have the *Fresno Bee* delivered to their house. Fifteen percent of the households in Visalia have the *Visalia Times Delta* and the *Fresno Bee* delivered to their house. Are the events "A household gets the *Times Delta* delivered to their house" and "A household gets the *Fresno Bee* delivered to their house" independent?

Let *D* be the event that a household has the *Times Delta* delivered to their house, and let *B* be the event that a household has the *Fresno Bee* delivered to their house. We know that $P(D) = 0.6$, $P(B) = 0.25$, and $P(D \cap B) = 0.15$. To check for independence, we can check to see whether $P(D) \cdot P(B) = P(D \cap B)$.

$$P(D) \cdot P(B) = (0.6) \cdot (0.25)$$
$$= 0.15$$
$$= P(D \cap B)$$

Thus the events "A household gets the *Times Delta* delivered to their house" and "A household gets the *Fresno Bee* delivered to their house" are independent. ■

The probability that a student passes their math class is 0.8, the probability that a student makes the Dean's List is 0.4, and the probability that a student passes their math class and makes the Dean's List is 0.25. Are the events "A student passes their math class" and "A student makes the Dean's List" independent events?

Let M be the event that the student passes their math class, and let D be the event that the student makes the Dean's List. We know that $P(M) = 0.8$, $P(D) = 0.4$, and $P(M \cap D) = 0.25$. To check for independence, we can check to see whether $P(M) \cdot P(D) = P(M \cap D)$.

$$P(M) \cdot P(D) = (0.8) \cdot (0.4)$$
$$= 0.32$$
$$\neq P(M \cap D)$$

Thus the events "A student passes their math class" and "A student makes the Dean's List" are not independent events. ■

EXAMPLE 3.63

Once again, let's return to the burger preference example. One hundred fifty students at the College of the Sequoias were asked their preference of burgers. Their responses were categorized as Burger King, McDonald's, In-N-Out, and Other. Here are the results.

	Burger King	*McDonald's*	*In-N-Out*	*Other*	*Total*
Male	15	18	35	15	83
Female	35	12	10	10	67
Total	50	30	45	25	150

For the students in this survey, are the events "A student is male" and "A student prefers Burger King" independent events?

Let M be the event that a student is male, and let B be the event that a student prefers Burger King. For these two events to be independent, we must show that $P(M) \cdot P(B) = P(M \cap B)$. Since 15 of the 150 students in the survey were male and preferred Burger King,

$$P(M \cap B) = \frac{15}{150}$$
$$= \frac{1}{10}$$

We know that $P(M) = \frac{83}{150}$, and $P(B) = \frac{50}{150}$ or $\frac{1}{3}$. Therefore,

$$P(M) \cdot P(B) = \frac{83}{150} \cdot \frac{1}{3}$$

$$= \frac{83}{450}$$

$$\approx 0.18$$

Since $P(M) \cdot P(B) \neq P(M \cap B)$, the two events are not independent. ∎

Later we will use such a survey to try and answer the broader question "Is a person's burger preference independent of the person's gender?" The events in this example were not independent, but that only relates to this sample. Were they close enough to independent to suggest that for all people burger preference is independent of a person's gender? We will examine this later.

Multiplication Rule

If two events A and B are independent, then the probability of A and B occurring is given by the following formula:

$$P(A \cap B) = P(A) \cdot P(B)$$

This is one form of the **Multiplication Rule.** Recall that we use intersection ($\cap$) in circumstances involving the word *and.* Also recall from our study of counting that we used multiplication when we were interested in the number of ways that one event *and* another event could occur. The Multiplication Rule is consistent with that idea.

EXAMPLE 3.64

A fair coin is flipped two times. Find the probability that both tosses result in heads.

Let H be the event that a toss results in heads. We know that $P(H) = \frac{1}{2}$. We are interested in the probability that both tosses result in heads, or, in other words, that the first toss is heads *and* the second toss is heads. The result of the second toss does not depend on the result of the first toss, so the two events are independent.

$$P(H \text{ first} \cap H \text{ second}) = P(H \text{ first}) \cdot P(H \text{ second})$$

$$= \frac{1}{2} \cdot \frac{1}{2}$$

$$= \frac{1}{4}$$

The probability that both tosses are heads is $\frac{1}{4}$. ∎

EXAMPLE 3.65

The probability that a randomly selected American has an unfavorable opinion of the president is 0.35. If two Americans are selected at random, find the probability that both have an unfavorable opinion of the president.

Let A be the event that a randomly selected American has an unfavorable opinion of the president. We know that $P(A) = 0.35$. The response of the second American does not depend on the response of the first American.

$$P(A \text{ first} \cap A \text{ second}) = P(A \text{ first}) \cdot P(A \text{ second})$$
$$= 0.35 \cdot 0.35$$
$$= 0.1225$$

The probability that both Americans have an unfavorable opinion of the president is 0.1225. ∎

We can extend the Multiplication Rule to cover three or more events, as in the following example.

The probability that an elementary school student selects soccer as his or her favorite sport is $\frac{1}{4}$. If four elementary school students are selected at random, find the probability that they all select soccer as their favorite sport.

Let A be the event that a student selects soccer. We are interested in the probability that all four students select soccer as their favorite sport. This is equivalent to the probability that the first student selects soccer *and* the second student selects soccer *and* the third student selects soccer *and* the fourth student selects soccer.

$$P(\text{All 4 select soccer}) = P(A) \cdot P(A) \cdot P(A) \cdot P(A)$$
$$= \frac{1}{4} \cdot \frac{1}{4} \cdot \frac{1}{4} \cdot \frac{1}{4}$$
$$= \frac{1}{256}$$

The probability that all four students select soccer is $\frac{1}{256}$. ∎

Multiplication Rule for Dependent Events

We can adapt the Multiplication Rule to cover events that are not independent. If A and B are any two events, then the probability of A and B occurring is given by

$$P(A \cap B) = P(A) \cdot P(B|A)$$

Instead of multiplying the probability of A occurring by the probability of B occurring, we instead multiply by the probability of B occurring if A has already occurred. The best way to remember this is to try to find the probability of the first event occurring, then multiply by the probability of the next event occurring. Take the events one at a time, and multiply their probabilities.

Two cards are drawn from a well-shuffled standard 52-card deck of playing cards. Find the probability that both cards are diamonds.

Let D be the event that a card is a diamond. We are interested in the probability that both cards are diamonds—that is, that the first and second cards are diamonds. What is the probability that the first card is a diamond? Since there are 13 diamonds and 52 cards in the deck, the probability is $\frac{13}{52}$. What is the probability that the second card is a diamond? There are now 12 diamonds and 51 cards left in the deck, so the probability is $\frac{12}{51}$. To find the probability that the first card is a diamond and the second card is a diamond, we multiply these two probabilities.

$$P(D \text{ first} \cap D \text{ second}) = P(D \text{ first}) \cdot P(D \text{ second})$$

$$= \frac{13}{52} \cdot \frac{12}{51}$$

$$= \frac{1}{17} \blacksquare$$

 EXAMPLE 3.68   Two cards are drawn from a well-shuffled standard 52-card deck of playing cards. Find the probability that both cards are face cards.

Let F be the event that a card is a face card. What is the probability that the first card is a face card? It is $\frac{12}{52}$, because there are 12 face cards in a standard 52-card deck. The probability that the second card is a face card is $\frac{11}{51}$, because there is one less face card in the deck on the second card, as well as one less card.

$$P(F \text{ first} \cap F \text{ second}) = P(F \text{ first}) \cdot P(F \text{ second})$$

$$= \frac{12}{52} \cdot \frac{11}{51}$$

$$= \frac{11}{221}$$

The probability that both cards are face cards is $\frac{11}{221}$. $\blacksquare$

EXAMPLE 3.69   Two cards are drawn from a well-shuffled standard 52-card deck of playing cards. Find the probability that one card is a king and the other card is a queen.

Let K be the event that a card is a king, and let Q be the event that a card is a queen. There are two ways that we could draw a king and a queen. We could draw a king and then a queen, or we could draw a queen and then a king. We will find the probability of each scenario, and then add. The probability that the first card is a king is $\frac{4}{52}$, and the probability that the second card is a queen is $\frac{4}{51}$. The probability that the first card is a queen is $\frac{4}{52}$, and the probability that the second card is a king is $\frac{4}{51}$. This gives us

$$P(K \text{ and a } Q) = P(K \text{ 1st}) \cdot P(Q \text{ 2nd}) + P(Q \text{ 1st}) \cdot P(K \text{ 2nd})$$

$$= \frac{4}{52} \cdot \frac{4}{51} + \frac{4}{52} \cdot \frac{4}{51}$$

$$= \frac{4}{663} + \frac{4}{663}$$

$$= \frac{8}{663}$$

The probability of drawing a king and a queen is $\frac{8}{663}$. $\blacksquare$

EXAMPLE 3.70   Three cards are drawn from a well-shuffled standard 52-card deck. Find the probability that two are aces and one is not.

Let A be the event that a card is an ace, and let N be the event that a card is not an ace. There are three different ways that two of the three cards could be aces: the first two cards could be aces, the first and third cards could be aces, or the last two cards could be aces. (In symbols: AAN, ANA, and NAA.) We must find the probability of each different way occurring, and then add those respective probabilities. Here is the probability of the first way, keeping in mind that there are four aces and 48 "non-aces" in a deck of 52 cards.

$$P(A \text{ first} \cap A \text{ second} \cap N \text{ third}) = P(A \text{ first}) \cdot P(A \text{ second}) \cdot P(N \text{ third})$$

$$= \frac{4}{52} \cdot \frac{3}{51} \cdot \frac{48}{50}$$

$$= \frac{24}{5525}$$

Now here is the second probability.

$$P(A \text{ first} \cap N \text{ second} \cap A \text{ third}) = P(A \text{ first}) \cdot P(N \text{ second}) \cdot P(A \text{ third})$$

$$= \frac{4}{52} \cdot \frac{48}{51} \cdot \frac{3}{50}$$

$$= \frac{24}{5525}$$

Note that the probability of the second way is exactly the same as the probability of the first. Here is the third way.

$$P(N \text{ first} \cap A \text{ second} \cap A \text{ third}) = P(N \text{ first}) \cdot P(A \text{ second}) \cdot P(A \text{ third})$$

$$= \frac{48}{52} \cdot \frac{4}{51} \cdot \frac{3}{50}$$

$$= \frac{24}{5525}$$

So the probability of drawing two cards that are aces and one that is not is found by adding the three probabilities.

$$P(2 \text{ aces and 1 "non-ace"}) = \frac{24}{2252} + \frac{24}{5525} + \frac{24}{5525}$$

$$= \frac{72}{5525}$$

We see that the probability is $\frac{72}{5525}$. ∎

The previous examples are examples of sampling **without replacement**, because the card that is drawn is not replaced in the deck before the next card is drawn. Occasionally the sampling is done **with replacement**. In such a situation, the card would be reinserted in the deck before the next card is drawn. In sampling with replacement, the events are independent because the sample space is not affected by the previous outcome.

EXAMPLE 3.71   Two cards are drawn, with replacement, from a well-shuffled standard 52-card deck of playing cards. Find the probability that both cards are diamonds.

Let D be the event that a diamond is drawn. We are interested in $P(D$ first$) \cap P(D$ second$)$. What is the probability that the first card is a diamond? Since there are 13 diamonds in a 52-card deck, the probability is $\frac{13}{52}$ or $\frac{1}{4}$. What is the probability that the second card is a diamond? The sample space has not been changed, so the probability is $\frac{13}{52}$ or $\frac{1}{4}$.

$$P(D \text{ first}) \cap P(D \text{ second}) = P(D \text{ first}) \cdot P(D \text{ second})$$

$$= \frac{1}{4} \cdot \frac{1}{4}$$

$$= \frac{1}{16}$$

The probability of drawing two diamonds with replacement is $\frac{1}{16}$. ■

Note that the probability of drawing two diamonds is higher when the sampling is done with replacement than without it. Can you explain why/how that happened? Think about it—you will get a chance to answer that question soon.

EXAMPLE 3.72   A fifth-grade class has 15 boys and 21 girls in it. Each week the teacher selects a student to clean the board at the end of the day. Find the probability that a boy will be selected two weeks in a row, if the same person cannot serve in consecutive weeks.

What we have here is essentially sampling without replacement, since the student selected one week is exempt the next week. Let B be the event that a boy is selected. We are interested in $P(B$ first$) \cap P(B$ second$)$. What is the probability that the helper will be a boy in the first week? Since there are 15 boys out of 36 students, the probability is $\frac{15}{36}$ or $\frac{5}{12}$. What is the probability that the helper will be a boy in the second week? Since that leaves only 14 boys out of 35 students, the probability is $\frac{14}{35}$ or $\frac{2}{5}$.

$$P(B \text{ first}) \cap P(B \text{ second}) = P(B \text{ first}) \cdot P(B \text{ second})$$

$$= \frac{15}{36} \cdot \frac{14}{35}$$

$$= \frac{5}{12} \cdot \frac{2}{5}$$

$$= \frac{1}{6}$$

The probability that a boy cleans the board for two straight weeks is $\frac{1}{6}$. ■

EXAMPLE 3.73   A fifth-grade class has 15 boys and 21 girls in it. Each week the teacher selects a student to clean the board at the end of the day. Find the probability that a boy will be selected two weeks in a row, if the same person can serve in consecutive weeks.

What we have here is essentially sampling with replacement, since the student selected one week can be selected the next week. Let B be the event that a boy is selected. We are interested in $P(B$ first$) \cap P(B$ second$)$. What is the probability that the helper will be a boy in the first week? Since there are 15 boys out of 36 students, the probability is $\frac{15}{36}$ or $\frac{5}{12}$. What is the probability that the helper will be a boy in the second week? Again, there are 15 boys out of 36 students, so the probability is $\frac{15}{36}$ or $\frac{5}{12}$.

$$P(B \text{ first}) \cap P(B \text{ second}) = P(B \text{ first}) \cdot P(B \text{ second})$$

$$= \frac{15}{36} \cdot \frac{15}{36}$$

$$= \frac{5}{12} \cdot \frac{5}{12}$$

$$= \frac{25}{144}$$

The probability that a boy cleans the board for two straight weeks is $\frac{25}{144}$. ■

Subjective Probability

When we introduced classical probability, it was necessary that all of the outcomes in the sample space were equally likely to occur. This is not always the case, however. Consider the following example.

The possible types of precipitation are rain, sleet, hail, snow, and none. Find the probability that it snows in Las Vegas on the next Fourth of July.

Let's look at the sample space. S = {Rain, Sleet, Hail, Snow, None}. Is the probability of snow $\frac{1}{5}$? It would be, according to classical probability, if each of the five outcomes were equally likely. I have relatives who live in Las Vegas who have told me that it is far too warm to snow in July, and this tells me that the probability should be a lot lower than $\frac{1}{5}$. ■

How do we find the probability of an event occurring when all of the outcomes are equally likely? This is the field of **subjective probability**. We use history, our own expertise, or some other method to estimate the probability in question.

Find the probability that a professional basketball player makes his next free throw, if he has made 221 out of his previous 425 attempts this season.

The best we can do is come up with an estimate of this probability. Does $\frac{221}{425}$ or 0.52 sound reasonable? If in the past the player has made 52% of his free throws, it stands to reason that there is a 52% chance on any single free throw. ■

Find the probability that a professional basketball player makes his next free throw, if he has made four out of his five attempts this game.

Does it seem reasonable that there is an 80% chance that the player will make his next free throw attempt? There is a little problem with this estimate. It is based on a small number of attempts (only five), and is not as reliable as the estimate in the previous example where there were 425 attempts. ∎

Law of Large Numbers

The free-throw examples lead us to the **Law of Large Numbers**.

> **LAW OF LARGE NUMBERS** *As the number of trials of an experiment increases, the proportion of successes in those trials approaches the probability of success on one trial.*

As far as the free throws go, 5 is not a large number but 425 is. The question is, how large a number is "large"? We will be able to calculate how many trials are appropriate in a later chapter, but for now we will have to use our intuition.

The following Excel project is the Law of Large Numbers in action. It simulates flipping a coin 1000 times. The probability that a coin lands on heads is 0.5, so the percentage of heads should get closer and closer to 50% as the number of tosses increases.

EXCEL EXPLORATION   The Law of Large Numbers

Before we begin, we must turn off the Excel feature that automatically recalculates functions. Click on **Tools,** and then **Options.** When the Options dialog box appears, click on the tab labeled **Calculation.** Click on **Manual,** and then **OK.**

We also must be sure that the Analysis Tool Pak has been added. Click on **Tools.** If you see a choice that says **Data Analysis,** then the Tool Pak is added and you may continue. If you do not see **Data Analysis,** click on **Add-ins** under **Tools.** When the dialog box opens up, click on the box for **Analysis Tool Pak,** and click on **OK.**

In cell **A1,** type Toss #.

In cell **B1,** type Heads(1), Tails(0).

In cell **C1,** type % Heads.

In cell **A2,** type 1.

In cell **B2,** type =RANDBETWEEN(0,1). This simulates the toss of a coin. The result will either be a 0, which stands for tails, or a 1, which stands for heads.

In cell **C2,** type =AVERAGE($A$2:A2).

Click on cell **A13.** Click on **Window,** and then **Split.** This will divide the screen horizontally.

Click on **Edit,** and then **Go To.** When the dialog box opens, type A1001. Hold down the **Shift** key and click on cell **A2.** This should highlight column A. Click on **Edit,** then **Fill,** and then **Series.** When the dialog box appears, click on **OK.** This will fill the first column with the number of the toss.

Highlight cells **B2** and **C2.** Click on **Edit,** and then **Copy.** Click on cell **B3.** Then hold down the **Shift** key and click on cell **C1001.** Press the **Enter** key. This will copy the information into columns **B** and **C.**

Press the **F9** key to recalculate the worksheet.

Here is an example of the results (graphically) of the first 50 tosses. Note how the graph appears to be drawn toward 0.5 as we move to the right.

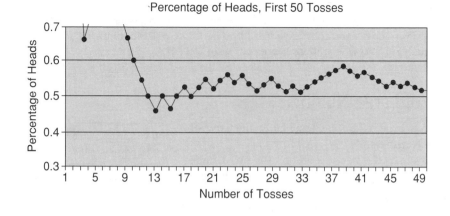

To produce a graph like this for your example, select **Chart** from the **Insert** menu. When the dialog box appears, select **XY (Scatter)** for the chart type and select the "Scatter with data points connected by smooth Lines" option for the chart subtype. Click on **Next** to move to step 2.

In the box labeled **Data Range**, type the following:

=Sheet1!$A$2:$A$51,Sheet1!$C$2:$C$51

Click on the **Series** tab. In the box labeled **X Values**, type =Sheet1!$A$2:$A$51. In the box labeled **Y Values**, type =Sheet1!$C$2:$C$51. Click on **Next** to move to step 3.

Click on the **Titles** tab. To add a label for the *x* axis, such as Number of Tosses, type it in the box labeled **Value (X) axis**. You can type % Heads in the box labeled **Value (Y) axis** to label the *y* axis. Click on **Finish** to see your graph.

Here are the results (graphically) of the first 200 tosses.

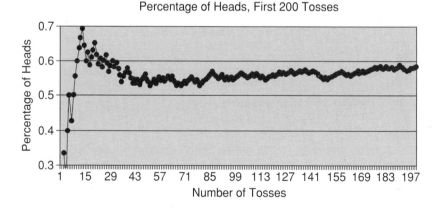

Note that the proportion is "jumpy" for the lower number of tosses, but as the number of tosses approaches 200, the proportion should be drawn toward 0.5. Now here are the results of the 1000 tosses. Note again that as the number of tosses increases, the proportion of heads is drawn in toward 0.5.

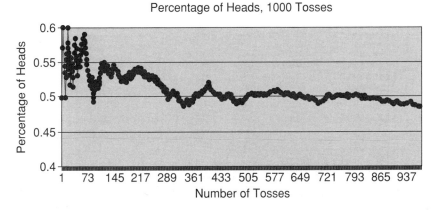

Percentage of Heads, 1000 Tosses

To produce these last two graphs, follow the above directions, extending the range of cells down to row 201 and row 1001, respectively. Print out your three graphs, and in your own words, comment on how they show the Law of Large Numbers at work.

EXERCISES 3.4

1. The probability that a community college student is female is 0.55. The probability that a community college student is on the Dean's List is 0.32. The probability that a community college student is female and on the Dean's List is 0.17. Is being female independent of being on the Dean's List for community college students?

2. The probability that a resident of Visalia attends the County Fair in June is 0.65. The probability that a resident of Visalia attends the Annual Fog Festival in January is 0.2. The probability that a resident of Visalia attends the County Fair and the Fog Festival is 0.13. Is attending the County Fair independent of attending the Fog Festival for Visalia residents?

3. If a fair six-sided die is rolled and a fair coin is tossed, the probability that the roll of the die is a 5 and the toss of the coin is heads is $\frac{1}{12}$. Are the events "A roll of a die is 5" and "A toss of a coin is heads" independent events?

4. List two events that you intuitively feel are independent events. Explain your reasoning in your own words. List two events that you intuitively feel are not independent events. Explain your reasoning in your own words.

A fair coin is flipped three times.

5. Find the probability that all three tosses are heads.
6. Find the probability that all three tosses are tails.
7. Find the probability that the first two tosses are heads and the last is tails.
8. Find the probability that two of the tosses are heads and the other is tails.

A fair six-sided die is rolled twice.

9. Find the probability that each roll is a 5.
10. Find the probability that neither roll is a 5.

11. The probability that an elementary school child lists basketball as his or her favorite sport is 0.18. If five elementary school children are selected at random, find the probability that they all name basketball as their favorite sport.

12. The probability that an adult lists blue as his or her favorite color is 0.15. If five adults are selected at random, find the probability that at least one fails to name blue as his or her favorite color.

13. In a certain experiment, the probability of an error is 5%, or 0.05. If the experiment is repeated six times, find the probability that at least one of the trials will result in an error.

14. In a certain experiment, the probability of an error is 1%, or 0.01. If the experiment is repeated four times, find the probability that at least one of the trials will result in an error.

Two cards are drawn from a well-shuffled standard 52-card deck of playing cards.

15. Find the probability that both cards are diamonds.
16. Find the probability that both cards are red.
17. Find the probability that both cards are kings.
18. Find the probability that the first card is red and the second card is black.
19. Find the probability that the first card is a spade and the second is a diamond.
20. Find the probability that the first card is a spade and the second card is not a spade.
21. Find the probability that one card is a spade and one card is a diamond.
22. Find the probability that one card is a face card and one card is an ace.

Three cards are drawn from a well-shuffled standard 52-card deck of playing cards.

23. Find the probability that all three cards are spades.
24. Find the probability that all three cards are face cards.
25. Find the probability that the first two cards are spades and the third card is a heart.
26. Find the probability that the first card is an ace and the last two cards are kings.
27. A card is drawn from a well-shuffled standard 52-card deck of playing cards, and then replaced in the deck. A second card is then drawn. Find the probability that both cards are diamonds. Now compare your answer to your answer from Exercise 15. Explain why the answer in this exercise (with replacement) is higher than the probability in Exercise 15 (without replacement).

A card is drawn from a well-shuffled standard 52-card deck of playing cards, and then replaced in the deck. A second card is then drawn.

28. Find the probability that both cards are aces.
29. Find the probability that one card is a diamond and one card is a heart.
30. Find the probability that one card is a spade and one card is not a spade.

31. An elementary school class has 16 girls and 11 boys. Each day the teacher selects a student to give a current events report to the class. The teacher posts a schedule for the entire week (Monday through Friday) each Friday. Find the probability that a boy is selected for each day next week, if no student can be selected more than once in a week.
32. A jar of M&M chocolate candies has 12 blue candies and 108 candies that are other colors. A child reaches into the jar, pulls out a candy, and then drops the candy back into the jar. If the child does this three times, find the probability that all three candies are blue.

MINI PROJECT

The purpose of this assignment is to estimate the probability that a randomly selected student on your campus is female—in other words, the percentage of the students at your school that are female. Visually survey 200 students on your campus. This can be done in your classes, in the library, in the center of campus, in the parking lot, or in some other fashion. Keep in mind that we want the sample to be representative of the population. Your best chance at this is to be as random as possible. Draw a graph showing the percentage of your survey that is female after 10 students, 25 students, 50 students, 100 students, 150 students, and 200 students. Here is an example of such a graph. (If you use Excel, type 1 for a female student and 0 for a male student.)

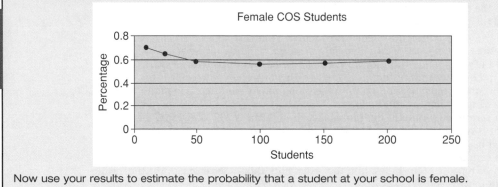

Now use your results to estimate the probability that a student at your school is female.

Overview

Unit 2 began with **counting**, which involves techniques used to determine the number of different outcomes an experiment may have. We began with **tree diagrams** and the "dice chart," but focused on three techniques: the **Multiplication Principle**, **permutations**, and **combinations**. We use multiplication when a choice consists of two or more steps. We multiply the number of different possibilities for the first step by the number of different possibilities for the second step, and then the third step, and so on. If we are selecting r objects, without repetition, from a set of n distinct objects, and the order of selection matters, then we use permutations. If the order of selection does not matter, we use combinations.

Probability is a measure of how likely it is that a particular outcome(s) occurs. To apply **classical probability**, all of the possible outcomes must be equally likely to occur. The **General Addition Rule** is used when we are interested in the probability of event A *or* event B occurring. We apply the **Complement Rule** when we are looking for the probability of the complement of an event occurring. The key word to the complement rule is often "not."

The rules of **conditional probability** are applied in situations where the original sample space has been restricted by some condition. This leads to the idea of **independent events**. If the probability of event *A* is unaffected by whether *B* has occurred or not, and vice versa, then the two events are independent. If we are interested in the probability of event *A* occurring *and* event *B* occurring, we often use the **Multiplication Rule**.

Review Exercises

1. Draw a tree diagram showing the results of three tosses of a coin. In how many ways is it possible to get no heads? one head? two heads? three heads?
2. Freebird's Pizza has four types of crust, three types of cheese, and 12 toppings. In how many ways can a customer order a three-topping pizza with one type of crust and one type of cheese?
3. Passwords at the computer lab consist of two letters of the alphabet, the first of which cannot be I, O, or Q, and three digits, the first of which cannot be 0. How many different passwords are possible with this scheme?
4. A multiple choice test consists of eight questions, each having four possible answers.
   (a) In how many ways can a student answer the eight questions?
   (b) In how many ways can a student answer the eight questions correctly?
   (c) In how many ways can a student answer the eight questions incorrectly?
5. A statistics instructor who has taken leave of his senses decides to randomly pick 5 of his 35 students and excuse them from the final exam. In how many ways can this be done?
6. A local club has 15 members. They decide to form a four-person committee to oversee the planning of meetings. In how ways can this be done under the following conditions?
   (a) There is to be one person in charge of notifying members, one person in charge of refreshments, one person to run the meeting, and one person to clean up.
   (b) There is no distinction between the jobs on the committee.
   (c) One person is to run the meeting, and there is no distinction between the other jobs on the committee.
7. Forty-eight of 60 dentists in a town recommend sugarless gum for their patients that chew gum. If one of these dentists is chosen at random, what is the probability that he or she recommends sugarless gum for their patients who chew gum?
8. Here is a table listing the probability of Tony Gwynn getting no, one, two, three, four, or five hits in a single game.

Number of hits	0	1	2	3	4	5
Probability	0.22	0.28	0.32	0.13	0.04	0.01

(a) Find the probability of Tony getting two or more hits in a game.
(b) Find the probability of Tony getting fewer than five hits in a game.
(c) Find the probability of Tony getting at least one hit in a game for three consecutive games.

9. The probability that a customer at a fast-food restaurant orders a beverage is 0.60. The probability that a customer orders French fries is 0.50. The probability that a customer orders both a beverage and French fries is 0.42. What is the probability that a customer at a fast-food restaurant orders a beverage or French fries?

10. One hundred students were asked their favorite fast-food restaurant: Taco Bell, McDonald's or neither. Here are the results.

	McDonald's	Taco Bell	Neither
Male	35	15	5
Female	10	20	15

(a) Find the probability that a student is male and prefers McDonald's.
(b) If a male student is selected at random, find the probability that he prefers McDonald's.
(c) If a student who prefers Taco Bell is selected at random, find the probability that the student is female.
(d) If a student who prefers neither is selected at random, find the probability that the student is female.
(e) If a female student is selected at random, find the probability that she prefers McDonald's or Taco Bell.

11. On a par 3 golf hole, the probability that a professional golfer hits his first shot onto the green and makes a birdie is 0.24. The probability that a professional golfer hits his first shot onto the green on a par 3 golf hole is 0.60. Find the probability that a professional golfer who hits his first shot onto the green will make a birdie.

12. If $P(A) = 0.30$, $P(B) = 0.40$ and $P(A \cap B) = 0.07$, are events A and B independent?

13. Find the probability of drawing two black cards from a deck
(a) with replacement
(b) without replacement

14. Six female students and six male students are nominated for a scholarship. Three scholarships are awarded. If all 12 students are equally qualified for the scholarship, find the probability that the three scholarships are awarded to females.

15. User ID's at the computer lab use four letters followed by four numbers. How many ID's are possible under the following conditions?
(a) The first letter cannot be O.
(b) The first letter cannot be O, and no letter can be repeated.

16. A student who is 20 years old and weighs 175 lb. owns four credit cards. He purchases three new shirts, two new pairs of shoes, and four new pairs of pants. In how many different ways can this student choose a new outfit to wear (one shirt, one pair of shoes, and one pair of pants)?

17. The "I Love Statistics Club" has 25 members. In how many ways can they elect
(a) a four-person leadership committee
(b) a president, a vice president, a treasurer, and a secretary
(c) a president, a vice president, and a two-person leadership committee

18. On a ten-question multiple choice test, if a student guesses on all ten questions, the probabilities of getting none, one, two, . . . , ten correct are listed below.

Number right	0	1	2	3	4	5	6	7	8	9	10
Probability	0.107	0.268	0.302	0.201	0.088	0.026	0.006	0.002	0	0	0

(a) Find the probability of getting at most two questions correct by guessing.
(b) Find the probability of getting at least five questions correct by guessing.

19. Find the probability of drawing two hearts from a well-shuffled deck of cards if the cards are drawn
    (a) without replacement
    (b) with replacement
20. Seven males and five females are to be interviewed for a job as a community college instructor. The top four candidates are sent forward to the president for a second interview. If all the candidates are equally qualified, find the probability that
    (a) four females get a second interview
    (b) at least one male gets a second interview
21. In the following table, 76 college students are classified according to their class standing and also according to their favorite pizza topping.

	Anchovies	Mushrooms	Onions	Pepperoni
Freshman	4	12	8	14
Sophomore	4	11	6	17

(a) Find the probability that a student is a sophomore or prefers mushrooms.
(b) If a freshman is selected at random, find the probability that the student prefers pepperoni.
(c) Find the probability that a student prefers onions and is a sophomore.
(d) If a student who prefers anchovies is selected at random, find the probability that the student is not a freshman.
(e) If a student who prefers anchovies or mushrooms is selected at random, find the probability that the student is a sophomore.

Formulas

Permutations: $_nP_r = \dfrac{n!}{(n-r)!}$

Combinations: $_nC_r = \dfrac{n!}{r! \cdot (n-r)!}$

Classical probability: $P(A) = \dfrac{n(A)}{n(S)}$

Special Addition Rule: $P(A \cup B) = P(A) + P(B)$

Complement Rule: $P(A) = 1 - P(A')$

General Addition Rule: $P(A \cup B) = P(A) + P(B) - P(A \cap B)$

Conditional probability: $P(A|B) = \dfrac{P(A \cap B)}{P(B)}$

Condition for independent events: $P(A|B) = P(A)$ or $P(A) \cdot P(B) = P(A \cap B)$

Multiplication Rule (independent events): $P(A \cap B) = P(A) \cdot P(B)$

Multiplication Rule (dependent events): $P(A \cap B) = P(A) \cdot P(B|A)$

unit
three

3

Probability Distributions

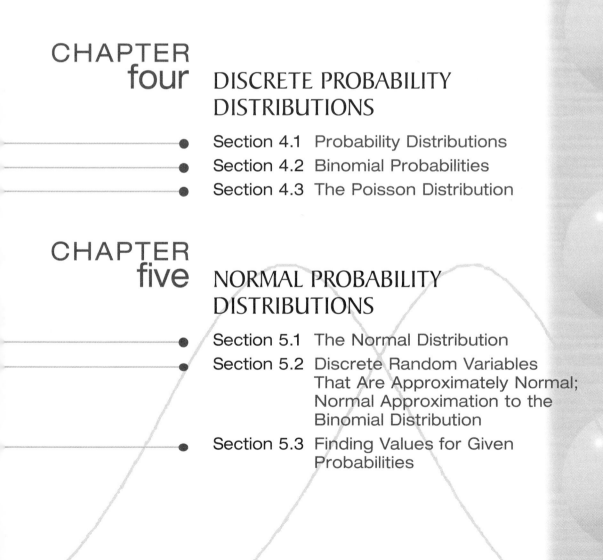

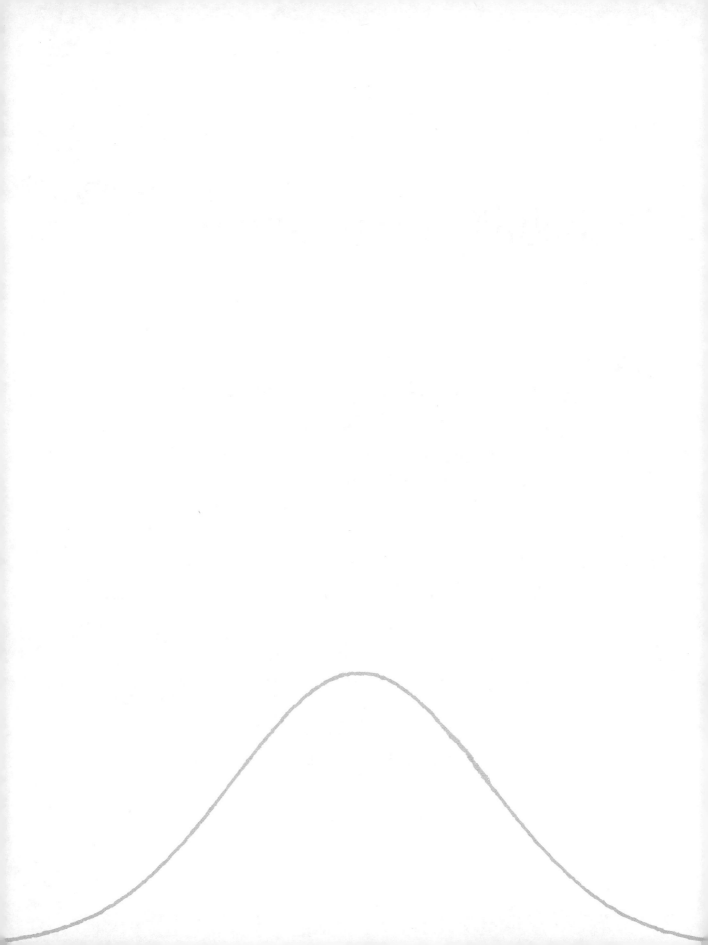

Discrete Probability Distributions

P robability is the tool that allows the study of inferential statistics. In this chapter we begin our study of probability distributions by focusing on two specific discrete probability distributions. In Section 4.2 we will learn about the binomial probability distribution. The binomial distribution is used to make inferences about population proportions, or percentages. In Section 4.3 we will study the Poisson distribution.

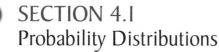

SECTION 4.1
Probability Distributions

In this section we continue our investigation into probability by introducing the concept of a probability distribution. A **probability distribution** associates a probability with each possible outcome of an experiment. For example, here is the probability distribution for flipping a fair coin.

Outcome	Probability
Heads	0.5
Tails	0.5

Another example is the probability distribution for the store at which a woman buys her New Year's Eve gown.

Store	Probability
Gottschalk's	0.4
JC Penney's	0.3
Mervyn's	0.1
Other	0.2

Before moving on, we introduce the idea of a *random variable*. A **random variable** is a variable that takes on a numerical value for each possible outcome of a probability experiment.

 EXAMPLE 4.1 A fair coin is tossed 3 times, and the number of heads is recorded.

What is the random variable?

The random variable *x* represents the number of heads in the 3 tosses.

What values can the random variable take on?

The number of heads in 3 tosses can be anywhere from zero through 3 (0, 1, 2, or 3). ■

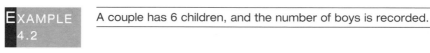 **EXAMPLE 4.2** A couple has 6 children, and the number of boys is recorded.

What is the random variable?

The random variable *x* represents the number of boys out of the 6 children.

What values can the random variable take on?

The number of boys out of 6 children can be anywhere from 0 through 6. ■

EXAMPLE
4.3

An adult male is selected at random, and his height is recorded.

What is the random variable?

The random variable *x* represents the height of the male.

What values can the random variable take on?

The height of an adult male can take on infinitely many positive values. ■

In the preceding example, the height of an adult male could take on infinitely many values. Recall from Chapter 1 that height was given as an example of continuous data. The random variable for the previous example is called a continuous random variable. There are two types of random variables: continuous and discrete. A **continuous random variable** is a variable that can assume infinitely many values from a continuous interval (there are no gaps between possible values). A **discrete random variable** is a variable that can assume either a finite number of values or a countable number of values.

Discrete random variables usually involve some sort of a count (0, 1, 2, 3, . . .). If the possible values of a discrete random variable were placed on a number line, there would be gaps between the values. Continuous random variables are usually associated with physical measurements.

EXAMPLE
4.4

Determine whether the random variables for the following experiments are continuous or discrete.

A fair coin is tossed 3 times, and the number of heads is recorded.

The random variable *x* (number of heads) is discrete because it can take on only 4 values: 0, 1, 2, and 3. Any variable that can assume only finitely many values is discrete.

A 6-month-old baby girl is selected at random, and her weight is recorded.

The random variable *x* (weight of the girl) is continuous. Recall that weight is an example of continuous data. The weight of a 6-month-old girl could take on values such as 15.3 pounds, 15.34 pounds, 15.347 pounds, and infinitely many other values in an interval that is centered at approximately 15.5 pounds.

A basketball player who is a poor free-throw shooter must make a free throw before he is allowed to leave practice. The number of free throws that he attempts until he makes one is recorded.

The random variable *x* (number of free throws attempted) is discrete. The random variable could take on any of the values 1, 2, 3, 4, ... Although there are an infinite number of possible values, this number is countable and the variable is discrete. ■

The remainder of this chapter is devoted to discrete random variables and their probability distributions. In Chapter 5 we will examine a continuous probability distribution called the normal distribution.

A probability distribution for a discrete random variable lists each possible value of the variable along with its probability. There are two basic rules that a probability distribution must follow. First, each probability must be between 0 and 1.

$$0 \leq P(x) \leq 1$$

Second, since we list all of the possible outcomes, the total of their probabilities must equal 1.

$$\sum P(x) = 1$$

 EXAMPLE 4.5

Are the following valid probability distributions?

x	P(x)
0	0.15
1	0.25
2	0.35
3	0.45

Although each individual probability is between 0 and 1, this is not a valid probability distribution because their sum is not equal to 1.

x	P(x)
0	0.2
1	0.3
2	1.5

This is not a valid probability distribution because $P(2)$ is not between 0 and 1.

x	P(x)
0	0.343
1	0.441
2	0.189
3	0.027

Since each probability is between 0 and 1, and the sum of all of the probabilities is equal to 1, this represents a valid probability distribution. ■

The use of a probability distribution makes it easier to calculate probabilities.

 EXAMPLE 4.6

Use the given probability distribution to find the following probabilities.

FIVE-QUESTION TRUE/FALSE QUIZ
(*x* = number correct)

x	P(x)
0	0.03125
1	0.15625
2	0.3125
3	0.3125
4	0.15625
5	0.03125

Find the probability that a randomly selected student gets 3 questions correct.

$P(x = 3) = 0.3125$

Find the probability that a randomly selected student gets at least 4 questions correct.

$P(\text{At least } 4) = P(x = 4) + P(x = 5)$

$= 0.15625 + 0.03125$

$= 0.1875$

Find the probability that a randomly selected student gets fewer than 4 questions correct.

$P(\text{Fewer than } 4) = P(x = 0) + P(x = 1) + P(x = 2) + P(x = 3)$

$= 0.03125 + 0.15625 + 0.3125 + 0.3125$

$= 0.8125$

We could have used the complement rule on the previous problem.

$P(\text{Fewer than } 4) = 1 - (P(x = 4) + P(x = 5))$

$= 1 - (0.15625 + 0.03125)$

$= 1 - 0.1875$

$= 0.8125$ ∎

Twenty-five percent of all Americans have a cellular phone. The probability distribution for the number of Americans with cell phones out of 10 randomly selected Americans is listed here.

Number with a Cell Phone (x)	Probability P(x)
0	0.0563
1	0.1877
2	0.2816
3	0.2503
4	0.1460
5	0.0584
6	0.0162
7	0.0031
8	0.0004
9	0.0000
10	0.0000

(Based on results of a study by Baskerville Communications.)

(a) If 10 Americans are selected at random, find the probability that 4 of them have a cellular phone.

Directly from the probability distribution,

$P(x = 4) = 0.1460$

(b) If 10 Americans are selected at random, find the probability that at least 7 of them have a cellular phone.

$P(\text{At least } 7) = P(x = 7) + P(x = 8) + P(x = 9) + P(x = 10)$

$= 0.0031 + 0.0004 + 0.0000 + 0.0000$

$= 0.0035$

(c) If 10 Americans are selected at random, find the probability that at least 2 of them have a cellular phone.

$P(\text{At least } 2) = 1 - (P(x = 0) + P(x = 1))$

$= 1 - (0.0563 + 0.1877)$

$= 1 - 0.2440$

$= 0.7560$ ■

If we construct a histogram for a probability distribution, we can use the area in the histogram to obtain a graphical representation of the probability distribution. Here is the histogram for the previous probability distribution.

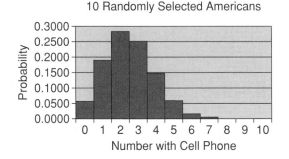

The probability that 2 of the 10 Americans have a cell phone is 0.2816, which is also the area (height) of the bar for $x = 2$. The association between area and probability will become very important when we investigate continuous probability distributions in Chapter 5.

Each probability distribution has a mean and standard deviation associated with it, just as a set of data has a mean and standard deviation. The mean is a typical value for the probability distribution and represents the average outcome that we can expect in the long run. Often the mean of a probability distribution is called its **expected value**.

The mean of a probability distribution can be calculated using the following formula:

$\mu = \Sigma x \cdot P(x)$

EXAMPLE
4.8

Find the mean of the following probability distribution.

FIVE-QUESTION TRUE/FALSE QUIZ
(*x* = number correct)

x	*P(x)*
0	0.03125
1	0.15625
2	0.3125
3	0.3125
4	0.15625
5	0.03125

There are two ways to calculate the mean. We can use the formula directly, or we can take a column approach.

μ = 0(0.03125) + 1(0.15625) + 2(0.3125) + 3(0.3125) + 4(0.15625) + 5(0.03125)

= 0 + 0.15625 + 0.625 + 0.9375 + 0.625 + 0.15625

= 2.5

The mean for this probability distribution is 2.5. If we repeated this experiment over and over, the mean number correct would be 2.5. This is also the number of questions that we could expect any single student to get correct, even though it is impossible for any student to get exactly 2.5 correct. Now we will use the column approach. We multiply each value of the variable by its probability, and write that result in the next column. When we total that column we will have the mean.

x	*P(x)*	*x* • *P(x)*
0	0.03125	0
1	0.15625	0.15625
2	0.3125	0.625
3	0.3125	0.9375
4	0.15625	0.625
5	0.03125	0.15625
		μ = 2.5

Does it make sense that the mean is 2.5? If there are 5 true/false questions, it stands to reason that we can expect to get half of the questions correct by guessing. ■

EXAMPLE 4.9

Don is a 65% free-throw shooter. This means that he makes 65% of his free throws, or the probability that he makes any individual free throw is 0.65. Given next is the probability distribution for the number of free throws made in 2 attempts by Don. Find the mean for this probability distribution.

x	*P(x)*
0	0.1225
1	0.455
2	0.4225

This time we will use the column approach.

x	P(x)	x · P(x)
0	0.1225	0
1	0.455	0.455
2	0.4225	0.845
		μ = 1.3

The mean for this probability distribution is 1.3. If Don repeated this experiment over and over, the mean number of free throws he would make is 1.3. Does it make sense that he would average 1.3 free throws made out of 2 attempts? 1.3 is 65% of 2, so if he makes 65% of his free throws he would make 1.3 out of 2 on average. ■

The standard deviation of a probability distribution measures the variability in the random variable. Here is the formula for standard deviation of a probability distribution.

$$\sigma = \sqrt{\Sigma((x - \mu)^2 \cdot P(x))}$$

We begin by finding μ, the mean of the probability distribution. We then subtract the mean from each value of x, and square those differences. After multiplying each of these squares by the probability of x occurring, we total these products. We finish by taking the square root of this total.

 EXAMPLE 4.10

Find the standard deviation of the following probability distribution.

FIVE-QUESTION TRUE/FALSE QUIZ
(x = number correct)

x	P(x)
0	0.03125
1	0.15625
2	0.3125
3	0.3125
4	0.15625
5	0.03125

As with the mean, we will use a "column approach." We begin with the mean.

x	P(x)	x · P(x)
0	0.03125	0
1	0.15625	0.15625
2	0.3125	0.625
3	0.3125	0.9375
4	0.15625	0.625
5	0.03125	0.15625
		μ = 2.5

Next we will calculate the standard deviation.

x	P(x)	x − μ	(x − μ)²	P(x) · (x − μ)²
0	0.03125	−2.5	6.25	0.19531
1	0.15625	−1.5	2.25	0.35156
2	0.3125	−0.5	0.25	0.07813
3	0.3125	0.5	0.25	0.07813
4	0.15625	1.5	2.25	0.35156
5	0.03125	2.5	6.25	0.19531
				1.25

Therefore,

$$\sigma = \sqrt{\Sigma((x - \mu)^2 \cdot P(x))}$$

$$= \sqrt{1.25}$$

$$= 1.118$$

The standard deviation of this distribution is 1.118 correct answers. ∎

Twenty-five percent of all Americans have a cellular phone. The probability distribution for the number of Americans with cell phones out of 10 randomly selected Americans is listed next.

Number with a Cell Phone (x)	Probability P(x)
0	0.0563
1	0.1877
2	0.2816
3	0.2503
4	0.1460
5	0.0584
6	0.0162
7	0.0031
8	0.0004
9	0.0000
10	0.0000

(Based on results of a study by Baskerville Communications.)

Find the mean and standard deviation for this distribution.

First, we find the mean.

x	P(x)	x · P(x)
0	0.0563	0.0000
1	0.1877	0.1877
2	0.2816	0.5632
3	0.2503	0.7509
4	0.1460	0.5840
5	0.0584	0.2920
6	0.0162	0.0972
7	0.0031	0.0217
8	0.0004	0.0032
9	0.0000	0.0000
10	0.0000	0.0000
		2.4999

The mean of this probability distribution is approximately 2.5 Americans. Next we calculate the standard deviation.

x	P(x)	x − μ	(x − μ)²	P(x) · (x − μ)²
0	0.0563	−2.5	6.25	0.3519
1	0.1877	−1.5	2.25	0.4223
2	0.2816	−0.5	0.25	0.0704

(continues)

(continued)

x	P(x)	x – μ	(x – μ)²	P(x) · (x – μ)²
3	0.2503	0.5	0.25	0.0626
4	0.1460	1.5	2.25	0.3285
5	0.0584	2.5	6.25	0.3650
6	0.0162	3.5	12.25	0.1985
7	0.0031	4.5	20.25	0.0628
8	0.0004	5.5	30.25	0.0121
9	0.0000	6.5	42.25	0.0000
10	0.0000	7.5	56.25	0.0000
				1.8741

Therefore,

$$\sigma = \sqrt{\Sigma((x - \mu)^2 \cdot P(x))}$$

$$= \sqrt{1.8741}$$

$$= 1.369$$

The standard deviation of this distribution is 1.369 Americans. ■

EXERCISES 4.1

1. Here is the probability distribution for the number of heads in 4 tosses of a fair coin.

x	P(x)
0	0.0625
1	0.2500
2	0.3750
3	0.2500
4	0.0625

(a) Find the probability that exactly 2 of the tosses are heads.
(b) Find the probability that at least 2 of the tosses are heads.
(c) Find the probability that at least 1 of the tosses is heads.
(d) How many heads can we expect in the 4 tosses?

2. A fair six-sided die is rolled 8 times. Here is the probability distribution for the number of 3s.

x	P(x)
0	0.2326
1	0.3721
2	0.2605
3	0.1042
4	0.0260
5	0.0042
6	0.0004
7	0.0000
8	0.0000

(a) Find the probability that between 3 and 5 rolls, inclusive, are 3s.
(b) Find the probability that more than 3 rolls are 3s.
(c) Find the probability of at least two 3s.
(d) How many 3s should we expect?

3. Here is the probability distribution for the number of candy bars sold at a house by a child selling the candy for a school fundraiser.

x	P(x)
0	0.4
1	0.25
2	0.2
3	0.15

(a) Find the probability that the child fails to sell a candy bar at a particular house.
(b) Find the probability that the child sells at least 2 candy bars at a particular house.
(c) How many candy bars should the child expect to sell at a house?

4. In California's Super Lotto game, the lottery draws 6 numbers from a drum containing 51 numbers (1–51). Players select 6 numbers for their ticket, and the payoff they receive depends on how many of their numbers match the 6 that were selected by the lottery. Here is a probability distribution for the number of numbers that a ticket matches.

x	P(x)
0	0.4522656426
1	0.4070390784
2	0.1240972800
3	0.0157583848
4	0.0008245666
5	0.0000149921
6	0.0000000555

(a) Players do not receive a payout unless they match at least 3 of the selected numbers. Find the probability that a ticket receives a payout.
(b) Find the probability that a ticket does not receive a payout.
(c) How many numbers should a player expect to match on a ticket?

5. Construct a probability distribution for x; the sum of 2 fair dice.

Use the probability distribution to answer the following.

(a) Find the probability of rolling a sum of at least 6.
(b) Find the probability of rolling a sum that is higher than 9.
(c) Find the probability of rolling a sum that is even.
(d) What sum should we expect when rolling 2 fair dice?

6. Complete the following probability distribution, and calculate its mean.

x	P(x)
0	0.28
1	0.31
2	0.19
3	
4	0.07
5	0.03
6	0.01

7. Complete the following probability distribution, and calculate its mean.

x	P(x)
0	0.04
1	0.05
2	0.07
3	0.15
4	0.22
5	
6	0.14
7	0.11
8	0.09

8. The mean for the following probability distribution is 3. Complete the probability distribution.

x	P(x)
0	0.1
1	0.1
2	0.15
3	
4	
5	0.2

For the following probability distributions, calculate their mean and standard deviations. (Note: Probabilities may not add to be 1 due to rounding.)

9. Eight randomly selected children, ages 12–17, x = number who smoke cigarettes

x	P(x)
0	0.2044
1	0.3590
2	0.2758
3	0.1211
4	0.0332
5	0.0058
6	0.0006
7	0.0000
8	0.0000

(Based on the 1998 Health and Human Services' National Household Survey on Drug Abuse.)

10. Six randomly selected Americans, ages 25–34, x = number who rent their home

x	P(x)
0	0.0905
1	0.2673
2	0.3292
3	0.2162
4	0.0799
5	0.0157
6	0.0013

(Based on the results of a study by the Census Bureau.)

11. Seven randomly selected residents of Mississippi, x = number who have health insurance

x	$P(x)$
0	0.2097
1	0.3670
2	0.2753
3	0.1147
4	0.0287
5	0.0043
6	0.0004
7	0.0000

(Source: Associated Press.)

12. Five randomly selected drivers, x = number who admit to running red lights

x	$P(x)$
0	0.0165
1	0.1049
2	0.2671
3	0.3400
4	0.2164
5	0.0551

(Based on the results of a study by the U.S. Department of Transportation.)

13. x = number of bicycle crashes on a university's bike paths on a given day

x	$P(x)$
0	0.1225
1	0.2572
2	0.2700
3	0.1890
4	0.0992
5	0.0417
6	0.0146
7	0.0044
8	0.0011
9	0.0003
10	0.0001

14. x = number of structure fires in a town on a given day

x	$P(x)$
0	0.7408
1	0.2222
2	0.0333
3	0.0033

15. x = number of cars stolen in a city on a given day

x	P(x)
0	0.4966
1	0.3476
2	0.1217
3	0.0284
4	0.0050
5	0.0007
6	0.0001

16. x = number of interceptions thrown by an NFL quarterback in a given game

x	P(x)
0	0.2725
1	0.3627
2	0.2321
3	0.1001
4	0.0325

SECTION 4.2
Binomial Probabilities

In this section we will examine one particular type of probability problem, which involves the binomial distribution. We could have done this problem in Chapter 3; now we will try to derive a formula to assist us. We begin with a solution that could have appeared in Chapter 3.

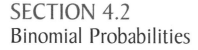

EXAMPLE 4.12   A 1996 study of first-time college freshmen by the Higher Education Research Institute shows that 20% of men planned to major in computer science or engineering. If 5 men who are first-time college freshmen are selected at random, find the probability that exactly 2 of them plan to major in computer science or engineering.

We must first decide in how many different ways 2 of the 5 freshmen could plan to major in computer science or engineering, with the other 3 planning to major in other than computer science or engineering. There are 10 different ways. One possible way is that the first 2 freshmen plan to major in computer science or engineering, and the last 3 plan on a different major. Here is the probability of that one way occurring. Keep in mind that the probability that a freshman male plans to major in computer science or engineering is 0.2, and the probability that a freshman male plans on another major is 0.8.

$$(0.2)(0.2)(0.8)(0.8)(0.8) = (0.2)^2(0.8)^3$$

$$= 0.02048$$

Here are the 10 different ways, along with their probabilities. We let Y represent a freshman male who plans to major in computer science or engineering, and we let N represent a freshman male who plans to major in something else.

1 and 2 do (YYNNN): $(0.2)(0.2)(0.8)(0.8)(0.8) = (0.2)^2(0.8)^3$

1 and 3 do (YNYNN): $(0.2)(0.8)(0.2)(0.8)(0.8) = (0.2)^2(0.8)^3$

1 and 4 do (YNNYN): $(0.2)(0.8)(0.8)(0.2)(0.8) = (0.2)^2(0.8)^3$

1 and 5 do (YNNNY):	$(0.2)(0.8)(0.8)(0.8)(0.2) = (0.2)^2(0.8)^3$
2 and 3 do (NYYNN):	$(0.8)(0.2)(0.2)(0.8)(0.8) = (0.2)^2(0.8)^3$
2 and 4 do (NYNYN):	$(0.8)(0.2)(0.8)(0.2)(0.8) = (0.2)^2(0.8)^3$
2 and 5 do (NYNNY):	$(0.8)(0.2)(0.8)(0.8)(0.2) = (0.2)^2(0.8)^3$
3 and 4 do (NNYYN):	$(0.8)(0.8)(0.2)(0.2)(0.8) = (0.2)^2(0.8)^3$
3 and 5 do (NNYNY):	$(0.8)(0.8)(0.2)(0.8)(0.2) = (0.2)^2(0.8)^3$
4 and 5 do (NNNYY):	$(0.8)(0.8)(0.8)(0.2)(0.2) = (0.2)^2(0.8)^3$

Note that the probability is the same for each way, so we may just multiply $(0.2)^2(0.8)^3$ by 10. The probability that 2 of the 5 freshman males plan to major in computer science or engineering is

$$10 \cdot (0.2)^2(0.8)^3 = 0.2048 \ \blacksquare$$

This problem is an application of the binomial distribution. For a problem to be a binomial problem there must be a fixed number of trials. We label the number of trials as n. In the above example there are 5 trials—each of 5 freshman males is asked about his major. Each outcome of a binomial trial must be classified as a *success* or a *failure*. The success is the outcome that we are interested in the probability of, and the failure is any other outcome. In the above example the success is that a freshman male plans to major in computer science or engineering, and the failure is that a freshman male plans to major in something other than computer science or engineering. Finally, each trial must be independent. This means that the probability of a success on any individual trial remains the same for each trial. In the above example, one freshman's choice of major does not affect another freshman's choice of major. These trials are independent, and the probability that any freshman male plans to major in computer science or engineering is 0.2.

As you read a problem, keep an eye out for the idea of "x out of n." Binomial problems involve phrases such as "2 out of 5," "between 3 and 6 out of 10," and "at least 1 out of 7." Also watch for a probability of success. Although most problems will list a probability of success on any individual trial, in some cases the probability of success will be assumed to be known (probability of heads, probability that a child is a girl, probability of getting a multiple choice question correct if there are 4 possible answers, and so on).

If a binomial experiment consists of n trials, and the probability of success is p, then the probability of x successes in n trials is given by the following formula:

$$P(x) = {}_nC_x \cdot p^x \cdot (1 - p)^{n-x}$$

Before we use the formula, let's analyze it. The number of different ways to have x successes in n trials is given by ${}_nC_x$. Since there are x successes, the probability of success p must show up x times, and this is represented in the formula by p^x. Similarly, since there are $n - x$ failures, the probability of failure $(1 - p)$ must show up $n - x$ times, and this is represented in the formula by $(1 - p)^{n-x}$.

Now we revisit the first example, this time applying the binomial formula.

EXAMPLE 4.13   A 1996 study of first-time college freshmen by the Higher Education Research Institute shows that 20% of men planned to major in computer science or engineering. If 5 men that are first-time college freshmen are selected at random, find the probability that exactly 2 of them plan to major in computer science or engineering.

The safest way to attack this problem is by listing the number of trials, the success, the number of successes that we are interested in, the probability of success, the failure, and the probability of failure.

Number of Trials: $n = 5$

Success: A freshman male plans to major in computer science or engineering.

Number of successes: $x = 2$

Probability of success: $p = 0.2$

Failure: A freshman male plans to major in something other than computer science or engineering.

Probability of failure: $1 - p = 0.8$

$$P(x) = {}_nC_x \cdot p^x \cdot (1 - p)^{n-x}$$
$$P(2) = {}_5C_2 \cdot (0.2)^2 \cdot (0.8)^3$$
$$= 0.2048$$

The probability that 2 of the 5 freshmen males plan to major in computer science or engineering is 0.2048. ■

 EXAMPLE 4.14   Find the probability that 4 of 7 randomly selected 18–25-year-olds used alcohol during the past 30 days, if the probability that any 18–25-year-old used alcohol during the past 30 days is 0.6. (According to data from the U.S. Department of Health and Human Services, 60% of all 18–25-year-olds reported using alcohol within the 30 days prior to the survey.)

Number of trials: $n = 7$

Success: An 18–25-year-old has used alcohol during the past 30 days.

Number of successes: $x = 4$

Probability of success: $p = 0.6$

Failure: An 18–25-year-old has not used alcohol during the past 30 days.

Probability of failure: $1 - p = 0.4$

$$P(x) = {}_nC_x \cdot p^x \cdot (1 - p)^{n-x}$$
$$P(4) = {}_7C_4 \cdot (0.6)^4 \cdot (0.4)^3$$
$$= 0.2903$$

When calculating binomial probabilities, we will always round to 4 decimal places. The probability that 4 of the 7 randomly selected 18–25-year-olds used alcohol during the past 30 days is 0.2903. ■

In the appendix, tables are provided with binomial probabilities that have already been calculated. If the number of trials (n) is 20 or less, and the probability of success (p) is 0.05, 0.1, 0.2, 0.3, 0.4, 0.5, 0.6, 0.7, 0.8, 0.9, or 0.95, then we can use these tables to find binomial probabilities. Here is the table for $n = 7$.

n = 7						P					
x	0.05	0.1	0.2	0.3	0.4	0.5	0.6	0.7	0.8	0.9	0.95
0	0.6983	0.4783	0.2097	0.0824	0.0280	0.0078	0.0016	0.0002	0	0	0
1	0.2573	0.3720	0.3670	0.2471	0.1306	0.0547	0.0172	0.0036	0.0004	0	0
2	0.0406	0.1240	0.2753	0.3177	0.2613	0.1641	0.0774	0.0250	0.0043	0.0002	0
3	0.0036	0.0230	0.1147	0.2269	0.2903	0.2734	0.1935	0.0972	0.0287	0.0026	0.0002
4	0.0002	0.0026	0.0287	0.0972	0.1935	0.2734	0.2903	0.2269	0.1147	0.0230	0.0036
5	0	0.0002	0.0043	0.0250	0.0774	0.1641	0.2613	0.3177	0.2753	0.1240	0.0406
6	0	0	0.0004	0.0036	0.0172	0.0547	0.1306	0.2471	0.3670	0.3720	0.2573
7	0	0	0	0.0002	0.0016	0.0078	0.0280	0.0824	0.2097	0.4783	0.6983

To use the table to find the probability from the above example, first locate the column for which the probability of success (p) is 0.6. This is the seventh column from the left. Now move down that column until you reach the row where the number of successes (x) is 4. You should see the value 0.2903, which is the same result that we got using the formula.

The tables are easier to use than the formula and we gain an advantage when we need to use the formula repeatedly, as in the following example.

EXAMPLE 4.15   Find the probability that between 4 and 6 out of 7 randomly selected 18–25-year-olds used alcohol during the past 30 days, if the probability that any 18–25-year-old used alcohol during the past 30 days is 0.6.

Number of trials: $n = 7$

Success: An 18–25-year-old has used alcohol during the past 30 days.

Number of successes: $x = 4, 5, 6$

Probability of success: $p = 0.6$

Failure: An 18–25-year-old has not used alcohol during the past 30 days.

Probability of failure: $1 - p = 0.4$

If we were to use the formula for this example, we would need to use the formula 3 times. However, we can use the tables because the number of trials (n) is 20 or less and the probability of success (p) is one that appears in the tables. It is much easier to add three probabilities from the table than to plug into the formula 3 times.

$$P(4 \text{ through } 6) = 0.2903 + 0.2613 + 0.1306$$

$$= 0.6822$$

The probability that between 4 and 6 out of 7 randomly selected 18–25-year-olds used alcohol during the past 30 days is 0.6822. ■

When you first determine that a problem is a binomial problem, you should decide whether you can use the tables. Next you should determine whether the complement rule would be helpful. Consider the following.

EXAMPLE 4.16   Find the probability that at least 1 of 7 randomly selected 18–25-year-olds used alcohol during the past 30 days, if the probability that any 18–25-year-old used alcohol during the past 30 days is 0.6.

Number of trials: $n = 7$

Success: An 18–25-year-old has used alcohol during the past 30 days.

Number of successes: $x = 1 - 7$

Probability of success: $p = 0.6$

Failure: An 18–25-year-old has not used alcohol during the past 30 days.

Probability of failure: $1 - p = 0.4$

To begin with, we can use the tables for this problem. We could begin with $x = 1$, and add our way down the table. However "at least one 18–25-year-old used alcohol during the past 30 days" is the complement of "zero 18–25-year-olds used alcohol during the past 30 days."

$$P(\text{At least 1}) = 1 - P(0)$$
$$= 1 - 0.0016$$
$$= 0.9984$$

If we had totaled the probabilities from $x = 1$ through $x = 7$, we would have obtained a probability of 0.9983. The error of 0.0001 is due to rounding. ■

The probability that a person has O-negative as his or her blood type is 0.06. If 15 people are selected at random, find the probability that at least one of them has O-negative blood.

Number of trials: $n = 15$

Success: A person has O-negative as his or her blood type.

Number of successes: $x = 1$ through 15 (at least 1)

Probability of success: $p = 0.06$

Failure: A person does not have O-negative as his or her blood type.

Probability of failure: $1 - p = 0.94$

Again, we have a problem that is best handled by using the complement rule. However, we cannot use the tables to assist us this time. We begin by finding the probability that none of the 15 people have O-negative blood.

$$P(x) = {}_nC_x \cdot p^x \cdot (1 - p)^{n-x}$$
$$P(0) = {}_{15}C_0 \cdot (0.06)^0 \cdot (0.94)^{15}$$
$$= 0.3953$$

To find the probability that at least 1 of the 15 people have O-negative blood, we subtract the above probability from 1.

$$P(\text{At least 1}) = 1 - P(0)$$
$$= 1 - 0.3953$$
$$= 0.6047$$

The probability that at least 1 of 15 randomly selected people has O-negative blood is 0.6047. ■

A basketball player makes 51% of his free-throw attempts (the probability that he makes any single free-throw attempt is 0.51).

> If he attempts 12 free throws in tonight's game, find the probability that he makes at least 9 of them.

Number of trials: $n = 12$

Success: The player makes a free throw.

Number of successes: $x = 9$ through 12

Probability of success: $p = 0.51$

Failure: The player misses a free throw.

Probability of failure: $1 - p = 0.49$

We cannot use the tables for this problem, because the probability of success (0.51) does not show up in the tables. Although we see the phrase "at least," we do not need to use the complement rule for this problem. A direct calculation uses the formula 4 times, whereas using the complement rule would require us to use the formula 9 different times (0 through 8).

$x = 9$

$P(x) = {}_nC_x \cdot p^x \cdot (1 - p)^{n-x}$

$P(9) = {}_{12}C_9 \cdot (0.51)^9 \cdot (0.49)^3$

$\quad = 0.0604$

$x = 10$

$P(x) = {}_nC_x \cdot p^x \cdot (1 - p)^{n-x}$

$P(10) = {}_{12}C_{10} \cdot (0.51)^{10} \cdot (0.49)^2$

$\quad = 0.0189$

$x = 11$

$P(x) = {}_nC_x \cdot p^x \cdot (1 - p)^{n-x}$

$P(11) = {}_{12}C_{11} \cdot (0.51)^{11} \cdot (0.49)^1$

$\quad = 0.0036$

$x = 12$

$P(x) = {}_nC_x \cdot p^x \cdot (1 - p)^{n-x}$

$P(12) = {}_{12}C_{12} \cdot (0.51)^{12} \cdot (0.49)^0$

$\quad = 0.0003$

$P(\text{At least } 9) = 0.0604 + 0.0189 + 0.0036 + 0.0003$

$\quad\quad\quad\quad\quad\; = 0.0832$

The probability that the player makes at least 9 of his 12 free throws is 0.0832. ■

EXAMPLE 4.19

A company makes a product to help reduce male hair loss. The company claims that 80% of the men that use their product will grow new hair. If 10 randomly selected men use the product, find the probability that at least 3 of them do not grow new hair.

In this example, the success is that a man does not grow new hair. However, the probability given in the problem (80% or 0.8) is not the probability of success. It is the probability of failure in this case, so we must use the complement rule to find the probability of success.

Number of trials: $n = 10$

Success: The man does not grow new hair.

Number of successes: $x = 3$ through 10

Probability of success: $p = 0.2$

Failure: The man does grow new hair.

Probability of failure: $1 - p = 0.8$

We can use the tables for this problem.

$$P(\text{At least } 3) = 1 - P(0, 1, 2)$$

$$= 1 - (0.1074 + 0.2684 + 0.3020)$$

$$= 1 - 0.6778$$

$$= 0.3222$$

The probability that at least three men do not grow new hair is 0.3222. ∎

Once again, here is the strategy you should employ when tackling binomial problems.

1. Use the tables whenever possible.
2. Determine whether the complement rule would make the problem easier.

There is one more topic that must be discussed when dealing with the binomial distribution. As with any probability distribution, the binomial distribution has a mean and a standard deviation. We can calculate these by the methods of Section 4.1, but this distribution has its own formulas for the mean and standard deviation.

Mean and Standard Deviation of the Binomial Distribution

$$\mu = n \cdot p$$

$$\sigma = \sqrt{n \cdot p(1 - p)}$$

 EXAMPLE
4.20

The probability that a man who is a first-time college freshman plans to major in computer science or engineering is 0.2. Five first-time college freshman males are selected at random, and asked to name their major. Find the mean and standard deviation for this probability distribution.

This is a binomial distribution with $n = 5$ and $p = 0.2$.

$$\mu = n \cdot p$$

$$= 5(0.2)$$

$$= 1$$

$$\sigma = \sqrt{n \cdot p(1 - p)}$$

$$= \sqrt{5(0.2)(0.8)}$$

$$= \sqrt{0.8}$$

$$= 0.89$$

This distribution has a mean of 1 success (1 first-time college freshman plans to major in computer science or engineering), with a standard deviation of 0.89 successes. ∎

It should seem reasonable that we should expect 1 of the 5 first-time freshman males to plan on majoring in computer science or engineering, as 20% of all first-time freshman males feel that way. The mean and standard deviation will take on importance in Section 5.2. For now, we will focus only on their calculation.

 EXAMPLE
4.21

The probability that a person has O-negative as his or her blood type is 0.06. Fifteen people are selected at random, and asked whether their blood type is O-negative or not. Find the mean and standard deviation of this probability distribution.

This is a binomial distribution with $n = 15$ and $p = 0.06$.

$$\mu = n \cdot p$$
$$= 15(0.06)$$
$$= 0.9$$

$$\sigma = \sqrt{n \cdot p(1 - p)}$$
$$= \sqrt{15(0.06)(0.94)}$$
$$= \sqrt{0.846}$$
$$= 0.92$$

This distribution has a mean of 0.9 success (0.9 people out of 15 people have this blood type), with a standard deviation of 0.92 successes. ■

MICROSOFT EXCEL   Binomial Probabilities

Binomial problems become more tedious as we have to use the formula more times. For instance, consider the following problem:

> In a certain county, 45% of registered voters are Republican. If 300 registered voters from that county are selected at random, find the probability that at least 155 are Republicans.

To calculate this probability directly, we would need to use the formula 146 times. The complement rule is no help here; we would need to use the formula 155 times. There are two tools that we may use to help us in such a situation. The first is Microsoft Excel and its built-in binomial probability functions. The second tool will be covered in Section 5.2.

To find the binomial probability of x successes in n trials, with probability of success p, we use Excel's built-in binomial function:

=BINOMDIST(x,n,p,FALSE)

There are four arguments in the parentheses. The first three are the number of successes (x), the number of trials (n), and the probability of success (p). The fourth argument allows you to determine whether you want the probability of exactly x successes (FALSE) or the probability of at most x successes (TRUE). Typing TRUE as the fourth argument finds the cumulative probability of 0 through x successes. We will use the cumulative option when we are interested in the probability of a range of successes. We begin by looking back to the first example of the section.

EXAMPLE 4.22   A 1996 study of first-time college freshmen by the Higher Education Research Institute shows that 20% of men planned to major in computer science or engineering. If 5 men that are first-time college freshmen are selected at random, find the probability that exactly 2 of them plan to major in computer science or engineering.

We are interested in exactly 2 successes in 5 trials, with the probability of success being 0.2. Since we are interested in exactly 2 successes, we do not need to use the cumulative option. Type the following in any cell.

=BINOMDIST(2,5,0.2,FALSE)

Excel's output is 0.2048. (The number of decimal places that you get depends on the width of the column. Be sure that there are at least four decimal places.) ■

 A basketball player makes 51% of his free-throw attempts (the probability that he makes any single free-throw attempt is 0.51). If he attempts 12 free throws in tonight's game, find the probability that he makes at most 8 of them.

This is a binomial problem with 12 trials, and the probability of success on any trial is 0.51. Since we are interested in the probability of *at most* 8 successes, we will use the cumulative option. Type the following in any cell.

=BINOMDIST(8,12,0.51,TRUE)

Excel tells us that the probability is 0.9168. ■

In general, to find the probability of *x* or fewer successes (or at most *x* successes) in *n* trials of a binomial experiment with probability of success *p*, use the following format.

=BINOMDIST(x,n,p,TRUE)

 The probability that a student passes a certain math exam is 0.7. If the exam is given to 15 students, find the probability that at least 12 of them pass.

In this problem, we are interested in the probability of *at least* 12 successes. We will need to use the complement rule to take advantage of Excel's cumulative option.

P(At least 12) 1 − P(11 or fewer)

Type the following in any cell.

=1-BINOMDIST(11,15,0.7,TRUE)

The result from Excel tells us that the probability of at least 12 students passing is 0.2969. ■

In general, to find the probability of at least *x* successes in *n* trials of a binomial experiment with probability of success *p*, use the following format.

=1-BINOMDIST(x-1,n,p,TRUE)

 The probability that a student passes a certain math exam is 0.7. If the exam is given to 15 students, find the probability that between 9 and 12 of them pass, inclusive.

To use Excel and its cumulative option, we begin by finding the probability of 12 or fewer successes. From that result, we subtract the probability of 8 or fewer successes. This will leave us with the probability of 9 through 12 successes. Type the following in any cell.

=BINOMDIST(12,15,0.7,TRUE)-BINOMDIST(8,15,0.7,TRUE)

From Excel, the probability is 0.7420. ∎

In general, to find the probability of *a* successes through *b* successes in *n* trials of a binomial experiment with probability of success *p,* use the following format:

= BINOMDIST(b,n,p,TRUE)-BINOMDIST(a-1,n,p,TRUE)

We will finish with the example from the beginning of this Excel section.

EXAMPLE 4.26   In a certain county, 45% of registered voters are Republican. If 300 registered voters from that county are selected at random, find the probability that at least 155 are Republicans.

We are looking for the probability of at least 155 successes in 300 trials of a binomial experiment with probability of success 0.45. Type the following in any cell:

=1-BINOMDIST(154,300,0.45,TRUE)

The probability that at least 155 are Republicans is 0.0120. ∎

TI-83   Binomial Probabilities

Binomial problems become more tedious as we have to use the formula more times. For instance, consider the following problem:

In a certain county, 45% of registered voters are Republican. If 300 registered voters from that county are selected at random, find the probability that at least 155 are Republicans.

To calculate this probability directly, we would need to use the formula 146 times. The complement rule is no help here; we would need to use the formula 155 times. There are two tools that we may use to help us in such a situation. The first uses technology, the TI-83, and its built-in binomial probability functions. The second tool will be covered in Section 5.2.

To find the binomial probability for *x* successes in *n* trials, with probability of success *p,* we must access the **DISTR** menu of the TI-83. To do this, press (2ND) (VARS). Select item 0: **binompdf(.** Inside the parentheses, type the number of trials, the probability of success, and the number of successes, all separated by commas. After closing the parentheses with ( )), press (ENTER).

We begin by looking back to the first example of the section.

EXAMPLE 4.27   A 1996 study of first-time college freshmen by the Higher Education Research Institute shows that 20% of men planned to major in computer science or engineering. If 5 men that are first-time college freshmen are selected at random, find the probability that exactly 2 of them plan to major in computer science or engineering.

We are interested in exactly 2 successes in 5 trials, with the probability of success being 0.2. Here is what the screen should look like after entering the proper values and pressing (ENTER):

```
binompdf(5,0.2,2
)
              .2048
```

The TI-83 tells us that the probability is 0.2048. ∎

By the way, if you leave off the number of successes, the TI-83 will calculate the probability of 0 successes, 1 success, . . . , and n successes. We can store these values to a list for convenient viewing. For the previous example, this is done by pressing (2ND)(VARS), selecting item 0, entering 5 (,) 0.2 ( ) (ENTER). After the calculations are done, press (STO→), then (2ND)(1)(ENTER) to store the values to list L_1. Recall that to view these values press (STAT) and select item 1 under the **EDIT** menu.

If we need to calculate probabilities for a range of successes, we use item A under the **DISTR** menu, **binomcdf(**. If we enter the number of trials n, the probability of success p, and a number of successes x, the TI-83 will calculate the probability that there are x or fewer successes.

A basketball player makes 51% of his free-throw attempts (the probability that he makes any single free-throw attempt is 0.51). If he attempts 12 free throws in tonight's game, find the probability that he makes at most 8 of them.

This is a binomial problem with 12 trials, and the probability of success on any trial is 0.51. Since we are interested in the probability of *at most* 8 successes, we will use the cumulative option. Press (2ND)(VARS) to access the **DISTR** menu, and select item **A**. Enter 12 (,) 0.51 (,) 8 ( ) and press (ENTER). Here is what the screen should look like:

```
binomcdf(12,0.51
,8)
        .9168416661
```

The TI-83 tells us that the probability is 0.9168. ∎

EXAMPLE 4.29 The probability that a student passes a certain math exam is 0.7. If the exam is given to 15 students, find the probability that at least 12 of them pass.

In this problem, we are interested in the probability of *at least* 12 successes. We will need to use the complement rule to take advantage of the TI-83's **binomcdf** function.

   P(At least 12) = 1 − P(11 or fewer)

Press 1 ⊖, then ⟨2ND⟩⟨VARS⟩ to access the **DISTR** menu, and select item **A.** Enter 15 ⟨,⟩ 0.7 ⟨,⟩ 11 ⟨)⟩ and press ⟨ENTER⟩. Here is what the screen should look like:

```
1-binomcdf(15,0.
7,11)
          .2968679279
```

The result from the TI-83 tells us that the probability of at least 12 students passing is 0.2969. ∎

EXAMPLE 4.30 The probability that a student passes a certain math exam is 0.7. If the exam is given to 15 students, find the probability that between 9 and 12 of them pass, inclusive.

To use the TI-83 and its **binomcdf** function, we begin by finding the probability of 12 or fewer successes. From that result, we subtract the probability of 8 or fewer successes. This will leave us with the probability of 9 through 12 successes. Press ⟨2ND⟩⟨VARS⟩ to access the **DISTR** menu, and select item **A.** Enter 15 ⟨,⟩ 0.7 ⟨,⟩ 12 ⟨)⟩. Press ⊖ for subtraction. Now press ⟨2ND⟩⟨VARS⟩ to access the **DISTR** menu, and select item **A.** Enter 15 ⟨,⟩ 0.7 ⟨,⟩ 8 ⟨)⟩. Finally, press ⟨ENTER⟩ to perform the calculation.

```
binomcdf(15,0.7,
12)-binomcdf(15,
0.7,8)
          .742029712
```

The TI-83 tells us that the probability is 0.7420. ∎

We will finish with the example from the beginning of this TI-83 section.

EXAMPLE 4.31   In a certain county, 45% of registered voters are Republican. If 300 registered voters from that county are selected at random, find the probability that at least 155 are Republicans.

We are looking for the probability of at least 155 successes in 300 trials of a binomial experiment with probability of success 0.45.

$$P(\text{At least 155}) = 1 - P(\text{154 or fewer})$$

Press 1 (−), then (2ND)(VARS) to access the **DISTR** menu, and select item **A.** Enter 300 (,) 0.45 (,) 154 ()) and press (ENTER). Here is what the screen should look like:

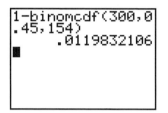

The TI-83 tells us that the probability at least 155 are Republicans is 0.0120. ∎

EXERCISES 4.2

1. Twenty-five percent of all Americans have a cellular phone. If 10 Americans are selected at random, find the probability that 4 of them have a cellular phone. (Based on results of a study by Baskerville Communications.)
2. According to the U.S. Department of Commerce, 62% of Alaskan households have a computer. If 15 Alaskan households are selected at random, find the probability that 9 of them have a computer.
3. Among women ages 15–19, 35% use the birth control pill to prevent unintended pregnancies, according to the Alan Guttmacher Institute. If 12 women between the ages of 15 and 19 are selected at random, find the probability that 3 of them use the birth control pill to prevent unintended pregnancies.
4. According to the 1998 Health and Human Services' National Household Survey on Drug Abuse, 18% of children between the ages of 12 and 17 are cigarette smokers. If 8 children between the ages of 12 and 17 are selected randomly, find the probability that 2 of them are cigarette smokers.
5. The probability that a college student works to earn extra money is 0.65. If 10 college students are selected at random, find the probability that at least 7 work to earn extra money. (Based on the results of the Social Science Research Center, Department of Psychology, Old Dominion University Campus Concepts.)
6. Eight percent of the students at a community college are at least 30 years old. If 40 students at that community college are selected at random, find the probability that between 2 and 4 students, inclusive, are at least 30 years old.
7. According to Cruise Lines International Corporation, 28% of the North Americans who have taken a cruise in the past five years were at least 60 years old. If 12 North Americans who have taken a cruise in the last five years are selected at random, find the probability that at least one of them was at least 60 years old at the time.
8. The U.S. Bureau of the Census reports that 33% of Americans between the ages of 25 and 34 rent their home. If 8 Americans between the ages of 25 and 34 are randomly selected, find the probability that at least 2 of them rent their home.

9. The probability that a high school student who takes the SAT has a combined score of 1290 or higher is 0.1. If 15 high school students who took the SAT are randomly selected, find the probability that between 2 and 5 students scored 1290 or above, inclusive.

10. The probability that a high school student who takes the SAT has a combined score of 610 or lower is 0.3. If 12 high school students who took the SAT are randomly selected, find the probability that between 3 and 7 students scored 610 or lower, inclusive.

11. A study by the U.S. Department of Health and Human Services revealed that 60% of 18–25-year-olds had drunk alcohol in the past 30 days. If twenty 18–25-year-olds are selected at random, find the probability that at least ten of them have drunk alcohol in the past 30 days.

12. According to the Associated Press, in 1998 20% of the people in Mississippi did not have health insurance. If 8 people from Mississippi were selected at random, find the probability that 1, 2 or 3 of them did not have health insurance.

13. A fair coin is flipped 9 times. Find the probability that it lands on heads at least twice.

14. A fair coin is flipped 14 times. Find the probability that it lands on heads at least 7 times.

15. A couple plans to have 6 children. Assuming the probability that a child is a girl is 0.5, find the probability that the couple has at least 3 girls.

16. A couple plans to have 5 children. Find the probability that they have at least 1 boy.

17. An instructor gives a five-question "True/False" quiz. A student must answer at least four questions correctly to pass the quiz. If a student randomly guesses on all five questions, find the probability that he passes the quiz.

18. An instructor gives a ten-question "True/False" quiz. A student must answer at least seven questions correctly to pass the quiz. If a student randomly guesses on all ten questions, find the probability that she passes the quiz.

19. An instructor gives an eight-question multiple-choice quiz. There are five possible answers for each question, of which only one is correct. A student must answer at least six questions correctly to pass the quiz. If a student randomly guesses on all eight questions, find the probability that he passes the quiz.

20. An instructor gives a six-question multiple-choice quiz. There are five possible answers for each question, of which only one is correct. A student must answer at least five questions correctly to pass the quiz. If a student randomly guesses on all six questions, find the probability that he passes the quiz.

21. According to the U.S. Transportation Department, 56% of all licensed drivers admit to running red lights. If 8 licensed drivers are selected at random, find the probability that at least 3 do not admit to running red lights.

22. A recent study by Market Data Retrieval showed that 89% of K–12 teachers feel that they are at least somewhat prepared to integrate technology into instruction. If nine K–12 teachers are selected at random, find the probability that at least two of them feel that they are not prepared to integrate technology into instruction.

23. A 1999 survey of teachers by *USA Today* revealed that 85% of teachers had not encountered a student with a knife, gun, or other deadly weapon in the last three years. If 30 teachers are selected at random, find the probability that at least 2 teachers had encountered a student with such a weapon in the last three years.

24. Data from the U.S. Department of Commerce shows that 19.5% of Americans age 18 and over hold a bachelor's or higher degree. If 7 Americans age 18 and over are selected at random, find the probability that at least 1 does not have a bachelor's or higher degree.

25. Information from the Census Bureau tells us that 28% of American men have a bachelor's degree or higher. If 15 American men are selected at random, find the mean and standard deviation of the resulting binomial probability distribution (x = number of men with a bachelor's degree or higher).

26. According to the March 2000 National Vital Statistics report, approximately 21% of all babies are delivered by Caesarian section. If 18 births are selected at random, find the mean and standard deviation of the resulting probability distribution (x = number of babies delivered by Caesarian section).

27. A coin is flipped 50 times, and the number of heads are recorded. Find the mean and standard deviation of this binomial probability distribution (x = number of heads).

28. A coin is flipped 500 times, and the number of heads are recorded. Find the mean and standard deviation of this binomial probability distribution (x = number of heads).

29. According to information from the Bureau of Labor Statistics, approximately 6% of all adult Americans have more than one job. If 225 adult Americans are selected at random, find the mean and standard deviation of the resulting probability distribution (x = number of adult Americans with more than one job).

30. According to the Morbidity and Mortality Weekly Report by the Centers for Disease Control and Prevention, approximately 70% of all high school students have tried smoking cigarettes. If 889 high school students are selected at random, find the mean and standard deviation of the resulting binomial probability distribution (x = number who have tried smoking).

EXTRA • The Geometric Distribution

There are discrete probability distributions other than the binomial distribution. One is the **geometric distribution**, which is closely related to the binomial distribution. The geometric distribution represents the probability that the first success of a binomial trial occurs on trial x. An example is finding the probability that the first time a coin lands on heads is on the fifth toss.

Let's develop the formula for the geometric distribution. Suppose we are conducting a binomial experiment with a probability of success p and probability of failure $1 - p$. The probability that the first trial is a success is p. For the first success to occur on the second trial, the first trial would have to be a failure, and the second trial would be a success. The probability that the second trial produces the first success is $(1 - p) \cdot p$. For the first success to occur on the third trial, there would have to be two failures followed by one success. The probability of this occurring is given by $(1 - p)^2 \cdot p$. We can now see that for the first success to occur on trial x, there must be $x - 1$ failures followed by one success. The probability of this occurring is given by $(1 - p)^{x-1} \cdot p$. This is the general formula for the geometric distribution.

$$P(x) = (1 - p)^{x-1} \cdot p$$

EXAMPLE
4.32

The probability that a basketball player makes a free throw is 0.8. Find the probability that the player makes his first free throw on his fourth attempt.

For this example, $p = 0.8$ and $x = 4$. The probability that the first success occurs on the fourth attempt is given by

$$P(x) = (1 - p)^{x-1} \cdot p$$
$$P(4) = (0.2)^3 \cdot 0.8$$
$$= 0.0064$$

The probability that the player makes his first free throw on his fourth attempt is 0.0064. ∎

When rolling a pair of fair six-sided dice, the probability that the sum is 8 or 9 is 0.25, because 9 of the 36 possible rolls have a sum that is 8 or 9. We will construct a geometric probability distribution for the roll containing the first 8 or 9.

Success: A roll is an 8 or a 9.

Probability of success: $p = 0.25$

Failure: A roll is not an 8 or a 9.

Probability of failure: $1 - p = 0.75$

The appropriate formula here is

$P(x) = (1 - p)^{x-1} \cdot p$

$\quad\quad = (0.75)^{x-1} \cdot 0.25$

Here is the probability distribution.

x	P(x)
1	0.2500
2	0.1875
3	0.1406
4	0.1055
5	0.0791
6	0.0593
7	0.0445
8	0.0334
9	0.0250
10	0.0188
11	0.0141
12	0.0106
13	0.0079
14	0.0059
15	0.0046
16	0.0033
17	0.0025
18	0.0019
19	0.0014
20	0.0011
21	0.0008
22	0.0006
23	0.0004
24	0.0003
25	0.0003
26	0.0002
27	0.0001
28	0.0001
29	0.0001
30	0.0001
31	0.0000

We could go on to answer any question regarding probabilities for this distribution, such as finding the probability that the fourth roll produces the first success.

EXTRA • EXERCISES

1. At the beginning of every football game, there is a coin flip to determine which team gets the coin flip. Each team has a 50% chance of winning the coin flip. Find the probability that a team does not win its first coin flip until the sixth game of the season.

2. A game show invites people across the country to dial a toll-free phone number to try to qualify for their show. To become eligible to be on the show, a caller must correctly answer 3 questions. The show estimates that 8% of all callers will correctly answer the 3 questions. Find the probability that it takes a caller 4 attempts to answer the 3 questions correctly.

3. At a racetrack, 32% of the races are won by the favorite. Find the probability that it takes six races for the first favorite to win a race.

4. The probability that the roll of a single die produces a 3 is $\frac{1}{6}$. Find the probability that it takes 8 rolls to produce the first 3.

EXCEL EXPLORATION   Binomial Probabilities through Simulation

In this exploration, we will use Excel to simulate the classic binomial experiment—flipping a coin. In this example, we will simulate flipping a coin 10 times, counting the number of heads. We will repeat this 100 times, and create a histogram to show our results.

We will use the Data Analysis Toolpak to simulate the tossing of the coin. Select the **Tools** menu. If you do not see **Data Analysis** as one of the choices, then you must add it. Select **Add-Ins**, and when the dialog box opens click the box next to **Analysis ToolPak** and click **OK.**

We will use Excel's built-in function RANDBETWEEN to simulate the coin flip. In cell **A1**, type the following:

=RANDBETWEEN(0,1)

After you press Enter, this will randomly generate either a 0 or a 1. We will treat a result of 1 as "heads," and a result of 0 as "tails." Click on cell **A1**, and then select **Copy** from the **Edit** menu. Highlight cells **A2** through **A10**, and select **Paste** from the **Edit** menu, which will repeat this random generation 10 times. To total the number of heads, type the following in cell **A11**:

=SUM(A1:A10)

Record the number of heads for later use. Then repeat the experiment by pressing the **F9** key on your keyboard. Record this new number of heads. Repeat this cycle a total of 100 times.

We will create a frequency distribution showing the number of times that we got 0, 1, 2, . . . , 10 heads. In cells **B1** through **B11** type the numbers from 0 through 10. In cell **C1** type the number of times that you got 0 heads. In cell **C2** type the number of times that you got 1 head. Repeat this until you put the number of times that you get 10 heads in cell **C11**.

Now we can create a histogram from this frequency distribution. Click on **Insert**, and then on **Chart**. A histogram is a column chart, which is the default selection. Click on **Next** to move from Step 1 to Step 2.

In Step 2, type **C1:C11** in the box labeled **Data range**. These are the heights of the bars in your histogram. Also, be sure that you click on columns instead of rows in the area that says **Series in:**. Click on the **Series** tab, and in the box labeled **Category (X) axis labels** type the following:

=Sheet1!$B$1:$B$11

These are the values that will appear on the *x* axis. Click on **Next** to move to Step 3.

Step 3 contains some ways to make your chart prettier. Click on the **Titles** tab to add a title to your graph (in the box **Chart Title**). To show the value for each class on the chart click on the **Data Labels** tab, then click on **Show Value**. Click on the **Legend** tab, and be sure the box labeled **Show Legend** is *not* checked. Click on **Next** to advance to Step 4.

Step 4 has to do with the location of the chart. Select the option **As object in**, and be sure that it says **Sheet 1**. Now click on **Finish**, and your histogram will appear.

We can also use this experiment to check the validity of the formulas for the mean and standard deviation of the binomial distribution. We have $n = 10$ and $p = 0.5$, so we can find the mean as follows:

$$\mu = n \cdot p$$
$$= 10(0.5)$$
$$= 5$$

Find the standard deviation as follows:

$$\sigma = \sqrt{n \cdot p \cdot (1 - p)}$$
$$= \sqrt{10(0.5)(0.5)}$$
$$= \sqrt{2.5}$$
$$= 1.58$$

Place the number of heads obtained from the 100 times you repeated this experiment in a column, and calculate the mean and standard deviation. Are they close to the values they were supposed to be?

After completing this, repeat the process for 50 flips instead of 10, and again for 100 flips.

SECTION 4.3
The Poisson Distribution

The next discrete probability distribution that we will study is the Poisson distribution. This distribution is named after Simeon Denis Poisson (1781–1840), a French mathematician. If we know the expected number of successes in a given interval, then we can use the Poisson distribution to calculate the probability of x successes occurring in the interval. We use the Greek letter λ (lambda) to represent the expected number of successes in the given interval. The given interval may be a measure of time (minute, hour, day, week, year) or something else such as a baseball game, a semester, or even a chocolate chip cookie.

If λ is the expected (or mean) number of successes in an interval, then the probability of having x successes in that interval is given by

$$P(x) = \frac{e^{-\lambda} \cdot \lambda^x}{x!}$$

where e is the base of the natural logarithm ($e \approx 2.71828$).

EXAMPLE 4.33 During a typical hour fishing at Lake Lotsafish, a fisher can expect to catch 3 fish. Find the probability that a fisher catches exactly 2 fish in his first hour fishing.

First, we must determine what the expected number of successes is for this situation. A fisher can expect to catch 3 fish during a typical hour, so we know that $\lambda = 3$. (The interval for this problem is one hour.) We are interested in the probability of exactly 2 fish being caught, so $x = 2$. As with binomial problems, it is wise to create a table of information before proceeding.

Success: A fish is caught.

Number of successes: $x = 2$

Expected number of successes: $\lambda = 3$

Interval: One hour

Now we may plug into the formula.

$$P(x) = \frac{e^{-\lambda} \cdot \lambda^x}{x!}$$

$$P(2) = \frac{e^{-3} \cdot 3^2}{2!}$$

$$= 0.2240$$

The probability that a fisher catches 2 fish in his first hour fishing at Lake Lotsafish is 0.2240. As with binomial probabilities, we round to four decimal places. ∎

As for the binomial distribution, there are tables in the appendix that give probabilities associated with certain values of λ. If λ is 0.1, 0.2, . . . , 0.8, 0.9, 1, 2, . . . , 8, or 9, then you may look up the probabilities in the tables. One of the tables follows. To find the probability of 2 successes when λ is 3, locate the column where $\lambda = 3$ and look down until you find the row where x is 2. The correct probability has a box around it, and is in bold.

					λ				
x	1	2	3	4	5	6	7	8	9
0	0.3679	0.1353	0.0498	0.0183	0.0067	0.0025	0.0009	0.0003	0.0001
1	0.3679	0.2707	0.1494	0.0733	0.0337	0.0149	0.0064	0.0027	0.0011
2	0.1839	0.2707	**0.2240**	0.1465	0.0842	0.0446	0.0223	0.0107	0.0050
3	0.0613	0.1804	0.2240	0.1954	0.1404	0.0892	0.0521	0.0286	0.0150
4	0.0153	0.0902	0.1680	0.1954	0.1755	0.1339	0.0912	0.0573	0.0337
5	0.0031	0.0361	0.1008	0.1563	0.1755	0.1606	0.1277	0.0916	0.0607
6	0.0005	0.012	0.0504	0.1042	0.1462	0.1606	0.1490	0.1221	0.0911
7	0.0001	0.0034	0.0216	0.0595	0.1044	0.1377	0.1490	0.1396	0.1171
8	0	0.0009	0.0081	0.0298	0.0653	0.1033	0.1304	0.1396	0.1318
9	0	0.0002	0.0027	0.0132	0.0363	0.0688	0.1014	0.1241	0.1318
10	0	0	0.0008	0.0053	0.0181	0.0413	0.0710	0.0993	0.1186
11	0	0	0.0002	0.0019	0.0082	0.0225	0.0452	0.0722	0.0970
12	0	0	0.0001	0.0006	0.0034	0.0113	0.0263	0.0481	0.0728
13	0	0	0	0.0002	0.0013	0.0052	0.0142	0.0296	0.0504
14	0	0	0	0.0001	0.0005	0.0022	0.0071	0.0169	0.0324
15	0	0	0	0	0.0002	0.0009	0.0033	0.009	0.0194
16	0	0	0	0	0	0.0003	0.0014	0.0045	0.0109
17	0	0	0	0	0	0.0001	0.0006	0.0021	0.0058
18	0	0	0	0	0	0	0.0002	0.0009	0.0029
19	0	0	0	0	0	0	0.0001	0.0004	0.0014
20	0	0	0	0	0	0	0	0.0002	0.0006
21	0	0	0	0	0	0	0	0.0001	0.0003
22	0	0	0	0	0	0	0	0	0.0001

Unlike the binomial distribution, there is no number of trials, so the number of successes in a given interval has no upper limit. Probabilities are calculated until they round to be equal to zero to four decimal places.

 EXAMPLE 4.34   During the typical statistics final exam, 2 students leave the room in tears. Find the probability that between 1 and 4 students, inclusive, leave your statistics final exam in tears.

We can expect 2 students to leave the room in tears during a typical final exam, so we know that $\lambda = 2$. (The interval for this problem is a final exam.) We are interested in the probability of between 1 and 4 students, inclusive, leaving in tears, so $x = 1, 2, 3,$ or 4.

Success: A student leaves in tears.

Number of successes: $x = 1, 2, 3,$ or 4

Expected number of successes: $\lambda = 2$

Interval: A final exam

We can use the tables for this problem, since $\lambda = 2$ and that is one of the values included in our table.

$$P(\text{Between 1 and 4 students}) = P(1) + P(2) + P(3) + P(4)$$
$$= 0.2707 + 0.2707 + 0.1804 + 0.0902$$
$$= 0.8120$$

The probability that between 1 and 4 students leave your final exam in tears is 0.8120. ∎

 EXAMPLE 4.35   During the typical statistics final exam, 2 students leave the room in tears. Find the probability that at least 1 student leaves your statistics final exam in tears.

Again, we know that $\lambda = 2$. We can use the tables once again, but instead of adding our way down the table from 1 success through 9 successes (the last nonzero probability listed for $\lambda = 2$), we can use the complement rule.

$$P(\text{At least 1 student}) = 1 - P(0)$$
$$= 1 - 0.1353$$
$$= 0.8647$$

The probability that at least 1 student leaves your final exam in tears is 0.8647. Does it make sense that this probability is higher than that of the previous example? ∎

EXAMPLE 4.36   During a typical baseball game, the Cleveland Indians score 6.4 runs. Find the probability that the Cleveland Indians do not get shut out in a game—in other words, that they score at least one run in a game.

The Cleveland Indians score 6.4 runs during a typical baseball game, so we know that $\lambda = 6.4$. (The interval for this problem is one baseball game.) We are interested in the probability of at least one run being scored, so the complement rule will come in handy again. This time we must use the formula, and this time the problem would be quite tedious without the complement rule.

Success: A run is scored.

Number of successes: x = at least 1

Expected number of successes: λ = 6.4

Interval: One baseball game

Now we may plug into the formula.

$$P(x) = \frac{e^{-\lambda} \cdot \lambda^x}{x!}$$

$$P(0) = \frac{e^{-6.4} \cdot 6.4^0}{0!}$$

$$= 0.0017$$

$$P(\text{At least 1}) = 1 - P(0)$$

$$= 1 - 0.0017$$

$$= 0.9983$$

The probability that the Cleveland Indians do not get shut out is 0.9983. ∎

Occasionally, the hardest part of figuring out a Poisson probability is determining how many successes we are interested in. Consider the following example.

 EXAMPLE 4.37   A juvenile detention center has enough room for 8 juveniles. On an average night, 5 juveniles are brought to the center to be kept overnight. Find the probability that the center will not have enough room on a given night.

In order for the center to not have enough room, more than 8 juveniles must be brought in. So, we are interested in the probability that at least 9 juveniles are brought to the center. On an average night, 5 juveniles are brought to the center to be kept overnight. This tells us that $\lambda = 5$. Since $\lambda = 5$, we may use the tables for this problem.

Success: A juvenile is brought to the center to be kept overnight.

Number of successes: x = at least 9

Expected number of successes: λ = 5

Interval: One night

$$P(\text{At least 9 juveniles}) = P(9) + P(10) + P(11) + P(12) + P(13) + P(14) + P(15)$$

$$= 0.0363 + 0.0181 + 0.0082 + 0.0034 + 0.0013 + 0.0005$$
$$+ 0.0002$$

$$= 0.0680$$

The probability that the center does not have enough room on a given night is 0.0680. ∎

Occasionally we will be given a typical value for a certain interval and will be asked for a probability involving another interval. For example, we could be told the number of runs that the Cleveland Indians score in a typical game, and be asked to find the probability that they score 10 runs in the next 2 games. The next example illustrates this idea.

EXAMPLE 4.38

On a typical day, a credit union opens 5 new accounts. Find the probability that the credit union opens exactly 6 accounts in the next two days.

If we can expect 5 new accounts on a given day, in two days we can expect twice as many, or 10 new accounts. This will be our value for λ. Now we proceed in the same fashion as before.

    Success: A new account is opened.

    Number of successes: $x = 6$

    Expected number of successes: $\lambda = 10$

    Interval: Two days

We must use the formula for this problem, since $\lambda = 10$ is too big for our tables.

$$P(x) = \frac{e^{-\lambda} \cdot \lambda^x}{x!}$$

$$P(6) = \frac{e^{-10} \cdot 10^6}{6!}$$

$$= 0.0631$$

The probability that the credit union opens exactly 6 accounts in the next two days is 0.0631. ∎

Like the binomial distribution, the Poisson distribution has a mean and a standard deviation.

Mean and Standard Deviation of the Poisson Distribution

$$\mu = \lambda$$

$$\sigma = \sqrt{\lambda}$$

It should come as no surprise that the mean of the Poisson distribution is λ. After all, that is the expected number of successes.

EXAMPLE 4.39

During a typical hour fishing at Lake Lotsafish, a fisher can expect to catch 3 fish. Find the mean and standard deviation of the probability distribution for the number of fish caught in an hour of fishing at Lake Lotsafish.

In this problem $\lambda = 3$, so

    $\mu = 3$

    $\sigma = \sqrt{3}$

    $= 1.73$

The mean for this distribution is 3 fish, with a standard deviation of 1.73 fish. ∎

EXAMPLE 4.40

During a typical baseball game, the Cleveland Indians score 6.4 runs. Find the mean and standard deviation of the probability distribution for the number of runs scored in a game by the Cleveland Indians.

In this problem $\lambda = 6.4$, so

$$\mu = 6.4$$
$$\sigma = \sqrt{6.4}$$
$$= 2.53$$

The mean for this distribution is 6.4 runs, with a standard deviation of 2.53 runs. ■

MICROSOFT EXCEL   Poisson Probabilities

To find the Poisson probability for *x* successes, with expected number of successes λ, we use Excel's built in Poisson function.

    =POISSON(x, λ,FALSE)

There are three arguments in the parentheses. The first two are the number of successes (*x*) and the expected number of successes (λ). The third argument allows you to determine whether you want the probability of exactly *x* successes (**FALSE**) or the probability of at most *x* successes (**TRUE**). Typing **TRUE** as the third argument finds the cumulative probability of 0 through *x* successes. We will use the cumulative option when we are interested in the probability of a range of successes.

We begin by using Excel to rework the first example.

EXAMPLE 4.41

During a typical hour fishing at Lake Lotsafish, a fisher can expect to catch 3 fish. Find the probability that a fisher catches exactly 2 fish in his first hour fishing.

Success: A fish is caught.
Number of successes: $x = 2$
Expected number of successes: $\lambda = 3$
Interval: One hour

Type the following in any cell.

    =POISSON(2,3,FALSE)

Excel's answer, to four decimal places, is 0.2240. This matches the answer that we calculated using the formula, as well as the result that we got from the Poisson table. ■

Excel's cumulative option for finding Poisson probabilities allows us to directly calculate the probability of *x* or fewer successes. Here is an example.

EXAMPLE 4.42

A small town has 1.6 cars stolen on an average day. Find the probability that there are 2 or fewer cars stolen on a given day.

Success: A car is stolen.

Number of successes: $x = 2$ or fewer

Expected number of successes: $\lambda = 1.6$

Interval: One day

Type the following in any cell.

=POISSON(2,1.6,TRUE)

Excel gives us 0.7834, rounded to four places. ■

In general, to find the probability of x or fewer successes (or at most x successes) in a Poisson distribution with expected number of successes λ, use the following format.

=POISSON(x, λ,TRUE)

We can extend the use of the cumulative probability option to calculate probabilities using the complement rule, as in the next example.

EXAMPLE 4.43

During a typical baseball game, the Cleveland Indians score 6.4 runs. Find the probability that the Cleveland Indians do not get shut out in a game—in other words, that they score at least one run in a game.

Success: A run is scored.

Number of successes: $x = $ At least 1

Expected number of successes: $\lambda = 6.4$

Interval: One baseball game

The probability that the Cleveland Indians score at least 1 run in a game is calculated by using the complement rule. The complement of "at least 1" is zero. Type the following in any cell.

=1-POISSON(0,6.4,TRUE)

Excel tells us that the probability is 0.9983. This result agrees with our calculations earlier in the section. ■

In general, to find the probability of x or more successes (or at least x successes) in a Poisson distribution with expected number of successes λ, use the following format.

=1-POISSON(x-1, λ,TRUE)

The last type of problem that we can use Excel's assistance on is finding the probability of between a and b successes. Here is an example.

EXAMPLE 4.44

During the typical statistics final exam, 2 students leave the room in tears. Find the probability that between 1 and 4 students, inclusive, leave your statistics final exam in tears.

Success: A student leaves in tears.

Number of successes: $x = 1, 2, 3,$ or 4

Expected number of successes: $\lambda = 2$

Interval: One final exam

To use Excel and its cumulative option, we begin by finding the probability of 4 or fewer successes. From that result, we subtract the probability of less than 1 success. This will leave us with the probability of 1 through 4 successes. Type the following in any cell.

=POISSON(4,2,TRUE)-POISSON(0,2,TRUE)

From Excel, the probability is 0.8120. Again, this is the same probability that we found earlier. ■

In general, to find the probability of a successes through b successes in a Poisson distribution with expected number of successes λ, use the following format.

= POISSON(b, λ,TRUE)-POISSON(a-1, λ,TRUE)

TI-83   Poisson Probabilities

The TI-83 has built-in functions to help us find the Poisson probability for x successes, with expected number of successes λ. Again, we will need to access the **DISTR** menu, which contains the functions for calculating Poisson probabilities. We access this menu by pressing (2ND)(VARS). Option B is **poissonpdf**, which we use for calculating the probability of exactly x successes. Option C is **poissoncdf**, which we use to find the probability of x or fewer successes.

We will begin by using the TI-83 to rework the first example.

EXAMPLE 4.45

During a typical hour fishing at Lake Lotsafish, a fisher can expect to catch 3 fish. Find the probability that a fisher catches exactly 2 fish in his first hour fishing.

Success: A fish is caught.

Number of successes: $x = 2$

Expected number of successes: $\lambda = 3$

Interval: One hour

Access the **DISTR** menu by pressing (2ND)(VARS). Select option B: **poissonpdf** because we are interested in exactly 2 successes instead of a range of successes. On the main screen enter 3 (,) 2 ( )), and press (ENTER).

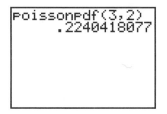

poissonpdf(3,2)
            .2240418077

The TI-83's answer, to four decimal places, is 0.2240. This matches the answer that we calculated using the formula, as well as the result that we got from the Poisson table. ∎

Here is an example using option C: **poissoncdf**. This calculates the probability of *x* or fewer successes. Again, we need to supply the expected number of successes (λ) and the number of successes that we are interested in (*x*).

 XAMPLE
4.46

A small town has 1.6 cars stolen on an average day. Find the probability that there are 2 or fewer cars stolen on a given day.

Success: A car is stolen.

Number of successes: *x* = 2 or fewer

Expected number of successes: λ = 1.6

Interval: One day

Access the **DISTR** menu by pressing (2ND)(VARS). Select option C: **poissoncdf** because we are interested in a range of successes (2 or fewer). On the main screen enter 1.6 (,) 2 ( )), and press (ENTER).

```
poissoncdf(1.6,2
)
            .78335849
```

The TI-83's answer is 0.7834, rounded to four places. ∎

We can extend the use of the Poisson functions, **poissonpdf** and **poissoncdf**, to calculate probabilities using the complement rule, as in the next example.

 XAMPLE
4.47

During a typical baseball game, the Cleveland Indians score 6.4 runs. Find the probability that the Cleveland Indians do not get shut out in a game—in other words, that they score at least one run in a game.

Success: A run is scored.

Number of successes: *x* = At least 1

Expected number of successes: λ = 6.4

Interval: One baseball game

The probability that the Cleveland Indians score at least 1 run in a game is calculated by using the complement rule. The complement of "*at least 1*" is zero.

P(At least 1) = 1 − P(0)

On the main screen, press 1 and (−). Access the **DISTR** menu by pressing (2ND) (VARS). Select option B: **poissonpdf** because we are interested in calculating the probability of exactly 0 successes and not a range of successes. On the main screen enter 6.4 (,) 0 ()), and press (ENTER).

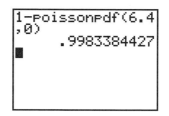

The TI-83's answer is 0.9983, rounded to four places. This result agrees with our calculations earlier in the section. ■

The last type of problem that we can use TI-83's assistance on is finding the probability of between *a* and *b* successes. Here is an example.

  During the typical statistics final exam, 2 students leave the room in tears. Find the probability that between 1 and 4 students, inclusive, leave your statistics final exam in tears.

Success: A student leaves in tears.

Number of successes: $x = 1, 2, 3,$ or 4

Expected number of successes: $\lambda = 2$

Interval: One final exam

To use the TI-83, we begin by finding the probability of 4 or fewer successes. From that result, we subtract the probability of less than 1 success. This will leave us with the probability of 1 through 4 successes.

Access the **DISTR** menu by pressing (2ND)(VARS). Select option C: **poissoncdf** because we are interested in a range of successes (four or fewer). On the main screen enter 2 (,) 4 ()). Now press (−) and return to the **DISTR** menu by pressing (2ND)(VARS). Select option B: **poissonpdf** because we are interested in exactly 0 successes. On the main screen enter 2 (,) 0 ()) and press (ENTER).

```
poissoncdf(2,4)-
poissonpdf(2,0)
        .8120116995
```

The TI-83 gives the answer as 0.8120, rounded to four places. Again, this is the same probability that we found earlier. ■

EXERCISES 4.3

1. A college baseball player averages 1.07 hits per game. Assuming that the number of hits per game follows a Poisson distribution with $\lambda = 1.07$, complete the following probability distribution.

Number of Hits	Probability
0	
1	
2	
3	
4	
5	
6	
7	

2. The number of bike crashes on a university's bike paths follows a Poisson distribution with a mean of 2.1 crashes per day. Complete the following probability distribution.

Number of Crashes	Probability
0	
1	
2	
3	
4	
5	
6	
7	
8	
9	
10	

3. The number of structure fires per day in a small city follows a Poisson distribution with a mean of 0.3 fires per day. Complete the following probability distribution.

Number of Fires	Probability
0	
1	
2	
3	
4	

4. The number of cars stolen per day in a city follows a Poisson distribution with 1.5 cars stolen per day. Complete the following probability distribution.

Number of Cars	Probability
0	
1	
2	
3	
4	
5	
6	
7	
8	

5. During a typical day at the Student Health Center, 4.2 students will come in complaining of stomach problems. Find the probability that 5 students will visit the center today to complain about stomach problems.

6. At a local gym, the number of people who are treated for being accidentally kicked in the head each week in an exercise class follows a Poisson distribution with a mean of 1.8 people treated per week. Find the probability that 3 people from exercise classes at the gym will need to be treated for such an injury this week.

7. During a semester at the university, there is an average of 1.8 students from the Lambda Lambda Lambda fraternity on the Dean's List. Find the probability that at least 2 students from the fraternity make the Dean's List this semester.

8. The number of interceptions thrown by college quarterbacks in a game follows a Poisson distribution with a mean of 1.7 interceptions per game. Find the probability of a college quarterback throwing at least 3 interceptions in a game.

9. The number of students absent from a statistics class follows a Poisson distribution with a mean of 3 students absent per day (not including test days).

   (a) Find the probability that no students are absent on a given day.
   (b) Find the probability that 2, 3, or 4 students are absent on a given day.
   (c) Find the probability that 6 or more students are absent on a given day.

10. The number of accounts opened at a small credit union follows a Poisson distribution with a mean of 2 new accounts per day.

   (a) Find the probability that at least 1 new account is opened on a given day.
   (b) Find the probability that 3, 4, or 5 new accounts are opened on a given day.
   (c) Find the probability that more than 4 accounts are opened on a given day.

11. The number of defects on an assembly line follows a Poisson distribution with a mean of 8 defects per hour.

   (a) Find the probability that 5 defects come off of the assembly line in the next hour.
   (b) Find the probability that fewer than 5 defects come off of the assembly line in the next hour.
   (c) Find the probability that 5 or fewer defects come off of the assembly line in the next hour.
   (d) Find the probability that the assembly line produces anywhere from 6 to 10 defects in the next hour.

12. The number of cars sold by a new car dealer follows a Poisson distribution with a mean of 4 cars sold per day.

   (a) Find the probability that at least 6 cars are sold today.
   (b) Find the probability that there is at least 1 sale today.
   (c) On what percentage of days are there fewer than 5 cars sold?
   (d) If we know that at least 2 cars were sold today, find the probability that exactly 4 cars were sold today.

13. On a typical day, 6 cats are brought in to a low-cost clinic to be spayed. If the clinic only has enough supplies to spay 8 cats per day, find the probability that the capacity will be exceeded today.

14. At a typical lunch meeting at a banquet hall, 5 people will request a vegan lunch. If the banquet hall only has 3 vegan lunches available for today's meeting, find the probability that this will not be enough for today's meeting.

15. The number of teachers absent at an elementary school follows a Poisson distribution with a mean of 0.3 absences per day. If the school has 2 substitute teachers available, find the probability that the school will be understaffed on a given day.

16. On an average day, a florist receives orders for 2 blooming plants. If there are 5 blooming plants in stock, find the probability that there will be enough blooming plants for today.

17. For a typical flight from Las Vegas to New York, 3 travelers do not show up. If the airline overbooked a flight by 5 passengers during advance sales, find the probability that it will have enough seats for this flight.

18. A door-to-door salesman averages 4 magazine subscriptions sold per evening. If the salesman needs at least 6 sales tonight to earn the weekly bonus, find the probability that he gets the bonus tonight.

19. The number of runs scored by a college baseball team follows a Poisson distribution with a mean of 4.4 runs per game. Find the probability that the team scores exactly 10 runs in the next two games.

20. A grocery store has an ATM machine inside. The number of customers who use the machine follows a Poisson distribution with a mean of 2.5 customers per hour. Find the probability that exactly 4 customers use the machine in the next three hours.

21. On a typical day at a small town hospital, 3 mothers give birth. Find the probability that there are at least 10 births in the next two days at the hospital.

22. A software company sells software on the Internet. The number of unsolicited sales follows a Poisson distribution with a mean of 5 sales per day. Find the probability that the company makes at least 15 unsolicited sales over the next two days.

23. Here are the number of accidents on a dangerous stretch of road on 40 randomly selected days.

Number of Accidents	Frequency
0	6
1	11
2	12
3	6
4	2
5	2
6	1
7	0
8	0

It is claimed that the number of accidents follows a Poisson distribution with a mean of 1.5 accidents per day. Does it? Complete the table that follows, and use it to help you determine whether the above table is close to what we would expect from a Poisson distribution whose mean is 1.5. To complete the table, first find the probability of 0 accidents, 1 accident, and so on for a Poisson distribution with a mean of 1.5 accidents per day. To find the expected number of days for each number of accidents, multiply the probabilities by 40 (for 40 days).

Number of Accidents	Probability	Expected
0		
1		
2		
3		
4		
5		
6		
7		
8		

24. Do the number of tornadoes that strike Utah per year follow a Poisson distribution with a mean of 1.7 tornadoes per year? Here is a table showing how many tornadoes struck Utah each year for the 46 years from 1950 through 1995.

Number of Tornadoes	Frequency
0	19
1	8
2	5

(continues)

(continued)

Number of Tornadoes	Frequency
3	5
4	3
5	3
6	3
7	0
8	0
9	0

Complete the following table, and use it to help you determine whether the number of tornadoes follows a Poisson distribution with a mean of 1.7 tornadoes per year.

Number of Tornadoes	Probability	Expected
0		
1		
2		
3		
4		
5		
6		
7		
8		
9		

25. At a local gym, the number of people who are treated for being accidentally kicked in the head each week in an exercise class follows a Poisson distribution with a mean of 1.8 people treated per week. Find the standard deviation for this distribution.

26. The number of teachers absent at an elementary school follows a Poisson distribution with a mean of 0.3 absences per day. Find the standard deviation of this distribution.

27. The number of penalties for a college football team follows a Poisson distribution with a mean of 7.2 penalties a game. Find the standard deviation for this distribution.

28. The number of errors that a Braille transcriber makes follows a Poisson distribution with a mean of 2.5 mistakes per page. Find the standard deviation for this distribution.

EXTRA • Poisson Distribution as a "Limit" for the Binomial Distribution

The Poisson distribution was initially developed as a limit for the binomial distribution. In other words, as n increased, the binomial distribution got closer and closer to the Poisson distribution. The fit was better for smaller values of p. The mean of the Poisson distribution can be found by using the formula $\lambda = n \cdot p$.

We will examine how the binomial distribution approaches the Poisson distribution as n becomes large and p becomes small. We will use four binomial distributions that have a mean of 1: $n = 10$ and $p = 0.1$; $n = 100$ and $p = 0.01$; $n = 1000$ and $p = 0.001$; and $n = 10,000$ and $p = 0.0001$. You will notice that as n becomes larger and p becomes smaller, the binomial probability distributions become closer to the Poisson probability distribution with $\lambda = 1$.

Complete the following probability distributions.

x	Binomial n = 10 p = 0.1	Binomial n = 100 p = 0.01	Binomial n = 1,000 p = 0.001	Binomial n = 10,000 p = 0.0001	Poisson λ = 1
0					0.3679
1					0.3679
2					0.1839
3					0.0613
4					0.0153
5					0.0031
6					0.0005
7					0.0001

What can we say as *n* is increasing?

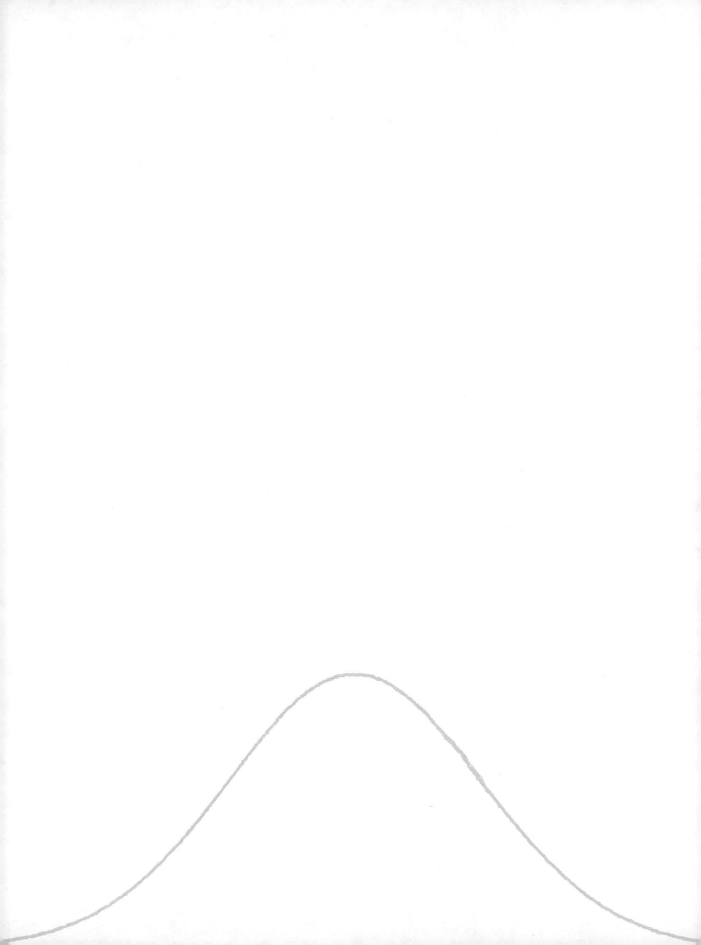

Normal Probability Distributions

I n this chapter we examine another probability distribution, the normal distribution.
The normal distribution differs from the
probability distributions in the previous
chapter because it is a continuous, rather than
discrete, distribution. The random variables used
in this chapter may take on infinitely many values, including decimal values. For the most part,
we will not be looking at counts of successes as
we did in the previous chapter. Another major
difference is the fact that the normal distribution is represented by a curve, and we will be
using areas to represent probabilities.

In Section 5.1, we will be introduced to the family of normal distributions and
their properties. We will also use the normal distribution to find probabilities.

In Section 5.2, we will learn how to make adjustments in order to handle probabilities for discrete random variables. Here we will learn how to approximate binomial probabilities using the normal distribution, under the correct conditions. The
normal approximation to the binomial distribution will give us the tools to make
inferences about a population proportion or percentage.

Finally, in Section 5.3 we will reverse the whole process. If given a probability,
we will learn how to find the value of the random variable associated with that
probability.

SECTION 5.1
The Normal Distribution

A normal distribution is a continuous probability distribution. In Chapter 4, we saw that the binomial and Poisson distributions were discrete because the random variables in those distributions could only take on values that were whole numbers. If asked to find the probability that at least 11.9 of 20 people waiting in line would not be able to register for the class that they wanted, we would see that the value 11.9 is not a possible value for the random variable. This is because we are counting the number of people that were not able to register for the class they wanted, which has to be a whole number. On the other hand, if we were asked to find the probability that a can of soda had at least 11.9 ounces of soda in it, we would see that 11.9 is a possible value—as are 11.93, 11.972, and infinitely many other decimal values.

In the first sentence of the previous paragraph, we referred to *a* normal distribution rather that *the* normal distribution. This is because there are infinitely many normal distributions. A normal distribution is determined by its mean and standard deviation.

Here is the graph of a normal distribution with mean μ and standard deviation σ.

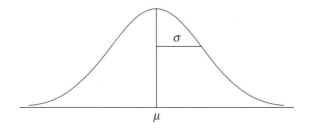

The graph of a normal distribution is bell-shaped and symmetric. The mean is the value on the horizontal axis where the peak of the graph occurs. The standard deviation, as you might expect, has to do with how spread out the distribution is. The heights of adult males are normally distributed with a mean of 69.0 inches and a standard deviation of 2.8 inches. The heights of adult females are normally distributed with a mean of 63.6 inches and a standard deviation of 2.5 inches. The graphs of these two distributions are shown next on the same set of axes.

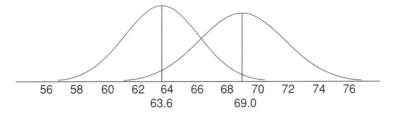

Note that the curve to the right is for male heights. The mean for male heights is greater than, or to the right of, the mean for female heights because the peak of the curve for male heights is to the right of the peak of the curve for female heights. The standard deviation for male heights is greater than the standard deviation for female heights. You can see that the curve for the male heights is flatter and more spread out.

EXAMPLE 5.1   Which of the following two normal distributions has the greater mean?

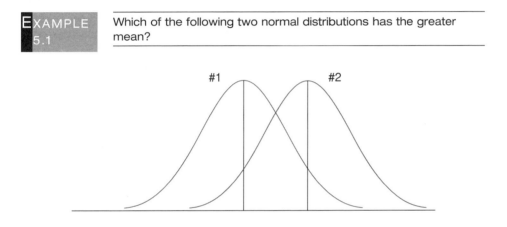

The distribution whose graph is to the right (#2) has the larger mean, because its peak occurs farther to the right. ■

EXAMPLE 5.2   The two normal distributions graphed here have the same mean. Which of the two has the greater standard deviation?

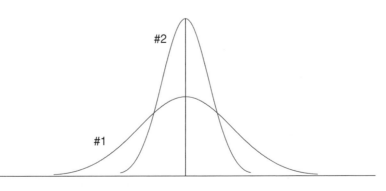

The distribution labeled as #1 has the greater standard deviation, because the curve is more spread out from the center. ■

A normal distribution's graph is bell-shaped. There are more values located near the center, or mean, of the distribution. There are fewer and fewer values as we move out from the center. The graph of the distribution continues forever in both directions, without ever touching the horizontal axis. The horizontal axis is an asymptote for the graph of a normal distribution. When drawing the graph of a normal distribution by hand, be sure that the curve never touches the horizontal axis.

Normal distributions are symmetric with respect to the mean. The area underneath the curve, but above the horizontal axis, is equal to 1. The fact that the total area is equal to 1 allows us to use the area underneath the curve to represent probabilities. Recall from the last chapter that the total of all probabilities for a probability distribution is equal to 1. Again, since a normal distribution is symmetric, the area underneath the right half of the curve is equal to 0.5, and the same can be said for the left half.

Before moving on to determine probabilities involving normal distributions, we must discuss one special normal distribution called the **standard normal distribution.** The standard normal distribution has a mean of 0 and a standard deviation of 1. Its graph is shown next. The standard normal distribution is sometimes referred to as the *z*-distribution. This has to do with the fact that we will later use *z*-scores to help find probabilities.

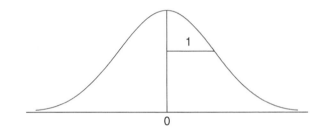

For any continuous probability distribution, including the standard normal distribution, we find the probability that the random variable is in a certain interval by finding the area underneath the curve in that interval. For example, if we wanted to find the probability that a standard normal variable (z) was between a and b, denoted $P(a < z < b)$, we would need to find the area underneath the curve between a and b. An example of such an area is shown next.

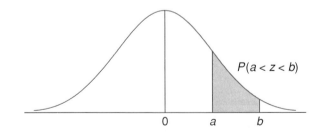

One important fact to take note of is that we can only find the probability that a random variable is in an interval. The probability that a continuous random variable is exactly equal to a certain value is 0.

Usually, to find the area underneath a curve, we would need to use calculus. However, there is a table (Table C) that will help us find the area underneath the curve for any region for the standard normal distribution. Table C gives us the area underneath the curve that lies to the left of any z-value (z*) for the standard normal distribution. For the graph shown here, our table gives us the shaded area.

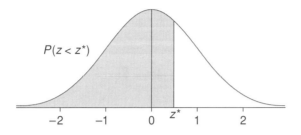

There are three different types of intervals that we will be finding probabilities for. The first is to the left of a certain value, the second is to the right of a value, and the third is in between two values.

Find the following probabilities.

(a) $P(z < 1.96)$

With all standard normal probability problems, begin by drawing the graph and shading the appropriate region. We draw a normal curve, with a mean of 0. We place 1.96 on the right side of the curve, draw a line segment above it, and then shade to the left. We shade to the left because we are looking for the probability that z is less than 1.96. The values that are less than 1.96 are to the left of 1.96. The graph is shown next, with the shaded area representing $P(z < 1.96)$.

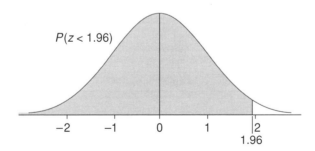

Since we are looking for the shaded region to the left of our value, we can use Table C directly to find our probability. We find the positive values of z on the second page of the table. We look down the column on the left side until we find 1.9, and then we move across until we find the column with a 6 above it (this is the second decimal place of our z-value). The value that you should see there is 0.9750. Therefore, $P(z < 1.96) = 0.9750$.

Would this problem have turned out differently if we were asked for $P(z \leq 1.96)$ instead of $P(z < 1.96)$? No. Recall that the probability of any particular value occurring is 0. Therefore,

$$P(z \leq 1.96) = P(z < 1.96) + P(z = 1.96)$$
$$= P(z < 1.96) + 0$$
$$= P(z < 1.96)$$

(b) $P(z < -1.25)$

Again, we begin with the graph.

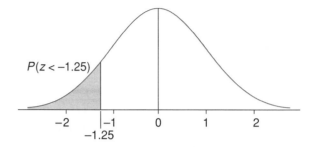

Since we are looking for the area of the region to the left of –1.25, we can get this area directly from the table. The area associated with –1.25 is 0.1056, so $P(z < -1.25) = 0.1056$. ∎

Often, we will be looking to find the area to the right of a value. Our table does not give us this value directly. As a matter of fact, it gives us the complement of the area that we are looking for. We will take the area from the table, and subtract it from 1 to find the area we are looking for. The following picture illustrates this idea.

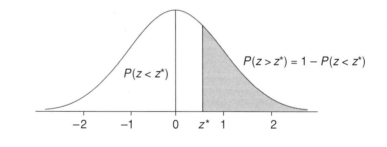

EXAMPLE 5.4   Find the following probabilities.

(a) $P(z > 1.28)$

Here is the graph for this region.

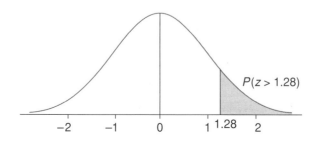

Our table gives us the area of the unshaded region for this graph, which is 0.8997. To find the area of the shaded region, we subtract this area from 1, which is the area underneath the entire curve. Therefore,

$P(z > 1.28) = 1 - 0.8997$

$= 0.1003$

(b) $P(z > -1.5)$

Again, we begin with the graph.

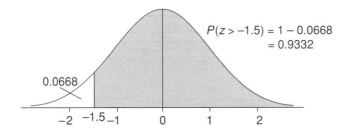

$$P(z > -1.5) = 1 - 0.0668$$
$$= 0.9332$$

Again, the table gives us the unshaded area to the left of –1.5, and we subtract that area from 1 to find the area of the shaded region.

$P(z > -1.5) = 1 - 0.0668$

$= 0.9332$

The probability that z is greater than –1.5 is 0.9332. ∎

The third possible type of shaded region occurs when we shade between two z-values. Consider the following graph, which is associated with $P(a < z < b)$, where a is less than b.

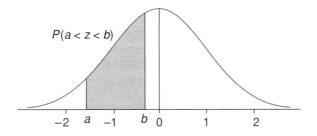

We can think of the shaded region as being the region to the left of b with the region to the left of a cut out of it. We can find the area of these two regions separately from the table, and then subtract the area to the left of a from the area to the left of b to give us the area of the region between a and b.

EXAMPLE
5.5

Find the following probabilities.

(a) $P(-1.23 < z < 0.97)$

First, sketch the shaded region. The area of the region to the left of 0.97 is 0.8340. The area of the region to the left of –1.23 is 0.1093. We subtract to find the area of the shaded region.

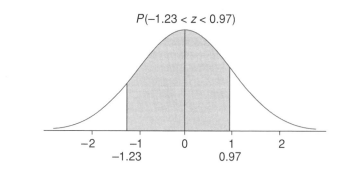

$$P(-1.23 < z < 0.97) = 0.8340 - 0.1093$$
$$= 0.7247$$

Therefore, $P(-1.23 < z < 0.97) = 0.7247$.

(b) $P(-1.53 < z < -0.75)$

Here is the graph.

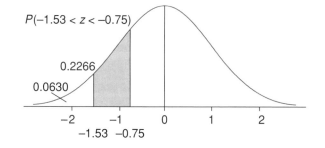

Using the table, we will subtract the area associated with the z-value on the right minus the area associated with the z-value on the left.

$$P(-1.53 < z < -0.75) = 0.2266 - 0.0630$$
$$= 0.1636$$

Therefore, $P(-1.53 < z < -0.75) = 0.1636$.

(c) $P(0.5 < z < 0.88)$

Here is the graph.

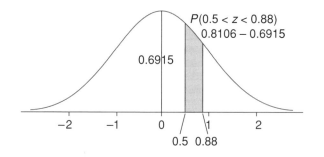

$$P(0.5 < z < 0.88) = 0.8106 - 0.6915$$
$$= 0.1191$$

Therefore, $P(0.5 < z < 0.88) = 0.1191$. ∎

A **nonstandard normal distribution** is a normal distribution that does not have a mean of 0 and a standard deviation of 1. An example of such a distribution is the height of adult females, which are normally distributed with a mean of 63.6 inches and a standard deviation of 2.5 inches. In order to be able to find probabilities associated with nonstandard normal distributions, we must first translate the distribution into a standard normal distribution. The formula for z-scores

$$z = \frac{x - \mu}{\sigma}$$

will assist us with that.

Suppose we want to find the probability that a randomly selected female is 5 feet tall (60 inches) or shorter. The graph for this probability looks like this:

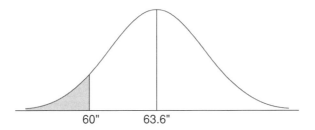

The values in *any* normal distribution are distributed in the same fashion. If we can translate this graph into one involving the standard normal (*z*) distribution, then we can find the area of the shaded region. As you may verify in the next example, a height of 60 inches corresponds to a *z*-value of –1.44, while a height of 63.6 inches corresponds to a *z*-value of 0. We can use those facts to create the following graph, from which we may find the area of the shaded region.

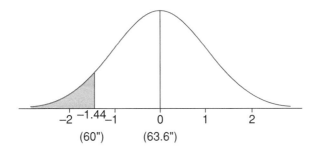

EXAMPLE 5.6 The heights of adult females are normally distributed with a mean of 63.6 inches and a standard deviation of 2.5 inches. Find the probability that a randomly selected female is 5 feet tall (60 inches) or shorter.

We begin by identifying the random variable in this problem. We let *x* represent the height of an adult female. Now, write the probability in terms of *x*.

$P(z \le 60)$

The next step is to change this probability into a probability involving the standard normal distribution by using the z-score formula $z = \frac{(x - \mu)}{\sigma}$, as follows.

$$P(z \le 60) = P\left(z \le \frac{60 - 63.6}{2.5}\right)$$

$$= P(z \le -1.44)$$

We finish by drawing the graph, and using our table.

Since we are looking for the area of the region to the left of z = –1.44, we can obtain this area directly from the table. The probability that an adult female is 5 feet tall or shorter is 0.0749. ■

The important idea to grasp from the previous example is that $P(x \le 60)$ is exactly the same as $P(z \le -1.44)$. All we have done is change our original problem into an equivalent problem involving the standard normal distribution.

We will take a look at two more examples involving the heights of adult females.

EXAMPLE 5.7   The heights of adult females are normally distributed with a mean of 63.6 inches and a standard deviation of 2.5 inches. Find the probability that a randomly selected female is between 5′ 2″ tall (62 inches) and 5′ 7″ tall (67 inches).

We will let x represent the height of an adult female, the random variable in this problem. We will write the probability that we are looking for in terms of x. Then we will convert it to a problem involving z, draw a graph, and use our table.

$$P(62 \le x \le 67) = P\left(\frac{62 - 63.6}{2.5} \le z \le \frac{67 - 63.6}{2.5}\right)$$

$$= P(-0.64 \le z \le 1.36)$$

Here is the graph.

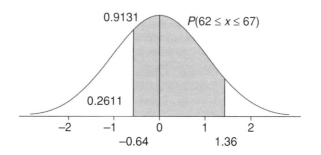

To find the area of the shaded region, we subtract the area of the region to the left of –0.64 from the area of the region to the left of 1.36.

$$P(-0.64 \leq z \leq 1.36) = 0.9131 - 0.2611$$

$$= 0.6520$$

The probability that an adult female is between 5′ 2″ and 5′ 7″ tall is 0.6520. ∎

 The heights of adult females are normally distributed with a mean of 63.6 inches and a standard deviation of 2.5 inches. Find the probability that a randomly selected female is 6′ 4″ tall or taller.

Before beginning here, we should be able to reason that this will be a small probability. As we walk down the street, it is a rare occasion that we see an adult female that is at least 6′ 4″ tall.

We will let x represent the height of an adult female.

$$P(x \geq 76) = P\left(z \geq \frac{76 - 63.6}{2.5}\right)$$

$$= P(z \geq 4.96)$$

Here is the graph.

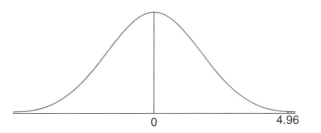

When we go to the table, we notice that the highest z-value is 3.50. For any z-value above 3.50, we will use the area associated with 3.50. (We will use a similar approach for negative z-values below –3.50.) The area of the unshaded region is 0.9998, so the probability that an adult female is at least 6′ 4″ tall is 1 – 0.9998 or 0.0002. This is an estimate of the probability; to find the actual probability we would need to use 4.96 for a z-value instead of 3.50. Later, we will see that the use of technology will allow us to find this probability, which is 0.000000353. ∎

 The lengths of human pregnancies are normally distributed with a mean of 268 days and a standard deviation of 15 days. Find the probability that a pregnancy lasts 280 days or longer.

We will let x represent the length of a human pregnancy.

$$P(x \geq 280) = P\left(z \geq \frac{280 - 268}{15}\right)$$

$$= P(z \geq 0.8)$$

Here is the graph.

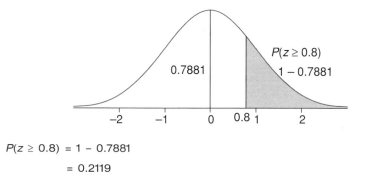

$P(z \geq 0.8) = 1 - 0.7881$

$= 0.2119$

The probability that a pregnancy lasts at least 280 days is 0.2119. ∎

MICROSOFT EXCEL   Normal Probabilities

Microsoft Excel provides us with two functions that we can use in this section. The first helps us to calculate probabilities involving the standard normal (z) distribution. The second helps us to calculate probabilities involving nonstandard normal distributions.

To find the probability that a standard normal variable is less than z, we use the function

=NORMSDIST(z)

We will look at the second function later.

Our first example involves the area of a region to the left of a z-value.

EXAMPLE 5.10   Find $P(z < 1.96)$.

As by hand, we begin with the graph.

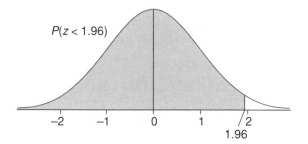

Now type the following in any cell.

=NORMSDIST(1.96)

Excel gives us 0.9750 to four decimal places, which agrees with our previous answer. ∎

In general, any time you need to find $P(z < a)$, type =NORMSDIST(a) in any cell. The next example deals with finding the area to the right of a z-value.

| EXAMPLE 5.11 | Find $P(z > -1.5)$. |

Again, we begin with the graph.

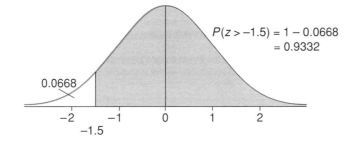

Since we have shaded to the right of $z = -1.5$, we will need to use the complement rule. This is because Excel gives us the area of the unshaded region to the left, not the shaded region to the right. Type the following in any cell.

=1-NORMSDIST(-1.5)

Excel tells us that the probability is 0.9332, which agrees with our previous answer. By the way, if you would like Excel to round your probabilities to four decimal places, type the following in any cell instead.

=ROUND(1-NORMSDIST(-1.5),4) ∎

In general, any time you need to find $P(z > a)$, type =1-NORMSDIST(a) in any cell. The next example deals with finding the area between two z-values.

| EXAMPLE 5.12 | Find $P(-1.23 < z < 0.97)$. |

Here is the graph.

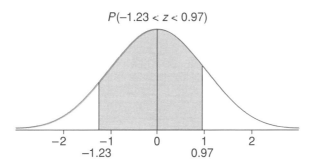

Recall that we want to subtract the area associated with the z-value on the left from the z-value on the right. Type the following in any cell.

=NORMSDIST(0.97)-NORMSDIST(-1.23)

Excel gives us 0.7246, which differs from our previous answer of 0.7247. This is due to the fact that the table rounded each probability before we subtracted, whereas with Excel we wait to round until after the subtraction. ∎

In general, any time you need to find $P(a \leq z \leq b)$, type

=NORMSDIST(b)-NORMSDIST(a)

When working with nonstandard normal distributions, we use the function

=NORMDIST(x,μ,σ,TRUE)

The first argument, x, is the x-value associated with the probability. The second and third arguments are the mean and standard deviation, respectively. The last argument must be **TRUE** if we are to find the probability to the left of x. (Typing **FALSE** for the last argument gives you the actual height of the normal curve and not a probability.)

We use this function in the same exact way that we used the first function. Here is one example of this function in use.

 EXAMPLE **5.13**   The heights of adult females are normally distributed with a mean of 63.6 inches and a standard deviation of 2.5 inches. Find the probability that a randomly selected female is 5 feet tall (60 inches) or shorter.

Here is the graph.

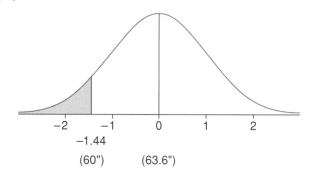

To find $P(x \leq 60)$, type the following in any cell.

=NORMDIST(60,63.6,2.5,TRUE)

Excel tells us that the probability that an adult female is 5 feet tall or shorter is 0.0749. This agrees with our previous answer. We could have also changed the problem to a standard normal distribution problem first, and then used Excel.

$$P(x \leq 60) = P\left(z \leq \frac{60 - 63.6}{2.5}\right)$$
$$= P(z \leq -1.44)$$

We would then type the following in any cell, using the first function.

=NORMSDIST(-1.44)

This gives the same result. ∎

TI-83   Normal Probabilities

The TI-83 uses its built-in function **normalcdf** for calculating probabilities involving a normal distribution. The basic format for using this function is as follows.

normalcdf(*lowerbound, upperbound, μ, σ*)

This will find the area underneath the curve between *lowerbound* and *upperbound* for a normal distribution with mean μ and standard deviation σ. If there is no lower bound (area to the left of a value), we enter –1E99 as the lower bound by pressing (–) 1 (2ND) (,) (to access **EE**) 99. If there is no upper bound (area to the right of a value), we enter 1E99 as the lower bound by pressing 1 (2ND) (,) (to access **EE**) 99. If we are using the standard normal distribution (μ = 0, σ = 1), then we may omit these values and enter only the lower bound and upper bound.

Our first example involves the area of a region to the left of a *z*-value.

EXAMPLE 5.14	Find $P(z < 1.96)$.

As by hand, we begin with the graph.

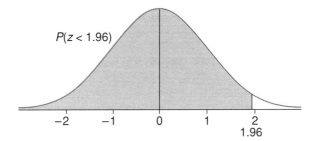

Press (2ND) (VARS) to access the **DISTR** menu, and select item 2: **normalcdf.** On the main screen, press (–) 1 (2ND) (,) 99 (,) 1.96 (,) (ENTER).

```
normalcdf(-1E99,
1.96)
        .9750021748
```

The TI-83 gives us 0.9750 to four decimal places, which agrees with our previous answer. ∎

The next example deals with finding the area to the right of a *z*-value.

EXAMPLE
5.15

Find $P(z > -1.5)$.

Again, we begin with the graph.

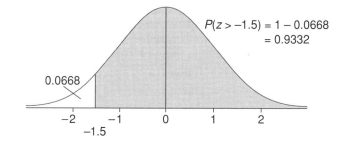

$$P(z > -1.5) = 1 - 0.0668$$
$$= 0.9332$$

0.0668

Since we have shaded to the right of $z = -1.5$, we have no upper bound. Press 2ND VARS to access the **DISTR** menu, and select item 2: **normalcdf.** On the main screen, press (–) 1.5 , 1 2ND , 99) ENTER .

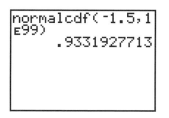

The TI-83 tells us that the probability is 0.9332, which agrees with our previous answer. ∎

The next example deals with finding the area between two z-values.

EXAMPLE
5.16

Find $P(-1.23 < z < 0.97)$.

Here is the graph.

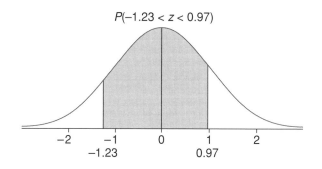

$$P(-1.23 < z < 0.97)$$

For such a case as this, we have both a lower bound and an upper bound. Press (2ND) (VARS) to access the **DISTR** menu, and select item 2: **normalcdf.** On the main screen, press ((–)) 1.23 ((,)) 0.97 (())(ENTER).

```
normalcdf(-1.23,
0.97)
          .7246281435
```

The TI-83 gives us 0.7246, which differs from our previous answer of 0.7247. This is due to the fact that the table rounded each probability before we subtracted, whereas the TI-83 waits to round until after the subtraction. ∎

Here is an example of how to work with nonstandard normal distributions using the TI-83.

 EXAMPLE 5.17 The heights of adult females are normally distributed with a mean of 63.6 inches and a standard deviation of 2.5 inches. Find the probability that a randomly selected female is 5 feet tall (60 inches) or shorter.

Here is the graph.

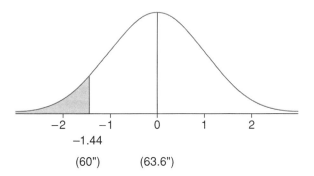

To find $P(x \leq 60)$, press (2ND) (VARS) to access the **DISTR** menu, and select item 2: **normalcdf.** On the main screen, press ((–)) 1 (2ND) ((,)) 99 ((,)) 60 ((,)) 63.6 ((,)) 2.5 (()) (ENTER).

```
normalcdf(-1E99,
60,63.6,2.5)
          .0749337427
```

The TI-83 tells us that the probability that an adult female is 5 feet tall or shorter is 0.0749. This agrees with our previous answer. We could have also changed the problem to a standard normal distribution problem first, and then omitted the mean and standard deviation. ∎

EXERCISES 5.1

Find the given probability.

1. $P(z < 1.25)$ 2. $P(z < 0.8)$ 3. $P(z < -2.16)$
4. $P(z \leq 0.85)$ 5. $P(z \leq 1.07)$ 6. $P(z < 3.75)$
7. $P(z > 2.33)$ 8. $P(z \geq 1.64)$ 9. $P(z \geq -0.43)$
10. $P(z > -1.96)$ 11. $P(z > 0.09)$ 12. $P(z > 4.17)$
13. $P(1.29 < z < 2.13)$ 14. $P(0.17 \leq z \leq 1.4)$ 15. $P(-2.05 < z < -1.13)$
16. $P(-1.57 \leq z \leq 0.79)$ 17. $P(-2 < z < 2)$ 18. $P(-1 < z < 1)$
19. $P(z = 1.65)$ 20. $P(z = 0)$

The heights of adult males are normally distributed with a mean of 69.0 inches and a standard deviation of 2.8 inches. Find the probability that an adult male is

21. 6′ or taller
22. 5′ 5″ or shorter
23. 5′ 10″ or shorter
24. 5′ 3″ or taller
25. Between 5′ 7″ and 5′ 9″ tall
26. Between 5′ 4″ and 6′ 4″ tall
27. Taller than 6′ 8″
28. Exactly 5′ 9″ tall

A computer software package generates random numbers that are normally distributed with a mean and standard deviation that are entered by the user. Suppose that the user enters a mean of 400 and a standard deviation of 100. Find the probability that a value will be

29. Between 525 and 650
30. Above 490
31. Below 535
32. Below 357
33. Above 619
34. Between 340 and 714

At a certain community college, the time that is required by students to complete the math competency exam is approximately a normal distribution with a mean of 57.6 minutes and a standard deviation of 8 minutes. Find the probability that a student takes

35. Longer than 1 hour to complete the test
36. Less than 1 hour and 15 minutes to complete the test
37. Between 55 minutes and 1 hour and 5 minutes to complete the test
38. At least 45 minutes to complete the test
39. No longer than 40 minutes to complete the test
40. Between 35 and 55 minutes to complete the test
41. Exactly 50 minutes to complete the test
42. If the college has a 90-minute time limit for this test, what percentage of all students do not finish the test?

The lengths of human pregnancies are normally distributed with a mean of 268 days and a standard deviation of 15 days. Find the probability that a human pregnancy lasts

43. At least 282 days
44. Less than 275 days
45. Between 254 days and 261 days
46. More than 240 days

47. 268 days or less
48. Between 253 days and 283 days

The heights of 12-month-old boys are approximately normally distributed with a mean of 29.8 inches and a standard deviation of 1.2 inches. Find the probability that a 12-month-old boy is

49. Less than 28 inches tall
50. Between 29 and 31 inches tall
51. Taller than 32 inches
52. Shorter than 33 inches
53. Between 25 and 27 inches tall
54. At least 31.5 inches tall

EXCEL EXPLORATION   Generating Random Normal Data

To use Excel to generate random data, we will use the Analysis ToolPak, which is an add-in to Microsoft Excel. To see whether it has already been added, select the **Tools** menu. If you do not see **Data Analysis** as one of the choices in this menu, then you have to add in the Analysis ToolPak. From the **Tools** menu, select **Add-Ins.** When the dialog box appears, click on the box to the left of **Analysis ToolPak** so that a check-mark appears, then click **OK.**

Select **Data Analysis** from the **Tools** menu. When the dialog box opens, scroll down and select **Random Number Generation** and click **OK.** This will open the dialog box for random number generation. We will be generating 1000 random normal values, and using them to investigate certain normal probabilities. To generate one column with 1000 values, enter 1 for **Number of Variables** and 1000 for **Number of Random Numbers.** For **Distribution,** select **Normal.** After selecting the normal distribution, enter 0 for **Mean** and 1 for **Standard Deviation** (this is the standard normal distribution). Enter any integer of your choice for **Random Seed.** Once this has all been done, click **OK** and Excel will generate your values on a new worksheet.

Now that these values have been generated, we will use them to investigate normal probabilities. First, it will help our efforts if the values have been sorted in ascending order. From the **Data** menu select **Sort.** A dialog box opens. Be sure that the box labeled **Sort by** says **Column A,** and select **Ascending** for the order. Click on **OK,** and your values will be sorted for you.

We found earlier in this section that $P(z < 1.96) = 0.9750$. This means that 97.5% of all z-values should be less than 1.96, or approximately 975 of our 1000 values should be less than 1.96. While attempting this simulation twice, I found that 975 and 972 values were less than 1.96. Go on to verify the following probabilities. Write down the percentage of your values that satisfy the given condition.

1. $P(z < -1.25) = 0.1056$
2. $P(z > -1.28) = 0.1003$
3. $P(z > -1.5) = 0.9332$
4. $P(-1.23 < z < 0.97) = 0.7247$
5. $P(-1.53 < z < -0.75) = 0.1636$
6. $P(0.5 < z < 0.88) = 0.1191$

SECTION 5.2

Discrete Random Variables That Are Approximately Normal; Normal Approximation to the Binomial Distribution

Although normal distributions are continuous, there are some discrete random variables that are approximately normal. In other words, their distribution roughly follows a normal distribution. If we were to look at a histogram for the distribution, we would see that it was symmetric and followed the same bell-shaped pattern that a normal distribution does. If we are to use a normal distribution to find probabilities associated with a discrete random variable, we must make a minor adjustment first to convert it to a continuous random variable. This minor adjustment is called a **continuity correction.**

Let's begin by taking a look at a discrete random variable that is approximately normally distributed. An IQ test is designed to produce a distribution that is approximately normally distributed with a mean score of 100 points and a standard deviation of 15 points. IQ scores are always whole numbers. If we let x represent the score on this IQ test, then this random variable is discrete because it can only take on whole numbers as values. It is impossible to have an IQ score of 103.7.

The problem we are faced with is that a normal distribution is continuous. If we look at a normal distribution with a mean of 100 and a standard deviation of 15, we see that 103.7 is a possible value. All of the values in between 103 and 104 are possible values. However, in a discrete distribution none of those values are possible. If we are to use a normal distribution, we must decide what to do with all of the space between the possible values. We will simply divide the values in the gap in half, assigning the first half to the previous possible value and the second half to the next possible value. An IQ score of 103 is associated with the interval $102.5 \leq x \leq 103.5$. The interval $102.5 \leq x \leq 103.5$ is the continuous equivalent of the discrete value $x = 103$. Changing a discrete value into a continuous interval is what we refer to as a continuity correction. In the following example, we will use the continuity correction and a normal distribution to calculate a probability involving a discrete random variable.

 EXAMPLE **5.18**   An IQ test is designed to produce a distribution that is approximately normally distributed with a mean score of 100 points and a standard deviation of 15 points. Find the probability that a randomly selected person has an IQ of 103 points.

Let x represent a person's IQ score. The random variable x is discrete, so we must make a continuity correction before we can find this probability.

$$P(x = 103) \overset{cc}{=} P(102.5 \leq x \leq 103.5)$$

$$= P\left(\frac{102.5 - 100}{15} \leq z \leq \frac{103.5 - 100}{15}\right)$$

$$= P(0.17 \leq z \leq 0.23)$$

In the first step, cc tells us that a continuity correction has been made. The probability $P(x = 103)$ involves a discrete random variable, whereas the probability $P(102.5 \leq x \leq 103.5)$ now involves a continuous random variable. We have changed a discrete problem into a continuous problem. We then finish the problem off with the process developed in the last section: make the transformation to z, draw the graph, and so on.

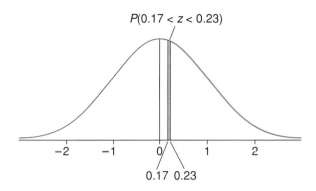

$P(0.17 < z < 0.23)$

We can tell from the graph that this probability must be small.

$P(0.17 < z < 0.23) = 0.5910 - 0.5675$

$= 0.0235$

The probability that a randomly selected person has an IQ of 103 is 0.0235. ∎

When we see a normal distribution problem, we must now first determine whether the continuity correction is necessary. Examine your random variable and determine whether only certain values are possible (discrete, make the continuity correction) or infinitely many values are possible (continuous, no correction).

The discrete value $x = 103$ is associated with the interval $102.5 \leq x \leq 103.5$, and the discrete value $x = 104$ is associated with the interval $103.5 \leq x \leq 104.5$. There seems to be a conflict, as 103.5 is part of the interval associated with $x = 103$ and $x = 104$. We should assign 103.5 to only one of the intervals, but the inclusion of both endpoints in the interval makes our continuity corrections easier to make. Also, recall that the probability of any one specific value of a continuous random variable occurring is 0, so the probabilities will be the same.

We will now focus on how to make the continuity correction if we are looking at a range of values, as opposed to a single value. Here are some possible discrete intervals, and their continuous equivalents.

Discrete	Continuous
$x \geq 103$	$x \geq 102.5$
$x \leq 103$	$x \leq 103.5$
$103 \leq x \leq 110$	$102.5 \leq x \leq 110.5$

For the first interval, $x \geq 103$, we begin at 103 and move to the right. As a discrete interval, this consists of the values 103, 104, 105, ... As a continuous interval, this consists of *all* values that are 103 and higher. The only values associated with 103 that are not included are the values to the left of 103—specifically, from 102.5 up. In general, we make the correction away from the interval that we are dealing with. Sometimes it is easier to look at the interval on the number line. Here is the original interval.

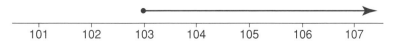

We will make our continuity corrections *away* from the given interval, like this.

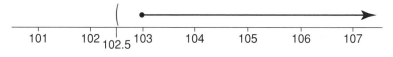

The interval $x \geq 103$ becomes $x \geq 102.5$ with a continuity correction.

How about the interval $x \leq 103$? This interval is to the left of 103, so we make our continuity correction of 0.5 to the right. This gives us the interval $x \leq 103.5$.

If we look at an interval between two values, such as $103 \leq x \leq 110$, all of the values inside of the interval are included. We make the continuity correction by extending the interval on the outside. We move 0.5 to the left of 103 and 0.5 to the right of 110, to produce the interval $102.5 \leq x \leq 110.5$.

Now we will look at a few examples, from start to finish.

 EXAMPLE **5.19**   The scores on a mathematics placement test are approximately normally distributed with a mean of 41.3 points and a standard deviation of 7.8 points. (Scores are whole numbers only.) Find the probability that a randomly selected student scores 55 or higher on the placement test.

Let x represent the score on the test. This random variable is discrete, since all scores are whole numbers. The probability that we are interested in is $P(x \geq 55)$. When we make the continuity correction, we make it to the left, since our interval goes to the right of 55.

$$P(x \geq 55) \overset{cc}{=} P(x \geq 54.5)$$

$$= P\left(z \geq \frac{54.5 - 41.3}{7.8}\right)$$

$$= P(z \geq 1.69)$$

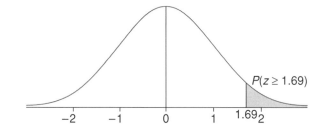

$$P(z \geq 1.69) = 1 - 0.9545$$

$$= 0.0455$$

The probability that a randomly selected student gets a score of 55 or higher is 0.0455. ∎

 EXAMPLE **5.20**   The number of fares for a Las Vegas taxi driver on Fridays is approximately normally distributed with a mean of 21.5 fares and a standard deviation of 6.3 fares. Find the probability that a driver has between 24 and 26 fares on a randomly selected Friday.

Let x represent the number of fares for a Las Vegas taxi driver on a Friday. The random variable in this problem is discrete, because a taxi driver can only have a whole number of fares for a day. The probability that we are interested in here is $P(24 \leq x \leq 26)$. Since our interval is between these two values, we make the continuity correction outside of this interval, changing it to $P(23.5 \leq x \leq 26.5)$.

$$P(24 \leq x \leq 26) \stackrel{cc}{=} P(23.5 \leq x \leq 26.5)$$

$$= P\left(\frac{23.5 - 21.5}{6.3} \leq z \leq \frac{26.5 - 21.5}{6.3}\right)$$

$$= P(0.32 \leq z \leq 0.79)$$

$$P(0.32 \leq z \leq 0.79) = 0.7852 - 0.6255$$

$$= 0.1597$$

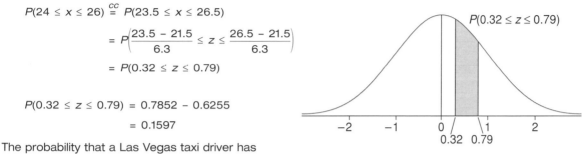

The probability that a Las Vegas taxi driver has between 24 and 26 fares is 0.1597. ∎

Normal Approximation to the Binomial Distribution

If we have a binomial distribution for which $n \cdot p \geq 5$ and $n(1 - p) \geq 5$, then we can use the normal distribution to approximate binomial probabilities. The reason is that, under these conditions, the binomial distribution is approximately normally distributed. The normal distribution has a mean $\mu = n \cdot p$ and a standard deviation $\sigma = \sqrt{n \cdot p(1 - p)}$. Recall that these formulas were provided in the section on the binomial distribution.

Consider a binomial distribution in which $n = 100$ and $p = 0.5$. A histogram showing the probability of having x successes is shown here.

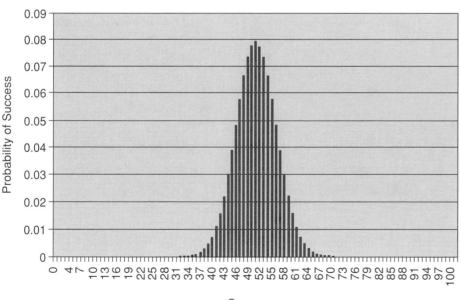

Binomial Distribution, $n = 100$, $p = 0.5$

According to the information above, this binomial distribution should be approximately normally distributed, and it appears to be. The mean of this normal distribution should be

$$\mu = n \cdot p$$
$$= 100(0.5)$$
$$= 50$$

The standard deviation of this normal distribution should be

$$\sigma = \sqrt{n \cdot p(1 - p)}$$
$$= \sqrt{100(0.5)(0.5)}$$
$$= \sqrt{25}$$
$$= 5$$

Here is the graph of a normal distribution with $\mu = 50$ and $\sigma = 5$.

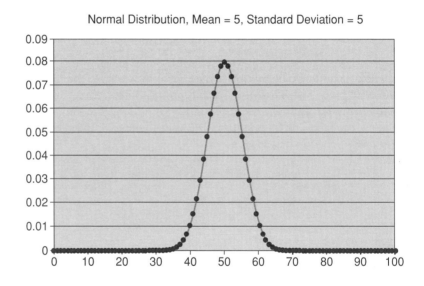

Here are the two distributions side by side.

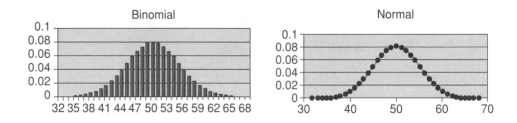

We see that the two distributions are essentially the same, except that the binomial distribution is discrete and the normal distribution is continuous.

Next is an example showing how to use the normal approximation to binomial probabilities. It will also show us how good an approximation it is.

| **EXAMPLE 5.21** | A fair coin is flipped 12 times. Find the probability that it lands on heads at least 5 times. |

This is a binomial problem, and we will do it first using the binomial tables. Then we will use the normal approximation and compare answers.

Number of trials: $n = 12$

Success: A toss is heads.

Number of successes: x = at least 5

Probability of success: $p = 0.5$

Failure: A toss is not heads.

Probability of failure: $1 - p = 0.5$

We can use the binomial tables here, since $n = 12$ and 0.5 is one of the probabilities given in the tables.

$$P(\text{At least 5}) = 1 - P(0 \text{ through } 4)$$

$$= 1 - (0.0002 + 0.0029 + 0.0161 + 0.0537 + 0.1208)$$

$$= 1 - 0.1937$$

$$= 0.8063$$

From the direct use of the binomial distribution, we see that the probability that at least 5 of the tosses are heads is 0.8063.

This problem satisfies the conditions required for the normal approximation to the binomial distribution as both $n \cdot p$ and $n(1 - p)$ are at least 5.

$$n \cdot p = 12(0.5) \qquad n(1 - p) = 12(0.5)$$

$$= 6 \qquad\qquad\qquad = 6$$

We may now approximate this binomial distribution with a normal distribution with the following mean and standard deviation.

$$\mu = n \cdot p \qquad \sigma = \sqrt{n \cdot p(1 - p)}$$

$$= 12(0.5) \qquad = \sqrt{12(0.5)(0.5)}$$

$$= 6 \qquad\qquad = \sqrt{3}$$

$$\approx 1.73$$

We are looking for $P(x \geq 5)$, but we must keep in mind that the binomial distribution is *discrete,* so a continuity correction is necessary.

$$P(x \geq 5) \overset{cc}{=} P(x \geq 4.5)$$

$$= P\left(z \geq \frac{4.5 - 6}{1.73}\right)$$

$$= P(z \geq -0.87)$$

Here is the graph.

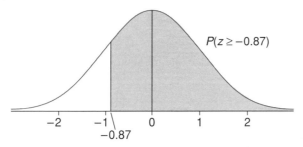

$$P(z \geq -0.87) = 1 - 0.1922$$

$$= 0.8078$$

The probability that the normal approximation gives us for at least 5 heads in 12 tosses is 0.8078. This is off, slightly, from our exact answer of 0.8063. This error decreases as the number of trials increases. ∎

 EXAMPLE 5.22 — A fair coin is flipped 1000 times. Find the probability that it lands on heads between 480 and 520 times, inclusive.

This is a binomial problem. First we should construct a table of information.

Number of trials: $n = 1000$

Success: A toss is heads.

Number of successes: $x = 480$ through 520

Probability of success: $p = 0.5$

Failure: A toss is not heads.

Probability of failure: $1 - p = 0.5$

We cannot use the binomial tables here, since $n = 1000$ is too large for our tables. Using the binomial formula would require us to plug into the formula 41 times. Let's check the conditions for the normal approximation.

$$n \cdot p = 1000(0.5) \qquad n(1 - p) = 1000(0.5)$$

$$= 500 \qquad\qquad\qquad = 500$$

$$\geq 5 \qquad\qquad\qquad\quad \geq 5$$

Since both conditions are satisfied, we continue by finding the mean and standard deviation for this distribution.

$$\mu = n \cdot p \qquad \sigma = \sqrt{n \cdot p(1 - p)}$$

$$= 1000(0.5) \qquad = \sqrt{1000(0.5)(0.5)}$$

$$= 500 \qquad\qquad = \sqrt{250}$$

$$\approx 15.81$$

Now, we begin with the probability $P(480 \leq x \leq 520)$.

$$P(480 \leq x \leq 520) \overset{cc}{=} P(479.5 \leq x \leq 520.5)$$

$$= P\left(\frac{479.5 - 500}{15.81} \leq z \leq \frac{520.5 - 500}{15.81}\right)$$

$$= P(-1.30 \leq z \leq 1.30)$$

Here's the graph.

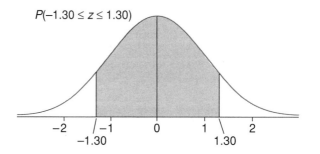

$P(-1.30 \leq z \leq 1.30)$

$P(-1.30 \leq z \leq 1.30) = 0.9032 - 0.0968$

$= 0.8064$

The approximation of the probability is 0.8064. The actual probability, which can be found using Microsoft Excel, is 0.8052. ∎

EXAMPLE 5.23   In a certain county, 45% of registered voters are Republican. If 300 registered voters from that county are selected at random, find the probability that at least 155 are Republicans.

This is a binomial problem, so we begin with a table of information.

   Number of trials: $n = 300$

   Success: A registered voter is Republican.

   Number of successes: $x =$ at least 155

   Probability of success: $p = 0.45$

   Failure: A registered voter is not Republican.

   Probability of failure: $1 - p = 0.55$

We cannot use the binomial tables for this problem, since the number of trials is too large. To use the binomial formula directly, we would need to plug into the formula 146 times (155 through 300). The complement rule is of no assistance, because it requires us to plug into the binomial formula 155 times (0 through 154). We check the conditions required to use the normal approximation.

$n \cdot p = 300(0.45) \qquad n(1 - p) = 300(0.55)$

$\qquad = 135 \qquad\qquad\qquad = 165$

$\qquad \geq 5 \qquad\qquad\qquad\quad \geq 5$

Since both conditions are satisfied, we continue by finding the mean and standard deviation for this distribution.

$\mu = n \cdot p \qquad\qquad \sigma = \sqrt{n \cdot p(1 - p)}$

$\quad = 300(0.45) \qquad\quad = \sqrt{300(0.45)(0.55)}$

$\quad = 135 \qquad\qquad\quad\; = \sqrt{74.25}$

$\qquad\qquad\qquad\qquad\quad\; \approx 8.62$

The probability that we are interested in is $P(x \geq 155)$.

$P(x \geq 155) \overset{cc}{=} P(x \geq 154.5)$

$\qquad = P\!\left(z \geq \dfrac{154.5 - 135}{8.62}\right)$

$\qquad = P(z \leq 2.26)$

Here is the graph.

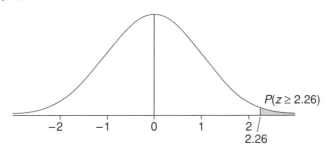

$$P(z \geq 2.26) = 1 - 0.9881$$

$$= 0.0119$$

The normal approximation to this probability is 0.0119. A direct calculation using the binomial formula shows that the exact probability is 0.0119. Our approximation is quite accurate in this situation. ∎

In the previous example, we see that it is rather unlikely that we draw a sample that has at least 155 Republicans if 45% of the registered voters are Republican. This could lead us to question whether the sample was truly random. Was there a bias in the way that people were selected to make Republicans more likely to be selected? We do not have enough information at this point to make a conclusion about this. Perhaps the 45% figure is not correct, and should actually be higher. In later chapters we will be able to make inferences about a claimed population proportion such as that 45% of registered voters in this county are Republicans.

EXERCISES 5.2

Scores for high school sophomores on the mathematics portion of the Preliminary SAT/National Merit Scholarship Qualifying Test (PSAT) are approximately normally distributed with a mean of 46.1 and a standard deviation of 10.5. Scores on the PSAT are whole numbers, with a low of 20 and a high of 80.

1. Find the probability that a sophomore's score on the mathematics portion of the PSAT is 60 or higher.
2. Find the probability that a sophomore's score on the mathematics portion of the PSAT is 42 or lower.
3. Find the probability that a sophomore's score on the mathematics portion of the PSAT is between 40 and 49.
4. Find the probability that a sophomore's score on the mathematics portion of the PSAT is between 65 and 75.
5. What percentage of sophomores earn a score of 52 or lower? The College Board reports that this percentage is 70%. How does your answer compare to the College Board's claim?
6. What percentage of sophomores earn a score of 68 or higher? The College Board reports that this percentage is 5%. How does your answer compare to the College Board's claim?

IQ scores are approximately normally distributed with a mean of 100 points, and a standard deviation of 15 points. IQ scores are always in whole numbers. Find the probability that a person has an IQ

7. 106 or lower
8. Between 85 and 95
9. At least 120
10. No higher than 80
11. Higher than 98
12. Lower than 92

A statistics final exam produces scores that are approximately normally distributed with a mean of 225 points and a standard deviation of 20 points. There are 300 points possible on the exam, and scores are always whole numbers. Find the probability that a student scores

13. 270 points or higher on this exam
14. 180 points or lower on this exam
15. Between 210 points and 255 points on this exam, inclusive
16. At least 200 points on this exam
17. No higher than 240 points on this exam

18. Between 180 points and 230 points on this exam, inclusive

SAT math scores are normally distributed with a mean of 512 points and a standard deviation of 112 points. SAT scores range from 200 to 800, and are always a multiple of 10. Because possible scores are 10 points apart, a continuity correction of 5 points is necessary instead of 0.5 points.

19. Find the probability that a high school student scores 400 points or lower on the math portion of the SAT.
20. What percentage of high school students score 640 points or lower on the math portion of the SAT?
21. Find the probability that a high school student scores 700 points or higher on the math portion of the SAT.
22. What percentage of high school students score 470 points or higher on the math portion of the SAT?
23. Find the probability that a high school student scores between 500 points and 600 points, inclusive, on the math portion of the SAT.
24. What percentage of high school students score between 350 points and 550 points, inclusive, on the math portion of the SAT?
25. Explain, in your own words, the conditions that require you to make a continuity correction.
26. Explain, in your own words, what the purpose of a continuity correction is.

A fair coin is tossed 500 times. Find the probability that

27. At least 270 tosses are heads
28. At most 240 tosses are heads
29. 225 or fewer tosses are heads
30. At least 285 tosses are heads
31. Between 230 tosses and 245 tosses, inclusive, are heads
32. Between 236 tosses and 264 tosses, inclusive, are heads
33. At least 255 tosses are heads
34. Exactly 255 tosses are heads
35. Between 49% and 51% of the tosses, inclusive, are heads
36. At least 55% of the tosses are heads

At a certain community college, 60% of the students are female. Three hundred students are randomly selected to be included in a survey. Find the probability that

37. At least 195 of the students are female
38. Between 150 and 190 students, inclusive, are female
39. 155 or fewer students are female
40. At least 70% of the students are female
41. At least 130 of the students are *male*
42. 150 or fewer students are *male*

43. On the night that Wilt Chamberlain scored 100 points in an NBA basketball game, he made 28 of his 32 free-throw attempts. Wilt was a notoriously poor free-throw shooter, making only approximately 51% of his attempts over his career. Assuming that attempting a free throw is an independent event, use the normal approximation to find the probability that a person who has a 51% chance of making any particular free throw makes at least 28 out of 32 attempts. Does your result suggest that Wilt was extremely fortunate to score 100 points in that game?
44. An election poll is printed in the newspaper, showing that the Republican candidate is well in front of his rival from the Democratic Party. Approximately 46% of the voters in the state that was surveyed are registered as Republicans. However, of the 1200 people polled in the survey, 602 were registered as Republicans. If 46% of the people in the state are registered as Republicans, find the probability that at least 602 of 1200 randomly selected voters from the state are registered as Republicans. What does your result suggest about the validity of the poll?
45. A mathematics competency exam is given at a community college as a requirement for graduation. The test is made up of 55 multiple-choice questions, each

with 5 possible answers, of which only 1 is correct. If a student guesses at all 55 questions, find the probability that the student gets 6 or fewer correct.

46. It is claimed that approximately 25% of Americans have a cellular phone. If 440 Americans are surveyed at random, find the probability that 90 or fewer have a cellular phone. (Based on results of a study by Baskerville Communications.)

47. While being interviewed about the status of education, a politician states that at least 20% of all new teachers leave the profession within the first three years. A random sample of 216 teachers hired three years ago revealed that 33 of them were no longer teaching. If 20% of all new teachers actually leave the profession within the first three years, find the probability that 33 or fewer of the 216 teachers were no longer teaching.

48. According to the U.S. Department of Commerce, 30% of Arkansas households have a computer. If 285 Arkansas households are selected at random, find the probability that 70 or fewer have a computer.

EXCEL EXPLORATION   Investigating the Normal Approximation

We will now examine the connection between the binomial distribution and the normal distribution, by using Excel to simulate tossing a coin 20 times and counting the number of heads. We will repeat this experiment 1000 times, and construct a histogram for the number of heads recorded.

We will use the Data Analysis Toolpak to simulate the tossing of the coins. Select the **Tools** menu. If you do not see **Data Analysis** as one of the choices, then you must add it. Select **Add-Ins,** and when the dialog box opens click the box next to **Analysis ToolPak** and click **OK.**

From the **Tools** menu, select **Data Analysis.** When the dialog box appears, select **Random Number Generation** from the list and click on **OK.** In the box labeled **Number of Variables** type 20. This will represent the 20 tosses of the coin. In the box labeled **Number of Random Numbers** type 1000. This will represent the 1000 times we repeat this experiment. In the box labeled **Distribution** select Bernoulli. This will generate a series of 1s (successes) and 0s (failures). Under **Parameters** enter 0.5 in the box labeled **p Value.** We enter 0.5 because the probability that any individual toss is heads is 0.5. Type any number in the box labeled **Random Seed;** this is a number that is used to start the random number generation. Click **OK.** The random numbers will be generated on another worksheet. (Be patient; this may take some time.)

Once the random generation is complete, type the following in cell **U1:**

    =SUM(A1:T1)

This will total the number of heads the first time we tossed the coin 20 times. Copy cell **U1** and paste it into cells **U2** through **U1000.** Now we have the number of heads for each of the 1000 times this experiment was repeated.

Now we will create a histogram for the number of heads. We will examine the shape of this histogram to determine whether the number of heads is approximately normally distributed. Before we do, we need to type a list of class boundaries for the histogram. Type the numbers from 0 through 20 in cells **V1** through **V21.** From the **Tools** menu, select **Data Analysis.** Select **Histogram** from the list of choices, and click **OK.** In the box labeled **Input Range** type U1:U1000. This is the range of cells containing the number of heads for each set of 20 tosses. In the box labeled **Bin Range,** type V1:V21. Again, this is the range of cells that we created to store the class boundaries. Be sure to select the box labeled **Chart Output,** and click **OK.** Excel will create a histogram on a new worksheet for you. Here is a sample of what your histogram might look like.

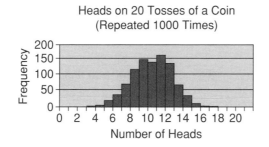

Note that it appears to be roughly symmetrical, and that it is almost bell-shaped.

When the conditions to use the normal approximation to the binomial distribution are met, the mean is given by $\mu = n \cdot p$ and the standard deviation is given by $\sigma = \sqrt{n \cdot p(1 - p)}$. For 20 tosses of a coin, $n = 20$ and $p = 0.5$. The mean and standard deviation are as follows.

$$\mu = n \cdot p \qquad \sigma = \sqrt{n \cdot p(1 - p)}$$
$$= 20(0.5) \qquad = \sqrt{20(0.5)(0.5)}$$
$$= 10 \qquad\qquad = \sqrt{5}$$
$$\approx 2.24$$

We will now examine the mean and standard deviation of our 1000 repetitions of this experiment. Go back to the worksheet that contains the randomly generated numbers. In cell **W1**, type =AVERAGE(U1:U1000). Below that, in cell **W2**, type =STDEV(U1:U1000). The data that generated the previous histogram had a mean of 9.995 and a standard deviation of 2.322. Both of these values are close to what they are supposed to be. Compare your mean and standard deviation to the values that we should get.

Now repeat this example for other values of n and p.

1. $n = 30$ and $p = 0.4$ 2. $n = 50$ and $p = 0.3$
3. $n = 50$ and $p = 0.2$ 4. $n = 100$ and $p = 0.1$

For each case, examine the histogram and comment on whether it appears to be symmetrical and bell-shaped. Also, calculate the mean and standard deviation for your simulation, and calculate the mean and standard deviation for this distribution according to the normal approximation to the binomial distribution. Compare these theoretical values to the values that you calculated from your random number generation.

SECTION 5.3
Finding Values for Given Probabilities

In this section we will examine problems that are essentially the opposite of the problems discussed earlier in this chapter. We will be given a certain probability, and we will try to find the value of the random variable associated with it. An example of such a problem is to find the height that separates the tallest 10% of all adult females from the rest. We would look for a height x^* such that $P(x > x^*)$ is approximately equal to 0.1 (10%).

We will "warm up" with problems involving the standard normal distribution, and then later we will look at problems involving nonstandard normal distributions.

 EXAMPLE 5.24   Find the z-value for which the area under the standard normal curve to the left of it is 0.01. (For this z-value, 1% of the observations fall below it.)

Now we begin with a graph.

We are looking for the z-value that has an area of 0.01 to the left of it. We look on the inside of the chart for an area of 0.0100. The closest we can find is 0.0099, which is associated with a z-value of –2.33. Therefore, the area to the left of the z-value –2.33 is approximately 0.01; that is, 1% of all z-values fall below –2.33. ■

EXAMPLE 5.25   Find the z-value for which the area under the standard normal curve to the right of it is 0.005. (For this z-value, 0.5% of the observations fall above it.)

We begin once again with the graph. We know that the area to the right of our z-value is 0.005. In order to use our tables, we need to know the area to the left of our z-value, which is 0.995 (1 – 0.005).

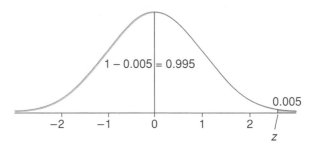

When we look for 0.9950 on the inside of the normal distribution table, we notice that it is exactly between 0.9949 and 0.9951. These two areas correspond to the z-values 2.57 and 2.58, respectively. In such a case where the desired area is exactly between two areas, we will use the mean of the two z-values.

$$z = \frac{2.57 + 258}{2}$$

$$= 2.575$$

The z-value that has an area of 0.005 to the right of it is 2.575; that is, 0.5% of z-values are higher than 2.575. ■

 EXAMPLE 5.26 Find the z-values that separate the middle 95% of the area under the standard normal curve from the rest of the curve. (For these z-values, 95% of all z-values lie between them.)

We are looking for two z-values in this problem. We will label them z_1 and z_2. Let's begin with a graph to get a sense of the big picture.

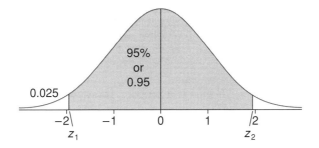

The first value, z_1, is the easiest to find, because it is the z-value with an area of 0.025 to the left of it. We look up 0.0250 in our normal table, and find that the z-value associated with it is –1.96. What is the area to the left of z_2? It is 0.975, which can be found by adding 0.95 to 0.025. We look at our normal table for 0.9750, and we find that the z-value associated with it is 1.96. This should come as no surprise, because the normal distribution is symmetric. Therefore, the z-values that separate the middle 95% of the area under the standard normal distribution from the rest are –1.96 and 1.96; that is, 95% of all z-values lie between –1.96 and 1.96. ■

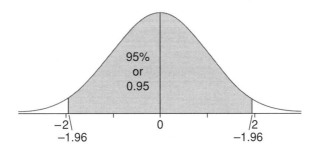

Now that we have discussed this idea for the standard normal distribution, we can extend it to nonstandard normal distributions. Reconsider the problem mentioned at the beginning of the section: Find the height that separates the tallest 10% of all adult females from the rest. We will begin by finding the z-value for which the area under the curve to the right of that value is 0.1. Then we will convert that z-value into the height of an adult female.

EXAMPLE 5.27   The heights of adult females are normally distributed with a mean of 63.6 inches and a standard deviation of 2.5 inches. Find the height that separates the tallest 10% of adult females from the rest.

We begin by finding the z-value for which the area to the right of it is 0.1 (10%). Here is the graph.

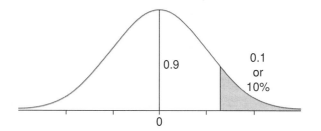

The area to the left of this z-value is 0.9. When we look at our normal distribution table for 0.9000, the closest we can find is 0.8997, which is associated with the z-value 1.28. The next step is to convert this z-value to the height of an adult female. This is done using the formula for z-scores. Since we know values for z, μ, and σ, we can solve for x, which in this problem represents the height of an adult female.

$$z = \frac{x - \mu}{\sigma}$$

$$1.28 = \frac{x - 63.6}{2.5}$$

$$1.28 \cdot 2.5 = x - 63.6$$

$$3.2 + 63.6 = x$$

$$66.8 = x$$

The height that separates the tallest 10% of adult females from the rest is 66.8 inches. ■

EXAMPLE 5.28   IQ tests are designed to produce a mean of 100 points, with a standard deviation of 15 points. Find the IQ score that separates the lowest 5% of IQ scores from the rest.

Again, we begin with the graph.

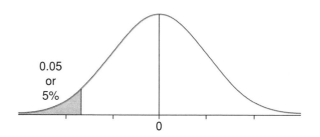

Since we have the area to the left of the z-value, we can go directly to the normal distribution table and look for 0.0500. It falls exactly between 0.4495 ($z = -1.64$) and 0.4505 ($z = -1.65$). We use the mean of these two values.

$$z = \frac{(-1.64) + (-1.65)}{2}$$

$$= -1.645$$

Now we transform that z-value to an IQ score.

$$z = \frac{x - \mu}{\sigma}$$

$$-1.645 = \frac{x - 100}{15}$$

$$-1.645 \cdot 15 = x - 100$$

$$-24.675 + 100 = x$$

$$75.325 = x$$

Therefore, 5% of all people have an IQ below 75.325. Since IQ scores are discrete, we could go on to say that 5% of all people have an IQ that is 75 or lower. ∎

Scores on the math portion of the SAT are approximately normally distributed with a mean of 512 points and a standard deviation of 112 points. Between what two scores are the middle 75% of all SAT math scores?

We will start with the graph and look for the two z-values, z_1 and z_2, that separate the middle 75% of z-values from the rest. The first z-value, z_1, has an area of 0.1250 to the left of it. Looking at our table, the closest z-value is -1.15. The second z-value, z_2, has an area of 0.8750 (0.1250 + 0.7500) to the left of it. Looking at our table, we see that this z-value is 1.15. We now must convert these two z-values into SAT scores.

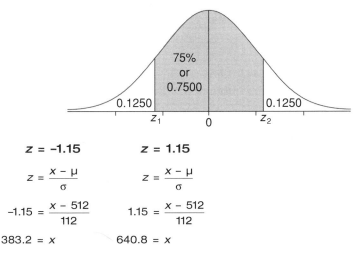

$z = -1.15$	$z = 1.15$
$z = \dfrac{x - \mu}{\sigma}$	$z = \dfrac{x - \mu}{\sigma}$
$-1.15 = \dfrac{x - 512}{112}$	$1.15 = \dfrac{x - 512}{112}$
$383.2 = x$	$640.8 = x$

Therefore, 75% of all SAT math scores are between 383.2 and 640.8. Since SAT scores are discrete (multiples of 10), we could go on to say that the middle 75% of all SAT math scores are between 390 and 640. ∎

Finding Values for Given
Probabilities

Microsoft Excel provides us with a built-in function to find a z-value that is associated
with a given probability. It is

=NORMSINV(probability)

Inside the parentheses we type a probability, and Excel returns the z-value for which
the area to the left of that value is the same as the probability we typed. For example,

=NORMSINV(0.9750)

returns 1.96, because the area to the left of $z = 1.96$ is equal to 0.9750.

EXAMPLE 5.30	Find the z-value for which the area under the standard normal curve to the left of it is 0.01.

We begin with a graph.

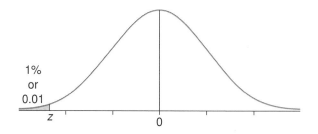

We are looking for the z-value that has an area of 0.01 to the left of it. Type the following in any cell.

=NORMSINV(0.01)

Excel gives us –2.32634, which rounds to –2.33 to two decimal places. ■

Microsoft Excel also provides us with a built-in function to use when the distribution is a nonstandard normal distribution. Here is the general formula.

=NORMINV(probability, mean, standard deviation)

Inside the parentheses we must provide three values. The first is the probability that
we are interested in, or the area to the left of the desired value. The last two values are
the mean and standard deviation of the distribution in question.

EXAMPLE 5.31	The heights of adult females are normally distributed with a mean of 63.6 inches and a standard deviation of 2.5 inches. Find the height that separates the tallest 10% of adult females from the rest.

We begin by drawing the graph, so that we know the area to the left of our desired value.

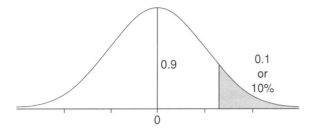

In this case the probability is 0.9. We can then type the following in any cell.

=NORMINV(0.9,63.6,2.5)

The value that Excel gives us is 66.8. Therefore, the tallest 10% of all adult females are taller than 66.8 inches. Note that this value agrees with the value found earlier in this section. ∎

 Finding Values for Given Probabilities

The TI-83 uses the built-in function **invNorm** to find the value associated with a given probability or area underneath the curve. The basic format for using this function is as follows.

invnorm(*area*, μ, σ)

It calculates the z-value associated with the *area* to the left of the value. If we are using the standard normal distribution (μ = 0, σ = 1), then we may omit these values and enter only the lower bound and upper bound. This function can be found in the **DISTR** menu.

EXAMPLE 5.32   Find the z-value for which the area under the standard normal curve to the left of it is 0.01.

We begin with a graph.

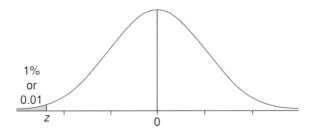

We are looking for the z-value that has an area of 0.01 to the left of it. Press (2ND) (VARS) to access the **DISTR** menu, and select item 3: **invNorm.** On the main screen, press 0.01 ( ) (ENTER).

The TI-83 gives us –2.326347877, which rounds to –2.33 to two decimal places. ∎

Here is an example involving a nonstandard normal distribution.

  The heights of adult females are normally distributed with a mean of 63.6 inches and a standard deviation of 2.5 inches. Find the height that separates the tallest 10% of adult females from the rest.

We begin by drawing the graph, so that we know the area to the left of our desired value.

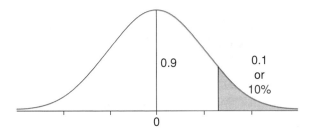

In this case the probability is 0.9. Press (2ND) (VARS) to access the **DISTR** menu, and select item 3: **invNorm.** On the main screen, press 0.9 (,) 63.6 (,) 2.5 ( )) (ENTER).

invNorm(0.9,63.6
,2.5)
          66.80387892

The value that the TI-83 gives us is 66.80387892. Therefore, the tallest 10% of all adult females are taller than 66.8 inches. Note that this value agrees with the value found earlier in this section. ∎

EXERCISES 5.3

Find the *z*-value that separates

1. The lower 1% of all *z*-values from the rest
2. The lower 5% of all *z*-values from the rest

3. The lower 10% of all z-values from the rest
4. The lower 31% of all z-values from the rest
5. The upper 3% of all z-values from the rest
6. The upper 2.5% of all z-values from the rest
7. The upper 15% of all z-values from the rest
8. The upper 20% of all z-values from the rest
9. The upper 60% of all z-values from the rest
10. The upper 72% of all z-values from the rest
11. The middle 80% of all z-values from the rest
12. The middle 77% of all z-values from the rest
13. The middle 93% of all z-values from the rest
14. The middle 50% of all z-values from the rest
15. Find the z-value that is associated with the first quartile of a normal distribution.
16. Find the z-value that is associated with the third quartile of a normal distribution.

The heights of adult males are normally distributed with a mean of 69.0 inches and a standard deviation of 2.8 inches.

17. What height separates the shortest 8% of adult males from the rest?
18. What height separates the shortest 25% of adult males from the rest?
19. Find the height that separates the tallest 15% of adult males from the rest.
20. Twenty percent of adult males are above what height?
21. What heights separate the middle 70% of adult males from the rest?
22. What heights separate the middle 50% of all adult males from the rest?
23. A builder wants to make the height of his doorways so that only 3% of adult males will have to duck on entering the room. How high should the doorways be?
24. A person of below-average height wants to form a social club for the "vertically challenged." He decides that adult males will have to be in the shortest 1% of all adult males to join. What is the maximum height that an adult male can have and still join this club?

The heights of 12-month-old boys are approximately normally distributed with a mean of 29.8 inches and a standard deviation of 1.2 inches.

25. What height separates the tallest 33% of all 12-month-old boys from the rest?
26. Sixty percent of 12-month-old boys are above what height?
27. Twenty-five percent of 12-month-old boys are shorter than what height?
28. What height separates the shortest 50% of all 12-month-old boys from the rest?

Scores for high school sophomores on the mathematics portion of the Preliminary SAT/National Merit Scholarship Qualifying Test (PSAT) are approximately normally distributed with a mean of 46.1 and a standard deviation of 10.5. Scores on the PSAT are whole numbers, with a low of 20 and a high of 80.

29. A university's engineering department decides to mail out brochures to students whose test scores were in the top 8%. What is the lowest score that will be sent a letter?
30. A high school decides to honor its students who scored in the top 20% of all students nationally. What is the lowest score that will be honored by the high school?
31. A high school decides to set up a program for students who struggled on the PSAT math exam, so that they may improve their performance. The school decides to invite students who scored in the lower 25% nationally. What is the highest score that will be invited into this program?
32. The College Board reports that 1% of all sophomores score below 25 points on this exam. Use the above mean and standard deviation to determine what score separates the lowest 1% of all scores from the rest. Compare your result to the College Board's claim. If there are differences, explain why they may occur.

IQ scores are approximately normally distributed with a mean of 100 points and a standard deviation of 15 points. IQ scores are always in whole numbers.

33. Forty percent of all people have an IQ above what score?
34. Thirty percent of all people have an IQ above what score?
35. Mensa, a group founded in England in 1946 by Roland Berrill and Dr. Lance Ware for people of high intelligence, requires that its members be in the top 2% on any intelligence test. What IQ is the lowest possible for admission to Mensa?

36. A new group is being founded for people who are of above-average intelligence, but fail to qualify for Mensa. The group has decided to accept people who have an IQ that is in the top 10%, but not in the top 2%. What are the lowest and highest IQ a person could have and belong to this group?

A statistics final exam produces scores that are approximately normally distributed with a mean of 225 points and a standard deviation of 20 points. There are 300 points possible on the exam, and scores are always whole numbers.

37. Seventy-five percent of all statistics students score above what score on this exam?
38. What scores separate the middle 40% of all scores from the rest?
39. What score is required to put a student in the top 1% of all scores?
40. Five percent of all students score below what score?

SAT math scores are normally distributed with a mean of 512 points and a standard deviation of 112 points. SAT scores range from 200 to 800, and are always a multiple of 10.

41. What score separates the top 3% of all scores from the rest?
42. Eighty percent of all students score below what score?
43. What scores separate the middle 60% of all scores from the rest?
44. An elite college refuses to accept students whose SAT math scores are not in the top 5%. What is the lowest score that a student could get and still be admitted to the school?

EXTRA • Is a Set of Data Approximately Normally Distributed?

If we have a set of data, how can we determine whether it is normally distributed? This task may be beyond the level of this text, but there are some techniques that we can use to get a good idea about a set of data and whether it is approximately a normal distribution.

The first aspect of the data set that we will examine is its mean and median. If the mean and median are close to each other, then the distribution is roughly symmetric. Second, we will take a look at a histogram for the set of data, to see whether it is bell-shaped like the normal distribution. Finally, we will compare the interquartile range and the standard deviation to each other. If a set of data is normally distributed, then the first quartile is associated with a z-value of -0.67 and the third quartile is associated with a z-value of 0.67. Thus, the interquartile range should be approximately equal to 1.34 standard deviations. If we multiply 1.34 by the standard deviation and find this product to be close to the interquartile range, then this is a good sign that the data follow a normal distribution.

The set of data that we will be examining follows. The values are the number of heads in 20 tosses of a fair coin. This experiment was carried out 100 times.

5	10	5	10	10	10	9	7	11	14
14	8	6	8	12	10	5	12	12	9
11	12	11	14	12	11	12	11	7	11
8	9	11	9	9	12	8	11	9	11
14	11	9	10	13	12	10	9	9	11
13	9	11	7	13	13	10	9	10	9
8	12	8	7	9	9	9	13	9	9
9	7	10	13	7	13	9	10	11	9
7	10	16	8	6	9	12	11	9	8
10	10	10	9	10	9	10	9	11	12

The first thing to examine is a histogram of the data. Is the histogram symmetric? Is it bell-shaped? Here is a histogram, generated by Microsoft Excel, for this set of data.

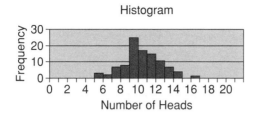

The set of data seems to be close to symmetric and bell-shaped, and may come from a population that is normally distributed.

We move on to compare the mean and median of the set of data. If the mean and median are vastly different, this is a sign that the distribution is not symmetric. The mean for the above set of data is 9.95, and the median is 10. The mean and median are somewhat close, although this suggests that the data may be negatively skewed rather than symmetric.

Finally, we compare the interquartile range and the standard deviation. The first quartile is 9, and the third quartile is 11. This means that the interquartile range is 2. The standard deviation is 2.13. If we multiply the standard deviation by 1.34, we get 2.8542, which is not very close to 2. This makes it difficult to conclude whether or not these values are approximately distributed. (Another procedure for testing the normality of a set of data is introduced in Section 9.3.)

Determine whether the following sets of data are approximately normal.

1. SAT math scores of 60 students

280	610	660	530	550	480	640	500	640	580
370	640	520	540	390	510	410	470	320	560
530	230	500	490	390	600	680	540	510	530
550	750	440	660	570	710	600	510	590	570
500	790	500	320	580	480	650	610	700	430
520	750	690	480	510	750	510	530	400	720

2. Number of pull-ups performed by 40 fourth-graders.

0	8	9	6	6	4	9	5	9	7
1	9	5	6	1	5	2	4	0	7
6	0	5	4	1	8	9	6	5	6
6	10	3	9	7	10	8	5	8	7

3. Number of rejects produced per day on an assembly line, for a 56-day period

4	3	2	0	0	1	1	3
2	4	2	4	3	2	2	2
1	3	2	0	3	3	1	1
1	0	1	0	0	1	1	0
3	0	1	0	1	3	1	1
2	0	0	1	1	0	2	2
1	1	0	0	3	1	3	2

4. Number of tornadoes that struck Utah for each year from 1950 through 1995

Year	Number	Year	Number	Year	Number	Year	Number	Year	Number
1950	0	1960	0	1970	5	1980	0	1990	4
1951	0	1961	1	1971	1	1981	2	1991	5
1952	0	1962	1	1972	0	1982	3	1992	4
1953	2	1963	1	1973	0	1983	0	1993	6
1954	1	1964	1	1974	0	1984	6	1994	0
1955	2	1965	5	1975	0	1985	0	1995	2
1956	0	1966	2	1976	0	1986	3		
1957	1	1967	2	1977	0	1987	3		
1958	0	1968	4	1978	0	1988	1		
1959	0	1969	3	1979	0	1989	6		

(Utah Disaster Center Web site)

Overview

Unit 3 began with discrete probability distributions. You should be familiar with how to calculate the mean and standard deviation of a probability distribution.

From there we shifted focus to the **binomial distribution**. Look for the idea of "*x* out of *n*." Also, we must have a probability of success. When trying to find a binomial probability, first try to use the binomial tables. If the tables are not applicable, we must use the formula except in the case where we may apply the normal approximation to the binomial distribution.

When reading a **Poisson distribution** problem, look for the expected number of successes. Poisson problems look similar to binomial problems, and that is no accident. Poisson problems lack the number of trials that are present in binomial probability problems. When calculating Poisson probabilities, try to use the tables first. If the tables are not applicable to your problem, then you must use the formula.

We next covered the **normal distribution**. The information required for normal probability problems is a mean and a standard deviation. Watch out for normal variables that are discrete, as they require a **continuity correction**. You should know when it is appropriate to use the normal approximation to the binomial distribution, and remember the continuity correction here as well.

When finding a value that goes with a given probability ("reverse normal"), you will always be asked for a value rather than a probability.

Review Exercises

1. During a typical week, an assembly line must be shut down 3 times due to accidents. Find the probability that the assembly line gets shut down at least once this week due to an accident.
2. Driving speeds on the 198 are normally distributed with a mean of 71 mph and a standard deviation of 5 mph. Find the probability that a car on the 198 is traveling faster than 75 mph.
3. IQ scores are approximately normally distributed with a mean of 100 and a standard deviation of 15. Find the probability that a randomly selected person has an IQ of at least 90.
4. Fifty-five percent of all Visalia households have cable TV. If 10 Visalia households are selected at random, find the probability that 5 of them have cable TV.
5. Fifty-five percent of all Visalia households have cable TV. If 100 Visalia households are selected at random, find the probability that at least 50 of them have cable TV.
6. The height of 12-month-old boys is normally distributed with a mean of 29.8 inches and a standard deviation of 1.2 inches. Ten percent of 12-month-old boys are above what height?
7. During the typical hour at Lake Lotsafish, I can expect to catch 2.5 fish. Find the probability that I catch 4 fish during my first hour fishing at the lake.
8. A quiz has 10 true/false questions. A student must answer at least 7 correctly in order to pass. If a student guesses on all 10 questions, find the probability that the student passes.
9. At the typical Academy Awards ceremony, 3 people cry as they accept their award. Find the probability that fewer than 3 people cry at this year's awards.
10. Fifty-eight percent of all Tulare County residents do not have health insurance. If 10 residents of Tulare County are selected at random, find the probability that at least 3 do not have health insurance.
11. The heights of adult women are normally distributed with a mean of 63.6 inches and a standard deviation of 2.5 inches. Find the probability that an adult woman is 5 feet tall or taller.

12. A statistics final exam is graded out of 300 possible points and each score must be a whole number. The scores are approximately normally distributed with a mean of 225 points and a standard deviation of 35 points. Find the probability that a student scores 209 or lower.

13. The heights of adult males are normally distributed with a mean of 69 inches and a standard deviation of 2.8 inches. What height separates the shortest 20% of adult males from the rest?

14. A restaurant has five rooms for private parties. On a typical Friday night, three of the private rooms are used. Find the probability that there will not be enough private rooms on a Friday night.

15. A 10-question multiple-choice exam is given in a math class. There are 5 possible answers to each question, of which only 1 is correct. Thus, the probability that a student gets an answer correct by guessing is 0.2. If a student guesses on all 10 questions, find the probability that he gets 6 or less correct.

16. A 100-question multiple-choice exam is given in a math class. There are 5 possible answers to each question, of which only 1 is correct. Therefore, the probability that a student gets an answer correct by guessing is 0.2. If a student guesses on all 100 questions, find the probability that he gets 25 or less correct.

17. On a typical Friday night, Joe D. Bouncer has to throw 2.5 people out of his club. Find the probability that he has to throw 5 people out this Friday night.

18. The lengths of human pregnancies are normally distributed with a mean of 268 days and a standard deviation of 15 days. Find the probability that a human pregnancy lasts between 255 and 275 days.

19. IQ scores are approximately normally distributed with a mean of 100 points and a standard deviation of 15 points. Find the probability that a person's IQ is between 110 and 125.

20. The probability of a woman giving birth to twins is approximately $\frac{1}{40}$ or 0.025. If 120 pregnant women are selected at random, find the probability that at least 5 of them have twins.

Formulas

Mean of a probability distribution: $\mu = \Sigma x \cdot P(x)$

Standard deviation of a probability distribution: $\sigma = \sqrt{\Sigma((x - \mu)^2 \cdot P(x))}$

Binomial probability: $P(x) = {}_nC_x \cdot p^x(1 - p)^{n-x}$

Poisson probability: $P(x) = \dfrac{e^{-\lambda} \cdot \lambda^x}{x!}$

Standard units: $z = \dfrac{x - \mu}{\sigma}$

Conditions for the normal approximation to the binomial distribution: $n \cdot p > 5$ and $n(1 - p) > 5$

Mean of binomial distribution: $\mu = n \cdot p$

Standard deviation of binomial distribution: $\sigma = \sqrt{n \cdot p(1 - p)}$

unit four

4

One-Sample Confidence Intervals and Hypothesis Tests

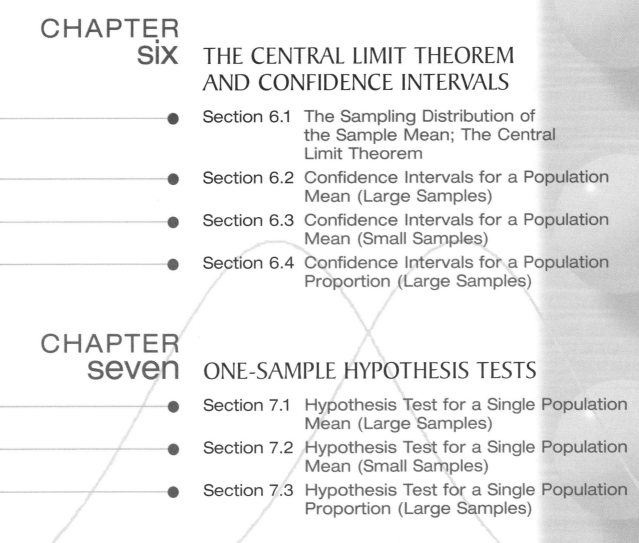

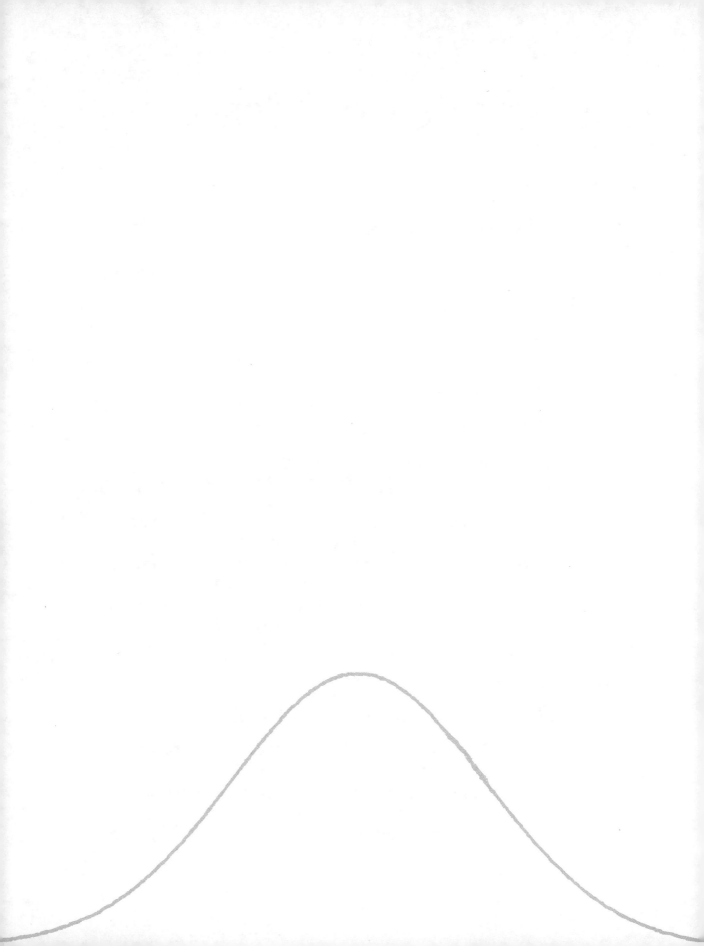

The Central Limit Theorem and Confidence Intervals

Recall that it is the goal of inferential statistics to make inferences about a population parameter, such as a mean or proportion, by taking a random sample and calculating the appropriate sample statistic. We can use this statistic to make an inference about the population parameter. For instance, suppose that we wanted to estimate the proportion of all women who smoke cigarettes while pregnant. It would be impossible to contact every pregnant woman to find out whether she smokes. A random sample of 400 pregnant women revealed that 60 (or 15%) are smoking during their pregnancy. Does this mean that *exactly* 15% of *all* pregnant women smoke during their pregnancy? Although it is possible, it is highly unlikely that exactly 15% of all women smoke while pregnant. How far off from the actual population proportion is our sample proportion? We do not know. The difference between the sample proportion and the population proportion is called **sampling error.**

In this chapter we will develop estimates for the sampling error, and use it to construct interval estimates for the population parameter. For example, we will conclude that "we are 95% sure that the percentage of all women who smoke is somewhere between 11.50% and 18.50%." In other words, we are 95% sure that our sample is off by no more than 3.50% from the population proportion.

Section 6.1 examines the sampling distribution of the sample mean and the Central Limit Theorem. This section provides us with the tools to construct interval estimates for population parameters.

In Section 6.2 we will construct an interval for a population mean based on a sample of size 30 or greater. We will also determine how large a sample we should take. In Section 6.3 we will construct an interval for a population mean when our sample is smaller than 30. Finally, in Section 6.4 we will construct an interval for a population proportion, like the one above for the proportion of all pregnant women who smoke.

SECTION 6.1
The Sampling Distribution of the Sample Mean; The Central Limit Theorem

Five students take an IQ test, and here are their scores:

Alan	Bob	Cari	Darla	Erin
102	84	108	112	98

We will treat these scores as a population. The mean score μ is 100.8, and the standard deviation σ is 9.68297475. Keep these measures in mind for later. We could take samples from this population in an attempt to estimate the mean of the population. Suppose we draw Alan's score and Bob's score at random. The mean of these two scores is 93. The sampling error is –7.8, since our sample mean is 7.8 below the actual population mean. Following is a table listing all of the different possible samples, drawn with replacement, of size 2.

Sample	Mean
Alan, Alan	102
Alan, Bob	93
Alan, Cari	105
Alan, Darla	107
Alan, Erin	100
Bob, Alan	93
Bob, Bob	84
Bob, Cari	96
Bob, Darla	98
Bob, Erin	91
Cari, Alan	105
Cari, Bob	96
Cari, Cari	108
Cari, Darla	110
Cari, Erin	103
Darla, Alan	107
Darla, Bob	98
Darla, Cari	110
Darla, Darla	112
Darla, Erin	105
Erin, Alan	100
Erin, Bob	91
Erin, Cari	103
Erin, Darla	105
Erin, Erin	98

The mean of these sample means is 100.8, and the standard deviation is 6.846897107. What we have listed is the sampling distribution of the sample mean. When all samples of a certain size are drawn from a population, and the sample means are calculated, the distribution that we get is the **sampling distribution of the sample mean.** Note that the mean of this distribution is exactly the same as the mean of the population. Also, the standard deviation is lower than the standard deviation of the population.

Let's examine the shape of this distribution. Here is a table of each sample mean, along with its frequency in the above distribution.

Mean	Frequency
84	1
91	2
93	2
96	2
98	3
100	2
102	1
103	2
105	4
107	2
108	1
110	2
112	1

Histogram of Sample Means

Note that the distribution does not look like a normal distribution. In general, as the sample size increases, the sampling distribution of the sample means does become approximately normal. This idea is summed up in the next theorem—the Central Limit Theorem.

As the sample size increases, the distribution of the sample means gets closer to a normal distribution. When the sample size is at least 30, a normal distribution approximates the sampling distribution of the sample means very well. In the Excel Exploration at the end of this section, you can see how the distribution of the sample means gets closer to a normal distribution as the sample size increases. We must note that if the original population was normally distributed, then the sampling distribution of sample means is normally distributed as well.

> **CENTRAL LIMIT THEOREM**   *If all samples of size n are drawn from a population with a mean μ and standard deviation σ, the distribution of the sample means $\bar{x}$ will be approximately a normal distribution. The mean of the sample means $(\mu_{\bar{x}})$ is the same as the mean of the population μ. The standard deviation of the sample means $(\sigma_{\bar{x}})$ is $\sigma/\sqrt{n}$. $\sigma_{\bar{x}}$ is called the **standard error of the mean,** or the **standard error.***

If we look back to our IQ means, we see that the standard deviation (6.846897107) is equal to the standard deviation of the population divided by $\sqrt{2}$, where 2 is the size of the samples we took. What standard deviation could we expect if our samples were of size 3?

$$\sigma_{\bar{x}} = \frac{\sigma}{\sqrt{n}}$$

$$= \frac{9.68297475}{\sqrt{3}}$$

$$= 5.590468078$$

The standard deviation would be 5.590468078. Note that as the sample size increases, the standard deviation decreases. In other words, the sample means have less variation as the sample size increases.

 The heights of adult females are normally distributed with a mean of 63.6 inches and a standard deviation of 2.5 inches. If we selected all possible samples of size 36 and calculated the mean of each sample, what would the mean of those sample means be? What would their standard deviation be?

The mean of the sample means $\mu_{\bar{x}}$ would be the same as the mean of the population—63.6 inches.

$$\mu_{\bar{x}} = \mu$$
$$= 63.6$$

The standard deviation of the sample means, or the standard error, is found as follows.

$$\sigma_{\bar{x}} = \frac{\sigma}{\sqrt{n}}$$
$$= \frac{2.5}{\sqrt{36}}$$
$$\approx 0.4167$$

Therefore, the standard error is approximately 0.4167. ∎

Probabilities Involving Sample Means

Now that we know that the sample means are approximately normally distributed, we can do probability problems involving sample means. First, let's begin with a problem from the last chapter.

 The heights of adult females are normally distributed with a mean of 63.6 inches and a standard deviation of 2.5 inches. Find the probability that a randomly selected adult female is 5′ 5″ (65 inches) or taller.

This is a typical normal probability problem with $\mu = 63.6$ and $\sigma = 2.5$.

$$P(x \geq 65) = P\left(z \geq \frac{65 - 63.6}{2.5}\right)$$
$$= P(z \geq 0.56)$$
$$= 1 - 0.7123$$
$$= 0.2877$$

The probability that a randomly selected adult female is 5′ 5″ or taller is 0.2877. ∎

The major difference for probability problems involving a sample mean is in the conversion to z. Here is the formula that we will use:

$$z = \frac{\bar{x} - \mu}{\sigma/\sqrt{n}}$$

It is very similar to the formula we had been using. We subtract the mean and then divide by the standard deviation. The difference is that the standard deviation for sample means, or standard error, is found by dividing the population's standard deviation by the square root of the sample size. Here is an example.

 EXAMPLE 6.3

The heights of adult females are normally distributed with a mean of 63.6 inches and a standard deviation of 2.5 inches. Find the probability that the mean height of 36 randomly selected adult females is 5′ 5″ (65 inches) or taller.

We are interested in the probability of a sample mean being at least 65 inches.

$$P(\bar{x} \geq 65) = P\left(z \geq \frac{65 - 63.6}{2.5/\sqrt{36}}\right)$$

$$= P(z \geq 3.36)$$

$$= 1 - 0.9996$$

$$= 0.0004$$

The probability that the mean height of 36 randomly selected adult females is 5′ 5″ or taller is 0.0004. ∎

Thus, we see that it is entirely possible that an adult female is at least 5′ 5″ tall (0.2877), but it is rather unlikely that a group of 36 women have a mean height of at least 5′ 5″ (0.0004). In an individual trial anything is possible, but as the sample size increases our sample mean will be drawn toward the population mean.

EXAMPLE 6.4

IQ scores are normally distributed with a mean of 100 points and a standard deviation of 15 points. If 100 people are randomly selected, find the probability that their mean score is below 97.

The mean of the sample means is 100, the same as the mean of the population. The standard error is 1.5.

$$\sigma_{\bar{x}} = \frac{\sigma}{\sqrt{n}}$$

$$= \frac{15}{\sqrt{100}}$$

$$= 1.5$$

Now we find the probability that the mean score is below 97.

$$P(\bar{x} < 97) = P\left(z < \frac{97 - 100}{15/\sqrt{100}}\right)$$

$$= P(z < -2)$$

$$= 0.0228$$

The probability of 100 randomly selected people having a mean IQ score below 97 is 0.0228. ∎

You may be wondering why the previous problem did not require a continuity correction—after all, IQ scores are discrete. The important thing to keep in mind is that we are interested in means of IQ scores, not IQ scores themselves. The means of IQ scores are continuous.

EXERCISES 6.1

1. Consider the population of possible rolls of a die: 1, 2, 3, 4, 5, 6.
   (a) Find the population mean μ.
   (b) Find the population standard deviation σ.
   (c) If all samples of size 2 are drawn from this population, what should the mean of the sample means ($\mu_{\bar{x}}$) be according to the Central Limit Theorem?
   (d) If all samples of size 2 are drawn from this population, what should the standard error ($\sigma_{\bar{x}}$) be according to the Central Limit Theorem?
   (e) Construct all 36 possible samples of size 2 from this population, and calculate their means. (Refer to the dice chart in Section 3.1.)
   (f) Calculate the mean and standard deviation of the sample means found in (e). Do these results agree with your answers in parts (c) and (d)?

2. IQ scores are normally distributed with a mean of 100 points and a standard deviation of 15 points.
   (a) If all samples of size 64 are drawn, what should the mean of the sample means ($\mu_{\bar{x}}$) be according to the Central Limit Theorem?
   (b) If all samples of size 64 are drawn, what should the standard error ($\sigma_{\bar{x}}$) be according to the Central Limit Theorem?

3. The heights of adult males are normally distributed with a mean height of 69.0 inches and a standard deviation of 2.8 inches.
   (a) If all samples of size 144 are drawn, what should the mean of the sample means ($\mu_{\bar{x}}$) be according to the Central Limit Theorem?
   (b) If all samples of size 144 are drawn, what should the standard error ($\sigma_{\bar{x}}$) be according to the Central Limit Theorem?

4. Community college students have a mean age of 23.5 years and a standard deviation of 7.5 years.
   (a) If all samples of size 30 are drawn, what should the mean of the sample means ($\mu_{\bar{x}}$) be according to the Central Limit Theorem?
   (b) If all samples of size 30 are drawn, what should the standard error ($\sigma_{\bar{x}}$) be according to the Central Limit Theorem?

5. SAT math test scores have a mean of 512 points with a standard deviation of 112 points.
   (a) If all samples of size 40 are drawn, what should the mean of the sample means ($\mu_{\bar{x}}$) be according to the Central Limit Theorem?
   (b) If all samples of size 40 are drawn, what should the standard error ($\sigma_{\bar{x}}$) be according to the Central Limit Theorem?

IQ scores are normally distributed with a mean of 100 points and a standard deviation of 15 points.

6. A group of 36 people is selected at random. Find the probability that their mean IQ score is at least 105.

7. A group of 49 people is selected at random. Find the probability that their mean IQ score is less than 110.

8. A group of 81 people is selected at random. Find the probability that their mean IQ score is between 95 and 105.

9. A group of 75 people is selected at random. Find the probability that their mean IQ score is 92 or higher.

10. A group of 50 people is selected at random. Find the probability that their mean IQ score is between 105 and 110.

The heights of adult males are normally distributed with a mean of 69 inches and a standard deviation of 2.8 inches.

11. A group of 64 adult males is selected at random. Find the probability that their mean height is below 67 inches.

12. A group of 225 adult males is selected at random. Find the probability that their mean height is between 68 inches and 70 inches.

13. A group of 121 adult males is selected at random. Find the probability that their mean height is 72 inches or more.

14. A group of 417 adult males is selected at random. Find the probability that their mean height is between 67.5 inches and 68.5 inches.

15. A group of 600 adult males is selected at random. Find the probability that their mean height is between 68.7 inches and 69.3 inches.

16. The mean age for a woman's first marriage is 25.0 years of age, and the standard deviation is 4.3 years. A random sample of 256 females who have been married at least one time was selected.

   (a) Find the probability that the mean age for this sample is 24.3 years old or younger.
   (b) If the mean age of the 256 females was 24.3 years old, is it likely that the true population mean age for a woman's first marriage is really 25.0 years?
   (c) If the mean age of the 256 females was 24.3 years old, is it likely that the sample is truly representative of the population?

17. It is widely known that people living in England drink a lot of tea—on average, more than 3 cups per day. Suppose that the mean number of cups of tea that people living in England drink per day is 3.4 cups, with a standard deviation of 1.3 cups per day.

   (a) Find the probability that a sample of 144 people produces a mean of more than 3.65 cups per day.
   (b) How likely is it that this sample is a representative sample of the English population?
   (c) What possible biases could have affected this study?

18. The heights of 12-month-old boys are approximately normally distributed with a mean of 29.8 inches and a standard deviation of 1.2 inches. A sample of thirty-six 12-month-old boys was taken.

   (a) Find the probability that the mean height of this sample is between 29.5 inches and 31.5 inches.
   (b) Find the probability that the mean height is below 28 inches.
   (c) Find the probability that the mean height is at least 30.2 inches.
   (d) Repeat parts (a)–(c) for a sample of 225 instead of thirty-six 12-month-old boys. Discuss, in your own words, what the increased sample size has done to these probabilities, and why.

19. The lengths of human pregnancies are normally distributed with a mean of 268 days and a standard deviation of 15 days. Three samples of women who have recently given birth are taken. One has a size of 49, the second has a size of 121, and the third has a size of 400. For each sample, find the probability that the mean length of pregnancy is (a) over 275 days; (b) less than 266 days; and (c) between 265 and 270 days.

Probability	n = 49	n = 121	n = 400
$P(\bar{X} > 275)$			
$P(\bar{X} < 266)$			
$P(265 < \bar{X} < 270)$			

20. The heights of adult males are normally distributed with a mean of 69.0 inches and a standard deviation of 2.8 inches. For samples of 36 adult males, 196 adult males, and 2025 adult males, find (a) $P(\bar{x} > 70.2)$; (b) $P(\bar{x} < 687)$; (c) $P(68.5 < \bar{x} < 69.5)$.

EXCEL EXPLORATION   Central Limit Theorem

In this computer project, we will explore the Central Limit Theorem using Excel. First we will take 1000 samples, each of size 5. For each sample, we will calculate the sample mean, $\bar{x}$. We will then look at the distribution of the sample means by constructing a histogram. We will go on to repeat this for samples of size 30 and 100.

Before we begin we must be sure that the Analysis Tool Pak has been added. Pull down the **Tools** menu. If you see a choice that says **Data Analysis,** then the Tool Pak is added and you may continue. If you do not see **Data Analysis,** click on **Add-ins** under **Tools.** When the dialog box opens up, click on the box for **Analysis Tool Pak,** and click on **OK.**

We will now select 1000 samples, each of size 5, from a population that has a uniform distribution. Pull down the **Tools** menu and then click on **Data Analysis.** When the dialog box appears, click on **Random Number Generation** and click **OK.** In the **Random Number Generation** dialog box, enter 5 in the box labeled **Number of Variables.** (This is our sample size.) Next, enter 1000 in the box labeled **Number of Random Numbers.** (This is the number of samples that we are going to take.) In the box labeled **Distribution,** click the arrow and select Uniform. In the boxes next to **Parameters,** type 0 and 100. **Click OK.** (This may take a few minutes for Excel to generate, so be patient.)

We need to split the window in order to see the top and bottom of our worksheet. Click on cell **A13.** Pull down the **Window** menu and then click on **Split.**

Pull down the **Edit** menu and click on **Go To.** When the dialog box appears, type **A1001** and click **OK.** Click on cell **A1.**

We will now calculate the mean for each of our samples. In cell **F1** type =AVERAGE(A1:E1). Click on cell **F1.** Hold down the SHIFT key and click on cell **F1000.** Pull down the **Edit** Menu; select **Fill** and then **Down.** This will calculate the mean for all 1000 samples. Pull down the **Window** menu and select **Remove Split.**

We now want to look at the distribution of the sample means. Pull down the **Tools** menu and select **Data Analysis.** Click on **Histogram** and click **OK.** When the dialog box appears, type F1:F1000 in the box labeled **Input Range.** Select **Chart Output,** and then click **OK.**

In new worksheets, repeat this for samples of size 30 and 100. For a sample of size 30, column **AD** is the thirtieth column, so the sample means need to be calculated in column **AE.** We will also use column **AE** for constructing the histogram. For a sample of size 100, column **CV** is the hundredth column, so the sample means need to be calculated in column **CW.** We will also use column **CW** for constructing the histogram.

According to the Central Limit Theorem, the sample means should follow a normal distribution. Here is a sample of the histogram that Excel generates for a sample size of 100. Note that the shape appears to be normal.

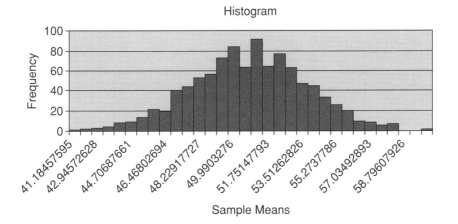

Compare the shape of your three histograms. As the sample size increases, does the shape of the histogram look more like a normal distribution?

SECTION 6.2

Confidence Intervals for a Population Mean (Large Samples)

We begin with a brief definition of a confidence interval for a population mean before we learn how to construct a confidence interval.

> **DEFINITION** A confidence interval for a population mean is an interval that we believe with some level of confidence contains the unknown population mean.

The level of confidence mentioned in the definition is usually 90%, 95%, or 99%. For a 95% confidence interval for a population mean, we are 95% sure that the interval actually contains the population mean. When we construct a 95% confidence interval, we are looking for two values for which there is a 95% chance that the population mean is between them. For now, we will call those values L (Low) and H (High). In symbols, we are looking for L and H such that

$$P(L \leq \mu \leq H) = 0.95$$

Thinking back to Chapter 5, we see that the z-values associated with the middle 95% are −1.96 and 1.96. We will start with a probability involving z, and work our way back.

$$0.95 = P(-1.96 \leq z \leq 1.96)$$

$$= P\left(-1.96 \leq \frac{\bar{x} - \mu}{\sigma/\sqrt{n}} \leq 1.96\right)$$

$$= P\left(-1.96 \cdot \frac{\sigma}{\sqrt{n}} \leq \bar{x} - \mu \leq 1.96 \cdot \frac{\sigma}{\sqrt{n}}\right)$$

$$= P\left(-\bar{x} - 1.96 \cdot \frac{\sigma}{\sqrt{n}} \leq -\mu \leq -\bar{x} + 1.96 \cdot \frac{\sigma}{\sqrt{n}}\right)$$

$$= P\left(\bar{x} + 1.96 \cdot \frac{\sigma}{\sqrt{n}} \geq \mu \geq \bar{x} - 1.96 \cdot \frac{\sigma}{\sqrt{n}}\right)$$

$$= P\left(\bar{x} - 1.96 \cdot \frac{\sigma}{\sqrt{n}} \leq \mu \leq \bar{x} + 1.96 \cdot \frac{\sigma}{\sqrt{n}}\right)$$

We now know that

$$P\left(\bar{x} - 1.96 \cdot \frac{\sigma}{\sqrt{n}} \leq \mu \leq \bar{x} + 1.96 \cdot \frac{\sigma}{\sqrt{n}}\right) = 0.95$$

This tells us that 95% of the intervals constructed in this fashion will contain the population mean μ. This also tells us that any interval constructed in this fashion has a 95% chance of containing the population mean. The probability that it does not contain the population mean is 0.05. This complement of the level of confidence is denoted by α (alpha).

Constructing a Confidence Interval for a Population Mean

From above, we see that if we multiply 1.96 by the standard error, and then move by that amount from the sample mean to the left and to the right, we get the interval that has a 95% chance of containing the sample mean.

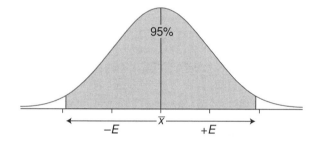

E in this picture stands for the **margin of error**. For a 95% confidence interval, we are really saying that there is a 95% chance that the sampling error (difference between the sample mean and population mean) is less than the margin of error. When we are trying to construct a confidence interval for a population mean, the first thing we should do is calculate the margin of error.

$$E = z_{\alpha/2} \cdot \frac{\sigma}{\sqrt{n}}$$

where $z_{\alpha/2}$ = critical value
   σ = population standard deviation
   $\sqrt{n}$ = sample size

We will work on the critical value first. Quite simply, it is the positive z-value that along with its opposite separates the middle percentage declared by the level of confidence from the rest of the curve. Suppose we were going to construct a 90% confidence interval. What z-value separates the middle 90% of the graph from the rest?

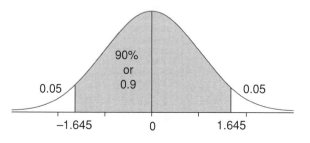

The critical value for a 90% confidence interval is 1.645. What is the critical value for a 99% confidence interval?

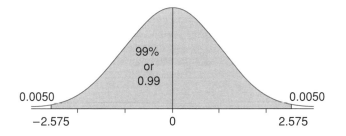

The critical value for a 99% confidence interval is 2.575.

Let's look at the formula for margin of error once again.

$$E = z_{\alpha/2} \cdot \frac{\sigma}{\sqrt{n}}$$

where $z_{\alpha/2}$ = critical value
σ = population standard deviation
$\sqrt{n}$ = sample size

Note that the formula contains the *population* standard deviation. If we are trying to estimate the population mean, we most likely will not know the population standard deviation either. If a sample has size 30 or greater, then we can use the sample standard deviation as a reliable estimate of the population standard deviation. In Section 6.3 we will learn how to handle a sample that has a size less than 30.

Thus, if the population standard deviation is unknown and the sample size is at least 30, we calculate the margin of error using the following.

$$E = z_{\alpha/2} \cdot \frac{s}{\sqrt{n}}$$

where $z_{\alpha/2}$ = critical value
s = sample standard deviation
$\sqrt{n}$ = sample size

Once the margin of error has been calculated, we plug into the following to obtain our confidence interval.

$$\bar{x} - E \leq \mu \leq \bar{x} + E$$

 EXAMPLE 6.5   A random sample of 350 male college students were asked for the number of units they were taking. The mean was 12.3 units, with a standard deviation of 2.50 units. Construct a 90% confidence interval for the mean number of units taken by a male college student.

First we calculate the margin of error. We need the critical value for a 90% confidence interval.

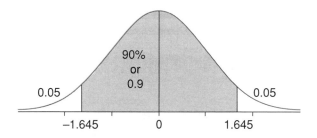

The critical value for a 90% confidence interval is $z_{\alpha/2}$ = 1.645. Now plug into the formula for margin of error.

$$E = z_{\alpha/2} \cdot \frac{s}{\sqrt{n}}$$

$$= 1.645 \cdot \frac{2.50}{\sqrt{350}}$$

$$= 0.22$$

When rounding the margin of error, round to one more decimal place than the sample mean. Our margin of error is 0.22. Now we add this to the sample mean, and subtract it from the sample mean, to obtain the confidence interval.

$$\bar{x} - E \le \mu \le \bar{x} + E$$

$$12.3 - 0.22 \le \mu \le 12.3 + 0.22$$

$$12.08 \le \mu \le 12.52$$

We are 90% confident that the mean number of units taken by all male college students is between 12.08 units and 12.52 units. ∎

 EXAMPLE 6.6   A random sample of 400 female college students were asked for the number of units they were taking. The mean was 11.5 units, with a standard deviation of 2.56 units. Construct a 90% confidence interval for the mean number of units taken by a female college student.

First we calculate the margin of error. We need the critical value for a 90% confidence interval.

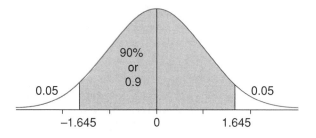

The critical value for a 90% confidence interval is $z_{\alpha/2}$ = 1.645. Now plug into the formula for margin of error.

$$E = z_{\alpha/2} \cdot \frac{s}{\sqrt{n}}$$

$$= 1.645 \cdot \frac{2.56}{\sqrt{400}}$$

$$= 0.21$$

Our margin of error is 0.21. Now we add this to the sample mean, and subtract it from the sample mean, to obtain the confidence interval.

$$\bar{x} - E \leq \mu \leq \bar{x} + E$$

$$11.5 - 0.21 \leq \mu \leq 11.5 + 0.21$$

$$11.29 \leq \mu \leq 11.71$$

We are 90% confident that the mean number of units taken by all female college students is between 11.29 units and 11.71 units. ∎

Let's analyze the two confidence intervals that we just constructed.

Male college students:	$12.08 \leq \mu \leq 12.52$
Female college students:	$11.29 \leq \mu \leq 11.71$

If someone claimed that the mean number of units taken by male college students is at least 12 units, does our confidence interval support such a statement? Yes—since we are 90% sure that the mean number of units is between 12.08 and 12.52, our interval is consistent with the claim.

If someone claimed that female students take an average of 12 units, what does our interval say about the claim? Our interval does not contain 12 units, so it is unlikely that the mean number of units taken by female students could be 12. There is a 10% chance that the true mean is not contained in our interval, so there is less than a 10% chance that this claim is accurate.

If someone claimed that the average number of units taken by male students is greater than the average number of units taken by female students, what do our intervals tell us? All of the values contained in the confidence interval for male students are greater than the values in the confidence interval for female students. This suggests that there is a good chance that the claim is true.

It is the goal of Chapters 7 and 8 to test the validity of such claims. We will revisit these claims then. We will develop formal procedures to test such claims, but they are all based upon confidence intervals.

If we increase the level of confidence, what happens to the confidence interval? It grows wider. Since we are trying to catch the unknown population mean in our interval, we need to increase the size of our interval if we are to be more likely to have the actual population mean in our interval. Maybe this next example will help. Suppose you are standing in the outfield with a glove, and you are blindfolded. We will hit fly balls to you, and you will try to catch them by just sticking out your glove and hoping for the best. To have a better chance of catching the ball, you need a bigger glove. To catch the unknown population mean, we need a bigger confidence interval.

We can make a mathematical argument for this as well. As the level of confidence increases, so does the critical value $z_{\alpha/2}$. This in turn increases the margin of error, which creates a wider confidence interval.

A random sample of 400 female college students were asked for the number of units they were taking. The mean was 11.5 units, with a standard deviation of 2.56 units. Construct a 95% confidence interval for the mean number of units taken by a female college student.

First we calculate the margin of error. We need the critical value for a 95% confidence interval.

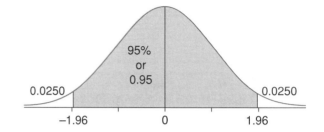

The critical value for a 95% confidence interval is $z_{\alpha/2} = 1.96$. Now plug into the formula for margin of error.

$$E = z_{\alpha/2} \cdot \frac{s}{\sqrt{n}}$$

$$= 1.96 \cdot \frac{2.56}{\sqrt{400}}$$

$$= 0.25$$

Our margin of error is 0.25. (Note that this is larger than the margin of error that we calculated for a 90% confidence interval.) Now we add this to the sample mean, and subtract it from the sample mean, to obtain the confidence interval.

$$\bar{x} - E \leq \mu \leq \bar{x} + E$$

$$11.5 - 0.25 \leq \mu \leq 11.5 + 0.25$$

$$11.25 \leq \mu \leq 11.75$$

We are 95% confident that the mean number of units taken by all female college students is between 11.25 units and 11.75 units. ■

A random sample of 400 female college students were asked for the number of units they were taking. The mean was 11.5 units, with a standard deviation of 2.56 units. Construct a 99% confidence interval for the mean number of units taken by a female college student.

First we calculate the margin of error. We need the critical value for a 99% confidence interval.

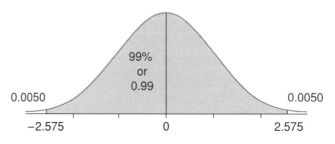

The critical value for a 99% confidence interval is $z_{\alpha/2}$ = 2.575. Now plug into the formula for margin of error.

$$E = z_{\alpha/2} \cdot \frac{s}{\sqrt{n}}$$

$$= 2.575 \cdot \frac{2.56}{\sqrt{400}}$$

$$= 0.33$$

Our margin of error is 0.33. Now we add this to the sample mean, and subtract it from the sample mean, to obtain the confidence interval.

$$\bar{x} - E \leq \mu \leq \bar{x} + E$$

$$11.5 - 0.33 \leq \mu \leq 11.5 + 0.33$$

$$11.17 \leq \mu \leq 11.83$$

We are 99% confident that the mean number of units taken by all female college students is between 11.17 units and 11.83 units. ■

As we stated, the margin of error increased as the level of confidence increased. This, in turn, created wider confidence intervals.

Level of Confidence	Margin of Error (E)	Lower Limit of Interval	Upper Limit of Interval
90%	0.21	11.29	11.71
95%	0.25	11.25	11.75
99%	0.33	11.17	11.83

We will look at one more example before moving on.

 EXAMPLE 6.9

Are movies too long? A random sample of 45 recent releases had a mean length of 111.1 minutes and a standard deviation of 22.6 minutes. Construct a 95% confidence interval for the mean length of all contemporary movies. Then use the interval to comment on the claim, "The average length of movies is at least 2 hours."

First we calculate the margin of error. We need the critical value for a 95% confidence interval.

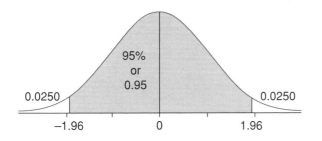

The critical value for a 95% confidence interval is $z_{\alpha/2}$ = 1.96. Now plug into the formula for margin of error.

$$E = z_{\alpha/2} \cdot \frac{s}{\sqrt{n}}$$

$$= 1.96 \cdot \frac{22.6}{\sqrt{45}}$$

$$= 6.60$$

Our margin of error is 6.60. Now we add this to the sample mean, and subtract it from the sample mean, to obtain the confidence interval.

$$\bar{x} - E \leq \mu \leq \bar{x} + E$$

$$111.1 - 6.60 \leq \mu \leq 111.1 + 6.60$$

$$104.50 \leq \mu \leq 117.70$$

We are 95% confident that the mean length of contemporary movies is between 104.50 minutes and 117.70 minutes. Since 120 minutes (2 hours) is not included in this interval, and all of the values in the interval are less than 120 minutes, it is not likely that the claim "The average length of movies is at least 2 hours" is valid. ■

MICROSOFT EXCEL   Finding Critical Values, Constructing Confidence Intervals

To find a critical value with Excel, we need to know what $\alpha/2$ is. To find α, subtract the level of confidence from 1. For example, if the level of confidence is 95%, α = 0.05. Divide α by 2 to obtain $\alpha/2$. For a 95% level of confidence, $\alpha/2$ = 0.025. Here are the Excel functions to find the left-tail and right-tail critical values.

Left-tail   **Right-tail**

$$=\text{NORMSINV}\left(\frac{\alpha}{2}\right) \qquad =\text{NORMSINV}\left(1 - \frac{\alpha}{2}\right)$$

For a 95% level of confidence, type the following in any two cells.

=NORMSINV(0.025)   =NORMSINV(1-0.025)

(We use the right-tail value for the critical value, so actually only the second formula is needed.) Here is an example using Excel.

 EXAMPLE 6.10

The superintendent of a golf course believes that the mean score for professional golfers on his course is above 72. He randomly samples the scores of 40 professional golfers from the last tournament:

69	67	69	71	75	78	70	71	72	74
72	76	69	76	70	77	78	79	72	71
75	75	75	78	71	64	71	68	72	71
73	72	73	71	71	75	72	72	78	75

Construct a 95% confidence interval for the mean score for all professional golfers on this course.

We will need the sample statistics (mean and standard deviation). Open a new Excel worksheet, and type the above values in column **A** (cells **A1** through **A40**). In cell **B1**, type Sample Mean. In cell **B2**, type Sample Standard Deviation. In cell **B3**, type Sample Size. In cell **B4**, type Critical Value.

Now that the framework is set up, we can begin the calculations. To calculate the sample mean, type =AVERAGE(A1:A40) in cell **C1**. To calculate the sample standard deviation, type =STDEV(A1:A40) in cell **C2**. Type the sample size (40) in cell **C3**. In cell **C4**, type =NORMSINV(1-0.025).

Now we calculate the margin of error. In cell **B6**, type Margin of Error. To calculate the margin of error in cell **C6**, type =C4*C2/SQRT(C3).

Finally, we can find the endpoints of the confidence interval. In cell **B8**, type Lower Limit, and type Upper Limit in cell **B9**. To find the lower limit, type =C1-C6 in cell **C8**. We find the upper limit in cell **C9** by typing =C1+C6.

The results of the worksheet appear to the right. The confidence interval that we obtain is $71.66 \le \mu \le 73.74$.

Is the superintendent's claim valid? Although most of our interval is above 72, we see that it is possible for the mean to be as low as 71.66, which does not support the claim. ■

Excel Worksheet

69	Sample Mean	72.7
67	Sample Standard Deviation	3.352764
69	Sample Size	40
71	Critical Value	1.959961
75		
78	Margin of Error	1.039012
70		
71	Lower Limit	71.66099
72	Upper Limit	73.73901
74		
72		
76		
69		
76		
70		
77		
78		
79		
72		
71		
75		
75		
75		
78		
71		
64		
71		
68		
72		
71		
73		
72		
73		
71		
71		
75		
72		
72		
78		
75		

TI-83	Finding Critical Values, Constructing Confidence Intervals

To find a critical value with the TI-83, we need to know what $\alpha/2$ is. To find α, subtract the level of confidence from 1. For example, if the level of confidence is 95%, $\alpha = 0.05$. Divide α by 2 to obtain $\alpha/2$. For a 95% level of confidence, $\alpha/2 = 0.025$. Here is how to find the left-tail and right-tail critical values using the TI-83.

To find the left-tail critical value, press (2ND)(VARS) to get to the **DISTR** menu.

```
DISTR DRAW
1:normalpdf(
2:normalcdf(
3:invNorm(
4:tpdf(
5:tcdf(
6:X²pdf(
7↓X²cdf(
```

Select option **3:invNorm(**. Type the value for $\alpha/2$. For a 95% confidence interval, type 0.025, and ( )). Press (ENTER), which will produce the following screen.

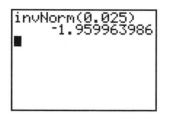

To two decimal places this is –1.96. Since the normal distribution is symmetric, the right-tail critical value is 1.96.

Here is an example using the TI-83.

 EXAMPLE **6.11**   The superintendent of a golf course believes that the mean score for professional golfers on his course is above 72. He randomly samples the scores of 40 professional golfers from the last tournament:

69	67	69	71	75	78	70	71	72	74
72	76	69	76	70	77	78	79	72	71
75	75	75	78	71	64	71	68	72	71
73	72	73	71	71	75	72	72	78	75

Construct a 95% confidence interval for the mean score for all professional golfers on this course.

Enter these 40 values in L_1. Press (STAT), and move to the **CALC** menu. Select option 1, and enter L_1 and press (ENTER).

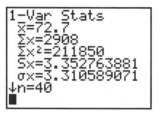

We need the sample mean (72.7) and the sample standard deviation (3.35). To construct the confidence interval press (STAT), and move to the **TESTS** menu.

```
EDIT CALC TESTS
1█Z-Test…
2: T-Test…
3: 2-SampZTest…
4: 2-SampTTest…
5: 1-PropZTest…
6: 2-PropZTest…
7↓ZInterval…
```

Select option **7:ZInterval.** For **Inpt,** select **Stats.** Enter 3.35 for the standard deviation, 72.7 for the mean, 40 for the sample size, and 0.95 for the confidence level.

```
ZInterval
 Inpt:Data Stats
 σ:3.35█
 x̄:72.7
 n:40
 C-Level:.95
 Calculate
```

Select **Calculate** and press (ENTER).

```
ZInterval
 (71.662,73.738)
 x̄=72.7
 n=40
```

The confidence interval that we obtain is 71.662 ≤ μ ≤ 73.738.

Is the superintendent's claim valid? Although most of our interval is above 72, we see that it is possible for the mean to be as low as 71.662, which does not support the claim. ∎

Determining the Appropriate Sample Size

When a statistical study is started, often the first question is "How large a sample should we take?" If we know

- how confident we want to be in our results (level of confidence)
- the amount of sampling error that is acceptable to us (margin of error)
- the standard deviation of the population

then we can calculate how large a sample is necessary to produce the desired results. We start with the formula for margin of error, and solve it for n.

$$E = z_{\alpha/2} \cdot \frac{\sigma}{\sqrt{n}}$$

$$E \cdot \sqrt{n} = z_{\alpha/2} \cdot \frac{\sigma}{\sqrt{n}} \cdot \sqrt{n}$$

$$E \cdot \sqrt{n} = z_{\alpha/2} \cdot \sigma$$

$$\frac{E \cdot \sqrt{n}}{E} = \frac{z_{\alpha/2} \cdot \sigma}{E}$$

$$\sqrt{n} = \frac{z_{\alpha/2} \cdot \sigma}{E}$$

$$(\sqrt{n})^2 = \left(\frac{z_{\alpha/2} \cdot \sigma}{E}\right)^2$$

$$n = \left(\frac{z_{\alpha/2} \cdot \sigma}{E}\right)^2$$

The formula for calculating how large a sample is necessary is

$$n = \left(\frac{z_{\alpha/2} \cdot \sigma}{E}\right)^2$$

As before, the population's standard deviation is rarely known, so we will need an estimate of the population's standard deviation. If no estimate is available, a preliminary sample must be taken so that we can come up with an estimate of the standard deviation.

 EXAMPLE 6.12

Suppose we want to estimate the mean home price in Visalia, California. From previous studies, we know that the standard deviation is approximately $60,000. If we want to be 90% sure that the mean home price in Visalia differs from our sample mean by no more than $5,000, how large a sample should we take?

First, let's list the important information.

Level of confidence: 90%

$z_{\alpha/2} = 1.645$

Standard deviation: $\sigma = \$60,000$

Error: $E = \$5,000$

$$n = \left(\frac{z_{\alpha/2} \cdot \sigma}{E}\right)^2$$

$$= \left(\frac{1.645 \cdot 60,000}{5,000}\right)^2$$

$$= 389.7$$

Since it is not possible to survey 389.7 homes, we must have a sample of at least 390 homes for the desired results.

We will always round our answers that are not whole numbers up to the next whole number. The reason for this is that our formula gives the minimum sample size necessary, and any size below that would not be adequate. ■

 EXAMPLE 6.13

A pediatrician wants to estimate the mean weight of firstborn babies. The standard deviation for all babies is approximately 1.15 pounds. If she wants to be 95% sure that the mean weight of firstborns differs from her sample mean by no more than 0.25 pounds, how large a sample is required?

First, let's list the important information.

Level of confidence: 95%

$z_{\alpha/2} = 1.96$

Standard deviation: $\sigma = 1.15$

Error: $E = 0.25$

$$n = \left(\frac{z_{\alpha/2} \cdot \sigma}{E}\right)^2$$

$$= \left(\frac{1.96 \cdot 1.15}{0.25}\right)^2$$

$$= 81.3$$

The pediatrician must sample at least 82 babies to produce the desired results. ∎

EXERCISES 6.2

1. A random sample of 150 college students regarding their work schedules produced a mean of 25.4 hours worked per week with a standard deviation of 12 hours. Construct a 95% confidence interval for the mean number of hours worked per week by college students.

2. A placement exam is given to 36 randomly selected college students. Their mean score is 63.6 points, with a standard deviation of 15.1 points. Construct a 90% confidence interval for the mean score on this placement exam for all students.

3. Thirty randomly selected college students were asked how many cavities they had. These students had a mean of 3.2 cavities with a standard deviation of 1.65 cavities.

   (a) Construct a 99% confidence interval for the mean number of cavities for all college students.

   (b) Explain how you would gather the data to construct this confidence interval, and list any biases that may be present using your method.

4. A random sample of 49 pepper seeds was planted. The length of time between planting and germination was measured. The mean time until germination was 10.4 days with a standard deviation of 2.13 days. Construct a 95% confidence interval for the true length of time for all pepper seeds to germinate.

5. How large are math classes in community colleges in California? A random sample of 100 math classes had a mean of 28.3 students, with a standard deviation of 4.5 students. Construct a 92% confidence interval for the mean class size for all community college math classes.

6. A random sample of 50 community college students produced a mean GPA of 3.18 with a standard deviation of 0.30.

   (a) Construct an 80% confidence interval for the mean GPA of all community college students.

   (b) Explain how you would gather the data to construct this confidence interval, and list any biases that may be present using your method.

7. A sample of 150 high school students who took the SAT test had a mean math score of 515 with a standard deviation of 125. Construct a 95% confidence interval for the mean SAT math score for all high school students who take the SAT.

8. A sample of 200 smokers were asked at what age they started smoking. The mean age was 18.3 years with a standard deviation of 4.35 years. Construct a 95% confidence interval for the mean age that all smokers begin smoking.

9. The president of a community college is interested in the mean distance driven to school by his students. A random sample of 40 students produced a mean distance of 11.9 miles, with a standard deviation of 12.29 miles.

   (a) Construct a 95% confidence interval for the mean distance driven to school by all students of the community college.

   (b) The president believes that the mean distance driven by all students of the community college is 15 miles. Is the president's belief consistent with your interval? Explain your answer in your own words.

10. The president of a community college claims that the mean age of the students at his school is less than 25. A random sample of 100 of his students had a mean age of 23.1 years old, with a standard deviation of 7.10 years.

(a) Construct a 99% confidence interval for the mean age of all students at his school.

(b) Is the president's claim consistent with your interval? Explain your answer in your own words.

11. A recent "Pop-Up Video" on VH1 claimed that the mean age of women at their first marriage is 25. A random sample of 104 married females had a mean age of 22.4 years old at their first marriage, with a standard deviation of 4.63 years.

(a) Construct a 90% interval for the mean age of all women at their first marriage.

(b) Is their claim consistent with your interval? Explain your answer in your own words.

12. A real estate agent claims that the mean sale price of a home in Visalia is $100,000. A random sample of 44 homes sold in Visalia produced a mean sales price of $115,661 with a standard deviation of $58,538.

(a) Construct a 95% confidence interval for the mean sale price of all Visalia homes.

(b) Is the agent's claim consistent with your interval? Explain your answer in your own words.

13. Construct a 95% confidence interval for the mean cost of all funerals. A survey of 60 funerals showed that they had a mean cost of $5235, with a standard deviation of $1739.

14. A random sample of 529 pregnancies lasted 38.8 weeks with a standard deviation of 2.33 weeks. Construct a 90% confidence interval for the mean length of all pregnancies.

15. A sample of 36 medium-sized oranges had a mean vitamin C content of 64.5 mg, with a standard deviation of 3.14 mg. Construct a 99% confidence interval for the mean vitamin C content of medium-sized oranges.

16. (a) Repeat Exercise 15 using a sample size of 360 rather than 36. How does this affect your interval? Explain why this happens, in your own words.

(b) Repeat Exercise 15 using a 95% confidence level instead of 99%. How does this affect your interval? Explain why this happens, in your own words.

17. How long is the typical honeymoon? A random sample of 180 couples who were recently married showed that their honeymoons had a mean length of 8.9 days with a standard deviation of 3.7 days. Construct a 95 % confidence interval for the mean length of all honeymoons. (Based on a survey by *Bride's* magazine.)

18. Many health experts stress the importance of drinking at least eight 8-oz. glasses of water a day. A random sample of 330 adults showed that they drank a mean of 4.6 glasses of water per day, with a standard deviation of 2.79 glasses. (Based on the results of a study by Opinion Research Corporation International for the Water Quality Association.)

(a) Construct a 95% confidence interval for the mean number of glasses that an adult drinks per day.

(b) Does it appear that we are falling short of the recommendations? Explain how your confidence interval answers that question.

19. A random sample of 40 tax returns of individuals who earned less than $40,000 showed a mean charitable contribution of $527 with a standard deviation of $167. Construct a 95% confidence interval for the mean charitable contribution of individuals who earn less than $40,000. (Based on the results of a survey by Gallup.)

20. A sample of 103 homeowners showed that they spend a mean of 4.3 hours per week working on their lawn and garden during spring and summer, with a standard deviation of 3.62 hours. Construct a 95% confidence interval for the mean number of hours worked per week by all homeowners on their lawn and garden during spring and summer. (Based on the results of a study by Bruskin/Goldring Research for Sears, Roebuck and Company.)

21. The public relations officer at a college wants to estimate the mean IQ of all college students. If she wants to be 95% confident that her sample mean is off by no

more than 2 points, how large a sample is necessary? The standard deviation for IQ scores is 15 points.

22. A pediatrician wants to estimate the mean height of boys on their first birthday. If she wants to be 90% sure that her sample mean is off by no more than 0.25 inches, how large a sample should she take? The standard deviation has been shown to be approximately 1.2 inches.

23. How many hours do college students work per week? How large a sample is required to be 99% sure that a sample mean will be off by no more than 2 hours? An initial study suggested that the standard deviation is approximately 8.3 hours.

24. A high school counselor wants to estimate the mean SAT combined score for high school students. He has been told that the standard deviation for all such scores is approximately 230 points. How large a sample is required to be 95% sure that his sample mean is off by no more than 20 points from the true mean SAT combined score?

25. A statistics instructor wants to determine what the mean score on a statistics exam would be. A preliminary sample yielded a standard deviation of 14.2 points. How large should his sample be in order to be 90% sure that the sample mean differs from the population mean by no more than 3 points?

26. How large a sample should a medical researcher take if he wants to estimate the mean pulse rate of healthy adults with 99% confidence within 3 beats per minute? An initial sample yielded a standard deviation of 6.2 beats per minute.

27. Explain, in your own words, how an increase in sample size affects the margin of error in a confidence interval for a population mean.

28. Explain, in your own words, how an increase in level of confidence affects the margin of error in a confidence interval for a population mean.

MINI PROJECT

What is the mean age of students at your school? Randomly sample at least 30 students at your college and find their age. Use these sample data to construct a 95% confidence interval for the mean age of all students at your school. In addition to your confidence interval, include

- Your raw data
- A histogram for the ages
- A pie chart showing the breakdown of ages. Set this up in any way that you feel accurately represents the students in your survey.

Finally, answer the following questions.

- How did you obtain your sample?
- Which type of sampling did you use?
- Was it truly random?
- What potential biases may show up in your sample?

EXCEL EXPLORATION Confidence Intervals

We will use Excel to explore the idea of confidence intervals and how to interpret them. First, we will take 100 samples, each of size 100. For each sample, we will calculate a confidence interval for μ. Finally, we will determine what percentage of the confidence intervals actually capture the true value of the population mean.

Before we begin we must be sure that the Analysis Tool Pak has been added. Pull down the **Tools** menu. If you see a choice that says **Data Analysis**, then the Tool Pak is added and you may continue. If you do not see **Data Analysis**, click on **Add-ins** under **Tools**. When the dialog box opens up, click on the box for **Analysis Tool Pak**, and click on **OK**.

We begin by selecting 100 samples, each of size 100, from a population that has a normal distribution. Pull down the **Tools** menu and click on **Data Analysis.**

When the dialog box appears, click on **Random Number Generation** and **OK.** When the next dialog box appears, enter 100 in the box labeled **Number of Variables.** (This is the number of samples that we are going to take.) Enter 100 in the box labeled **Number of Random Numbers.** (This is the sample size.) In the box labeled **Distribution**, click on **Normal.** For **Parameters**, enter 50 for the mean and 5 for the standard deviation. Usually, the population parameter μ is unknown. Therefore, we try to estimate it using a confidence interval. However, for the purpose of this simulation it is imperative that we know the true mean so that we can find out how many of our confidence intervals contain μ. Click **OK.** (This may take a few minutes, so be patient.)

In cell **A101** type =AVERAGE(A1:A100). After pressing the Enter key, click on cell **A101** again. Hold down the Shift key and click on cell **CV101.** Pull down the **Edit** Menu; select **Fill** and then **Right.**

In row **102** we will calculate the lower limit of each confidence interval. In cell **A102** type =(A101 - 1.96*(STDEV(A1:A100)/(SQRT(100)))). After pressing the Enter key, click on cell **A102** again. Hold down the Shift key and click on cell **CV102.** Pull down the **Edit** Menu; click on **Fill** and then **Right.**

In row **103** we will calculate the upper limit of each confidence interval. In cell **A103** type =(A101 + 1.96*(STDEV(A1:A100)/(SQRT(100)))). After pressing the Enter key, click on cell **A103** again. Hold down the Shift key and click on cell **CV103.** Pull down the **Edit** Menu; click on **Fill** and then **Right.**

Now we will determine what percentage of our intervals contain the actual population mean (50) in row **104.** Type =IF((A102<50)*AND(A103>50),1,0) in cell **A104.** After pressing the Enter key, the cell will contain a 1 if the interval contains 50 and a 0 if it does not. Click on cell **A104.** Hold down the Shift key and click on cell **CV104.** Pull down the **Edit** Menu; click on **Fill** and then **Right.** To determine what percentage of our intervals contain the actual population mean (50), type =SUM(A104:CV104)/100 in cell **A105.** Did 95% of your intervals contain the population mean? If not, was it close?

EXTRA • One-Sided Confidence Intervals

Occasionally, there is a need for a one-sided confidence interval. For example, we may want to be 95% sure that the population mean is below a certain value, as opposed to being between two values. Or we may want to be 99% sure that the population mean is above a certain value. The procedure is similar to the confidence interval procedure developed in this section. One difference is when we find the critical value. The second difference is that we only find one endpoint, as opposed to two.

We begin with an example.

EXAMPLE 6.14

The president of a community college claims that the mean age of the students at his school is less than 25. A random sample of 100 of his students had a mean age of 23.1 years, with a standard deviation of 7.10 years. Construct a 99% left-sided confidence interval for the mean age of all students at his school.

We are looking for the age for which we are 99% sure that the population mean age is no higher than it. We begin by finding the *z*-value that separates the lower 99% from the rest.

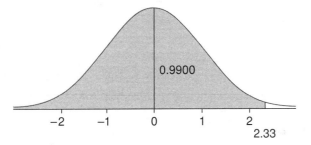

0.9900

The value that we are looking for is $z = 2.33$. Now we proceed to find what we considered to be the right endpoint of our confidence interval. First, the margin of error.

$$E = z_\alpha \cdot \frac{s}{\sqrt{n}}$$

$$= 2.33 \cdot \frac{7.10}{\sqrt{100}}$$

$$= 1.65$$

Now we add the margin of error to the sample mean to obtain the confidence interval.

$$\mu \le \bar{x} + E$$

$$\mu \le 23.1 + 1.65$$

$$\mu \le 24.75$$

We are 99% confident that the mean age is 24.75 or lower. The president's claim appears to be valid. ∎

A right-sided interval works in a similar fashion. We begin by finding the critical value, for which we will use a positive value. Next, we find the margin of error. Finally, we subtract the margin of error from our sample mean.

A television commercial claims that the mean cost of funerals is above $5000. A survey of 60 funerals showed that they had a mean cost of $5235, with a standard deviation of $1739. Construct a 95% right-sided confidence interval for the mean cost of all funerals.

First, we find the critical value.

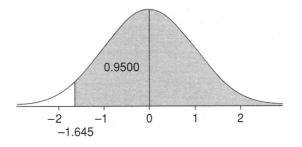

0.9500

We will use $z = 1.645$.

$$E = z_\alpha \cdot \frac{s}{\sqrt{n}}$$

$$= 1.645 \cdot \frac{1739}{\sqrt{60}}$$

$$= 369.31$$

Now we subtract the margin of error from the sample mean to obtain the confidence interval.

$$\mu \geq \bar{x} - E$$

$$\mu \geq 5235 - 369.31$$

$$\mu \geq 4865.69$$

We are 95% confident that the mean cost is $4865.69 or higher. The commercial's claim may be false. ∎

SECTION 6.3
Confidence Intervals for a Population Mean (Small Samples)

In the last section, we saw that the sampling distribution of the sample means followed the z-distribution if either the population standard deviation was known or the sample size was at least 30. If neither of those conditions are true, then the distribution follows **Student's t-distribution.** William Gosset, who was a chemist for the Guinness Brewery, developed this distribution that is appropriate for small samples. The Guinness Brewery did not permit its employees to publish research results, so Gosset published under the pen name Student.

The t-distribution is similar to the z-distribution in some ways. The t-distribution is symmetric and bell-shaped with a mean of $t = 0$. There are different t-distributions; they depend on the sample size. Actually, they depend on the *degrees of freedom*. The degrees of freedom for a t-distribution are equal to 1 less than the sample size $(n - 1)$. As the size of the sample (n) increases, the t-distribution approaches the z-distribution. Therefore, if

- the sample size is less than 30
- the population mean is unknown
- we have reason to believe that the population follows a normal distribution

then we can construct a confidence interval for the population mean using the t-distribution. Here is the formula for the margin of error.

$$E = t_{\alpha/2} \cdot \frac{s}{\sqrt{n}}$$

where $t_{\alpha/2}$ = critical value of t
s = sample standard deviation
n = sample size

We construct the interval in the same fashion as with large samples.

$$\bar{x} - E \leq \mu \leq \bar{x} + E$$

We begin our work by learning how to find the critical *t*-value.

EXAMPLE 6.16   Find the critical value of *t* necessary to construct a 90% confidence interval for a population mean, if there are 11 degrees of freedom.

A 90% confidence interval goes along with $\alpha = 0.10$. On your *t* chart, look for the column under the heading 0.10 for a two-tailed value. We use the two-tailed value because our confidence interval focuses on the "middle" 90%, excluding the two tails. Go down that column until you reach 11 degrees of freedom. Here is the correct row from the chart.

	Two-Tailed Value					
Degrees of freedom	0.01	0.02	0.05	0.1	0.2	0.5
11	3.106	2.718	2.201	1.796	1.363	0.697

The critical value is 1.796. ∎

EXAMPLE 6.17   Find the critical value of *t* necessary to construct a 99% confidence interval for a population mean, if there are 21 degrees of freedom.

A 99% confidence interval goes along with $\alpha = 0.01$. On your *t* chart, look for the column under the heading 0.01 for a two-tailed value. Go down that column until you reach 21 degrees of freedom. Here is the correct row from the chart.

	Two-Tailed Value					
Degrees of freedom	0.01	0.02	0.05	0.1	0.2	0.5
21	2.831	2.518	2.080	1.721	1.323	0.686

The critical value is 2.831. ∎

EXAMPLE 6.18   Find the critical value of *t* necessary to construct a 95% confidence interval for a population mean, if there are 6 degrees of freedom.

A 95% confidence interval goes along with $\alpha = 0.05$. On your *t* chart, look for the column under the heading 0.05 for a two-tailed value. Go down that column until you reach 6 degrees of freedom. Here is the correct row from the chart.

	Two-Tailed Value					
Degrees of freedom	0.01	0.02	0.05	0.1	0.2	0.5
6	3.707	3.143	2.447	1.943	1.440	0.718

The critical value is 2.447. ∎

We are now ready to construct some confidence intervals. With the exception of finding the critical value of *t*, everything else is exactly like the confidence intervals for a population mean based on a large sample.

Mark, an avid golfer, claims that his average drive is greater than 250 yards. To prove this, he measured his 14 drives from a round of golf. Here are the distances, in yards.

262	247	255	239	275	257	260
242	256	281	265	248	255	263

Construct a 95% confidence interval for the mean length of Mark's drives. Based on your confidence interval, does Mark's claim appear to be valid?

We begin by calculating the mean and standard deviation for this sample. The mean is 257.5 yards and the standard deviation is 11.67 yards.

Next, we need to find the critical value. For a 95% confidence interval, $\alpha = 0.05$. We have 13 degrees of freedom.

	Two-Tailed Value					
Degrees of freedom	0.01	0.02	0.05	0.1	0.2	0.5
13	3.012	2.650	2.160	1.771	1.350	0.694

The critical value is 2.160. Now we can calculate the margin of error.

$$E = t_{\alpha/2} \cdot \frac{s}{\sqrt{n}}$$

$$= 2.160 \cdot \frac{11.67}{\sqrt{14}}$$

$$= 6.74$$

Finally, we construct the confidence interval.

$$\bar{x} - E \leq \mu \leq \bar{x} + E$$

$$257.5 - 6.74 \leq \mu \leq 257.5 + 6.74$$

$$250.76 \leq \mu \leq 264.24$$

We are 95% confident that the mean length of Mark's drives is between 250.76 yards and 264.24 yards. This interval is consistent with Mark's claim that his average drive is greater than 250 yards. ∎

A report on coins states that the mean weight of pennies minted before 1982 is 3 grams. A random sample of 4 such pennies had a mean weight of 3.11 grams with a standard deviation of 0.027 grams. Construct a 99% confidence interval for the mean weight of all pennies minted before 1982. Based on your interval, does the report's claim seem valid?

We need to find the critical value. For a 99% confidence interval, $\alpha = 0.01$. We have 3 degrees of freedom.

			Two-Tailed Value			
Degrees of freedom	0.01	0.02	0.05	0.1	0.2	0.5
3	5.841	4.541	3.182	2.353	1.638	0.765

The critical value is 5.841. Now we can calculate the margin of error.

$$E = t_{\alpha/2} \cdot \frac{s}{\sqrt{n}}$$

$$= 5.841 \cdot \frac{0.027}{\sqrt{4}}$$

$$= 0.0789$$

Finally, we construct the confidence interval.

$$\bar{x} - E \leq \mu \leq \bar{x} + E$$
$$3.11 - 0.0789 \leq \mu \leq 3.11 + 0.0789$$
$$3.0311 \leq \mu \leq 3.1889$$

We are 99% confident that the mean weight of pre-1982 pennies is between 3.0311 and 3.1889. It appears that the report is in error because 3 is not part of this interval. ■

It is important that our sample seems to have come from a population that is normally distributed, or we cannot use the t-distribution to construct a confidence interval for the population mean. Look for outliers, because outliers may suggest a distribution that is skewed and not normal. There is a procedure in Section 9.3 to test whether a set of data follows a normal distribution, but until then you can use the material presented in "Extra—Is a Set of Data Approximately Normally Distributed?" which follows Section 5.3.

MICROSOFT EXCEL Two-Tailed Critical Values for the t-Distribution

As with the standard normal distribution, Excel can help us find critical values for constructing confidence intervals with the t-distribution. We have to supply Excel with α and the number of degrees of freedom. Here is the format.

=TINV(α, degrees of freedom)

For example, to find the critical value for a 90% confidence interval with 11 degrees of freedom (sample size of 12), type the following in any cell.

=TINV(0.10,11)

Excel returns a value of 1.795884. (To three decimal places this rounds to 1.796.)

Here is an example that uses Excel.

EXAMPLE 6.21	A calculus exam on applications of the derivative was given to 19 randomly selected first-semester calculus students. Here are their results.

53	85	93	36	63	74	76	67	72	66
31	50	86	41	53	74	92	88	74	

Construct a 95% confidence interval for the mean score for all first-semester calculus students.

We will need the sample statistics (mean and standard deviation). Open a new Excel worksheet, and type the above values in column **A** (cells **A1** through **A19**). In cell **B1,** type Sample Mean. In cell **B2,** type Sample Standard Deviation. In cell **B3,** type Sample Size. In cell **B4,** type Critical Value.

Now that the framework is set up, we can begin the calculations. To calculate the sample mean, type =AVERAGE(A1:A19) in cell **C1**. To calculate the sample standard deviation, type =STDEV(A1:A19) in cell **C2**. Type the sample size (**19**) in cell **C3**. In cell **C4**, type =TINV(0.05,18).

Now we calculate the margin of error. In cell **B6,** type Margin of Error. To calculate the margin of error in cell **C6,** type =C4*C2/SQRT(C3).

Finally, we can find the endpoints of the confidence interval. In cell **B8,** type Lower Limit, and type Upper Limit in cell **B9.** To find the lower limit, type =C1-C6 in cell **C8.** We find the upper limit in cell **C9** by typing =C1+C6.

Your worksheet should look like this:

53	Sample Mean	67.05263
85	Sample Standard Deviation	18.72453
93	Sample Size	19
36	Critical Value	2.100924
63		
74	Margin of Error	9.024944
76		
67	Lower Limit	58.02769
72	Upper Limit	76.07758
66		
31		
50		
86		
41		
53		
74		
92		
88		
74		

By our calculations, the confidence interval is $58.03 \leq \mu \leq 76.08$. ■

TI-83   Two-Tailed Critical Values for the t-Distribution

As with the standard normal distribution, the TI-83 can help us to construct confidence intervals for a population mean using the t-distribution. We will use the following example to walk through the steps.

EXAMPLE 6.22

A calculus exam on applications of the derivative was given to 19 randomly selected first-semester calculus students. Here are their results.

53	85	93	36	63	74	76	67	72	66
31	50	86	41	53	74	92	88	74	

Construct a 95% confidence interval for the mean score for all first-semester calculus students.

First, we need to put the 19 values in list L_1. You may need to clear the values in L_1 first. Once the values are in the list, press (STAT) and move to the **TESTS** menu. Select option **8:TInterval**. The screen should look like this:

```
TInterval
 Inpt:DATA Stats
 List:L₁
 Freq:1
 C-Level:.95█
 Calculate
```

For **Inpt:** select **Data.** This will automatically calculate the sample mean and standard deviation that we need to construct the confidence interval. Next to **List:** be sure that it says L_1, which is the list that contains our values. **Freq:** should have a 1 next to it. Next to **C-Level:** enter 0.95. Once you have **Calculate** highlighted, press (ENTER). This should produce the following screen.

```
TInterval
 (58.028,76.078)
 x̄=67.05263158
 Sx=18.72453436
 n=19
█
```

The 95% confidence interval for the mean score for all first-semester calculus students is $58.028 \le \mu \le 76.078$. ■

EXERCISES 6.3

1. A major company has a large fleet of cars. A vice president feels that gas costs are too high, which is because the cars in the fleet get poor gas mileage. A random sample of 15 cars produced a mean mileage of 17.9 miles per gallon with a standard deviation of 3.9 miles per gallon. Construct a 95% confidence interval for the mean mileage for all of the cars in the fleet.

2. Do college students eat enough fruits and vegetables? A random sample of 24 college students produced a mean of 1.7 servings of fruits and vegetables per day with a standard deviation of 2.3 servings per day.

   (a) Construct a 90% confidence interval for the mean number of servings of fruits and vegetables eaten daily by all college students.

(b) Explain how you would gather the data to construct this confidence interval, and list any biases that may be present using your method.

3. A random sample of 10 NFL football games produced a mean of 39.7 total points per game with a standard deviation of 8.3 points. Construct a 99% confidence interval for the mean total points for all NFL games.

4. A certain brand of candy makes a package that they claim contains at least 110 candies. A random sample of 18 packages had a mean of 123.5 candies with a standard deviation of 7.6 candies. Construct a 95% confidence interval for the mean number of candies contained in all packages of this brand.

5. Six artichoke plants at a farm were selected at random. Here are the number of artichokes produced by each plant last year.

   | 32 | 17 | 51 | 40 | 36 | 34 |

   Construct a 99% confidence interval for the mean number of artichokes produced by all artichoke plants.

6. Here are 10 randomly selected blood sugar levels from a laboratory. (Levels were measured in mg/DL after a 12-hour fast.)

   | 105 | 89 | 96 | 135 | 94 | 91 | 111 | 107 | 141 | 83 |

   Construct a 90% confidence interval for the mean blood sugar level of all people after a 12-hour fast.

7. The Dean of Registration at a community college has implemented a new telephone registration system. The company that sold the system to the college claims that the mean length of phone calls is less than 10 minutes. To check this, the Dean randomly selected 20 calls, which had a mean length of 7.1 minutes and a standard deviation of 1.89 minutes.

   (a) Construct a 95% confidence interval for the mean length of all phone calls using this system.
   (b) Is the company's claim consistent with your interval? Explain your answer in your own words.

8. How old are the students at a local community college? The president of the college claims that the mean age of the students at her school is 25. A random sample of 25 students produced a mean age of 25.8 years with a standard deviation of 9.83 years.

   (a) Construct a 95% confidence interval for the mean age of all students at her community college.
   (b) Is the president's claim consistent with your interval? Explain your answer in your own words.

9. It is claimed that the mean height of adult females is 65 inches (5' 5"). A random sample of 11 adult females had a mean height of 63.7 inches and a standard deviation of 3.00 inches.

   (a) Construct a 99% confidence interval for the mean height of all adult females.
   (b) Is the claim consistent with your interval? Explain your answer in your own words.
   (c) Explain how you would gather the data to construct this confidence interval, and list any biases that may be present using your method.

10. A cotton farmer is considering growing a new type of cotton. Before he switches all of his fields to the new type, he must be sure that the mean yield per acre is above 850 pounds per acre, which is the yield of his current type of cotton. He randomly selects 7 acres in different fields, and the mean yield for these fields is 941.4 pounds with a standard deviation of 195.5 pounds.

    (a) Construct a 90% confidence interval for the mean yield per acre for this type of cotton.
    (b) Do you believe that the mean yield per acre is above 850 pounds per acre? Explain your answer in your own words.

11. Construct a 95% confidence interval for the mean number of trades per day on the New York Stock Exchange. Here are the number of trades reported on the New York Stock Exchange for 8 randomly selected days in 1999.

713,908	698,287	733,686	746,847
583,053	854,294	606,022	690,561

12. Construct a 90% confidence interval for the mean winning mutual payout for horse races at Santa Anita Park. Here are the winning mutual payouts for 20 randomly selected horse races at Santa Anita Park.

$3.40	$14.60	$3.40	$5.40	$44.40
$4.40	$8.00	$13.00	$4.20	$10.60
$14.80	$23.80	$5.20	$37.20	$23.80
$5.80	$4.80	$7.20	$7.00	$10.00

13. These are the listed home prices, in thousands of dollars, of 24 homes for sale in Visalia, California.

$92.9	$59.5	$97.9	$125.9	$115	$97.9
$239.9	$49.9	$179.9	$74.9	$63.9	$329
$69.9	$125	$229	$79.5	$121	$98.5
$84.9	$175	$89.9	$112.9	$185	$185

Construct a 95% confidence interval for the mean sale price of homes in Visalia, California.

14. Here are the lengths of eight Bruce Springsteen concerts in minutes.

180	175	165	163	150	158	175	165

Construct a 95% confidence interval for the mean length of a Bruce Springsteen concert.

15. Over a one-week period, 18 customers were randomly selected and the length of time that they had to wait at the check-in counter at an airport had a mean of 11.3 minutes with a standard deviation of 3.22 minutes.

   (a) Construct a 90% confidence interval for the mean waiting time at an airport check-in counter.
   (b) Explain how you would gather the data to construct this confidence interval, and list any biases that may be present using your method.

16. These are the scores in 18 randomly selected games by NBA teams.

103	114	98	93	99	104	107	109	81
79	89	85	90	98	81	105	108	111

Construct a 95% confidence interval for the mean number of points scored per game by an NBA team.

17. Ten cups of fresh orange juice were purchased at 10 different grocery stores and analyzed for vitamin C content. The mean content of vitamin C was 120.9 mg, with a standard deviation of 5.06 mg. Construct a 95% confidence interval for the mean vitamin C content of fresh orange juice.

18. A random sample of 23 men who first married in 1998 had a mean age of 26.7 years with a standard deviation of 5.03 years. Construct a 95% confidence interval for the mean age at which men get married. (Based on data from the U.S. Census Bureau.)

19. The results of a study on the effect of smoking on blood pressure appeared in the *American Journal of Hypertension* (1998). Ambulatory 24-hour blood pressure monitoring was conducted in 29 healthy middle-aged smokers to study the influence of smoking on blood pressure and heart rate. Their daytime heart rates had a mean of 74 beats per minute, with a standard deviation of approximately 34.2 beats per minute. Construct a 95% confidence interval for the mean daytime heart rate for all healthy middle-aged smokers. (The study was conducted by G. Bolinder and U. deFaire.)

20. Twenty cows from a college dairy farm were selected at random. Records showed that the cows' first lactation had a mean length of 307.2 days with a standard deviation of 46.03 days. Construct a 95% confidence interval for the mean length of first lactations for all cows.

MINI PROJECT

What is the mean number of units taken by students at your school this term? Randomly sample fewer than 30 students at your college and find out how many units they are taking this term. Use these sample data to construct a 95% confidence interval for the mean number of units being taken this term by all of the students at your school. In addition to your confidence interval, include

- Your raw data
- A histogram for the number of units being taken
- A pie chart showing the breakdown of units. Set this up in any way that you feel accurately represents the students in your survey.

Finally, answer the following questions.

- How did you obtain your sample?
- Which type of sampling did you use?
- Was it truly random?
- What potential biases may show up in your sample?

SECTION 6.4
Confidence Intervals for a Population Proportion (Large Samples)

Quite often, the population parameter that we are interested in is not a mean or average, but a proportion or percentage. In this section, we will learn how to construct a confidence interval for a population proportion. We will use a sample proportion (p) to estimate the population proportion (π). Creating a confidence interval is based on our work with the normal approximation to the binomial distribution. The conditions for the normal approximation must be satisfied [$np \geq 5$ and $n(1 - p) \geq 5$], so we must have at least 5 successes and 5 failures in our n independent trials (sample).

The basic idea is the same as it was in the previous two sections. We will calculate a margin of error, and then subtract it from the sample proportion p to get the left endpoint of the confidence interval. We add the margin of error to the sample proportion to find the right endpoint of the confidence interval. The major difference is that the formula for margin of error is different.

$$E = z_{\alpha/2} \cdot \sqrt{\frac{p \cdot (1 - p)}{n}}$$

where $z_{\alpha/2}$ = critical value
p = sample proportion
n = sample size

We then construct the confidence interval using

$$p - E \leq \pi \leq p + E$$

EXAMPLE 6.23

A random sample of 400 pregnant women revealed that 60 are smoking during their pregnancy. Construct a 95% confidence interval for the proportion of all pregnant women who smoke while pregnant.

In this problem, the sample proportion is 60 out of 400, so p = 60/400 or 0.15. The sample size n is 400. Now we find the critical value, which is done in the exact same way as in Section 6.2.

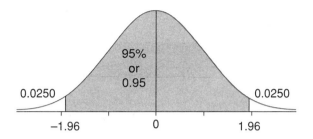

Thus, $z_{\alpha/2}$ = 1.96. Now we calculate the margin of error.

$$E = z_{\alpha/2} \cdot \sqrt{\frac{p \cdot (1 - p)}{n}}$$

$$= 1.96 \cdot \sqrt{\frac{0.15 \cdot 0.85}{400}}$$

$$= 0.0350$$

When rounding the margin of error for a proportion, round to four decimal places. Now construct the confidence interval.

$$p - E \leq \pi \leq p + E$$

$$0.15 - 0.0350 \leq \pi \leq 0.15 + 0.0350$$

$$0.1150 \leq \pi \leq 0.1850$$

We are 95% confident that the proportion of pregnant women who smoke while pregnant is between 0.1150 (11.50%) and 0.1850 (18.50%). ∎

EXAMPLE 6.24

A random sample of 65 community college statistics students revealed that 57 of them were under 30 years old. Construct a 95% confidence interval for the proportion of all community college statistics students who are under 30 years old.

In this problem, the sample proportion is 57 out of 65, so p = 57/65. The sample size n is 65. Now we find the critical value.

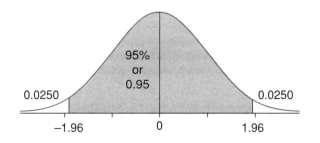

Thus, $z_{\alpha/2}$ = 1.96. Now calculate the margin of error.

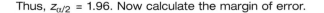

$$E = z_{\alpha/2} \cdot \sqrt{\frac{p \cdot (1 - p)}{n}}$$

$$= 1.96 \cdot \sqrt{\frac{\frac{57}{65} \cdot \frac{8}{65}}{65}}$$

$$= 0.0799$$

When rounding the margin of error for a proportion, round to four decimal places. Next construct the confidence interval.

$$p - E \le \pi \le p + E$$

$$\frac{57}{65} - 0.0799 \le \pi \le \frac{57}{65} + 0.0799$$

$$0.7970 \le \pi \le 0.9568$$

We are 95% confident that the proportion of college statistics students who are under 30 is between 0.7970 (79.70%) and 0.9568 (95.68%). ■

We will now take a look at how the size of the sample affects the confidence interval.

The university is considering adding a sushi restaurant to their food court. They will only add it if they are sure that more than half of the students are in favor of it. A sample of 30 university students revealed that 18 of them were in favor of adding a sushi restaurant. Construct a 99% confidence interval for the true proportion of university students who are in favor of adding a sushi restaurant. Based on the university's criteria and your confidence interval, will the university add a sushi restaurant?

The sample proportion is 18/30, or 0.6. The level of confidence is 99%.

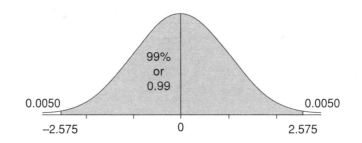

The critical value is 2.575. Now we calculate the margin of error.

$$E = z_{\alpha/2} \cdot \sqrt{\frac{p \cdot (1 - p)}{n}}$$

$$= 2.575 \cdot \sqrt{\frac{0.6 \cdot 0.4}{30}}$$

$$= 0.2303$$

Next, the confidence interval.

$$p - E \leq \pi \leq p + E$$

$$0.6 - 0.2303 \leq \pi \leq 0.6 + 0.2303$$

$$0.3697 \leq \pi \leq 0.8303$$

Based on this interval, we are not 99% confident that the proportion of all university students in favor of adding a sushi restaurant to the food court is more than 0.5. ■

We will leave the results as they are, but will base the interval on a sample size of 300 instead of 30.

EXAMPLE
6.26

The university is considering adding a sushi restaurant to their food court. They will only add it if they are sure that more than half of the students are in favor of it. A sample of 300 university students revealed that 180 of them were in favor of adding a sushi restaurant. Construct a 99% confidence interval for the true proportion of university students who are in favor of adding a sushi restaurant. Based on the university's criteria and your confidence interval, will the university add a sushi restaurant?

The sample proportion is 180/300, or 0.6. The level of confidence is 99%.

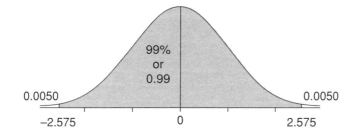

The critical value is 2.575. Now we calculate the margin of error.

$$E = z_{\alpha/2} \cdot \sqrt{\frac{p \cdot (1 - p)}{n}}$$

$$= 2.575 \cdot \sqrt{\frac{0.6 \cdot 0.4}{300}}$$

$$= 0.0728$$

Next, the confidence interval.

$$p - E \leq \pi \leq p + E$$

$$0.6 - 0.0728 \leq \pi \leq 0.6 + 0.0728$$

$$0.5272 \leq \pi \leq 0.6728$$

Based on this interval, we are 99% confident that the proportion of all university students in favor of adding a sushi restaurant to the food court is more than 0.5. ■

The major difference in the above two examples is the size of the sample. As the sample size increases, the margin of error grows smaller. Let's look at the same example, based on a sample of size 3000.

EXAMPLE 6.27

The university is considering adding a sushi restaurant to their food court. They will only add it if they are sure that more than half of the students are in favor of it. A sample of 3000 university students revealed that 1800 of them were in favor of adding a sushi restaurant. Construct a 99% confidence interval for the true proportion of university students who are in favor of adding a sushi restaurant. Based on the university's criteria and your confidence interval, will the university add a sushi restaurant?

The sample proportion is 1800/3000, or 0.6. The level of confidence is 99%.

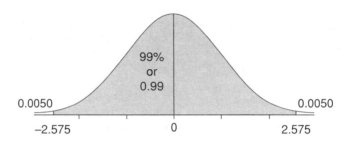

The critical value is 2.575. Now we calculate the margin of error.

$$E = z_{\alpha/2} \cdot \sqrt{\frac{p \cdot (1 - p)}{n}}$$

$$= 2.575 \cdot \sqrt{\frac{0.6 \cdot 0.4}{3000}}$$

$$= 0.0230$$

Next, the confidence interval.

$$p - E \le \pi \le p + E$$

$$0.6 - 0.0230 \le \pi \le 0.6 + 0.0230$$

$$0.5770 \le \pi \le 0.6230$$

Based on this interval, we are 99% confident that the proportion of all university students in favor of adding a sushi restaurant to the food court is more than 0.5. ■

Here's one more example that you may find interesting. It was done by a student of mine as his semester research project.

 EXAMPLE 6.28

A student, interested in personal hygiene, was curious about the percentage of men who wash their hands after using the bathroom. At a local mall, the student observed that 9 of 30 men did not wash their hands after using the men's room. Construct a 90% confidence interval for the percentage of men who do not wash their hands after using a public bathroom.

The sample proportion is 9/30, or 0.3. The level of confidence is 90%.

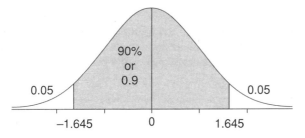

The critical value is 1.645. Now we calculate the margin of error.

$$E = z_{\alpha/2} \cdot \sqrt{\frac{p \cdot (1 - p)}{n}}$$

$$= 1.645 \cdot \sqrt{\frac{0.3 \cdot 0.7}{30}}$$

$$= 0.1376$$

Next, the confidence interval.

$$p - E \leq \pi \leq p + E$$

$$0.3 - 0.1376 \leq \pi \leq 0.3 + 0.1376$$

$$0.1624 \leq \pi \leq 0.4376$$

Thus, we are 90% sure that between 16.24% and 43.76% of all men do not wash their hands after using a public bathroom. ∎

MICROSOFT EXCEL Finding Critical Values, Constructing Confidence Intervals for Population Proportions

To find a critical value with Excel for confidence intervals for a population proportion, we use the same procedure that we applied when constructing confidence intervals for a population mean. This is because both require the use of the z-distribution. Again, we need to know what $\alpha/2$ is. To find α, subtract the level of confidence from 1. For example, if the level of confidence is 95%, $\alpha = 0.05$. Divide α by 2 to obtain $\alpha/2$. For a 95% level of confidence, $\alpha/2 = 0.025$. Here are the Excel functions to find the left-tail and right-tail critical values.

Left-tail   **Right-tail**

=NORMSINV$\left(\dfrac{\alpha}{2}\right)$ =NORMSINV$\left(1 - \dfrac{\alpha}{2}\right)$

For a 95% level of confidence, type the following in any two cells.

=NORMSINV(0.025)   =NORMSINV(1-0.025)

(We use the right-tail value for the critical value, so actually only the second formula is needed.)

The following example uses Excel to construct a confidence interval.

EXAMPLE
6.29

A company that makes a drug to reverse hair loss in males claims that 60% of the men who try their drug will grow new hair. A test on 40 randomly selected males experiencing hair loss produced the following results (Yes = grew new hair; No = no new hair).

Yes	No	No	No	Yes	No	No	No	Yes
No	Yes	Yes	Yes	Yes	No	No	Yes	No
Yes	Yes	Yes	Yes	Yes	Yes	No	Yes	Yes
Yes	No	No	No	Yes	No	No	Yes	Yes
No	Yes	No	Yes					

Construct a 95% confidence interval for the proportion of all men experiencing hair loss that will grow new hair by taking this drug.

Yes	1	Number: Yes	22
No	0	Sample Size	40
No	0	Sample Proportion	0.55
No	0	Critical Value	1.959961
Yes	1		
No	0	Margin of Error	0.154172
No	0		
No	0	Lower Limit	0.395828
Yes	1	Upper Limit	0.704172
No	0		
Yes	1		
Yes	1		
Yes	1		
Yes	1		
No	0		
No	0		
Yes	1		
No	0		
Yes	1		
Yes	1		
Yes	1		
Yes	1		
Yes	1		
Yes	1		
No	0		
Yes	1		
Yes	1		
Yes	1		
No	0		
No	0		
No	0		
Yes	1		
No	0		
No	0		
Yes	1		
Yes	1		
No	0		
Yes	1		
No	0		
Yes	1		

Type the results in column **A** (cells **A1** through **A40**). In this case, the success is a "Yes". We need to change all of the successes into 1s, and the failures into 0s. We do this in order to be able to calculate the proportion of successes.

In cell **B1,** type =IF(A1="Yes",1,0). This will change a "Yes" into a 1, and anything else into a 0. After you press the Enter key, click on cell **B1** again. Next, hold down the Shift key and click on cell **B40.** Pull down the **Edit** menu, and click on **Fill** and then **Down.** This action converts the rest of column **A.**

In cell **C1,** type Number: Yes. Next to that in cell **D1,** type =SUM(B1:B40). This will total column **B,** and tell us how many men grew new hair.

In cell **C2,** type Sample Size. Type the sample size (40) in cell **D2.**

In cell **C3,** type Sample Proportion. In cell **D3** type =D1/D2 to calculate the sample proportion.

In cell **C4,** type Critical Value. In cell **D4,** type =NORMSINV(1-0.025) to calculate the critical value.

The next step is to calculate the margin of error. In cell **C6** type Margin of Error. In cell **D6,** type =D4*SQRT(D3*(1-D3)/D2) to calculate the margin of error.

Finally, we can find the endpoints of the confidence interval. In cell **C8,** type Lower Limit, and type Upper Limit in cell **C9.** To find the lower limit, type =D3-D6 in cell **D8.** We find the upper limit in cell **D9** by typing =D3+D6.

The confidence interval is $0.3958 \leq \pi \leq 0.7042$. ∎

The Excel worksheet is shown on the left.

TI-83 | Finding Critical Values, Constructing Confidence Intervals for Population Proportions

We can use the TI-83 to construct a confidence interval for a population proportion with ease. Here is an example for which we will construct a confidence interval.

EXAMPLE 6.30

A company that makes a drug to reverse hair loss in males claims that 60% of the men who try their drug will grow new hair. A test on 40 randomly selected males experiencing hair loss produced the following results (Yes = grew new hair; No = no new hair).

Yes	No	No	No	Yes	No	No	No	Yes
No	Yes	Yes	Yes	Yes	No	No	Yes	No
Yes	Yes	Yes	Yes	Yes	Yes	No	Yes	Yes
Yes	No	No	No	Yes	No	No	Yes	Yes
No	Yes	No	Yes					

Construct a 95% confidence interval for the proportion of all men experiencing hair loss that will grow new hair by taking this drug.

Begin by pressing (STAT), and moving to the **TESTS** menu. Select option **A: 1-PropZInt.** Next to **x:** enter the number of successes (22). Next to **n:** enter the sample size (40). After entering the level of confidence, highlight **Calculate** and press (ENTER). Here is the screen that you will see.

The confidence interval is $0.39583 \le \pi \le 0.70417$. ∎

Appropriate Sample Size

We can determine how large our sample needs to be for estimating a population proportion just as we did for estimating a population mean. We begin with the formula for margin of error, and solve for n.

$$E = z_{\alpha/2} \cdot \sqrt{\frac{p \cdot (1 - p)}{n}}$$

$$E^2 = \left(z_{\alpha/2} \cdot \sqrt{\frac{p \cdot (1 - p)}{n}}\right)^2$$

$$E^2 = z_{\alpha/2}{}^2 \cdot \frac{p \cdot (1 - p)}{n}$$

$$E^2 \cdot n = z_{\alpha/2}{}^2 \cdot \frac{p \cdot (1 - p)}{n} \cdot n$$

$$E^2 \cdot n = z_{\alpha/2}^2 \cdot p \cdot (1 - p)$$

$$\frac{E^2 \cdot n}{E^2} = \frac{z_{\alpha/2}^2 \cdot p \cdot (1 - p)}{E^2}$$

$$n = \frac{z_{\alpha/2}^2 \cdot p \cdot (1 - p)}{E^2}$$

$$n = \left(\frac{z_{\alpha/2}}{E}\right)^2 \cdot p \cdot (1 - p)$$

The formula for the appropriate sample size to estimate a proportion is

$$n = \left(\frac{z_{\alpha/2}}{E}\right)^2 \cdot p \cdot (1 - p)$$

If there is no estimate of the sample proportion given, then we replace p by 0.5. This makes the biggest possible result, which will have us sample too many rather than too few. When substituting for the margin of error, be sure to use a decimal value and not a percentage for E.

 EXAMPLE 6.31   A campaign manager is interested in finding out what percentage of registered voters will vote for her candidate in November. She knows that her candidate has been getting approximately 40% of the voters in other polls. How large a sample is required to be 95% confident that the percentage of all registered voters in favor of her candidate is within 2% of the sample percentage?

We have an estimate of the sample proportion p in this problem. We will use 40% for p. The level of confidence is 95%.

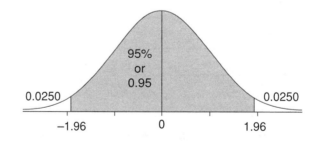

Thus, the critical value is 1.96. Finally, we use 0.02 for E.

$$n = \left(\frac{z_{\alpha/2}}{E}\right)^2 \cdot p \cdot (1 - p)$$

$$= \left(\frac{1.96}{0.02}\right)^2 \cdot 0.4 \cdot 0.6$$

$$= 2304.96$$

The rule for rounding is the same as when trying to estimate a mean—if your result is not a whole number, round your answer up to the next whole number. Thus, she must sample at least 2305 registered voters. ∎

EXAMPLE 6.32   A financial aid officer is curious to find out what percentage of college students who are parents are single parents. There is no known estimate of the sample proportion. How large a sample is required to be 95% confident that the percentage of single parents is within 4% of the sample percentage?

Since there is no known estimate of the sample proportion, we will use 0.5 for *p*. The level of confidence is 95%.

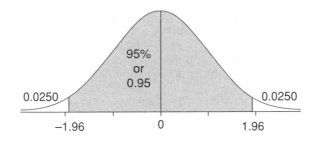

The critical value is 1.96. We will use 0.04 for *E*.

$$n = \left(\frac{z_{\alpha/2}}{E}\right)^2 \cdot p \cdot (1 - p)$$

$$= \left(\frac{1.96}{0.04}\right)^2 \cdot 0.5 \cdot 0.5$$

$$= 600.25$$

The financial officer must sample at least 601 college students. ■

EXERCISES 6.4

1. In a random sample of 400 registered voters in New York State during the summer of 1999 who were asked whether they thought that Hillary Clinton should run for the U.S. Senate, 220 of the respondents said that she should run. Construct a 95% confidence interval for the proportion of all registered voters in New York State that felt that Hillary Clinton should run for the U.S. Senate.

2. A community college is considering offering an intensive algebra course to replace a two-semester algebra sequence. Of the 120 students who were randomly placed into the intensive course, 72 passed the course. Construct a 90% confidence interval for the proportion of all algebra students who could pass this course.

3. What percentage of Americans are not exposed to secondhand smoke? A sample of 200 nonsmokers revealed that only 22 of them did not show traces of a chemical that appears in the blood of people exposed to secondhand smoke. Construct a 99% confidence interval for the percentage of all Americans that are not exposed to secondhand smoke.

4. During the 1998 election, an exit poll of 4400 registered voters revealed that 2376 approved of the job that Bill Clinton was doing as president. Construct a 99% confidence interval for the proportion of all registered voters who approved of the job that Bill Clinton was doing as president. (Based on the results of a *USA Today/Gallup Poll*.)

5. A random sample of 250 registered voters revealed that 63 of them feel that education is the most important issue when deciding on a candidate. Construct a 90% confidence interval for the proportion of all registered voters who feel that education is the most important issue when deciding on a candidate.

6. A sixth-grade student claims to have a rigged coin—it is supposed to land on heads more often than tails. He flips it 30 times, and 18 of the tosses are heads. Construct a 95% confidence interval for the proportion of times that this coin will land on heads.

7. A local buffet restaurant has many patrons that are senior citizens. The manager of the restaurant claims that at least 60% of her customers are senior citizens. A random sample of 50 diners revealed that 33 of them are senior citizens.

   (a) Construct a 95% confidence interval for the true proportion of customers that are senior citizens.

   (b) Is the manager's claim consistent with your interval? Explain your answer in your own words.

8. A nurse at a community college claims that less than 25% of college students smoke. A random sample of 30 college students revealed that 6 of them smoke.

   (a) Construct a 90% confidence interval for the true proportion of college students that smoke.

   (b) Is the nurse's claim consistent with your interval? Explain your answer in your own words.

   (c) Explain how you would gather the data to construct this confidence interval, and list any biases that may be present using your method.

9. A magazine article claimed that half of all people list pepperoni as their favorite pizza topping. A random sample of 150 people revealed that 69 of them listed pepperoni as their favorite pizza topping.

   (a) Construct a 99% confidence interval for the true proportion of people that list pepperoni as their favorite pizza topping.

   (b) Is the article's claim consistent with your interval? Explain your answer in your own words.

10. A 911 operator claims that half of all 911 calls can be classified as "nonemergency" calls. Thirty-four of 64 randomly selected 911 calls were nonemergency calls.

    (a) Construct a 95% confidence interval for the true proportion of 911 calls that are nonemergency calls.

    (b) Is the operator's claim consistent with your interval? Explain your answer in your own words.

11. A *USA Today*/CNN/Gallup Poll of 1005 adults nationally conducted October 21–24, 1999, asked "If Al Gore were the Democratic presidential nominee and George W. Bush were the Republican nominee, whom would you support?" Five hundred fifty-three of those polled said they would support George W. Bush. Construct a 95% confidence interval for the proportion of all voters that supported George W. Bush for president versus Al Gore at this time.

12. A poll of 240 registered voters at least 40 years old revealed that 158 of them voted in the last presidential election.

    (a) Construct a 90% confidence interval for the proportion of registered voters at least 40 years old that voted in the last presidential election.

    (b) Explain how you would gather the data to construct this confidence interval, and list any biases that may be present using your method.

13. A sample of 75 renters in a city showed that 40 were under the age of 35. Construct a 99% confidence interval for the percentage of all renters that are under the age of 35.

14. A survey of 200 people aboard a cruise ship showed that 76 of them had children under age 18 at home. Construct a 95% confidence interval for the proportion of all cruise-goers that have children under age 18 at home. (Based on the results of a study by Cruise Line International.)

15. An ABC News poll of 452 "leaned Democrats" (Democrats and independents who lean toward the Democratic party) was conducted during the period of August 10–15, 1999. This was right after rumors began to circulate about actor Warren Beatty seeking the Democratic presidential nomination. The poll showed that 9% of those surveyed supported Beatty for the nomination. Construct a 90% confidence interval for the proportion of all leaned Democrats that supported Warren Beatty for the nomination.

16. A nonscientific Internet poll showed that 73% of 2430 drivers feel that they are "above average" drivers. Construct a 95% confidence interval for the proportion

of all drivers that feel they are above average drivers. (Based on the results of a poll on abcnews.com.)

17. What percentage of young drivers run red lights? A survey of 124 drivers ages 18 to 25 showed that 89 of them run red lights. (Based on the results of a study by the Social Science Research Center, Department of Psychology, Old Dominion University.)

    (a) Construct a 90% confidence interval for the proportion of all drivers ages 18 to 25 that run red lights.

    (b) Explain how you would gather the data to construct this confidence interval, and list any biases that may be present using your method.

18. A sample of 412 adults showed that 93 of them are connected to the Internet from home. Construct a 95% confidence interval for the proportion of all adults that have Internet access at home.

19. From a random sample of 139 incoming freshmen, 79 ended up graduating within six years. Construct a 95% confidence interval for the proportion of all college students that graduate within six years. (Based on results published by the NCAA.)

20. A random sample of 47 people showed that 19 of them get more than 10 e-mails per day at work. Construct a 90% confidence interval for the proportion of all people that get more than 10 e-mails per day at work. (Based on a survey by the John J. Heldrich Center for Workforce Development.)

21. In a survey of 320 women, 64 said that Valentine's Day was their favorite occasion to receive flowers. Construct a 99% confidence interval for the proportion of all women who feel that Valentine's Day is their favorite occasion to receive flowers. (Based on a study by the International Communications Research for the Society of American Florists.)

22. One hundred parents of children age 4 and younger were asked whether they were "very concerned" about their child's safety and 89 said that they were. Construct a 95% confidence interval for the proportion of all parents of children age 4 and younger that are "very concerned" about their safety. (Based on a study by Yankelovich Partners for Gerber Products.)

23. A sample of 340 Americans were asked whether they had ever seen a UFO, and 20 said that they had. Construct a 90% confidence interval for the proportion of all Americans who have seen a UFO. (Based on a study by Yankelovich Partners for *Life* magazine.)

24. When asked whether they drink alcohol or not, 79 of 441 college students said that they did not. Construct a 95% confidence interval for the proportion of all college students that do not drink alcohol. (Based on the Harvard School of Public Health College Alcohol Study.)

25. A study of 619 twelfth-graders showed that 74 of them smoked at least a half-pack of cigarettes daily. Construct a 99% confidence interval for the proportion of all twelfth-graders that smoke a half-pack or more daily. (Based on results of a study by the American Academy of Pediatric Dentistry.)

26. A random sample of 436 workers by the Society of Financial Service showed that 47% of them felt that it was "seriously unethical" to inspect lockers or work areas, whereas 41% of 121 senior-level bosses felt the same way.

    (a) Construct a 95% confidence interval for the proportion of all workers who view this as "seriously unethical."

    (b) Construct a 95% confidence interval for the proportion of all bosses who view this as "seriously unethical."

    (c) Does it appear that the proportion of all workers who view this as "seriously unethical" is the same as the proportion of all bosses who view this as "seriously unethical"? Using your confidence intervals, explain why or why not.

27. A random sample of 250 mobile phone users showed that 105 had taken an incoming call while dining out. (Based on a study by the National Restaurant Association.)

    (a) Construct a 95% confidence interval for the proportion of all mobile phone users who have taken an incoming call while dining out.

    (b) Based on your confidence interval, does it appear that fewer than 50% of all mobile phone users have taken an incoming call while dining out?

28. A random sample of 1477 high school students revealed that 489 of them had ridden in a car at least once in the preceding 30 days with a driver who had been drinking alcohol. Construct a 90% confidence interval for the proportion of all

high school students who have ridden in a car at least once in the past 30 days with a driver who had been drinking alcohol. (Based on the Morbidity and Mortality Weekly Report by the Centers for Disease Control and Prevention.)

29. A recent survey showed that 802 of 4155 high school students had seriously considered attempting suicide in the past 12 months. Construct a 99% confidence interval for the proportion of all high school students that have seriously considered attempting suicide in the past 12 months. (Based on the Morbidity and Mortality Weekly Report by the Centers for Disease Control and Prevention.)

30. A random sample of 250 high school students that smoke cigarettes revealed that 174 of them were not asked to show proof of age the last time they purchased cigarettes. Construct a 95% confidence interval for the proportion of all high school students that were not asked to show proof of age the last time they purchased cigarettes. (Based on the Morbidity and Mortality Weekly Report by the Centers for Disease Control and Prevention.)

31. A researcher wants to determine what proportion of all high school students have Internet access at home. He has no idea what the sample proportion will be. How large a sample is required to be 95% sure that the sample proportion is off by no more than 5%?

32. A news agency is planning a poll. The agency wants to determine what proportion of American citizens supported NATO involvement in Kosovo. If the agency wants to be 99% sure that the sample proportion differs from the true population proportion by no more than 3%, how large a sample is necessary?

33. A student wants to determine what percentage of college students smoke. How large a sample should she take to be 90% confident that her sample proportion is off by no more than 4.5%?

34. An ambitious health researcher wants to determine the proportion of American teenagers that are overweight. The researcher wants to be within 1% of the actual proportion. How large a sample is necessary to be 95% confident that the sample proportion is within the researcher's desired margin of error?

35. A parents' organization is trying to determine the percentage of high school athletes that make the Dean's List. The leader of the organization believes that the percentage is in the neighborhood of 30%. How large a sample will the organization have to take to be 99% confident that their sample proportion is within 4% of the actual proportion?

36. An orange grower has several hundred orchards that were replanted this year. He suspects that many of the new trees have a root disease. The nursery has agreed to reimburse the grower for the trees that are diseased. The disease can only be detected by digging up the tree, which will kill the tree. The highest known percentage of trees to have this disease is 5%. How many trees must be dug up to be 90% confident that the sample proportion of diseased trees is within 2% of the actual proportion of trees that have the root disease?

MINI PROJECT

Randomly sample at least 100 vehicles in your school's parking lots, and record how many are convertibles. Use these sample data to construct a 95% confidence interval for the proportion of all vehicles at your school that are convertibles. You may choose some other classification of vehicles that interests you (for example, vans or foreign-made cars). In addition to your confidence interval, include

- Your raw data
- A pie chart showing the percentage of vehicles at your school that are convertibles and the percentage that are not.

Finally, answer the following questions.

- How did you obtain your sample?
- Which type of sampling did you use?
- Was the sample truly random?
- What potential biases may show up in your sample?

Confidence Intervals for
Population Proportions

We will use Excel to explore the idea of confidence intervals for population proportions and how to interpret them. We will take 100 samples, each of size 100. Each observation will be either a 1 (success) or a 0 (failure). For each sample, we will calculate a confidence interval for π. We will then determine what percentage of the confidence intervals actually capture the true value of the population proportion.

Before we begin we must be sure that the Analysis Tool Pak has been added. Pull down the **Tools** menu. If you see a choice that says **Data Analysis**, then the Tool Pak is added and you may continue. If you do not see **Data Analysis**, click on **Add-ins** under **Tools**. When the dialog box opens up, click on the box for **Analysis Tool Pak**, and click on **OK**.

We begin by selecting 100 samples, each of size 100. Pull down the **Tools** menu and click on **Data Analysis**. When the dialog box appears, click on **Random Number Generation** and OK. When the next dialog box appears, enter 100 in the box labeled **Number of Variables**. (This is the number of samples that we are going to take.) Enter 100 in the box labeled **Number of Random Numbers**. (This is the sample size.) In the box labeled **Distribution**, click on **Bernoulli**. For **Parameters**, enter 0.4 for the p-value. Usually, the population parameter π is unknown. Therefore, we try to estimate it using a confidence interval. However, for the purpose of this simulation it is imperative that we know the true population proportion so that we can find out how many of our confidence intervals contain π. Our value for the population proportion is 0.4. To create the samples, click **OK**. (This may take a few minutes, so be patient.)

In cell **A101**, type =SUM(A1:A100)/100. This will calculate the proportion of successes in the first sample. Click on cell **A101**, then hold the Shift key and click on cell **CV101** to highlight the 101st row. Pull down the **Edit** Menu; select **Fill** and then **Right**. This will calculate the proportion of successes for all 100 samples.

In row 102 we will calculate the lower limit of each confidence interval. In cell **A102**, type =A101-1.96*SQRT(A101*(1-A101)/100). Click on cell **A102**, then hold the Shift key and click on cell **CV102** to highlight the 102nd row. Pull down the **Edit** Menu, select **Fill** and then **Right**.

In row 103 we will calculate the upper limit of each confidence interval. In cell **A103**, type =A101+1.96*SQRT(A101*(1-A101)/100). Click on cell A103. Hold down the Shift key and click on cell **CV103**. Pull down the **Edit** Menu; click on **Fill** and then **Right**.

Now we will determine exactly how many intervals contain the true population proportion, 0.4. In cell **A104**, type =IF((A102<0.4)*AND(A103>0.4),1,0). If the first confidence interval contains 0.4, the value in cell **A104** will be 1; otherwise, it will be 0. To determine whether the other intervals contain 0.4, click on cell **A104**. Hold down the Shift key and click on cell **CV104**. Pull down the **Edit** Menu; click on **Fill** and then **Right**. To determine how many intervals contain 0.4, type =SUM(A104:CV104) in cell **A105**. If things go according to plan, approximately 95 of our intervals should contain the population proportion.

Repeat this process for sample sizes of 250 and 1000. Does a different sample size dramatically affect the number of intervals that contain the population proportion? Why or why not?

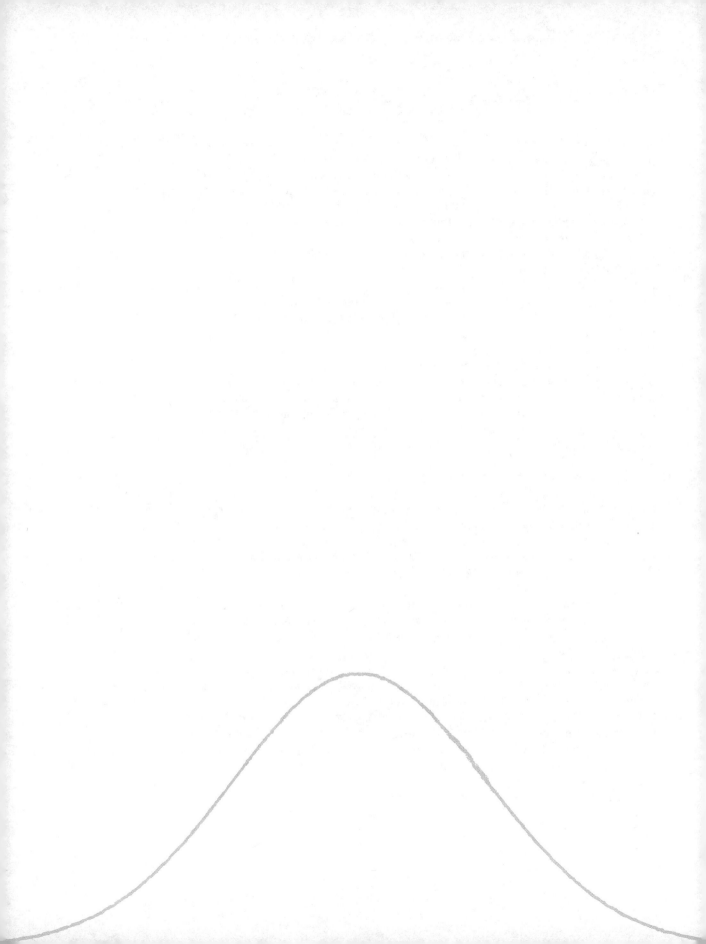

One-Sample Hypothesis Tests

A hypothesis test is a procedure that we use to determine whether a claim about a population parameter is valid, invalid, or somewhere in between. In this chapter we will be performing three different tests that involve using a single sample to make an inference about a population parameter.

In Section 7.1 we look at the test for a population mean based on a large sample. This test involves comparing the mean of a population to a certain number, such as "The mean age of college statistics students is 24." It is in this section that the procedure for testing a claim is established.

In Section 7.2 we will cover the small-sample test for a single population mean. This test, using the *t*-distribution, is used when the sample size is below 30.

Finally, in Section 7.3, we cover the large-sample single proportion test. This test is used when a population proportion is compared to a certain proportion. An example of such a claim would be "13.6% of pregnant women smoke while they are pregnant."

SECTION 7.1
Hypothesis Test for a Single Population Mean (Large Samples)

A college statistics instructor claims that the mean age of college statistics students is 24. Do we just assume that the instructor is right, or is there a way we can find out whether the claim is true? We will develop a five-step procedure to determine the validity of such a claim. The procedure is called a **hypothesis test**. The test is based on the confidence intervals that we developed in Chapter 6, and in many ways mirrors our legal system. Let's take a look at a confidence interval example to develop some intuition for this process.

 EXAMPLE 7.1   A college statistics instructor claims that the mean age of college statistics students is 24. A random sample of 116 college statistics students revealed a mean age of 22.7 years with a standard deviation of 5.68 years. Construct a 95% confidence interval for the true mean age of all college statistics students. Use the interval to determine the validity of the instructor's claim.

First, we find the critical value $z_{\alpha/2}$ for a 95% confidence interval ($\alpha = 0.05$).

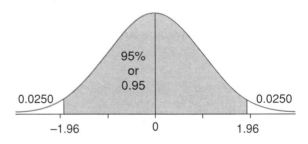

We see that $z_{\alpha/2} = 1.96$. Now we determine the margin of error.

$$E = z_{\alpha/2} \cdot \frac{s}{\sqrt{n}}$$

$$= 1.96 \cdot \frac{5.68}{\sqrt{116}}$$

$$= 1.03$$

Next we construct the confidence interval.

$$\bar{x} - E \leq \mu \leq \bar{x} + E$$

$$22.7 - 1.03 \leq \mu \leq 22.7 + 1.03$$

$$21.67 \leq \mu \leq 23.73$$

The confidence interval tells us that we are 95% sure that the true mean age for all college statistics students is between 21.67 and 23.73 years old. There is a 5% chance that the mean is not in this interval, which tells us that there is less than a 5% chance that the mean is actually 24. The instructor's claim appears to be invalid. ■

Now we develop the five-step procedure for a hypothesis test. We will come back to this example several times to show how a hypothesis test goes hand in hand with a confidence interval. The five steps are as follows.

- State the claim and the hypothesis
- Choose a level of significance
- Choose a test statistic
- State a decision rule
- Do the calculations and draw conclusions

Step 1: State the Claim and the Hypotheses

The **claim** is the statement that we are interested in investigating. Each hypothesis test should begin by identifying the claim to be tested, and writing it in words. For the previous example, the claim would be "The mean age of college statistics students is 24."

Once the claim has been identified, we need to write it in symbolic form. In our claim, we are comparing a mean of a population to the number 24. We let μ represent the mean age of all college statistics students, and our claim becomes $\mu = 24$.

The next thing that we do in Step 1 is to write the complement of the claim in symbolic form. What is the complement of $\mu = 24$? It is $\mu \neq 24$.

Suppose the claim had been $\mu < 24$? What is the complement of "less than 24"? At first, we might be tempted to say "greater than 24," but this would not be correct. The mean could be less than 24, greater than 24, or equal to 24. If the claim only uses one of these, the complement combines the other two. Therefore, the complement of $\mu < 24$ is $\mu \geq 24$.

 EXAMPLE 7.2   Here are the six different claims that could be made comparing the mean age of all college statistics students to 24.

1. The mean age of all college statistics students is 24.
   Claim: $\mu = 24$ Complement: $\mu \neq 24$

2. The mean age of all college statistics students is not 24.
   Claim: $\mu \neq 24$ Complement: $\mu = 24$

3. The mean age of all college statistics students is less than 24.
   Claim: $\mu < 24$ Complement: $\mu \geq 24$

4. The mean age of all college statistics students is at least 24.
   Claim: $\mu \geq 24$ Complement: $\mu < 24$

5. The mean age of all college statistics students is greater than 24.
   Claim: $\mu > 24$ Complement: $\mu \leq 24$

6. The mean age of all college statistics students is at most 24.
   Claim: $\mu \leq 24$ Complement: $\mu > 24$ ■

Once the claim and its complement have been established, we must decide which of those two is the **null hypothesis (H_0)** and which is the **alternate hypothesis (H_A)**. Of the claim and its complement, the null hypothesis is the statement containing equality. In other words, we are looking for the symbols =, ≤ , or ≥. We assume that the null hypothesis is true, and look for sample evidence to suggest that it is false. In our example, we are looking for a sample mean that is significantly lower than 24 or significantly higher than 24 to suggest that the true mean is not 24.

The assumption that the null hypothesis is true is equivalent to assuming that a person on trial is innocent (until proven guilty). If there is sufficient evidence to prove that the assumption of the person's innocence is false beyond a reasonable doubt, then we reject our assumption that the person is innocent and declare him guilty. The evidence in a hypothesis test comes from a sample. If there is enough sample evidence to suggest that the null hypothesis is false, we reject the null hypothesis and accept the alternate hypothesis.

When we assume that the null hypothesis is true, we are actually assuming that the population mean is equal to the value in the claim. We assume that $\mu = 24$ whether the null hypothesis is $\mu = 24$, $\mu \leq 24$, or $\mu \geq 24$. We say that the claimed value of the mean is 24.

After the null hypothesis has been determined, the alternate hypothesis is the statement that does not contain equality. If we reject the null hypothesis as false, then we accept the alternate hypothesis as true. If we fail to reject the null hypothesis, we are not saying that the null hypothesis is true. We can only say that there is not enough sample evidence to convince ourselves that it is false. In this case, we fail to support the alternate hypothesis as true. When there is not enough evidence to find a defendant guilty, is the defendant then found innocent? No, the defendant is found "not guilty." This means that it was not proven that the person was innocent. There was simply not enough evidence to conclude that the person was guilty.

Reject null hypothesis: support alternate hypothesis
Fail to reject null hypothesis: fail to support alternate hypothesis

 EXAMPLE 7.3   Let's take a look at three claims, and try to determine what the null and alternate hypotheses are.

1. The mean number of hours spent studying per week by college students is less than 10 hours.

    Claim: $\mu < 10$ H_0: $\mu \geq 10$
    Complement: $\mu \geq 10$ H_A: $\mu < 10$

2. The mean price for a home sold in Visalia, California is $100,000.

    Claim: $\mu = 100{,}000$ H_0: $\mu = 100{,}000$
    Complement: $\mu \neq 100{,}000$ H_A: $\mu \neq 100{,}000$

3. The mean starting salary for a college graduate is at most $30,000.

    Claim: $\mu \leq 30{,}000$ H_0: $\mu \leq 30{,}000$
    Complement: $\mu > 30{,}000$ H_A: $\mu > 30{,}000$ ∎

There are two types of errors that we can make in a hypothesis test. A Type I error is rejecting a null hypothesis that is actually true. This is equivalent to convicting an innocent man. A Type II error is failing to reject a null hypothesis that is actually false.

This is equivalent to letting a guilty man go free. We try to minimize Type I errors in hypothesis testing. We control the chances of making a Type I error by establishing a *level of significance.*

Step 2: Choose a Level of Significance (α)

We want to reject the null hypothesis only if the probability that it is true is very small. By "very small" we usually use 5% or 1%. In the case of 5%, we reject the null hypothesis if there is less than a 5% chance that the null hypothesis is true. In other words, if the chances that the null hypothesis is true are low enough, then we will declare it false. This 5%, or 1%, is referred to as the **level of significance.** We use the Greek letter alpha (α) to stand for the level of significance. This is equivalent to the phrase "reasonable doubt" in a trial. We will discuss the use of α in confidence intervals and hypothesis tests when we get to Step 4.

The level of significance that is used most frequently is 0.05. However, if the consequences of rejecting a null hypothesis that is true are serious, it is common to use 0.01 instead. Suppose a company is testing a new drug that they believe lowers high blood pressure. A lower level of significance is appropriate here, because of the serious health implications.

Step 3: Choose a Test Statistic

The **test statistic** is a formula that will be used to determine the probability that the null hypothesis is true based on the sample evidence. For the large-sample, single-mean test the formula is

---

Test Statistic for a Large-Sample, Single-Mean Test
(Population Standard Deviation Known)

$$z = \frac{\bar{x} - \mu}{\sigma/\sqrt{n}}$$

where $\bar{x}$ = sample mean
  μ = claimed population mean
  σ = population standard deviation
  n = sample size

As mentioned earlier in the text, the population standard deviation is often unknown. If the sample size is 30 or greater, the sample standard deviation is an acceptable estimate of the population standard deviation.

Test Statistic for a Large-Sample, Single-Mean Test
(Population Standard Deviation Unknown)

$$z = \frac{\bar{x} - \mu}{s/\sqrt{n}}$$

where $\bar{x}$ = sample mean
  μ = claimed population mean
  s = population standard deviation
  n = sample size

This test statistic is the same formula we used in the previous chapter when trying to find probabilities involving a sample mean. If we are assuming that the value of μ is correct, this test statistic will allow us to calculate the probability of obtaining a sample mean that is as extreme as the mean of our sample.

The test statistic will change for different types of hypothesis tests. The above formula applies only to a test for a single population mean when the sample size is at least 30. In this chapter there will be a second test statistic for a small-sample, single-mean test, and a third for the large-sample, single-proportion test.

Step 4: State the Decision Rule

When we state the **decision rule**, we are establishing the guidelines for rejecting the null hypothesis. We will be finding the critical value(s) for our test statistic. If the value of the test statistic is beyond the critical value for the test, we reject the null hypothesis. The decision rule for this test depends on two things: the alternate hypothesis and the level of significance.

We return to our example to explore the establishment of a decision rule.

 EXAMPLE **7.4**   A college statistics instructor claims that the mean age of college statistics students is 24. Test his claim at the 0.05 level of significance. A random sample of 116 college statistics students revealed a mean age of 22.7 years with a standard deviation of 5.68 years.

Step 1

Claim in words: The mean age of all college statistics students is 24.
Claim: $\mu = 24$
Complement: $\mu \neq 24$
H_0: $\mu = 24$
H_A: $\mu \neq 24$

Step 2

Level of significance: $\alpha = 0.05$

Step 3

Test statistic: $z = \dfrac{\bar{x} - \mu}{s/\sqrt{n}}$

To establish our decision rule, we must first decide what types of sample means suggest that the null hypothesis is false and, more important, that the alternate hypothesis is true. There are two ways that the population mean could be not equal to 24: it could be less than 24, and it could be greater than 24. This is an example of a **two-tailed test.** It is called a two-tailed test because the sample means that suggest the null hypothesis is false are found in the right tail of the z-distribution ($\mu > 24$) and the left tail of the z-distribution ($\mu < 24$).

To set up our decision rule, we look for the critical values that separate the outer 5% of the z-distribution from the middle 95%. We do this by dividing the level of significance in half ($\alpha/2$) and placing that amount in each of the tails. We then look for the z-values that go with such a graph. We need to look for the z-value that is associated with an area of 0.0250 ($z = -1.96$), and the z-value that is associated with an area of 0.9750 ($z = 1.96$).

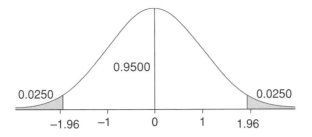

The shaded regions on the outside are called the **regions of rejection.** If the value of the test statistic falls into one of these regions, we reject the null hypothesis. Otherwise, we fail to reject the null hypothesis. All that remains to be done is to formally state our decision rule.

Decision rule: Reject H_0 if $z < -1.96$ or if $z > 1.96$. ■

Two-Tailed Tests   A test is two-tailed if our alternate hypothesis involves the symbol ≠. To find our decision rule for a two-tailed test, divide the level of significance and place that amount in the left and right tails. Find the critical values of z for the graph that we have, and reject the null hypothesis if the value of the test statistic is outside of these critical values.

We used the symbol $z_{\alpha/2}$ for the critical value in a confidence interval in the previous chapter. Here is why we did that. Suppose we were going to construct a 95% confidence interval for a population mean. The "middle" 95% leaves 5% of the area under the curve in the two outer tails.

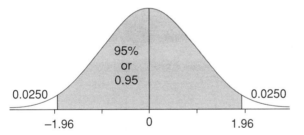

That 5% we call alpha (α). If we divided α in half, that would put $\alpha/2$ in each tail. Thus, our critical values go along with the critical values that we would obtain by putting $\alpha/2$ in each tail to begin with.

EXAMPLE 7.5

Find the decision rule for a two-tailed test at the 0.01 level of significance.

First we divide α by 2, which gives us 0.005. We put 0.005 in the left tail and right tail, which produces the graph shown here.

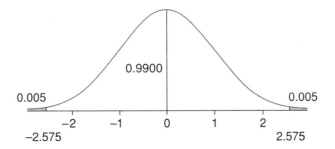

We look for the z-values associated with areas of 0.0050 (left tail) and 0.9950 (right tail). From the graph, the decision rule is

Reject H_0 if $z < -2.575$ or if $z > 2.575$ ■

Right-Tailed Tests   Suppose we had the following hypotheses.

H_0: $\mu \leq 30,000$
H_A: $\mu > 30,000$

What type of sample mean would be required to suggest that the null hypothesis was false and that the alternate hypothesis was true? To conclude that $\mu > 30,000$, we would have to have a sample mean that was significantly higher than 30,000, which is the claimed value of μ. We find the sample means that are significantly higher than 30,000 in the right tail.

The decision rule for a right-tailed test differs from a two-tailed test because the entire level of significance α is put in the right tail. It is not divided in half, because there are not two tails to put it in.

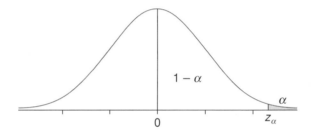

Once α is placed in the right tail, we find the critical value in the same fashion as we did with a two-tail test.

 EXAMPLE 7.6   Find the decision rule for a right-tailed test at the 0.01 level of significance.

We put 0.01 in the right tail, which produces the graph shown here.

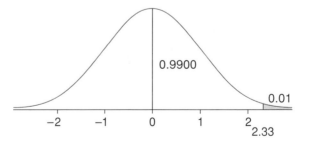

We look for the z-value associated with an area of 0.9900. From the graph, the decision rule is

Reject H_0 if $z > 2.33$ ■

Left-Tailed Tests   Suppose we had the following hypotheses.

$H_0: \mu \geq 10$
$H_A: \mu < 10$

What type of sample mean would be required to suggest that the null hypothesis was false and that the alternate hypothesis was true? To conclude that $\mu < 10$, we would have to have a sample mean that was significantly lower than 10, which is the claimed value of μ. We find the sample means that are significantly lower than 10 in the left tail.

The decision rule for a left-tailed test differs from the previous tests because the entire level of significance α is put in the left tail. It is not divided in half, because there are not two tails to put it in.

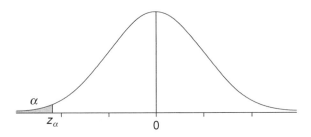

Once α is placed in the left tail, we find the critical value in the same fashion as before.

 Find the decision rule for a left-tailed test at the 0.05 level of significance.

We put 0.05 in the left tail, which produces the graph shown here.

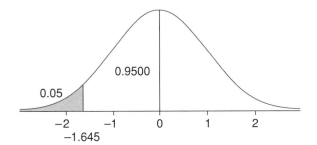

We look for the z-value associated with an area of 0.0500. From the graph, the decision rule is

Reject H_0 if z < –1.645 ∎

Step 5: Do the Calculations and Draw Conclusions

The first task in Step 5 is to calculate the test statistic using the sample information. Once that has been accomplished, we check our decision rule to make our decision about the null hypothesis H_0. Then we must draw our final conclusion about the claim being tested. Let's begin with the calculation.

Recall our example: A college statistics instructor claims that the mean age of college statistics students is 24. Test his claim at the 0.05 level of significance. A random sample of 116 college statistics students revealed a mean age of 22.7 years with a standard deviation of 5.68 years.

Let's calculate the test statistic for our example. There are four values that need to be plugged into this statistic: the sample mean ($\bar{x}$), the claimed population mean (μ), the sample standard deviation (s), and the sample size (n). Here is what we get when we plug the correct values into the test statistic.

$$z = \frac{\bar{x} - \mu}{s/\sqrt{n}}$$

$$= \frac{22.7 - 24}{5.68/\sqrt{116}}$$

$$= -2.47$$

If the value of our test statistic is -2.47, what is our decision about the null hypothesis H_0? Recall the decision rule:

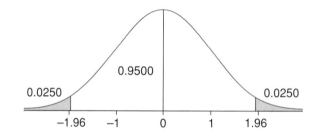

Reject H_0 if $z < -1.96$ or if $z > 1.96$

Since -2.47 falls in a rejection region ($-2.47 < -1.96$), our decision is to reject H_0.

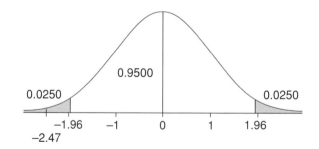

We still must say something about our claim; after all, we were testing the instructor's claim, not the null hypothesis. Our conclusion about the claim depends on two things: whether our claim was the null or alternate hypothesis, and whether the null hypothesis was rejected or not.

If the claim is the null hypothesis, then our conclusion is easy to state. In such a case, whatever our decision is about the null hypothesis, our conclusion about the claim is the same. If the claim is the null hypothesis and the null hypothesis is rejected, then the claim is rejected as well. We have proved the claim to be false. If the claim is the null hypothesis and we fail to reject the null hypothesis, then we fail to reject the claim. In this case we have failed to prove the claim to be false. Please note that this is not the same as proving the claim to be true. When the claim is the null hypothesis, either the claim is proved to be false or we fail to prove that it is false.

If the claim is the alternate hypothesis, then we either support it as being true or we fail to support it. If the claim is the alternate hypothesis and the null hypothesis is rejected, then the claim is supported. In other words, if we throw the null hypothesis out as being false, then we accept the alternate hypothesis (the claim) as being true. We have proved the claim to be true. If the claim is the alternate hypothesis and we fail to reject the null hypothesis, then we fail to support the claim. In this case we have failed to prove the claim to be true.

The possible conclusions are illustrated in the following chart. You may find it handy to use as a guide as you work through the homework exercises, but you must understand why each conclusion is the way it is. If you cannot understand and be able to draw your own conclusion, then you will never fully understand the process of conducting a hypothesis test.

	Reject H_0	**Fail to Reject H_0**
Claim is H_0	Reject the claim	Fail to reject the claim
Claim is H_A	Support the claim	Fail to support the claim

In our example, the claim was the null hypothesis. Since our decision was to reject the null hypothesis, then our conclusion is that there is sufficient sample evidence to reject the claim that the mean age of all college statistics students is 24.

P-Values

One more step that can be added to a hypothesis test is the calculation of a **p-value**. The p-value for a hypothesis test is the probability of obtaining a sample as extreme as our sample if the null hypothesis is actually true. We will find the probability of obtaining a value as extreme as the test statistic, or more extreme, if the null hypothesis is indeed true. When calculating a p-value, we need to know two things: what tailed test we have and the value of the test statistic.

Two-Tailed P-Values   We find a p-value for a two-tailed hypothesis test by putting the value of the test statistic on a normal distribution graph, finding the area in the tail that it creates, and multiplying by 2. Multiplying by 2 takes into consideration the tail on the opposite side.

For our example, we found a test statistic of –2.47. Put it on a normal distribution graph, and find the area in the left tail.

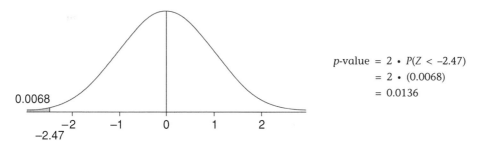

$$p\text{-value} = 2 \cdot P(Z < -2.47)$$
$$= 2 \cdot (0.0068)$$
$$= 0.0136$$

This tells us that the probability of obtaining a sample mean as extreme as ours if the null hypothesis is actually true is only 0.0136. This is a much stronger statement than saying that there is less than a 0.05 probability that it is true.

The decision rule can be rewritten in terms of the *p*-value and the level of significance. Recall that we want to reject the null hypothesis only if the probability that it is true is less than the level of significance. This translates to the following:

Reject H_0 if *p*-value < α

In our example, since the *p*-value was less than 0.05, the correct decision is to reject the null hypothesis. What if the level of significance had been 0.01? We would have failed to reject the null hypothesis, because our *p*-value is greater than 0.01.

Left-Tailed P-Values   We find a *p*-value for a left-tailed hypothesis test by putting the value of the test statistic on a normal distribution graph, and finding the area to the left of that value. If we had a left-tailed hypothesis test, and the test statistic was calculated to be –1.57, here is how the *p*-value would be calculated.

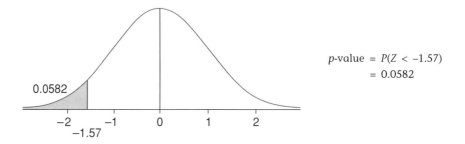

$$p\text{-value} = P(Z < -1.57)$$
$$= 0.0582$$

Right-Tailed P-Values   We find a *p*-value for a right-tailed hypothesis test by putting the value of the test statistic on a normal distribution graph, and finding the area to the right of that value. If we had a right-tailed hypothesis test, and the test statistic was calculated to be 2.29, here is how the *p*-value would be calculated.

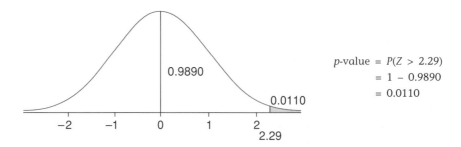

$$p\text{-value} = P(Z > 2.29)$$
$$= 1 - 0.9890$$
$$= 0.0110$$

Our example has been spread out over the entire section. Before moving on, here it is from start to finish, all in one example.

EXAMPLE 7.8

A college statistics instructor claims that the mean age of college statistics students is 24. Test his claim at the 0.05 level of significance. A random sample of 116 college statistics students revealed a mean age of 22.7 years with a standard deviation of 5.68 years.

Step 1

Claim in words: The mean age of all college statistics students is 24.
Claim: $\mu = 24$
Complement: $\mu \neq 24$
$H_0: \mu = 24$
$H_A: \mu \neq 24$

Step 2

Level of significance: $\alpha = 0.05$

Step 3

Test statistic: $z = \dfrac{\bar{x} - \mu}{s/\sqrt{n}}$

Step 4

Two-tailed test, $\alpha = 0.05$

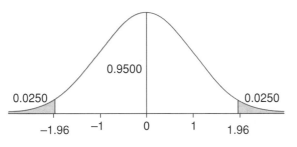

Decision rule: Reject H_0 if $z < -1.96$ or if $z > 1.96$.

Step 5

$$z = \frac{\bar{x} - \mu}{s/\sqrt{n}}$$

$$= \frac{22.7 - 24}{5.68/\sqrt{116}}$$

$$= -2.47$$

Decision: Reject H_0.

Conclusion: There is sufficient sample evidence to reject the claim that the mean age of all college statistics students is 24.

***p*-value**

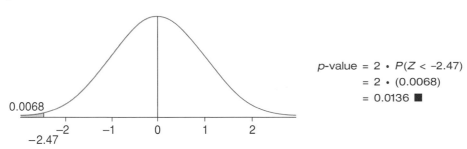

p-value $= 2 \cdot P(Z < -2.47)$
$= 2 \cdot (0.0068)$
$= 0.0136$ ■

Now we will look at three more examples in their entirety. Please use them as a reference while you do the homework exercises. Although this process may seem foreign and difficult, it will be used throughout most of the remainder of the book, so you need to get comfortable with it now. There will be small changes from section to section (different test statistics, different ways of finding the decision rule, . . .) but the general process will remain the same.

 Test the claim that the mean price for a dozen roses at a California florist is higher than $45 at the 0.01 level of significance. A random sample of 30 California florists had a mean price of $48.33 per dozen with a standard deviation of $6.07.

Step 1

Claim in words: The mean price for a dozen roses at a California florist is higher than $45.
Claim: $\mu > 45$
Complement: $\mu \leq 45$
H_0: $\mu \leq 45$
H_A: $\mu > 45$

Note that the claim is the alternate hypothesis, so our conclusion will be either to support the claim or fail to support the claim. Also note that the alternate hypothesis involves the symbol >. This tells us that our test is a right-tailed test.

Step 2

Level of significance: $\alpha = 0.01$

Step 3

Test statistic: $z = \dfrac{\bar{x} - \mu}{s/\sqrt{n}}$

Since we are comparing the mean of a population to a certain number ($45) and the sample size is at least 30, the above test statistic is the appropriate one.

Step 4

Right-tailed test, $\alpha = 0.01$

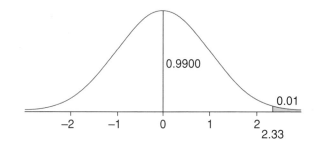

Decision rule: Reject H_0 if $z > 2.33$.

Step 5

$$z = \frac{\bar{x} - \mu}{s/\sqrt{n}}$$

$$= \frac{48.33 - 45}{6.07/\sqrt{30}}$$

$$= 3.00$$

Decision: Reject H_0.

Conclusion: There is sufficient sample evidence to support the claim that the mean price for a dozen roses at a California florist is higher than $45.

p-value

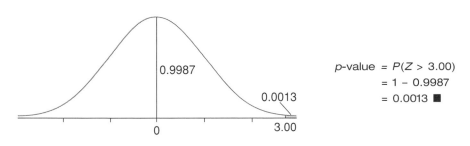

$$p\text{-value} = P(Z > 3.00)$$
$$= 1 - 0.9987$$
$$= 0.0013 \blacksquare$$

 EXAMPLE 7.10 Test the claim that the mean amount spent by a customer at Dylan's Place, a local fast-food restaurant, is less than $5 at the 0.05 level of significance. A random sample of 34 customers had spent a mean of $4.43 and a standard deviation of $3.23.

Step 1

Claim in words: The mean amount spent by a customer at Dylan's Place is less than $5.
Claim: $\mu < 5$
Complement: $\mu \geq 5$
H_0: $\mu \geq 5$
H_A: $\mu < 5$

Note that the claim is the alternate hypothesis, so our conclusion will be either to support the claim or fail to support the claim. Also note that the alternate hypothesis involves the symbol <. This tells us that our test is a left-tailed test.

Step 2

Level of significance: $\alpha = 0.05$

Step 3

Test statistic: $z = \dfrac{\bar{x} - \mu}{s/\sqrt{n}}$

Since we are comparing the mean of a population to a certain number ($5) and the sample size is at least 30, the above test statistic is the appropriate one.

Step 4

Left-tailed test, $\alpha = 0.05$

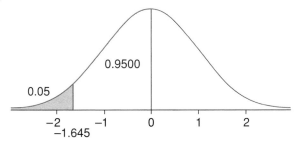

Decision rule: Reject H_0 if $z < -1.645$.

Step 5

$$z = \frac{\bar{x} - \mu}{s/\sqrt{n}}$$

$$= \frac{4.43 - 5}{3.23/\sqrt{34}}$$

$$= -1.03$$

Decision: Fail to reject H_0.

Conclusion: There is not sufficient sample evidence to reject the claim that the mean amount spent by a customer at Dylan's Place is less than $5.

p-value

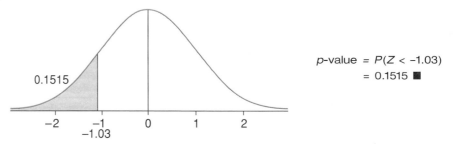

0.1515

$$p\text{-value} = P(Z < -1.03)$$
$$= 0.1515 \ \blacksquare$$

EXAMPLE 7.11   Test the claim that the mean time Americans spend watching TV per day is 4.4 hours at the 0.05 level of significance. A random sample of 80 people had a mean time of 2.225 hours viewing per day and a standard deviation of 1.088 hours.

Step 1

Claim in words: The mean time spent watching TV by Americans per day is 4.4 hours.
Claim: $\mu = 4.4$
Complement: $\mu \neq 4.4$
H_0: $\mu = 4.4$
H_A: $\mu \neq 4.4$

Note that the claim is the null hypothesis, so our conclusion will be either to reject the claim or fail to reject the claim. Also note that the alternate hypothesis involves the symbol $\neq$. This tells us that our test is a two-tailed test.

Step 2

Level of significance: $\alpha = 0.05$

Step 3

Test statistic: $z = \frac{\bar{x} - \mu}{s/\sqrt{n}}$

Since we are comparing the mean of a population to a certain number (4.4 hours) and the sample size is at least 30, the above test statistic is the appropriate one.

Step 4

Two-tailed test, $\alpha = 0.05$

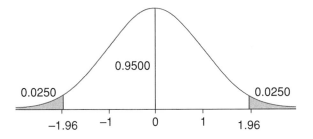

Decision rule: Reject H_0 if $z < -1.96$ or if $z > 1.96$.

Step 5

$$z = \frac{\bar{x} - \mu}{s/\sqrt{n}}$$

$$= \frac{2.225 - 4.4}{1.088/\sqrt{80}}$$

$$= -17.88$$

Decision: Reject H_0.

Conclusion: There is sufficient sample evidence to reject the claim that the mean time Americans spend watching TV per day is 4.4 hours.

p-value

Since -17.88 does not come close to being on our z-table, we will use the biggest value we have—namely, -3.50.

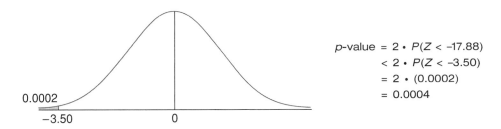

p-value $= 2 \cdot P(Z < -17.88)$
$< 2 \cdot P(Z < -3.50)$
$= 2 \cdot (0.0002)$
$= 0.0004$

The p-value for this test is (much) less than 0.0004. ■

MICROSOFT EXCEL   Hypothesis Tests for a Single Population Mean

TWO-TAILED CRITICAL VALUES

To use Excel to find two-tailed critical values, we first must find $\alpha/2$. If the level of significance (α) is 0.05, then $\alpha/2 = 0.025$. To find the critical value for the left tail, type the following in any cell.

=NORMSINV(0.025)

This Excel function returns the z-value for which the area to the left of it is equal to 0.025. Round Excel's result to two decimal places. To find the critical value for the right tail, type the following in any cell.

=NORMSINV(1-0.025)

We must subtract 0.025 from 1 to obtain the area to the left of the right tail. Again, round the result to two decimal places. For values of $\alpha/2$ other than 0.025, replace 0.025 by the appropriate value.

RIGHT-TAILED CRITICAL VALUES

To use Excel to find a right-tailed critical value, we need α. Suppose the level of significance was 0.01; in other words, $\alpha = 0.01$. To find the critical value for the right tail, type the following in any cell.

=NORMSINV(1-0.01)

Round Excel's result to two decimal places.

LEFT-TAILED CRITICAL VALUES

To use Excel to find a left-tailed critical value, we need α. Suppose that $\alpha = 0.05$. To find the critical value for the left tail, type the following in any cell.

=NORMSINV(0.05)

Round Excel's result to two decimal places.

TWO-TAILED p-VALUES

If the calculated value of the test statistic is negative, we need to multiply the area to the left of the test statistic by 2. In any cell, we type the following.

=2*NORMSDIST(z-value)

If the test statistic was calculated to be $z = -2.47$, then we would type

=2*NORMSDIST(-2.47)

Round all p-values to four decimal places.

We can also use Excel to calculate two-tailed p-values if the calculated value of the test statistic is positive. In such a case, we type the following in any cell.

=2*(1-NORMSDIST(z-value))

For example, if we calculated the test statistic to be $z = 1.23$, then we would type the following in any cell.

=2*(1-NORMSDIST(1.23))

We must subtract the result from 1 because Excel's **NORMDIST** function gives us the area to the left of the z-value, but we need the area in the right tail.

LEFT-TAILED p-VALUES

Since we are looking for the area to the left of a particular z-value, NORMSDIST will work without any adjustments. In any cell, we type the following.

=NORMSDIST(z-value)

If the test statistic was calculated to be $z = -1.57$, then we would type

=NORMSDIST(-1.57)

Round that answer to four decimal places.

RIGHT-TAILED *p*-VALUES

Because we are looking for the area to the right of a particular *z*-value, we must subtract the value that NORMSDIST gives us from 1. In any cell, we type the following.

=1-NORMSDIST(*z*-value)

If the test statistic was calculated to be $z = 2.29$, then we would type

=1-NORMSDIST(2.29)

Round that answer to four decimal places.

Here is an example that uses these tools in performing a hypothesis test.

EXAMPLE 7.12

An algebra instructor has designed a final exam that he believes will produce a mean score of 70. Mr. Thomas, one of his colleagues, disagrees, claiming that the mean score for all algebra students on this exam will be below 70. Mr. Thomas randomly selects 38 algebra students, and gives them the exam. Here are their scores.

32	41	43	44	46	46	47	48	49	50
52	54	54	56	57	58	58	60	60	61
65	66	66	67	67	68	68	68	68	71
72	76	83	87	92	92	94	100		

Use these sample data to test Mr. Thomas' claim at the 0.01 level of significance.

We will need the sample statistics (mean and standard deviation).

- Open a new Excel worksheet, and type the sample values in column **A** (cells **A1** through **A38**).
- In cell **B1**, type Sample Mean. In cell **B2**, type Sample Standard Deviation. In cell **B3**, type Sample Size.

Now that the framework is set up, we can begin the calculations.

- To calculate the sample mean, type =AVERAGE(A1:A38) in cell **C1.**
- To calculate the sample standard deviation, type =STDEV(A1:A38) in cell **C2.**
- Type the sample size (38) in cell **C3.**

We begin with the hypothesis test by hand, and then will turn to Excel when needed.

Step 1

Claim in words: The mean score for all algebra students on this exam is below 70.
Claim: $\mu < 70$
Complement: $\mu \geq 70$
H_0: $\mu \geq 70$
H_A: $\mu < 70$

Note that the claim is the alternate hypothesis, so our conclusion will be either to support the claim or fail to support the claim. Also note that the alternate hypothesis involves the symbol <. This tells us that our test is a left-tailed test.

- In cell **B4,** type Claimed Mean. Type the claimed value of the mean (70) in cell **C4.**

Step 2

Level of significance: $\alpha = 0.01$

Step 3

Test statistic: $z = \dfrac{\bar{x} - \mu}{s/\sqrt{n}}$

Since we are comparing the mean of a population to a certain number (70) and the sample size is at least 30, the above test statistic is the appropriate one.

Step 4

Left-tailed test, $\alpha = 0.01$

- Type Critical Value in cell **B6.** We will calculate the critical value for this test in cell **C6.** Since this is a left-tailed test, type =NORMSINV(0.01).

The value Excel gives us is -2.32634. Our decision rule is then

Reject H_0 if $z < -2.32634$

Here is the critical value using our usual method.

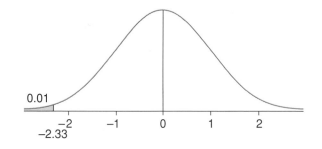

Decision rule: Reject H_0 if $z < -2.33$.

Step 5

- In cell **B7,** type Test Statistic.
- Next we calculate the test statistic in cell **C7.** Type =(C1-C4)/(C2/SQRT(C3)). The result Excel gives us is -2.7704. Here is the formula worked out.

$$z = \frac{\bar{x} - \mu}{s/\sqrt{n}}$$

$$= \frac{62.8 - 70}{16.04/\sqrt{38}}$$

$$= -2.77$$

Decision: Reject H_0.

Conclusion: There is sufficient sample evidence to support the claim that the mean score for all algebra students on this exam is below 70.

p-value

- Type p-value in cell **B9.** We will calculate the p-value for this test in cell **C9.** Since we are looking for a left-tailed p-value, type =NORMSDIST(C7). The result that Excel gives us is 0.002799.

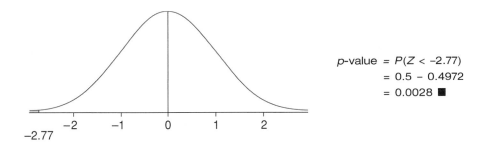

$$p\text{-value} = P(Z < -2.77)$$
$$= 0.5 - 0.4972$$
$$= 0.0028 \ \blacksquare$$

Here is what your Excel worksheet should look like.

32	Sample Mean	62.78947
41	Sample Standard Deviation	16.04412
43	Sample Size	38
44	Claimed Mean	70
46		
46	Critical Value	-2.32634
47	Test Statistic	-2.7704
48		
49	*p*-value	0.002799
50		
52		
54		
54		
56		
57		
58		
58		
60		
60		
61		
65		
66		
66		
67		
67		
68		
68		
68		
68		
71		
72		
76		
83		
87		
92		
92		
94		
100		

TI-83 Hypothesis Tests for a Single Population Mean

The TI-83 can help us to perform a hypothesis test, as shown in the following example.

EXAMPLE 7.13

An algebra instructor has designed a final exam that he believes will produce a mean score of 70. Mr. Thomas, one of his colleagues, disagrees, claiming that the mean score for all algebra students on this exam will be below 70. Mr. Thomas randomly selects 38 algebra students, and gives them the exam. Here are their scores.

32	41	43	44	46	46	47	48	49	50
52	54	54	56	57	58	58	60	60	61
65	66	66	67	67	68	68	68	68	71
72	76	83	87	92	92	94	100		

Use these sample data to test Mr. Thomas' claim at the 0.01 level of significance.

Put these 38 scores in list L_1. We will need sample statistics before continuing on to the hypothesis test. Press (STAT), then use the (→) key to move to the **CALC** menu. Select item **1: 1-Var** Stats. On the main screen, press (2nd)(1) to access list L_1, and press (ENTER). Here is the screen of sample information.

```
1-Var Stats
 x̄=62.78947368
 Σx=2386
 Σx²=159340
 Sx=16.04412479
 σx=15.83161044
↓n=38
■
```

```
1-Var Stats
↑n=38
 minX=32
 Q₁=50
 Med=60.5
 Q₃=68
 maxX=100
```

We can use 62.8 for the sample mean, and 16.04 for the sample standard deviation. Since the sample size of 38 is at least 30, we may use the sample standard deviation as an estimate of the population standard deviation. To begin the hypothesis test, Press (STAT), then use the (→) key to move to the **TESTS** menu. Select item **1: Z-Test.** Fill in the screen in the following manner.

- **Inpt:** Highlight **Stats.** We use this option when we type in the standard deviation, mean, and sample size manually. We use the **Data** option to take these statistics from a list.
- **μ_0:** Enter the claimed value of the mean, which is 70.
- **σ:** Enter the value that will be used for the population standard deviation. Again, in this example, we will use the sample standard deviation, which is 16.04.
- **$\bar{x}$:** Enter the value of the sample mean. You may find that it has already been entered for you. If it has not, use 62.8.
- **n:** Enter the sample size. Again, you may find that it has already been entered for you. If it has not, use 38.
- **μ:** Select the form of the alternate hypothesis. The claim for this test is that the mean score for all algebra students on this exam is below 70, so we are testing H_0: $\mu \geq 70$ against H_A: $\mu < 70$. Enter the second option, $<\mu_0$.
- Highlight **Calculate** and press (ENTER).

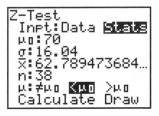

You will see the following screen, which lists the alternate hypothesis, calculated value of the test statistic, *p*-value, sample mean, and sample size. ■

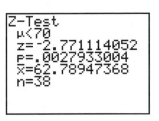

EXERCISES 7.1

Explain the following, in your own words.

1. How to write the claim in symbols
2. How to write the complement of the claim in symbols
3. How to identify the null hypothesis
4. How to identify the alternate hypothesis
5. The meaning of "level of significance"
6. The purpose of a test statistic
7. How to determine whether a test is left-tailed, right-tailed, or two-tailed
8. How to formulate the decision rule for a right-tailed test
9. How to formulate the decision rule for a left-tailed test
10. How to formulate the decision rule for a two-tailed test
11. The difference between the decision about the null hypothesis and the conclusion about the claim
12. The meaning of a *p*-value
13. A skeptical student believes that the mean height of adult males is actually greater than 69 inches. He randomly samples 45 adult males. The sample has a mean height of 70.4 inches with a standard deviation of 2.44 inches. Test the student's claim at the 0.05 level of significance.

    Step 1:
    (a) State the claim in words.
    (b) State the claim in symbols.
    (c) State the complement of the claim in symbols.
    (d) State the null hypothesis (H_0).
    (e) State the alternate hypothesis (H_A).

    Step 2:
    (f) State the level of significance (α).

    Step 3:
    (g) State the appropriate test statistic.

    Step 4:
    (h) What tailed test is this test?
    (i) State the decision rule.

Step 5:
(j) Calculate the test statistic.
(k) State your decision about the null hypothesis (H_0).
(l) State your conclusion about the claim.

For the remaining hypothesis tests, use the five-step procedure developed in this section. Refer to Exercise 13 for help.

14. A counselor at a community college claims that the mean GPA for students who have transferred to State University is higher than 3.0. A random sample of 45 such students had a mean GPA of 3.12 with a standard deviation of 0.31.

(a) Test the counselor's claim at the 0.05 level of significance.
(b) Explain how you would gather the data to test this claim, and list any biases that may be present using your method.

15. A student is interested in the mean life span of males in his town. He has heard claims that the mean age at death for males in his town is 70 years. He randomly selects 88 male deaths from the past year. The mean age at death for these 88 men was 68.9 years, with a standard deviation of 16.71 years. Test the claim that the mean age at death for males in his town is 70 years, at the 0.05 level of significance.

16. A student is concerned with the lack of effort among students at her school. She claims that students spend 5 or fewer hours a week studying. A random sample of 100 students produces a mean study time of 5.73 hours per week, with a standard deviation of 3.464 hours.

(a) Test the claim that the mean study time per week for these students is 5 or fewer hours at the 0.05 level.
(b) Explain how you would gather the data to test this claim, and list any biases that may be present using your method.

17. I recently heard someone say "Movies are getting too long! It seems like there are a lot of movies that are over two hours long!" I claim that the mean length of movies is less than 2 hours (120 minutes). A random sample of 90 recent releases produced a mean length of 108.44 minutes with a standard deviation of 17.952 minutes. Test my claim at the 0.01 level of significance.

18. At the 0.05 level of significance, test the claim that the mean mileage for cars is less than 20 miles per gallon. A sample of 48 cars had a sample mean of 16.9 miles per gallon, with a standard deviation of 6.83 miles per gallon.

19. A TV commercial claims that the mean cost of a funeral today is above $5000. A survey of 60 funerals showed that they had a mean cost of $5235, with a standard deviation of $1739.

(a) Test the commercial's claim at the 0.05 level of significance.
(b) Explain how you would gather the data to test this claim, and list any biases that may be present using your method.

20. A random sample of 100 women who smoked while pregnant showed that they smoked a mean of 7.4 cigarettes per day with a standard deviation of 4.95 cigarettes. (Based on data contained in Centers for Disease Control and Prevention Monthly Vital Statistics Report.)

(a) Test the claim that women who smoke while pregnant smoke a mean of less than 10 cigarettes per day, at the 0.01 level of significance.
(b) Explain how you would gather the data to test this claim, and list any biases that may be present using your method.

21. A sample of 212 women who gave birth showed that they gained a mean of 26.2 pounds, with a standard deviation of 10.48 pounds. (Based on data contained in Centers for Disease Control and Prevention Monthly Vital Statistics Report.)

(a) Test the claim that the mean weight gain while pregnant is higher than 25 pounds, at the 0.05 level of significance.
(b) Explain how you would gather the data to test this claim, and list any biases that may be present using your method.

22. Test the claim that the mean number of prenatal visits for pregnant women is higher than 10 visits at the 0.05 level of significance. A random sample of 441

women who gave birth showed that they had a mean of 11.2 prenatal doctor visits, with a standard deviation of 3.76. (Based on data contained in Centers for Disease Control and Prevention Monthly Vital Statistics Report.)

23. A racing publication reports that the average winning time for a six-furlong race is 1 minute and 10.8 seconds (70.8 seconds). These are the winning times, in seconds, for 63 six-furlong races at Santa Anita Park. (Six furlongs are equal to three-quarters of a mile.)

71.4	72.6	71.0	70.0	71.8	72.0	70.0
71.0	70.8	69.4	70.0	71.0	68.8	71.0
69.2	70.2	72.2	71.0	71.6	69.8	70.4
69.8	70.0	70.2	69.4	69.2	70.6	69.0
70.6	72.2	71.2	71.4	70.2	72.2	72.2
70.6	71.6	69.6	71.2	70.4	69.0	68.6
70.8	69.4	69.6	69.4	70.2	69.4	70.0
68.8	69.8	70.6	71.0	70.8	69.2	70.8
71.0	71.0	70.8	71.0	71.0	69.2	71.6

Test the publication's claim at the 0.01 level of significance.

24. A random sample of 120 fillets from farm-raised Atlantic salmon had a mean fat content of 10.5 grams and a standard deviation of 1.12 grams. Use this sample to test the claim that the mean fat content of farm-raised Atlantic salmon fillets is higher than 8.5 grams at the 0.05 level of significance.

25. A random sample of 67 fillets from wild Atlantic salmon had a mean fat content of 8.1 grams and a standard deviation of 1.48 grams. Use this sample to test the claim that the mean fat content of wild Atlantic salmon fillets is lower than 8.5 grams at the 0.05 level of significance.

26. A survey of 65 beer drinkers revealed a mean consumption of 25.2 beers per month, with a standard deviation of 4.96 beers. (Based on the results of a survey by Maritz AmeriPoll.)

   (a) Test the claim that the mean number of beers consumed per month by beer drinkers is 24 beers (1 case) at the 0.05 level of significance.
   (b) Explain how you would gather the data to test this claim, and list any biases that may be present using your method.

27. A random sample of 300 incoming flights at an airport showed that travelers had to wait in the baggage claim area for a mean of 18.9 minutes, with a standard deviation of 5.73 minutes. Test the claim that the mean waiting time in the baggage claim area is less than 20 minutes at the 0.01 level of significance.

28. Test the claim that the mean vitamin C content of medium-sized oranges is 66 mg at the 0.05 level of significance. A sample of 36 medium-sized oranges had a mean vitamin C content of 64.5 mg, with a standard deviation of 4.14 mg.

29. A politician claims that the mean number of hours Americans worked per week in a typical week is greater than 40 hours. A sample of 324 American workers produced a mean of 41.3 hours per week with a standard deviation of 10.63 hours per week. Test the politician's claim at the 0.05 level of significance. (Based on a study by the Center for Survey Research and Analysis at University of Connecticut.)

30. A random sample of 74 recently engaged men showed that they had spent a mean of $2982 on the engagement ring, with a standard deviation of $1691. At the 0.05 level of significance, test the claim that the mean amount spent on engagement rings by all men is above $2500. (Based on the results of a survey by *Bride's* magazine.)

31. How many e-mails does a person get at work per day? A random sample of 47 people showed that they receive a mean of 10.7 e-mails at work per day, with a standard deviation of 10.53 e-mails. At the 0.05 level of significance, test the claim that the mean number of e-mails received per day is higher than 5. (Based on a survey by the John J. Heldrich Center for Workforce Development.)

32. Test the claim that the mean cost of a honeymoon is over $3000 at the 0.05 level of significance. A random sample of 180 couples spent a mean of $3657 on their

honeymoons, with a standard deviation of $2109. (Based on a survey by *Bride's* magazine.)

33. A sample of 225 parents were asked how much time they spent per week on school work or school-related activities. The sample produced a mean of 5.6 hours per week, with a standard deviation of 4.4 hours. At the 0.01 level of significance, test the claim that the mean number of hours spent by parents on school work or school-related activities is 5 hours per week. (Based on a study by Yankelovich Partners for LearningPays.com.)

34. One hundred one cruise ship passengers were selected at random, and they had a mean age of 50.1 years old, with a standard deviation of 16.41 years. At the 0.05 level of significance, test the claim that the mean age of a passenger on a cruise ship is below 55 years old. (Based on a study by Polk.)

35. A study of 42 restaurants showed that the mean cost for a sit-down dinner for one person at these restaurants, not including tip, was $27.05 with a standard deviation of $7.20. At the 0.05 level of significance, test the claim that the mean cost for a sit-down dinner for one person at all restaurants is $25. (Based on a study by the National Restaurant Association.)

36. Television networks and movie studios are always interested in the age of the audience they are reaching. A random sample of 104 people at a showing of *Toy Story 2* had a mean age of 26.4 years with a standard deviation of 14.82 years. At the 0.05 level of significance, test the claim that the mean age of people going to see *Toy Story 2* is less than 30 years. (Based on the results of an informal survey by Disney in six cities.)

37. A television golf analyst claims that the mean drive for professional golfers is 270 yards. Here are the lengths of 63 randomly selected drives from the 100th U.S. Open at Pebble Beach.

305	269	289	287	293	289	289	256
288	283	300	229	262	312	274	286
268	282	290	270	281	296	306	256
280	273	293	268	284	266	254	242
283	270	277	293	262	270	273	261
275	276	279	263	273	261	281	241
262	258	274	261	264	260	266	234
252	259	284	257	270	281	250	

At the 0.05 level of significance, test the television golf analyst's claim. (*Source:* On-line scoring for U.S. Open at lycos.com.)

What is the appropriate conclusion about the claim under the following conditions, if the level of significance is 0.05?

38. Claim to be tested: Mean is less than 25.

    p-value: 0.0885

39. Claim to be tested: Mean is greater than 17.1.

    p-value: 0.0322

40. Claim to be tested: Mean is equal to 129.7.

    p-value: 0.0039

41. Claim to be tested: Mean is not 25.

    p-value: 0.0668

42. A colleague tells you that he is trying to verify the claim that households that subscribe to a particular magazine have a mean income that is above $45,000. He is elated that the mean income of 225 randomly selected households that subscribe to the magazine is $49,245. Explain to him why this may not be enough to support his claim, and what other information may be necessary.

? ## What Is Wrong with This Picture?

The following statistical project contains a major error. Potential errors could be choosing the wrong hypothesis test or not meeting the assumptions required for this test. Find the error, and explain what could be done to correct it.

A student randomly selects 6 elementary schools from his county. He visits these schools and obtains the count of students in 12 different third-grade classes. He uses these data to test the claim that the mean number of students in all third-grade classes in his county is above 20 students, using the hypothesis test introduced in this section.

What is wrong with this picture?

MINI PROJECT

I claim that the mean height for adult females is 5′ 5″ (65 inches). You are to test my claim at the 0.05 level of significance. Randomly sample 50 adult females, and find out their height in inches. Use these sample data to test the claim. In addition to your complete hypothesis test, include

- Your raw data
- A histogram for the heights
- A pie chart showing the percentages that are 5′ 2″ or shorter, between 5′ 3″ and 5′ 7″, and 5′ 8″ or taller

Explain your technique of sampling. Which type of sampling did you use? Was it truly random? What potential biases may show up in your sample?

EXCEL EXPLORATION   The Effect of Sample Size

In this exploration, we will use Microsoft Excel to examine how the sample size affects a hypothesis test. We will be testing the null hypothesis that a population mean is equal to 50 by randomly generating samples from a population whose mean is 48 and whose standard deviation is 15. We know that the mean is not equal to 50 (it is equal to 48), but will our hypothesis-testing procedure be able to reach that conclusion? We will see that one factor that may play a part is the sample size.

If we are testing the null hypothesis that $\mu = 50$ at the 0.05 level of significance, the two-tailed decision rule is to reject the null hypothesis if $z < -1.96$ or $z > 1.96$. We will use Excel to generate a sample, and then calculate the test statistic for that sample.

Before we begin we must be sure that the Analysis Tool Pak has been added. Pull down the **Tools** menu. If you see a choice that says **Data Analysis**, then the Tool Pak is added and you may continue. If you do not see **Data Analysis**, click on **Add-ins** under **Tools**. When the dialog box opens up, click on the box for **Analysis Tool Pak**, and click on **OK**.

We begin by generating 100 samples of size 30. Pull down the **Tools** menu and click on **Data Analysis**. When the dialog box appears, click on **Random Number Generation** and **OK**. When the next dialog box appears, enter 100 in the box labeled **Number of Variables**. (This is the number of samples that we are going to take.) Enter

30 in the box labeled **Number of Random Numbers.** (This is the sample size.) In the box labeled **Distribution** click on **Normal.** For **Parameters,** enter **48** in the box labeled **Mean,** and **15** in the box labeled **Standard Deviation.**

Now we will calculate the test statistic. In cell **A31,** type =AVERAGE(A1:A30). This is the mean of the sample. To calculate the standard deviation of the sample, type =STDEV(A1:A30) in cell **A32.** Finally, to calculate the test statistic, type the following in cell **A33.**

=(A31-50)/(A32/SQRT(30))

In the above formula, **A31** represents the sample mean, **50** is the claimed population mean, **A32** is the sample standard deviation, and **30** is the sample size. Highlight cells **A31** through **A33,** then select **Copy** from the **Edit** menu. Highlight the cells in rows **31, 32,** and **33** through column **CV,** then select **Paste** from the **Edit** menu. This will calculate the test statistic for each of the 100 samples.

In a trial run, using 100 samples, the null hypothesis was not rejected a single time. The results of your 100 samples will most likely be the same. Repeat this procedure for samples of size 50, 100, 250, and 1000. Summarize your results, stating how the increase in sample size affects the test statistic.

SECTION 7.2
Hypothesis Test for a Single Population Mean (Small Samples)

Recall that the sampling distribution of sample means follows the *t*-distribution for samples of size less than 30. When we are trying to test a claim about a population mean using a small sample ($n < 30$) and the population standard deviation σ is unknown, it is appropriate to use the *t*-distribution as long as the sample comes from a population that is normally distributed. With small samples, outliers suggest that the population is skewed and not normally distributed. There is a hypothesis test in Section 9.3 to test whether a set of data follows a normal distribution, but until then you can use the material presented in "Extra—Is a Set of Data Approximately Normally Distributed?" which follows Section 5.3. If it appears that the population is not normally distributed, we may instead use a hypothesis test from Section 11.1. Here is the test statistic for the small-sample test for a single population mean.

$$t = \frac{\bar{x} - \mu}{s/\sqrt{n}} \qquad \text{degrees of freedom} = n - 1$$

The test statistic is identical to the one that we use for a large sample test when the population standard deviation is unknown, but the decision rule will be different. Before we take a look at a few hypothesis tests, we will review how to obtain the decision rule.

EXAMPLE 7.14   Find the decision rule for a right-tailed test at the 0.01 level of significance with 22 degrees of freedom.

Since this is a one-tailed test, we look for 0.01 as a one-tailed value. When we locate that column (the second column from the left), we go down until we reach 22 degrees of freedom.

One-Tailed Value						
Degrees of Freedom	0.005	0.01	0.025	0.05	0.1	0.25
22	2.819	2.508	2.074	1.717	1.321	0.686

The critical value is 2.508. This value is positive, because we are dealing with the right tail. Now for the decision rule. For a right-tailed test, we want to reject the null hypothesis if the value of the test statistic is greater than the critical value.

Decision rule: Reject H_0 if $t > 2.508$ ■

EXAMPLE 7.15   Find the decision rule for a left-tailed test at the 0.05 level of significance with 9 degrees of freedom.

Since this is a one-tailed test, we look for 0.05 as a one-tailed value. When we locate that column (the fourth column from the left), we go down until we reach 9 degrees of freedom.

One-Tailed Value						
Degrees of Freedom	0.005	0.01	0.025	0.05	0.1	0.25
9	3.250	2.821	2.262	1.833	1.383	0.703

The value in the chart is 1.833, so the critical value is –1.833. This value is negative, because we are dealing with the left tail. For a left-tailed test, we want to reject the null hypothesis if the value of the test statistic is less than the critical value.

Decision rule: Reject H_0 if $t < -1.833$ ■

EXAMPLE 7.16   Find the decision rule for a two-tailed test at the 0.05 level of significance with 19 degrees of freedom.

Since this is a two-tailed test, we look for 0.05 as a two-tailed value. When we locate that column (the third column from the left), we go down until we reach 19 degrees of freedom.

Two-Tailed Value						
Degrees of Freedom	0.01	0.02	0.05	0.1	0.2	0.5
19	2.861	2.539	2.093	1.729	1.328	0.688

The value in the table is 2.093, so the critical values are –2.093 and 2.093. We use both the positive and negative values because we are dealing with both the right and left tails. For a two-tailed test, we want to reject the null hypothesis if the value of the test statistic is greater than the positive critical value or less than the negative critical value.

Decision rule: Reject H_0 if $t < -2.093$ or if $t > 2.093$ ■

Now we will perform some actual tests, using the same five-step procedure as in the last section. The major difference is the decision rule.

EXAMPLE **7.17** Test the claim that the mean number of books read per month by community college students is less than 2 at the 0.05 level of significance. A random sample of 28 community college students had read a mean of 1.2 books with a standard deviation of 2.14 books.

Step 1

Claim in words: The mean number of books read per month by community college students is less than 2.
Claim: $\mu < 2$
Complement: $\mu \geq 2$
H_0: $\mu \geq 2$
H_A: $\mu < 2$

Since the alternate hypothesis involves <, we have a left-tailed test. Also, since our claim is the alternate hypothesis, we will either support or fail to support our claim.

Step 2

Level of significance: $\alpha = 0.05$.

Step 3

Test statistic: $t = \dfrac{\bar{x} - \mu}{s/\sqrt{n}}$ degrees of freedom = $n - 1$

This is the appropriate test statistic to use when testing a claim about a single population mean based on a small sample ($n < 30$) when the population standard deviation is unknown.

Step 4

This is a left-tailed test, $\alpha = 0.05$, with 27 degrees of freedom.

	One-Tailed Value					
Degrees of Freedom	0.005	0.01	0.025	0.05	0.1	0.25
27	2.771	2.473	2.052	1.703	1.314	0.684

Decision rule: Reject H_0 if $t < -1.703$.

Step 5

$$t = \frac{\bar{x} - \mu}{s/\sqrt{n}}$$

$$= \frac{1.2 - 2}{2.12/\sqrt{28}}$$

$$= -1.997$$

Decision: Reject H_0.

Conclusion: There is sufficient sample evidence to support the claim that the mean number of books read per month by community college students is less than 2.

***p*-value**

Finding *p*-values for *t* tests are tricky when using the table. We need to go to the chart, and look at the entire row for 27 degrees of freedom.

	One-Tailed Value					
Degrees of Freedom	0.005	0.01	0.025	0.05	0.1	0.25
27	2.771	2.473	2.052	1.703	1.314	0.684

Our calculated value for the test statistic (–1.997) falls between 1.703 (α = 0.05) and 2.052 (α = 0.025). Our p-value is between 0.025 and 0.05. Using technology, we will be able to obtain a more precise p-value later. ∎

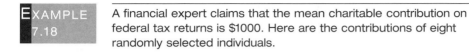

EXAMPLE 7.18   A financial expert claims that the mean charitable contribution on federal tax returns is $1000. Here are the contributions of eight randomly selected individuals.

$329	$495	$552	$734
$951	$1379	$1394	$2550

At the 0.01 level of significance, test the expert's claim. (Data based on the results of a survey by Gallup for *Independent Sector.*)

First we need to calculate the mean and standard deviation for this sample. The mean is $1048.00, and the standard deviation is $723.29.

Step 1

Claim in words: The mean charitable contribution on federal tax returns is $1000.
Claim: $\mu = 1000$
Complement: $\mu \neq 1000$
H_0: $\mu = 1000$
H_A: $\mu \neq 1000$

Since the alternate hypothesis involves $\neq$, we have a two-tailed test. Also, since our claim is the null hypothesis, we will either reject or fail to reject our claim.

Step 2

Level of significance: $\alpha = 0.01$

Step 3

Test statistic: $t = \dfrac{\bar{x} - \mu}{s/\sqrt{n}}$ degrees of freedom = $n - 1$

Step 4

This is a two-tailed test, $\alpha = 0.01$, with 7 degrees of freedom.

	Two-Tailed Value					
Degrees of Freedom	0.01	0.02	0.05	0.1	0.2	0.5
7	3.499	2.998	2.365	1.895	1.415	0.711

Decision rule: Reject H_0 if $t < -3.499$ or if $t > 3.499$.

Step 5

$$t = \frac{\bar{x} - \mu}{s/\sqrt{n}}$$

$$= \frac{1048 - 1000}{723.29/\sqrt{8}}$$

$$= 0.188$$

Decision: Fail to reject H_0.

Conclusion: There is not sufficient sample evidence to reject the claim that the mean charitable contribution on federal tax returns is $1000.

p-value

We go to the chart, and look at the entire row for 7 degrees of freedom.

Degrees of Freedom	*Two-Tailed Value*					
	0.01	0.02	0.05	0.1	0.2	0.5
7	3.499	2.998	2.365	1.895	1.415	0.711

Our calculated value for the test statistic (0.188) falls below 0.711 ($\alpha = 0.5$). Our p-value is greater than 0.5. ∎

Mark, an avid golfer, claims that his average drive is greater than 250 yards. To prove this, he measured his 14 drives from a round of golf. Here are the distances, in yards.

262	247	255	239	275	257	260
242	256	281	265	248	255	263

Test Mark's claim that his average drive is greater than 250 yards at the 0.05 level.

Again, we begin by calculating the mean and standard deviation for this sample. The mean is 257.5 yards and the standard deviation is 11.67 yards.

Step 1

Claim in words: The mean length for all of Mark's drives is greater than 250 yards.
Claim: $\mu > 250$
Complement: $\mu \leq 250$
H_0: $\mu \leq 250$
H_A: $\mu > 250$

Since the alternate hypothesis involves >, we have a right-tailed test. Also, since our claim is the alternate hypothesis, we will either support or fail to support our claim.

Step 2

Level of significance: $\alpha = 0.05$

Step 3

Test statistic: $t = \dfrac{\bar{x} - \mu}{s/\sqrt{n}}$ degrees of freedom = $n - 1$

Step 4

This is a right-tailed test, $\alpha = 0.05$, with 13 degrees of freedom.

Degrees of Freedom	*One-Tailed Value*					
	0.005	0.01	0.025	0.05	0.1	0.25
13	3.012	2.650	2.160	1.771	1.350	0.694

Decision rule: Reject H_0 if $t > 1.771$.

Step 5

$$t = \frac{\bar{x} - \mu}{s/\sqrt{n}}$$

$$= \frac{257.5 - 250}{11.67/\sqrt{14}}$$

$$= 2.405$$

Decision: Reject H_0.

Conclusion: There is sufficient sample evidence to support the claim that the mean length for all of Mark's drives is greater than 250 yards.

p-value

We go to the chart, and look at the entire row for 13 degrees of freedom.

	One-Tailed Value					
Degrees of Freedom	0.005	0.01	0.025	0.05	0.1	0.25
13	3.012	2.650	2.160	1.771	1.350	0.694

Our calculated value for the test statistic (2.405) falls between 2.160 ($\alpha = 0.025$) and 2.650 ($\alpha = 0.01$). Our *p*-value is therefore between 0.01 and 0.025. ■

MICROSOFT EXCEL   Hypothesis Tests for a Single Population Mean (Small Samples)

RIGHT-TAILED CRITICAL VALUES (*t*-DISTRIBUTION)

As with the standard normal distribution, Excel can help us find critical values for the *t*-distribution. With a one-tailed test, we must multiply the level of significance by 2 to use Excel's built-in function. This function returns critical values for two-tailed tests, for which the level of significance is twice that of a one-tailed test. Here is the format.

    =TINV(2*level of significance, degrees of freedom)

For example, for a right-tailed test at the 0.01 level of significance with 22 degrees of freedom, type the following in any cell.

    =TINV(2*0.01,22)

Excel returns a value of 2.508323. The decision rule is

    Reject H_0 if $t > 2.508323$

LEFT-TAILED CRITICAL VALUES (*t*-DISTRIBUTION)

The difference that arises with a left-tailed critical value, as compared to a right-tailed critical value, is that Excel gives us a positive value. It is up to us to remember that this value must become negative. We can take care of this by including a negative sign before the function's name. Again, with a one-tailed test, we must multiply the level of significance by 2 to use Excel's built-in function.

For example, for a left-tailed test at the 0.05 level of significance with 9 degrees of freedom, type the following in any cell.

    = -TINV(2*0.05,9)

Excel returns a value of –1.83311. The decision rule is

    Reject H_0 if $t < -1.83311$

TWO-TAILED CRITICAL VALUES (*t*-DISTRIBUTION)

With a two-tailed test we only have to supply Excel with the level of significance (not 2 times the level of significance) and the number of degrees of freedom. Here is the format.

=TINV(level of significance, degrees of freedom)

For example, for a two-tailed test at the 0.05 level of significance with 19 degrees of freedom, type the following in any cell.

=TINV(0.05,19)

Excel returns a value of 2.093025. The decision rule is

Reject H_0 if $t > 2.093025$ or if $t < -2.093025$

RIGHT-TAILED *p*-VALUES

Microsoft Excel can help us with the *p*-value. Excel will give us a precise *p*-value instead of a range of values. We need to supply Excel with the value of the test statistic (use positive values only), the degrees of freedom, and the number of tails. If the test statistic is negative, you need to type it as a positive value. (There is no difference in the way you enter a left-tailed test or a right-tailed test.) Here is the format.

=TDIST(test statistic, degrees of freedom, number of tails)

If we had a right-tailed test, with 13 degrees of freedom, and the test statistic calculated to be 2.405, we would find the *p*-value by typing the following in any cell.

=TDIST(2.405,13,1)

The result that we get is 0.0159, which is our *p*-value. Using our tables, the *p*-value is between 0.01 and 0.025, but Excel's *p*-value is more precise and more useful.

LEFT-TAILED *p*-VALUES

We follow the same format as we did for a right-tailed *p*-value. However, we do not type a negative sign in front of our test statistic.

If we had a left-tailed test, with 27 degrees of freedom, and the test statistic calculated to be −1.997, we would find the *p*-value by typing the following in any cell.

=TDIST(1.997,27,1)

The result that we get is 0.0280, which is our *p*-value. Using our tables, the *p*-value is between 0.025 and 0.05, but Excel's *p*-value is more precise and more useful.

TWO-TAILED *p*-VALUES

The major difference for calculating a two-tailed *p*-value is that we type a 2 instead of a 1 for the number of tails.

If we had a two-tailed test, with 11 degrees of freedom, and the test statistic calculated to be 2.038, we would find the *p*-value by typing the following in any cell.

=TDIST(2.038,11,2)

The result that we get is 0.0663, which is our *p*-value. The *p*-value is between 0.05 and 0.1, but Excel's *p*-value is more precise and more useful.

We conclude with an example that uses Excel to perform a hypothesis test.

EXAMPLE 7.20   A math instructor has written a new placement exam. He is concerned that the exam may be too long. The math department will use this new exam if the mean time required to finish the exam for all students who take it is at most 50 minutes. The exam is given to 20 randomly selected students. Here are the times (in minutes) required to complete the exam for these students.

43	44	45	46	49	50	51	53	54	55
57	57	58	59	59	60	60	60	60	60

Use these sample data to test the claim that the mean length of time required to finish the exam for all students who take it is at most 50 minutes at the 0.05 level of significance.

We will need the sample statistics (mean and standard deviation).

- Open a new Excel worksheet, and type the above values in column **A** (cells **A1** through **A20**).
- In cell **B1,** type Sample Mean. In cell **B2,** type Sample Standard Deviation. In cell **B3,** type Sample Size. In cell **B4,** type Claimed Mean.

Now that the framework is set up, we can begin the calculations.

- To calculate the sample mean, type =AVERAGE(A1:A20) in cell **C1.**
- To calculate the sample standard deviation, type =STDEV(A1:A20) in cell **C2.**
- Type the sample size (20) in cell **C3.**
- Type the claimed value of the mean (50) in cell **C4.**

We begin with the hypothesis test by hand, and then will turn to Excel when needed.

Step 1

Claim in words: The mean length of time required to finish the exam for all students who take it is at most 50 minutes.
Claim: $\mu \leq 50$
Complement: $\mu > 50$
H_0: $\mu \leq 50$
H_A: $\mu > 50$

Note that the claim is the null hypothesis, so our conclusion will be either to reject the claim or fail to reject the claim. Also note that the alternate hypothesis involves the symbol >. This tells us that our test is a right-tailed test.

Step 2

Level of significance: $\alpha = 0.05$

Step 3

Test statistic: $t = \dfrac{\bar{x} - \mu}{s/\sqrt{n}}$

This is the appropriate test statistic, since this is a one-mean test with a sample size that is smaller than 30.

Step 4

Right-tailed test, $\alpha = 0.05$

- Type Critical Value in cell **B6.**
- We will calculate the critical value for this test in cell **C6.** Since this is a left-tailed test, type =TINV(2*0.05,19). (Since this is a one-tailed test, we must multiply the level of significance by 2.)

The value Excel gives us is 1.729131. Our decision rule is then

Reject H_0 if $t > 1.729131$

Here is the critical value using our usual method.

	One-Tailed Value					
Degrees of Freedom	0.005	0.01	0.025	0.05	0.10	0.25
19	2.861	2.539	2.093	1.729	1.328	0.688

Decision rule: Reject H_0 if $t > 1.729$.

Step 5

- In cell **B7,** type Test Statistic.
- Next we will calculate the test statistic in cell **C7.** Type =(C1-C4)/(C2/SQRT(C3)).

The result Excel gives us is 2.985792.

Here is the formula worked out.

$$t = \frac{\bar{x} - \mu}{s/\sqrt{n}}$$

$$= \frac{54 - 50}{5.99/\sqrt{20}}$$

$$= 2.986$$

Decision: Reject H_0.

Conclusion: There is sufficient sample evidence to reject the claim that the mean length of time required to finish the exam for all students who take it is at most 50 minutes.

p-value

- Type p-value in cell **B9.**
- We will calculate the p-value for this test in cell **C9.** Since we are looking for a left-tailed p-value, type =TDIST(C7,19,1).

The result that Excel gives us is 0.003798.

Using the t-distribution chart, we can see that the p-value is less than 0.005. This is because the test statistic (2.986) is greater than 2.861, which is the highest value in the table for 19 degrees of freedom. ∎

Here is what your Excel worksheet should look like when you are done.

43	Sample Mean	54
44	Sample Standard Deviation	5.991222
45	Sample Size	20
46	Claimed Mean	50
49		
50	Critical Value	1.729131
51	Test Statistic	2.985792
53		
54	p-value	0.003798
55		
57		
57		
58		
59		
59		
60		
60		
60		
60		
60		

TI-83   Hypothesis Tests for a Single Population Mean (Small Samples)

The TI-83 can help us perform hypothesis tests for a single mean based on a small sample, and find *p*-values for these tests.

RIGHT-TAILED *p*-VALUES

The TI-83 can help us with calculating the *p*-value. The calculator will give us a precise *p*-value instead of an interval that it falls in. Suppose we have a right-tailed test, with 13 degrees of freedom, and the test statistic calculated to be 2.405. Press (2ND) (VARS) to access the **DISTR** menu. Select option 5: **tcdf(**. We need to enter three values. First, enter the value of the test statistic, which is **2.405**. After pressing (,), press 1, then (2ND) and the (,) key to access **EE**, and then finally press 99 followed by a (,). Finally, we need to type in the degrees of freedom, **13**, followed by (). Here is what the screen should look like. To calculate the *p*-value, press (ENTER).

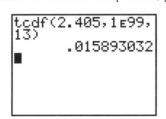

The result that we get is 0.0159, which is our *p*-value. Using our tables, the *p*-value is between 0.01 and 0.025, but the TI-83's *p*-value is more precise and more useful.

LEFT-TAILED *p*-VALUES

We follow the same format as we did for a right-tailed *p*-value.

tcdf(-1E99, test statistic, degrees of freedom)

Here is what the screen should look like for a left-tailed p-value for which the test statistic is calculated to be –1.997, with 27 degrees of freedom.

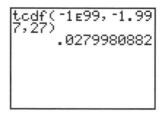

The result that we get is 0.0280, which is our p-value. Using our tables, the p-value is between 0.025 and 0.05, but the TI-83's p-value is more precise and more useful.

TWO-TAILED p-VALUES

To calculate a two-tailed p-value, we calculate the appropriate one-tailed p-value, and then double that value. If the test statistic is positive, double the right-tailed p-value. If the test statistic is negative, double the left-tailed p-value.

If we had a two-tailed test, with 11 degrees of freedom, and the test statistic calculated to be 2.038, here is the screen that calculates the p-value.

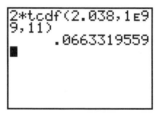

The result that we get is 0.0663, which is our p-value. The p-value is between 0.05 and 0.1, but the TI-83's p-value is more precise and more useful.

We conclude this section with an example that uses the TI-83 to perform a hypothesis test.

EXAMPLE 7.21

A math instructor has written a new placement exam. He is concerned that the exam may be too long. The math department will use this new exam if the mean time required to finish the exam for all students who take it is at most 50 minutes. The exam is given to 20 randomly selected students. Here are the times (in minutes) required to complete the exam for these students.

| 43 | 44 | 45 | 46 | 49 | 50 | 51 | 53 | 54 | 55 |
| 57 | 57 | 58 | 59 | 59 | 60 | 60 | 60 | 60 | 60 |

Use these sample data to test the claim that the mean length of time required to finish the exam for all students who take it is at most 50 minutes at the 0.05 level of significance.

Put these 20 times in list L_1. To begin the hypothesis test, press (STAT), then use the (→) key to move to the **TESTS** menu. Select item **2: T-Test.** Fill in the screen in the following manner.

- **Inpt:** Highlight **Data.** We use this option, rather than **Stats,** when we want to take the sample statistics from a list.
- **μ_0:** Enter the claimed value of the mean, which for this example is 50.
- **List:** Enter the list that contains the sample data. In this example, we will use L_1.
- **Freq:** Be sure that this says 1.
- **μ:** Select the form of the alternate hypothesis. The claim for this test is that the mean length of time required to finish the exam for all students who take it is at most 50 minutes.

  This becomes a test of H_0: $\mu \le 50$ versus H_A: $\mu > 50$.

  Select option 3: $>\mu_0$.
- Highlight **Calculate** and press (ENTER).

You will see the following screen, which lists the alternate hypothesis, calculated value of the test statistic, p-value, sample mean, sample standard deviation, and sample size. ■

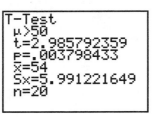

```
T-Test
μ>50
t=2.985792359
p=.003798433
x̄=54
Sx=5.991221649
n=20
```

EXERCISES 7.2

1. A random sample of 25 households without health insurance had a mean annual income of $37,850 with a standard deviation of $25,554. At the 0.05 level of significance, test the claim that the mean annual income of households without insurance is less than $50,000. (Based on results of a study by the Census Bureau.)

2. A mortgage broker believes that the typical family that rents their home owns fewer than 2 cars on average. A random sample of 26 families that rent their home owned a mean of 1.29 cars with a standard deviation of 0.97 cars. At the 0.05 level of significance, test the claim that the mean number of cars owned by families that rent their home is less than 2. (Based on the results of the Census Bureau American Housing Brief.)

3. In testing the effect of a new zinc spray on the common cold, 27 people were randomly selected and given the spray to take home. At the first sign of a cold they were to begin using the spray. Their first colds lasted a mean of 1.5 days, with a standard deviation of 2.1 days. At the 0.05 level of significance, test the claim that the mean length of a cold being treated by this spray is equal to 2 days. (Based on the results of a study funded by Gel-Tech.)

4. Fifteen households were randomly selected and asked how much they spend per week on groceries. The mean grocery bill per week was $73.25 with a standard deviation of $29.30. At the 0.05 level of significance, test the claim that the mean grocery bill per week for all households is less than $100 per week.

5. A health researcher is interested in the age that males begin smoking. She claims that the mean age is less than 19 years old. A random sample of 17 male smokers had a mean starting age of 18.2 years with a standard deviation of 2.60 years.
   (a) At the 0.05 level of significance, test the researcher's claim.

(b) Explain how you would gather the data to test this claim, and list any biases that may be present using your method.

6. A statistics instructor, when asked how students do on the first statistics exam, claims that the mean score for all students on the first exam is above 75. To test this claim, 22 students are selected at random. Their scores from the first exam had a mean of 78.2 with a standard deviation of 13.95. Test the instructor's claim at the 0.05 level of significance.

7. How many pairs of shoes do female college students own? A random sample of 15 female college students produced a sample mean of 8.7 pairs of shoes, with a standard deviation of 0.85 pairs. Use these data to test the claim that the mean number of pairs of shoes owned by female college students is less than 10 at the 0.05 level of significance.

8. An experimental algebra exam is given to 20 randomly selected college algebra students. Here are their scores.

68	94	50	60	92	14	49	68	41	67
76	71	32	15	54	54	65	92	58	68

At the 0.05 level of significance, test the claim that the mean score for all college algebra students on this exam is below 60.

9. A random sample of 24 high school students were found to be working a mean of 11.6 hours per week with a standard deviation of 11.54 hours. At the 0.05 level of significance, test the claim that the mean number of hours worked per week by high school students is less than 15 hours per week. (Based on results in the National Assessment of Educational Progress survey for the National Center for Education Statistics.)

10. How often do teen mothers read to their children per week? A random sample of 22 teenage mothers read a mean of 2.7 days per week with a standard deviation of 1.64 days per week. At the 0.05 level of significance, test the claim that teenage mothers read to their children more than twice a week.

11. Here are the heights of 11 randomly selected professional women's basketball players.

78	69	71	69	75	71	77	71	77	73	68

At the 0.05 level of significance, test the claim that the mean height of professional women's basketball players is more than 6 feet.

12. Here are the weights of 11 randomly selected professional women's basketball players.

195	148	168	142	170	165
170	165	180	172	132	

At the 0.05 level of significance, test the claim that the mean weight of professional women's basketball players is above 150 pounds.

13. These are the number of packages handled by a shipping office on 17 randomly selected days.

1103	1488	1713	1536	1037	1462
1625	1627	1080	1216	1639	1539
1545	907	1307	1387	1547	

Test the claim that the shipping office handles more than 1200 packages per day at the 0.01 level of significance.

14. A random sample of 13 accountants showed that they had a mean salary of $46,328 and a standard deviation of $17,298.

(a) Use this sample to test the claim that the mean accountant salary is higher than $40,000 at the 0.05 level of significance.

(b) Explain how you would gather the data to test this claim, and list any biases that may be present using your method.

15. These are the total scores from 15 randomly selected NFL football games.

41	26	51	38	30	46	63	31
45	27	9	37	42	40	37	

Use this sample to test the claim that the mean number of points scored in an NFL game is 34 points, at the 0.05 level of significance.

16. A random sample of 28 symphony musicians were asked how many hours they practiced their instrument per week. The survey produced a mean practice time of 7.2 hours with a standard deviation of 3.62 hours per week. At the 0.05 level of significance, test the claim that the mean time that symphony musicians practice per week is greater than 5 hours per week.

17. A random sample of 20 workers who log on to Yahoo! while at work were asked how long they stayed logged on while at work. The mean length was 106 minutes, with a standard deviation of 42 minutes. Use this sample to test the claim that the mean time spent on Yahoo! is greater than 90 minutes at the 0.05 level of significance. (Based on the results of Nielsen/Net Ratings, January 2000.)

18. A math instructor has developed a test to determine a college student's math competency. He claims that the test is designed to produce a mean score of 70. The test is given to 12 randomly selected college students, and here are their scores.

76	73	81	63	71	82
68	75	66	79	90	70

    Test the instructor's claim that the mean score for all students is 70 at the 0.05 level of significance.

19. Tiger Woods has been dominating professional golf. One of his greatest assets is the length of his drives, which makes every hole "shorter" for him. Here are the lengths, in yards, of eight of his drives from the 100th U.S. Open at Pebble Beach.

305	290	308	289	302	309	302	289

    Use these drives to test the claim that the mean length of all of Tiger's drives is longer than 290 yards at the 0.05 level of significance. (*Source:* On-line scoring for U.S. Open at lycos.com.)

20. An agriculture student is interested in the mean length of a cow's lactation (number of days producing milk). She randomly selects 20 of the university's cows and looks at their records. Here are the lengths, in days, of each cow's first lactation.

263	279	284	249	335	326	399
283	285	293	329	338	266	286
229	387	387	317	294	315

    Test the claim that the mean length of a cow's first lactation is greater than 290 days at the 0.05 level of significance.

21. Major league baseball managers keep a close eye on the "pitch count" of their starting pitcher, because they believe that a pitcher loses his effectiveness after a certain number of pitches. Here are the pitch counts of 16 randomly selected starting pitchers.

117	94	137	66	103	100	81	98
86	77	94	95	115	79	93	86

    At the 0.05 level of significance, test the claim that the mean pitch count for starting pitchers is 100 pitches.

What Is Wrong with This Picture?

The following statistical project contains a major error. Potential errors could be choosing the wrong hypothesis test or not meeting the assumptions required for this test. Find the error, and explain what could be done to correct it.

A student randomly selects 10 eighth-grade boys and counts the number of pull-ups that they can do. Here are the results.

5	6	6	7	7	7	8	8	9	52

Sue uses these results to test the claim that the mean number of pull-ups that eighth-grade boys can do is higher than 5 pull-ups, using the hypothesis test introduced in this section. What is wrong with this picture?

SECTION 7.3
Hypothesis Test for a Single Population Proportion (Large Samples)

In this section we will learn to test claims about a single population proportion. These claims will compare the proportion of a population to a certain value. For example, "13.6% of all pregnant women smoke while pregnant" compares the proportion of all pregnant women who smoke during their pregnancy to 13.6%.

In similar fashion to our work in constructing a confidence interval for a population proportion (Section 6.4), we must expect at least 5 successes and 5 failures in our sample. This is due to the fact that we are essentially applying the normal approximation to the binomial distribution, which can only be done when the expected number of successes and failures are both at least 5. If n is the size of our sample and π is the claimed population proportion, then both $n \cdot \pi$ and $n(1 - \pi)$ must be at least 5, or we cannot use the z-distribution for this hypothesis test. (We could test the claim using the binomial distribution, however.)

This test uses the z-distribution. A major difference between this test and the large-sample test for a single mean is the test statistic. Here it is.

$$z = \frac{p - \pi}{\sqrt{\dfrac{\pi(1 - \pi)}{n}}}$$

where p = sample proportion
π = claimed population proportion
n = sample size

If we examine the numerator, we notice that it is again the difference between the sample proportion and the claimed population proportion. The denominator is the standard error of the distribution of sample proportions.

The test for a single population proportion follows the same five-step procedure. Let's look at some examples.

EXAMPLE 7.22	A physician claims that 13.6% of all pregnant females smoke during their pregnancy. A random sample of 400 pregnant women revealed that 60 of them are smoking while pregnant. Test the physician's claim at the 0.05 level of significance.

Step 1

Claim in words: 13.6% of all pregnant females smoke during their pregnancy.
Claim: $\pi = 0.136$ (We use π to represent the population proportion.)
Complement: $\pi \neq 0.136$
H_0: $\pi = 0.136$
H_A: $\pi \neq 0.136$

Step 2

Level of significance: $\alpha = 0.05$

Step 3

Test statistic: $z = \dfrac{p - \pi}{\sqrt{\dfrac{\pi(1 - \pi)}{n}}}$

Step 4

$\alpha = 0.05$, two-tailed test

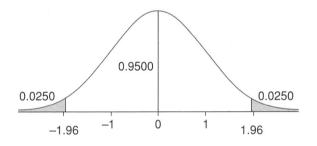

Decision rule: Reject H_0 if $z < -1.96$ or if $z > 1.96$.

Step 5

$$z = \frac{p - \pi}{\sqrt{\dfrac{\pi(1 - \pi)}{n}}}$$

$$= \frac{0.15 - 0.136}{\sqrt{\dfrac{(0.136)(0.864)}{400}}}$$

$$= 0.82$$

Decision: Fail to reject H_0.

Conclusion: There is not sufficient sample evidence to reject the claim that 13.6% of all pregnant women smoke while pregnant.

p-value

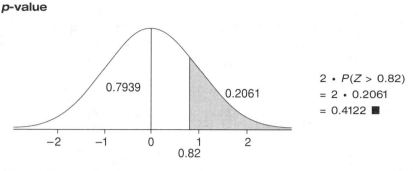

$2 \cdot P(Z > 0.82)$
$= 2 \cdot 0.2061$
$= 0.4122$ ∎

 Test the claim that more than 60% of all college statistics students are female at the 0.05 level of significance. A random sample of 114 college statistics students revealed that 81 of them were female.

Step 1

Claim in words: More than 60% of all college statistics students are female.
Claim: $\pi > 0.6$
Complement: $\pi \leq 0.6$
H_0: $\pi \leq 0.6$
H_A: $\pi > 0.6$

Step 2

Level of significance: $\alpha = 0.05$

Step 3

Test statistic: $z = \dfrac{p - \pi}{\sqrt{\dfrac{\pi(1 - \pi)}{n}}}$

Step 4

$\alpha = 0.05$, right-tailed test

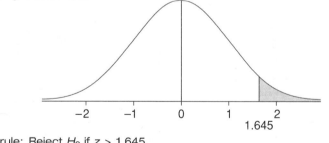

Decision rule: Reject H_0 if $z > 1.645$.

Step 5

$$z = \dfrac{p - \pi}{\sqrt{\dfrac{\pi(1 - \pi)}{n}}}$$

$$= \dfrac{\dfrac{81}{114} - 0.6}{\sqrt{\dfrac{(0.6)(0.4)}{114}}}$$

$$= 2.41$$

Decision: Reject H_0.

Conclusion: There is sufficient sample evidence to support the claim that more than 60% of all college statistics students are female.

p-value

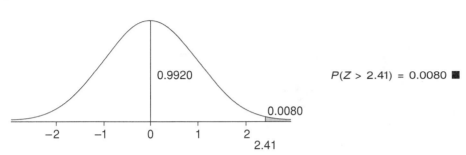

0.9920

0.0080

$P(Z > 2.41) = 0.0080$ ∎

2.41

EXAMPLE
7.24

Test the claim that less than 20% of all college students use prepaid phone cards, at the 0.05 level of significance. A random sample of 250 students revealed that 41 of them used prepaid phone cards.

Step 1

Claim in words: Less than 20% of all college students use prepaid phone cards.
Claim: $\pi < 0.2$
Complement: $\pi \geq 0.2$
H_0: $\pi \geq 0.2$
H_A: $\pi < 0.2$

Step 2

Level of significance: $\alpha = 0.05$

Step 3

Test statistic: $z = \dfrac{p - \pi}{\sqrt{\dfrac{\pi(1 - \pi)}{n}}}$

Step 4

$\alpha = 0.05$, left-tailed test

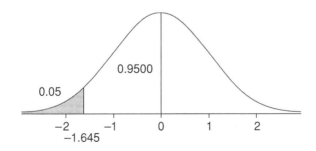

0.9500

0.05

−1.645

Decision rule: Reject H_0 if $z < -1.645$.

Step 5

$$z = \dfrac{p - \pi}{\sqrt{\dfrac{\pi(1 - \pi)}{n}}}$$

$$= \dfrac{\dfrac{41}{250} - 0.2}{\sqrt{\dfrac{(0.2)(0.8)}{250}}}$$

$$= -1.42$$

Decision: Fail to reject H_0.

Conclusion: There is not sufficient sample evidence to support the claim that less than 20% of all college students use prepaid phone cards.

p-value

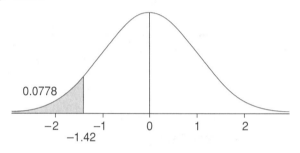

$P(Z < -1.42) = 0.0778$ ∎

0.0778

| MICROSOFT EXCEL | **Hypothesis Tests for a Single Population Proportion** |

Here is an example using Excel.

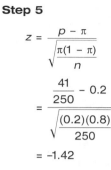

EXAMPLE 7.25

While at a dice table in a casino, a gambler comes up with the idea that the dice may be loaded. Particularly, he believes that they are set up so that the sum 7 will come up more often than it is supposed to. He recalls from his probability training that the probability of rolling a sum of 7 is $\frac{1}{6}$. He writes down the results from the next 50 rolls, and here they are.

5	9	9	6	6	7	8	9	6	7
11	6	10	8	10	5	2	7	6	7
8	9	3	7	6	8	7	3	3	9
5	8	7	7	9	4	9	7	4	12
3	5	9	7	9	10	8	9	8	7

Use these sample data to test the claim that the proportion of 7s is greater than $\frac{1}{6}$ at the 0.05 level of significance.

Enter the 50 values in column **A** (cells **A1** through **A50**). Now we will let Excel determine whether a roll was 7 or not. In cell **B1,** type

```
=IF(A1=7,1,0)
```

If cell **A1** is 7, Excel will place a 1 in cell **B1.** Otherwise, it will place a 0 in cell **B1.** Now copy cell **B1** and paste it in cells **B2** through **B50.**

In cell **C1,** type Number of Sevens. Next to that in cell **D1,** type

```
=SUM(B1:B50)
```

This will total column **B,** and tell us how many rolls were 7s.

In cell **C2,** type Sample Size. Type the sample size (50) in cell **D2.**

In cell **C3,** type Sample Proportion. In cell **D3** type the following to calculate the sample proportion.

```
=D1/D2
```

In cell **C4,** type Claimed Proportion. Next to that, type =1/6 in cell **D4.**

Now that the sample information is set up, we will begin the hypothesis test and use Excel where needed.

Step 1

Claim in words: The proportion of 7s is greater than $\frac{1}{6}$.

Claim: $\pi > \frac{1}{6}$

Complement: $\pi \leq \frac{1}{6}$

H_0: $\pi \leq \frac{1}{6}$

H_A: $\pi > \frac{1}{6}$

Step 2

Level of significance: $\alpha = 0.05$

Step 3

Test statistic: $z = \dfrac{p - \pi}{\sqrt{\dfrac{\pi(1 - \pi)}{n}}}$

Step 4

$\alpha = 0.05$, right-tailed test

- In cell **C6,** type Critical Value.
- To calculate the critical value for this test, type the following in cell **D6.**

```
=NORMSINV(1-0.05)
```

Excel returns a value of 1.644853.

Decision rule: Reject H_0 if $z > 1.644853$.

Here is how the decision rule is formulated without Excel.

Decision rule: Reject H_0 if $z > 1.645$.

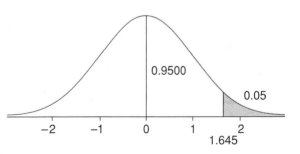

Step 5

- In cell **C7,** type Test Statistic.
- To calculate the test statistic, type the following in cell **D7**.

```
=(D3-D4)/SQRT(D4*(1-D4)/D2)
```

Excel returns a value of 1.011929 for the test statistic.

Now, here is the calculation by hand using the formula.

$$z = \frac{p - \pi}{\sqrt{\dfrac{\pi(1 - \pi)}{n}}}$$

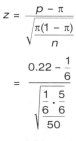

$$= \frac{0.22 - \dfrac{1}{6}}{\sqrt{\dfrac{\dfrac{1}{6} \cdot \dfrac{5}{6}}{50}}}$$

$$= 1.01$$

Decision: Fail to reject H_0.

Conclusion: There is not sufficient sample evidence to support the claim that the proportion of 7s is greater than $\frac{1}{6}$.

p-value

● In cell **C9,** type *p*-value. In cell **D9,** we calculate the *p*-value by typing

   =1-NORMSDIST(D7)

Excel returns a value of 0.155786.

Here is the "non-Excel" calculation.

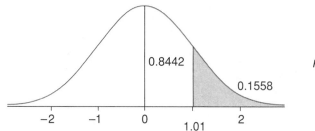

$$P(Z > 1.01) = 1 - 0.8442$$
$$= 0.1558 \ \blacksquare$$

Below is the Excel worksheet for this example.

5	0	Number of Sevens	11	8	0
9	0	Sample Size	50	7	1
9	0	Sample Proportion	0.22	3	0
6	0	Claimed Proportion	0.166667	3	0
6	0			9	0
7	1	Critical Value	1.644853	5	0
8	0	Test Statistic	1.011929	8	0
9	0			7	1
6	0	p-value	0.155786	7	1
7	1			9	0
11	0			4	0
6	0			9	0
10	0			7	1
8	0			4	0
10	0			12	0
5	0			3	0
2	0			5	0
7	1			9	0
6	0			7	1
7	1			9	0
8	0			10	0
9	0			8	0
3	0			9	0
7	1			8	0
6	0			7	1

TI-83 Hypothesis Tests for a Single Population Proportion

Here is an example for the TI-83.

EXAMPLE 7.26

While at a dice table in a casino, a gambler comes up with the idea that the dice may be loaded. Particularly, he believes that they are set up so that the sum 7 will come up more often than it is supposed to. He recalls from his probability training that the probability of rolling a sum of 7 is $\frac{1}{6}$. He writes down the results from the next 50 rolls, and here they are.

5	9	9	6	6	7	8	9	6	7
11	6	10	8	10	5	2	7	6	7
8	9	3	7	6	8	7	3	3	9
5	8	7	7	9	4	9	7	4	12
3	5	9	7	9	10	8	9	8	7

Use these sample data to test the claim that the proportion of 7s is greater than $\frac{1}{6}$ at the 0.05 level of significance.

Single-proportion tests are also quite easy on the TI-83. Before we begin, we must know what the claimed population proportion is, what the sample proportion is, and what form the alternate hypothesis takes. The claim is that the proportion of 7s is greater than $\frac{1}{6}$. This turns into a test of H_0: $\pi \leq \frac{1}{6}$ versus H_A: $\pi > \frac{1}{6}$. Since there were 11 sevens out of 50 rolls, the sample proportion is $\frac{11}{50}$.

Press (STAT), then use the (→) key to move to the **TESTS** menu. Select option 5: **1-Prop Ztest.** Here is how to fill out the next screen.

- **p₀:** Type 1 (÷) 6 here, as this is the claimed population proportion.
- **x:** Type 11 here, the number of successes.
- **n:** Type 50 here, the sample size.
- **prop:** Choose the form of the alternate hypothesis here. The correct option this time is >**p₀.**
- Highlight **Calculate** and press (ENTER).

Here is the screen that you should see. It lists the claimed population proportion, the test statistic, the *p*-value, the sample proportion, and the sample size. ■

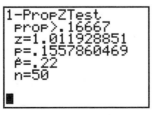

```
1-PropZTest
prop>.16667
z=1.011928851
p=.1557860469
p̂=.22
n=50
```

EXERCISES 7.3

1. At a campus-wide meeting, the president of a community college claims that more than 50% of all community college students are female. In a random sample of 155 community college students, 88 were female. Test the president's claim at the 0.01 level of significance.

2. In a current events survey, college students were asked to name the star of the hit movie *The Water Boy*. Twenty-eight out of 35 randomly selected college students were able to name Adam Sandler as the star of the movie. At the 0.05 level of significance, test the claim that more than 70% of all college students can name the star of the movie *The Water Boy*.

3. In a current events survey, college students were asked to name the U.S. senator who went into space in 1998. Twenty-three out of 35 randomly selected college students were able to name John Glenn as the senator who went into space. At the 0.05 level of significance, test the claim that less than 70% of all college students can name the U.S. senator who went into space in 1998.

4. A counselor claims that more than one-third of college students have divorced parents. A random sample of 40 college students revealed that 18 had divorced parents.
   (a) At the 0.01 level of significance, test the counselor's claim.
   (b) Explain how you would gather the data to test this claim, and list any biases that may be present using your method.

5. The author of a blackjack strategy book claims that the dealer wins 52.5% of all hands dealt. A blackjack player played 100 hands against a computer dealer, and the computer dealer won 58 of the hands. Use these data to test the author's claim at the 0.05 level of significance.

6. An algebra instructor believes that a cumulative midterm exam will help his students. His colleagues feel otherwise. One colleague claims that, in general, at most 25% of students would pass such a test. The cumulative midterm is given to 40 randomly selected students, and only 9 pass with a C or better. Test the colleague's claim at the 0.05 level of significance.

7. Pete Rose, who was elected to the All-Century major league baseball team, is banned for life from baseball. This prevents the all-time major league hit leader from being inducted into the Baseball Hall of Fame. A random sample of 200 baseball fans revealed that 121 want Pete Rose reinstated so that he may be elected into Baseball's Hall of Fame. Test the claim that more than half of all baseball fans feel that Pete Rose should be reinstated, at the 0.05 level of significance.

8. A doctor claims that at least 75% of the patients that he operates on for carpal tunnel syndrome return to work within a year of the operation. A sample of 55 patients that were operated on by the doctor for carpal tunnel syndrome showed that only 35 of them returned to work within a year of the surgery. Test the doctor's claim at the 0.05 level of significance.

9. It is claimed that 60% of all 18–25-year-olds have used alcohol in the past 30 days. A survey of 125 students on campus who are between the ages of 18 and 25 showed that 83 have used alcohol in the past 30 days. Test the claim at the 0.05 level of significance. (Based on the results of a study by the U.S. Department of Health and Human Services.)

10. It is claimed that more than 25% of all 18–25-year-olds have used illicit drugs in the past 30 days. A sample of 300 people between the ages of 18 and 25 revealed that 85 had used illicit drugs in the past 30 days. Test the claim at the 0.05 level of significance. (Based on the results of a study by the U. S. Department of Health and Human Services.)

11. A magazine article claimed that more than half of all Americans who work at home are men. A survey of 1200 Americans who work at home revealed that 635 were men. (Based on the results of a study by the U. S. Bureau of Labor Statistics.)
    (a) Test the magazine's claim at the 0.05 level of significance.
    (b) Explain how you would gather the data to test this claim, and list any biases that may be present using your method.

12. A political pundit claims that at least 55% of all female voters over the age of 60 vote for the Democratic Party candidate. A survey of 445 female voters who were 60 years old or older showed that 241 of them voted for the Republican candidate in the 1998 House of Representatives election. Test the pundit's claim at the 0.01 level of significance. (Based on the results of a poll conducted for the American Association of Health Plans.)

13. It has been claimed that at least 2 out of every 3 pet owners allow their pets to sleep in bed with them. A survey of 295 pet owners at a local pet shop showed that

236 of them allowed their pet to sleep in bed with them. At the 0.05 level of significance, test the claim that at least $\frac{2}{3}$ of all pet owners allow their pets to sleep in bed with them. (Based on the results of a study by CommSciences for PETsMART.)

14. A group of 630 adults were asked whether they take vitamins. Three hundred eighty-four of the people asked said yes. Test the claim that less than 75% of adults take vitamins, using the 0.01 level of significance. (Based on the results of a study by Market Facts for Vitamins.com.)

15. A survey of 595 college students showed that 240 did not use a condom the last time they had intercourse. Test the claim that less than half of all college students did not use a condom the last time they had intercourse at the 0.05 level of significance.

16. An NBC News/*Wall Street Journal* survey conducted of 2011 adult Americans revealed that 83% felt that "parents not paying enough attention to what's going on in their children's lives" has become a "very serious problem." Use this sample to test the claim that more than 80% of all adult Americans feel that "parents not paying enough attention to what's going on in their children's lives" has become a "very serious problem." Use the 0.05 level of significance.

17. An NBC News/*Wall Street Journal* survey conducted of 2011 adult Americans revealed that 63% identified "the high rate of divorce and the breakup of families" as a very serious problem for American society. Use this sample to test the claim that 60% of all adult Americans believe that "the high rate of divorce and the breakup of families" is a very serious problem for American society at the 0.05 level of significance.

18. Test the claim that more than 25% of all people who suffer strokes are smokers, at the 0.05 level of significance. The New Zealand Auckland Stroke Study compared 521 stroke patients with 1851 healthy people. Of the stroke patients, 164 were active smokers.

19. Test the claim that more than 20% of all teachers feel that the value of classroom technology is overrated. Use the 0.05 level of significance. A survey of 190 teachers showed that 49 of them felt that the value of classroom technology is overrated. (Based on the results of a survey by *USA Today*.)

20. A random sample of 611 Californians revealed that 140 did not have health insurance. Use this sample to test the claim that more than 20% of Californians do not have health insurance. Use the 0.05 level of significance. (Based on the results of a study by the Census Bureau.)

21. Test the claim that more than 1% of all pregnant women drink alcohol while pregnant at the 0.05 level of significance. A random sample of 325 women who gave birth showed that 6 of them drank alcohol while pregnant. (Based on data contained in Centers for Disease Control and Prevention Monthly Vital Statistics Report.)

22. Test the claim that more than half of all babies born are boys, at the 0.01 level of significance. During 1994, there were 2,022,589 male births and 1,930,178 female births. (Based on data contained in Centers for Disease Control and Prevention Monthly Vital Statistics Report.)

23. Test the claim that more than 60% of all college graduates use the Internet at the 0.05 level of significance. An alumni survey of 495 college graduates showed that 307 of them currently use the Internet.

24. A recent survey showed that 112 out of 180 two-parent households with children had a computer at home. Use this survey to test the claim that more than half of all households that have two parents living in the home have a computer at home. Use the 0.05 level of significance. (Based on the results of a study by the Henry J. Kaiser Family Foundation.)

25. A recent survey showed that 23 out of 72 households that were headed by a single female had a computer at home. Use this survey to test the claim that less than half of all households that are headed by a single female have a computer at home. Use the 0.05 level of significance. (Based on the results of a study by the Henry J. Kaiser Family Foundation.)

26. A random sample of 1401 women, financed by the Centers for Disease Control and Prevention, showed that 55% of them had been victims of some type of abuse (from physical assault to psychological battering) by an intimate male partner. At the 0.01 level of significance, test the claim that more than half of all women have been victims of such abuse from an intimate male partner.

27. A random sample of 118 adult women revealed that 70 were satisfied with their weight. At the 0.05 level of significance, test the claim that more than half of all adult women are satisfied with their weight. (Based on a survey conducted by Maritz Poll.)

28. According to research done by the John J. Heldrich Center for Workforce Development, here are the ways that U.S. workers learned to use a computer.

Self-Taught	Learned at School	Learned at Work	From Family or Friend	Other
41%	26%	23%	6%	4%

If these percentages were based on a sample of 300 workers, test the claim that more than one-third of all U.S. workers taught themselves to use a computer at the 0.01 level of significance.

29. A sample of 225 pregnant women who are smoking while pregnant showed that 70 of them smoke more than half a pack of cigarettes (10 cigarettes) per day. At the 0.05 level of significance, test the claim that more than one-quarter of all pregnant women who smoke while pregnant smoke more than half a pack per day. (Based on the results of a Gallup Poll.)

30. Test the claim that 25% of all couples spend their honeymoon in Hawaii at the 0.05 level of significance. A random sample of 180 couples showed that 35 had visited Hawaii on their honeymoon. (Based on a survey by *Bride's* magazine.)

31. Eight hundred forty-nine births were randomly selected, and 179 were by Caesarian section. At the 0.01 level of significance, test the claim that less than one-fourth of all babies are born by Caesarian section. (Based on data in the March 2000 National Vital Statistics report.)

32. A sample of 340 Americans were asked whether they thought that UFOs are real, and 146 said they thought that UFOs are real. At the 0.05 level of significance, test the claim that less than half of all Americans think that UFOs are real. (Based on a study by Yankelovich Partners for *Life* magazine.)

33. A random sample of 225 adult Americans showed that 13 of them had more than one job. At the 0.05 level of significance, test the claim that 5% of adult Americans have more than one job. (Based on data from the Bureau of Labor Statistics.)

34. A survey given to 441 college students showed that 247 either drink responsibly or abstain from drinking alcohol. At the 0.05 level, test the claim that more than half of all college students either drink responsibly or abstain from drinking alcohol. (Based on the Harvard School of Public Health College Alcohol Study.)

35. A study of 745 eighth-graders showed that 30 of them smoked at least half a pack of cigarettes daily. At the 0.01 level of significance, test the claim that 5% of all eighth-graders smoke half a pack or more daily. (Based on results of a study by the American Academy of Pediatric Dentistry.)

36. A random study of 1424 high school freshmen showed that 1054 went on to obtain their high school diploma. At the 0.01 level of significance, test the claim that fewer than 80% of high school students go on to earn their high school diploma. (Based on the results of a study by the Organization for Economic Cooperation and Development.)

37. Many women talk about being a "June bride," but how many women actually are? A random sample of 315 recent marriages showed that 35 occurred in the month of June. At the 0.05 level of significance, test the claim that 10% of all marriages take place during the month of June. (Based on a study by *Bride's* magazine.)

38. A random sample of 1374 high school students revealed that 238 of them had carried a weapon to school during the preceding 30 days. At the 0.05 level of significance, test the claim that more than 10% of all high school students have carried a weapon to school in the last 30 days. (Based on the Morbidity and Mortality Weekly Report by the Centers for Disease Control and Prevention.)

39. A random sample of 3610 high school students showed that 300 had attempted suicide at least once in the 12 months preceding the survey. At the 0.05 level of significance, test the claim that 10% of all high school students have attempted suicide in the past 12 months. (Based on the Morbidity and Mortality Weekly Report by the Centers for Disease Control and Prevention.)

40. A random sample of 1954 high school students showed that 186 of them have tried cocaine at least once. At the 0.01 level of significance, test the claim that 10% of all high school students have ever tried cocaine at least once. (Based on the Morbidity and Mortality Weekly Report by the Centers for Disease Control and Prevention.)

? **What Is Wrong with This Picture?**

The following statistical project contains a major error. Potential errors could be the choice of the wrong hypothesis test or that the assumptions required for this test are not met. Find the error, and explain what could be done to correct it.

A student believes that most college students feel that tuition is too high. He asks 42 randomly selected students on his campus and 38 of them feel that tuition is too high. He uses these results to test the claim that 90% of all college students feel that tuition is too high, using the hypothesis test introduced in this section.

What is wrong with this picture?

MINI PROJECT

I claim that more than 50% of the students at your school are female. You are to test my claim at the 0.05 level of significance. Randomly sample at least 100 students and record their gender. Use the sample data to test the claim. In addition to your complete hypothesis test, include

- Your raw data
- A pie chart showing the percentage of students at your school that are female and the percentage that are male

Finally, answer the following questions.

- How did you obtain your sample?
- Which type of sampling did you use?
- Was the sample truly random?
- What potential biases may show up in your sample?

EXCEL EXPLORATION   *p*-Values

We will use Excel to explore the idea of *p*-values. We will generate several random samples based on a known population proportion. We will then see what proportion of these randomly generated sample proportions are at least as extreme as our given sample proportion. This is the *p*-value for the given sample.

Before we begin we must be sure that the Analysis Tool Pak has been added. Pull down the **Tools** menu. If you see a choice that says **Data Analysis**, then the Tool Pak is added and you may continue. If you do not see **Data Analysis**, click on **Add-ins** under **Tools**. When the dialog box opens up, click on the box for **Analysis Tool Pak**, and click on **OK**.

We will use the following hypothesis test as an example.

EXAMPLE 7.27

A doctor claims that at least 75% of the patients that he operates on for carpal tunnel syndrome return to work within a year of the operation. A sample of 55 patients that were operated on by the doctor for carpal tunnel syndrome showed that only 35 of them returned to work within a year of the surgery. Test the doctor's claim at the 0.05 level of significance.

This is a left-tailed test, with the alternate hypothesis being $\pi < 0.75$. To estimate the p-value for this test, we will assume that the actual population proportion π is equal to 0.75. We will then generate a random sample of size 55 based on this value of π. We will repeat this process 1000 times. Our estimate of the p-value will be equal to the proportion of samples that contain 35 or fewer successes, $p \leq \frac{35}{55}$.

In a new Excel worksheet, select **Data Analysis** from the **Tools** menu. When the dialog box opens, select **Random Number Generation** and click on **OK.** In the box labeled **Number of Variables,** type 1. In the box labeled **Number of Random Numbers,** enter 1000. In the box labeled **Distribution,** select **Binomial.** Enter 0.75 for the **p-value** (this is the claimed population proportion π, not the p-value we are trying to estimate), and enter 55 in the box labeled **Number of Trials.** Type in a positive integer of your choice for the **Random Seed.** For the **Output Options,** select **Output Range** and type A1. Finally, click on **OK.**

The data will show up in column **A.** To sort the data, select **Sort** from the **Data** menu. When the dialog box appears, be sure that **Column A** is in the box labeled **Sort By,** and the **Ascending** option is selected. Press **OK** to sort the values from lowest to highest.

Keep in mind that this is just an estimate of the p-value. The actual p-value is 0.0256, whereas estimates of this p-value ranged from 0.024 to 0.055.

If the test is a right-tailed test, we count the proportion of sample proportions that are as high as or higher than the given sample proportion. For a two-tailed test, count the proportion of sample proportions that are at least as far away (above or below) the population proportion π as the given sample proportion p is. ■

Use Excel to estimate the p-values for the following hypothesis tests from the exercises following Section 7.3.

1. A local buffet restaurant has many patrons that are senior citizens. The manager of the restaurant claims that at least 60% of her customers are senior citizens. A random sample of 50 diners revealed that 33 of them were senior citizens. At the 0.05 level of significance, test the manager's claim. (**Actual p-value: 0.1922**)
2. A nurse at a community college claims that less than 25% of college students smoke. A random sample of 30 college students revealed that 6 of them smoke. Test the nurse's claim at the 0.01 level of significance. (**Actual p-value: 0.2643**)
3. A magazine article claimed that half of all people list pepperoni as their favorite pizza topping. A random sample of 150 people revealed that 69 of them listed pepperoni as their favorite pizza topping. Test the article's claim at the 0.05 level of significance. (**Actual p-value: 0.3270**)
4. A 911 operator claims that half of all 911 calls can be classified as "nonemergency" calls. Thirty-four of 64 randomly selected 911 calls were nonemergency calls. Test the operator's claim at the 0.05 level of significance. (**Actual p-value: 0.6170**)

5. At a campus-wide meeting, the president of a community college claims that more than 50% of all community college students are female. In a random sample of 155 community college students, 88 were female. Test the president's claim at the 0.01 level of significance. (**Actual p-value: 0.0455**)

Choosing the Appropriate Tool

One of the most difficult things for introductory statistics students to do is deciding which hypothesis test or confidence interval is appropriate for problems that do not come directly from a textbook. In this section, we establish guidelines for determining the appropriate course of action.

The first big question that must be answered is whether we are interested in a population mean or a proportion. If your data are categorical, then you will be working with a population proportion. If your data are numerical, then a population mean is the parameter of interest. Think about the question(s) or measurement(s) that lead to the data, which will help you determine whether the data are numerical or categorical. For instance, if you ask a person in your sample which presidential candidate he plans to vote for in November, the response will be the name of one of the candidates. The data are categorical, so we will be interested in the proportion of all voters that plan to vote for each candidate. Look for the conditions that are present in a binomial experiment—each response can be classified as a success (plan to vote for the candidate) or a failure (do not plan to vote for the candidate), and each trial is independent.

If you ask a smoker in your sample at what age she started smoking, then you will most likely be interested in the mean age at which all smokers begin smoking. It is not always so clear-cut; sometimes we collect data that are numerical but use them as if they were categorical. For example, you may be interested in the proportion of all smokers who began smoking before they were 13 years old. In this case, your data are treated as categorical data: either the person was younger than 13 or was not.

The decision of whether to create a confidence interval or perform a hypothesis test depends on what we are interested in. A confidence interval is used whenever we are interested in estimating a population mean or proportion. We perform a hypothesis test whenever we want to evaluate a claim about a population mean or proportion.

Suppose we want to construct a confidence interval for a population mean, or perform a hypothesis test to test a claim about a population mean. If our sample size is at least 30, then we may use the z-distribution. If the sample size is below 30 and the sample is drawn from a population that appears to be normally distributed, then we use the t-distribution. If the sample size is below 30 and the sample data are skewed, this suggests that the population is not normally distributed and that we should use methods that will be introduced in Chapter 11.

If we want to construct a confidence interval for a population proportion, or perform a hypothesis test to test a claim about a population proportion, we must verify certain conditions first. In order to construct a confidence interval, our sample must contain at least 5 successes and 5 failures, since our confidence interval is based on the normal approximation to the binomial distribution. (If there are fewer than 5 successes or failures, we could use the binomial distribution to construct a

(continues)

Choosing the Appropriate Tool *(continued)*

crude confidence interval or perform a hypothesis test. These procedures are not covered in this text.) For a hypothesis test, we need the expected number of successes and the expected number of failures to be 5 or higher, rather than the actual number of successes and failures in the sample.

Confidence Intervals

We construct a confidence interval when we want to estimate a population mean or proportion.

Confidence Interval	Section	Margin of Error	Interval	Conditions
Population mean (large samples)	6.2	$E = z_{\alpha/2} \cdot \dfrac{s}{\sqrt{n}}$	$\bar{x} - E \leq \mu \leq \bar{x} + E$	$n \geq 30$
Population mean (small samples)	6.3	$E = t_{\alpha/2} \cdot \dfrac{s}{\sqrt{n}}$ degrees of freedom = $n - 1$	$\bar{x} - E \leq \mu \leq \bar{x} + E$	• $n < 30$ • Sample comes from a population that is normally distributed
Population proportion	6.4	$E = z_{\alpha/2} \cdot \sqrt{\dfrac{p \cdot (1 - p)}{n}}$	$p - E \leq \pi \leq p + E$	• Number of successes in the sample is at least 5 • Number of failures in the sample is at least 5

Hypothesis Tests

We perform a hypothesis test when we want to evaluate a claim about a population mean or proportion.

Test	Section	Test Statistic	Conditions
Population mean (large samples)	7.1	$z = \dfrac{\bar{x} - \mu}{s/\sqrt{n}}$	$n \geq 30$
Population mean (small samples)	7.2	$t = \dfrac{\bar{x} - \mu}{s/\sqrt{n}}$ degrees of freedom = $n - 1$	• $n < 30$ • Sample comes from a population that is normally distributed
Population proportion	7.3	$z = \dfrac{p - \pi}{\sqrt{\dfrac{\pi(1 - \pi)}{n}}}$	• $n \cdot \pi \geq 5$ • $n \cdot (1 - \pi) \geq 5$

Overview

In this unit there are three major types of problems. The first is determining how large a sample must be taken. The second is constructing a confidence interval for a population parameter. The third is testing a claim using the five-step formal procedure for hypothesis tests.

How large a sample we must take depends first on whether we are trying to estimate a population mean or a population proportion. If we are estimating a population mean, we need to have a level of confidence (which gives us a value for z), an acceptable margin of error (E), and an estimate of the population's standard deviation (σ). If we are estimating a population proportion, we need to have a level of confidence, an acceptable margin of error, and an estimate of what the sample proportion (p) will be. If no estimate of the sample proportion is available, we use 0.5 for p.

Again, when constructing a confidence interval, it depends on whether we are trying to estimate a population mean or proportion. The basic steps are the same for both. First, we calculate the margin of error. Then we subtract it from the sample statistic to obtain the left endpoint of our interval, and add it to the sample statistic to obtain the right endpoint of our interval. If we are constructing a confidence interval for a population mean and our sample size is at least 30, then we use the z-distribution. Otherwise, we use Student's t-distribution.

We learned three types of hypothesis tests in Chapter 7. We are able to test claims about a population mean (large sample or small sample) or a population proportion.

The **null hypothesis** is a statement that is assumed to be true for testing purposes. If there is enough sample evidence to suggest that the null hypothesis is false, we reject it and support the **alternate hypothesis** as true. Recall that the null hypothesis always contains equality.

Each test has a **level of significance**. The level of significance is the probability of committing a Type I error, which means that we reject a null hypothesis that is actually true.

Each type of hypothesis test has its own **test statistic**. This is a formula that we use to determine the probability of obtaining sample information as extreme, or more extreme, than our sample if the null hypothesis is actually true.

We establish a **decision rule**, which tells us when to reject the null hypothesis and when to fail to reject it. The decision rule depends on the level of significance and the tail of the test. In the case of the small sample test for a single population mean, we also depend on the number of degrees of freedom for our decision rule.

We complete the test by calculating the test statistic, deciding whether to reject the null hypothesis or not, making a conclusion about our claim, and calculating a p-value for the test. The **p-value** is the probability of obtaining a sample as extreme as our sample, or more extreme than our sample, if the null hypothesis were actually true.

Unit 4 Review Exercises

1. A sample of 14 elementary school teachers revealed that they had spent a mean of $606 out of their own pockets during the last school year for classroom supplies, with a standard deviation of $744. Construct a 90% confidence interval for the mean annual out-of-pocket expense for elementary school teachers.
2. Construct a 95% confidence interval for the mean number of magazines read per week by women. A random sample of 34 women produced a mean of 2.647 magazines with a standard deviation of 1.889 magazines.
3. At the 0.05 level, test the claim that more than half of all shoes purchased at a mall shoe store are for children. A random sample of 150 purchases revealed that 80 had been for children.

4. Construct a 95% confidence interval for the proportion of community college students that attend church regularly. A random sample of 168 students revealed that 67 of them attend church regularly.

5. A researcher wants to find out what percentage of people prefer to do their grocery shopping at a bulk warehouse store rather than a traditional grocery store. How many people must she survey to be 95% sure that her estimate is off by no more than 5%?

6. At the 0.05 level, test the claim that the mean number of units completed by community college students is greater than 10. A random sample of 28 students produced a mean of 12.071 units with a standard deviation of 3.661 units.

7. A researcher wants to estimate the mean number of hours that a member of a local health club spends working out per week. How large a sample should she take to be 95% confident that her estimate is off by no more than 0.2 hours? Past studies suggest a standard deviation of 3 hours.

8. At the 0.05 level, test the claim that the hours worked by community college students is greater than 15 hours per week. A random sample of 30 students produced a sample mean of 20.83 hours per week with a standard deviation of 14.20 hours per week.

Formulas

Confidence interval for a population mean (large sample):

$$E = z_{\alpha/2} \cdot \frac{s}{\sqrt{n}} \qquad \bar{x} - E \le \mu \le \bar{x} + E$$

Appropriate sample size for estimating a population mean:

$$n = \left(\frac{z_{\alpha/2} \cdot \sigma}{E}\right)^2$$

Confidence interval for a population mean (small sample):

$$E = t_{\alpha/2} \cdot \frac{s}{\sqrt{n}} \qquad \bar{x} - E \le \mu \le \bar{x} + E$$

Confidence interval for a population proportion (large sample):

$$E = z_{\alpha/2} \cdot \sqrt{\frac{p \cdot (1 - p)}{n}} \qquad p - E \le \pi \le p + E$$

Appropriate sample size for estimating a population proportion:

$$n = \left(\frac{z_{\alpha/2}}{E}\right)^2 \cdot p \cdot (1 - p)$$

Hypothesis test for a population mean (large sample):

$$z = \frac{\bar{x} - \mu}{s/\sqrt{n}}$$

Hypothesis test for a population mean (small sample):

$$t = \frac{\bar{x} - \mu}{s/\sqrt{n}} \qquad \text{degrees of freedom} = n - 1$$

Hypothesis test for a population proportion (large sample):

$$z = \frac{p - \pi}{\sqrt{\dfrac{\pi(1 - \pi)}{n}}}$$

unit
five

5

Two-Sample Inferences, *F* Tests, Chi-Square Tests

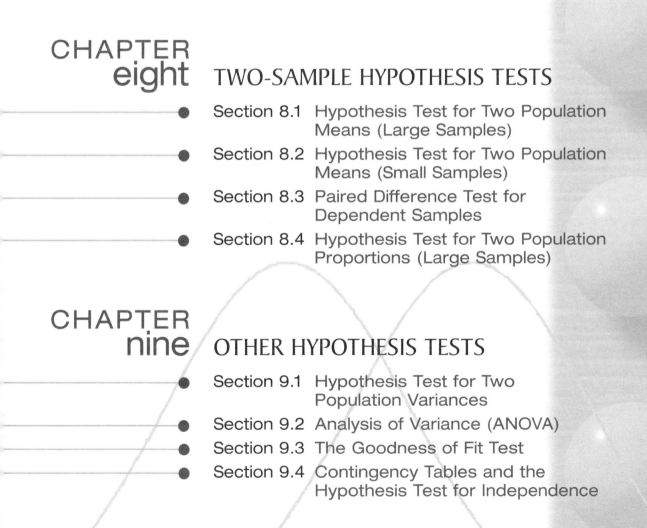

Two-Sample Hypothesis Tests

I n this chapter we continue our work from Chapter 7 by examining inferences made from using two samples. Instead of comparing a single population parameter (mean, proportion) to a single claimed value, we will be comparing a parameter from one population to a parameter from another population. As an example, instead of testing the claim that the mean serum cholesterol level of men over 60 years old is above 200 mg/dL, we will be testing the claim that the mean serum cholesterol level of men over 60 years old is less than the mean serum cholesterol level of women over 60 years old.

The last claim mentioned is an example of a "two-mean" test, which compares the mean of one population to the mean of another population. In Section 8.1 we will use the z-distribution for this test when both samples are of size 30 or greater. If one of the samples has a size smaller than 30, we use the t-distribution instead. This is covered in Section 8.2.

Testing two means using dependent samples is the focus of Section 8.3. This test is known as the paired difference test, and is used when there is a direct relation between a value in the first sample and a particular value in the second sample. An example is to test whether a certain diet is effective by taking the weights of people before the diet begins and again one month later. Each person's weight in the first sample is related to that person's weight one month after beginning the diet.

We finish the chapter with a test concerning two population proportions. The "two-proportion" test compares the proportion of one population to the proportion of another population. For instance, we could test the claim that the proportion of female high school graduates who go on to attend college is greater than the proportion of male high school graduates who go on to attend college.

SECTION 8.1
Hypothesis Test for Two Population Means (Large Samples)

In this section we develop a hypothesis test to use when we are comparing the mean of one population to the mean of another population. Such a test is often referred to as a "two-sample test," because two samples are gathered to test a claim that compares the means of two different populations. For this test, each sample size must be at least 30. Comparing two means when one or both of the sample sizes is lower than 30 will be handled in Section 8.2, using the t-distribution. Also, the two samples must be independent. A hypothesis test involving two dependent samples is presented in Section 8.3. In general, the procedure for hypothesis testing developed in Chapter 7 still applies here.

When performing a two-sample hypothesis test, we must label one population as population 1 and the other as population 2. The mean of population 1, denoted by μ_1, always appears first in the null and alternate hypotheses. We denote the mean of the second population by μ_2, and it always appears last in the null and alternate hypotheses. If you fail to do this, it will become impossible to determine whether a test is left-tailed or right-tailed. It does not matter which of the two populations is labeled as population 1, as long as you are consistent throughout the test. A good idea is to make the first population referred to in the claim population 1. This makes it easier to write the claim in symbols.

Here is the test statistic for comparing two population means, based on two large samples.

$$z = \frac{\bar{x}_1 - \bar{x}_2}{\sqrt{\dfrac{s_1^2}{n_1} + \dfrac{s_2^2}{n_2}}}$$

The mean of the first sample is represented by $\bar{x}_1$, s_1 is the standard deviation of the first sample, and n_1 is the size of the first sample. The subscript "1" identifies a statistic associated with sample 1, and we use a subscript "2" to denote the second sample. If for some reason the population standard deviations are known, they may be used instead of the sample standard deviations. We can use the sample standard deviation as an estimate of the population standard deviation for samples of size 30 or larger. If one or both samples have a size that is smaller than 30, we will have to use Student's t-distribution, which is covered in the next section. Many statisticians prefer to use the t-distribution even for sample sizes that are 30 and larger, but we will use the z-distribution.

We begin with an example that was included originally in the Unit 1 review.

| EXAMPLE 8.1 | A math instructor wrote two versions of the same test, which he believed to be of equal difficulty. He gave the first version (A) to 36 students, and the second version (B) to 41 students. We will consider the two groups to be random samples from the population of all community college statistics students. The 36 students who took version A had a mean score of 79.3 with a standard deviation of 11.26. The 41 students who took version B had a mean score of 84.1 with a standard deviation of 9.40. At the 0.05 level of significance, test the claim that the two versions were of equal difficulty. |

Since one version was not specifically mentioned first in the claim ("the two versions were of equal difficulty"), we will let version A be population 1 and version B be population 2.

Step 1

Population 1: version A; population 2: version B
Claim in words: The two versions are of equal difficulty.
Claim: $\mu_1 = \mu_2$
Complement: $\mu_1 \neq \mu_2$
H_0: $\mu_1 = \mu_2$
H_A: $\mu_1 \neq \mu_2$

Step 2

Level of significance: $\alpha = 0.05$

Step 3

Test statistic: $z = \dfrac{\bar{x}_1 - \bar{x}_2}{\sqrt{\dfrac{s_1^2}{n_1} + \dfrac{s_2^2}{n_2}}}$

Step 4

$\alpha = 0.05$, two-tailed test

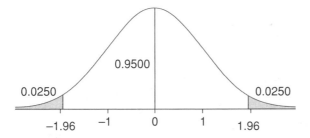

0.9500

0.0250   0.0250

-1.96 -1 0 1 1.96

Decision rule: Reject H_0 if $z < -1.96$ or if $z > 1.96$.

Step 5

$$z = \frac{\bar{x}_1 - \bar{x}_2}{\sqrt{\dfrac{s_1^2}{n_1} + \dfrac{s_2^2}{n_2}}}$$

$$= \frac{79.3 - 84.1}{\sqrt{\dfrac{11.26^2}{36} + \dfrac{9.40^2}{41}}}$$

$$= -2.01$$

Decision: Reject H_0.

Conclusion: There is sufficient sample evidence to reject the claim that the two exams were of equal difficulty.

p-value

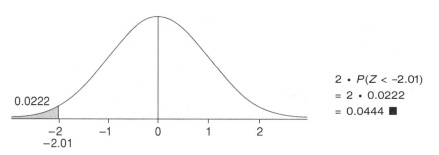

0.0222

$$2 \cdot P(Z < -2.01)$$
$$= 2 \cdot 0.0222$$
$$= 0.0444 \ \blacksquare$$

You may be wondering why we did not use the one-mean test developed in the last chapter. For instance, we could have tested the claim that the mean score for version A is below 82, and then tested the claim that the mean score for version B is above 82. The first problem with this approach is that it is very difficult to come up with that single number to test each mean against ahead of time. The second problem associated with using two separate tests involves the level of significance. Recall that using a level of significance of 0.05 means that the probability of rejecting a null hypothesis that is actually true is 0.05. However, if we perform two separate hypothesis tests at the 0.05 level of significance, then there is a 0.0975 probability of rejecting at least one null hypothesis that is actually true. This is calculated as follows.

$$P(\text{Rejecting at least 1 true null hypothesis}) = 1 - P(\text{not rejecting a true null hypothesis})$$
$$= 1 - (0.95)^2$$
$$= 0.0975$$

Not only is performing two hypothesis tests more tedious and time consuming, it also increases our chances of making a Type I error.

We continue with two more examples.

 EXAMPLE 8.2   A reading group claims that Americans read more as they grow older. A random sample of 115 Americans age 60 or older read for a mean length of 62.8 minutes per day, with a standard deviation of 18.3 minutes per day. A random sample of 88 Americans between the ages of 50 and 59 read for a mean length of 54.2 minutes per day, with a standard deviation of 23.1 minutes per day. At the 0.01 level of significance, test the claim that the mean time spent reading per day by Americans age 60 and older is longer than the mean time spent reading per day by Americans between the ages of 50 and 59.

We will let Americans age 60 and older be population 1 as they were the first group mentioned in the claim. We will let Americans between the ages of 50 and 59 be population 2.

Step 1

Population 1: 60 or older; population 2: 50 to 59 years old

Claim in words: The mean time spent reading per day by Americans age 60 and older is longer than the mean time spent reading per day by Americans between the ages of 50 and 59.

Claim: $\mu_1 > \mu_2$

Complement: $\mu_1 \leq \mu_2$

H_0: $\mu_1 \leq \mu_2$

H_A: $\mu_1 > \mu_2$

Step 2

Level of significance: $\alpha = 0.01$

Step 3

Test statistic: $z = \dfrac{\bar{x}_1 - \bar{x}_2}{\sqrt{\dfrac{s_1^2}{n_1} + \dfrac{s_2^2}{n_2}}}$

Step 4

$\alpha = 0.01$, right-tailed test

Decision rule: Reject H_0 if $z > 2.33$.

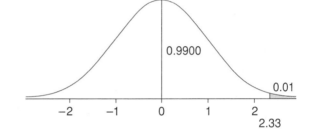

Step 5

$z = \dfrac{\bar{x}_1 - \bar{x}_2}{\sqrt{\dfrac{s_1^2}{n_1} + \dfrac{s_2^2}{n_2}}}$

$= \dfrac{62.8 - 54.2}{\sqrt{\dfrac{18.3^2}{115} + \dfrac{23.1^2}{88}}}$

$= 2.87$

Decision: Reject H_0.

Conclusion: There is sufficient sample evidence to support the claim that the mean time spent reading per day by Americans age 60 and older is longer than the mean time spent reading per day by Americans between the ages of 50 and 59.

p-value

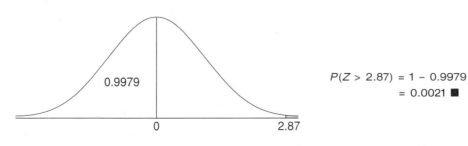

$P(Z > 2.87) = 1 - 0.9979$
$= 0.0021$ ∎

EXAMPLE
8.3

Many baseball fans prefer the National League to the American League because they believe that the National League has better pitchers, which produces lower scoring games. A random sample of 71 National League baseball games had a mean combined score of 9.7 runs per game, with a standard deviation of 4.94 runs per game. A random sample of 62 American League baseball games had a mean combined score of 10.2 runs per game, with a standard deviation of 4.98 runs per game. At the 0.05 level of significance, test the claim that the mean number of runs scored in a National League baseball game is less than it is in the American League.

We will let the National League games be population 1 as they were the first group mentioned in the claim. We will let the American League games be population 2.

Step 1

Population 1: National League games; population 2: American League games
Claim in words: The mean number of runs scored in a National League baseball game is less than it is in the American League.
Claim: $\mu_1 < \mu_2$
Complement: $\mu_1 \geq \mu_2$
H_0: $\mu_1 \geq \mu_2$
H_A: $\mu_1 < \mu_2$

Step 2

Level of significance: $\alpha = 0.05$

Step 3

Test statistic: $z = \dfrac{\bar{x}_1 - \bar{x}_2}{\sqrt{\dfrac{s_1^2}{n_1} + \dfrac{s_2^2}{n_2}}}$

Step 4

$\alpha = 0.05$, left-tailed test

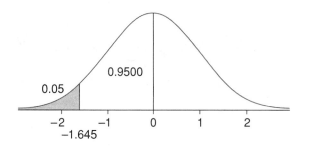

Decision rule: Reject H_0 if $z < -1.645$.

Step 5

$z = \dfrac{\bar{x}_1 - \bar{x}_2}{\sqrt{\dfrac{s_1^2}{n_1} + \dfrac{s_2^2}{n_2}}}$

$$= \frac{9.7 - 10.2}{\sqrt{\dfrac{4.94^2}{71} + \dfrac{4.98^2}{62}}}$$

$$= -0.58$$

Decision: Fail to reject H_0.

Conclusion: There is not sufficient sample evidence to support the claim that the mean number of runs scored in a National League baseball game is less than it is in the American League.

p-value

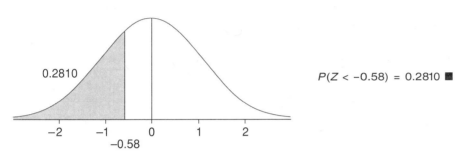

$P(Z < -0.58) = 0.2810$ ■

Confidence Intervals

Recall that the idea of hypothesis testing is based on confidence intervals. There is a confidence interval that can be constructed for the difference between two population means, when both sample sizes are large. We begin, as we did before, by calculating the margin of error. Here is the formula.

$$E = z_{\alpha/2} \sqrt{\frac{s_1^2}{n_1} + \frac{s_2^2}{n_2}}$$

We then subtract this margin of error from $\bar{x}_1 - \bar{x}_2$ to find the left endpoint of the confidence interval, and we add the margin of error to $\bar{x}_1 - \bar{x}_2$ to find the right endpoint of the confidence interval. (If you want to create a one-sided confidence interval, see the "Extra" material at the end of the Section 6.2 exercises.)

$$(\bar{x}_1 - \bar{x}_2) - E \leq (\mu_1 - \mu_2) \leq (\bar{x}_1 - \bar{x}_2) + E$$

A math instructor wrote two versions of the same test that he believed to be of equal difficulty. He gave the first version (A) to 36 students and the second version (B) to 41 students. We will consider the two groups to be random samples from the population of all community college statistics students. The 36 students who took version A had a mean score of 79.3 with a standard deviation of 11.26. The 41 students who took version B had a mean score of 84.1 with a standard deviation of 9.40. Construct a 95% confidence interval for the difference between the mean score of version A and the mean score of version B.

We will let population 1 be the scores on version A. We begin by finding the value for $z_{\alpha/2}$. Recall that for a 95% confidence interval, $z_{0.025} = 1.96$.

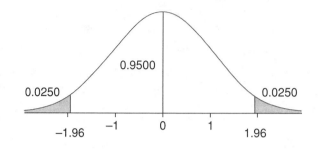

Next we calculate the margin of error E.

$$E = z_{\alpha/2} \sqrt{\frac{s_1^2}{n_1} + \frac{s_2^2}{n_2}}$$

$$= 1.96 \sqrt{\frac{11.26^2}{36} + \frac{9.40^2}{41}}$$

$$= 4.67$$

Now we construct the interval.

$$(\bar{x}_1 - \bar{x}_2) - E \le (\mu_1 - \mu_2) \le (\bar{x}_1 - \bar{x}_2) + E$$
$$(79.3 - 84.1) - 4.67 \le (\mu_1 - \mu_2) \le (79.3 - 84.1) + 4.67$$
$$-4.8 - 4.67 \le (\mu_1 - \mu_2) \le -4.8 + 4.67$$
$$-9.47 \le (\mu_1 - \mu_2) \le -0.13$$

We are 95% confident that the difference between the two means is between –0.13 points and –9.47 points. Note that this interval does not contain 0; this is strong evidence that the two population means are not equal. This conclusion can be affirmed by checking the hypothesis test of the first example of the section. ∎

MICROSOFT EXCEL   Hypothesis Test for Two Population Means (Large Samples)

Microsoft Excel's Data Analysis ToolPak has a built-in procedure for a hypothesis test comparing two population means. Before using it, we must calculate the variance for each sample, because their procedure assumes that both population variances are known.

We will step through an example that demonstrates how to use Excel for this hypothesis test.

A mathematics instructor has changed his instruction style to include more group activities, both inside and outside of class. He wants to know whether the results of such an instruction style have any effect on students' performance on a standard department final exam. Here are the final exam scores of 30 students from the semester before the change in styles, and 32 students from the semester after the change. Scores are out of a maximum of 50 points. At the 0.05 level of significance, test the claim that the mean score for both styles is the same.

Before

34	21	36	28	46	31	45	40
48	42	27	41	39	29	37	37
29	46	34	38	43	46	35	34
32	45	27	40	47	41		

After

35	39	50	25	25	36	34	31
45	35	37	32	36	31	44	38
34	23	31	36	21	30	44	45
30	39	39	42	28	47	40	27

In column **A,** enter the scores for the semester before the change in cells **A1** through **A30**. In column **B,** enter the scores for the semester after the change in cells **B1** through **B32**.

In cell **C1,** type Variance #1, and type Variance #2 in cell **D1**. To calculate the variance of the scores in sample 1, type the following in cell **C2**.

    =VAR(A1:A30)

To calculate the variance of the scores in sample 2, type the following in cell **D2**.

    =VAR(B1:B32)

Excel tells us that the variance of the first sample is 50.48 and the variance of the second sample is 52.53.

We proceed with the test by selecting **Data Analysis** from the **Tools** menu. When the dialog box appears, select **z-Test: Two Sample for Means** and click on **OK**. When the z-Test dialog box opens, type A1:A30 in the box labeled **Variable 1 Range** and type B1:B32 in the box labeled **Variable 2 Range**. In the box labeled **Hypothesized Mean Difference,** enter 0. (We would enter 5 if we claimed that the first mean was 5 points higher than the second mean.) In the box labeled **Variable 1 Variance (known)** enter 50.48, which was the variance that we calculated earlier. In the box labeled **Variable 2 Variance (known)** enter 52.53. In the box labeled **Alpha,** enter 0.05, which is the level of significance for this test. Click on **OK**.

Here is a copy of the output, which appears on another worksheet.

z-Test: Two Sample for Means

	Variable 1	Variable 2
Mean	37.26666667	35.28125
Known Variance	50.48	52.53
Observations	30	32

(continues)

(continued)

	Variable 1	Variable 2
Hypothesized Mean Difference	0	
z	1.088945603	
P(Z<=z) one-tail	0.138088975	
z Critical one-tail	1.644853	
P(Z<=z) two-tail	0.276177949	
z Critical two-tail	1.959961082	

Excel lists the mean for each sample. The value labeled as *z* is the test statistic, which is 1.09. Then Excel lists the *p*-values and critical values for a one-tailed test and a two-tailed test. Since our test is a two-tailed test, the *p*-value is 0.2762 and the critical value to be used in the decision rule is 1.96.

With this information we can write the hypothesis test.

Step 1

Population 1: previous teaching style; population 2: new teaching style
Claim in words: The mean score for both styles is the same.
Claim: $\mu_1 = \mu_2$
Complement: $\mu_1 \neq \mu_2$
H_0: $\mu_1 = \mu_2$
H_A: $\mu_1 \neq \mu_2$

Step 2

Level of significance: $\alpha = 0.05$

Step 3

Test statistic: $z = \dfrac{\bar{x}_1 - \bar{x}_2}{\sqrt{\dfrac{s_1^2}{n_1} + \dfrac{s_2^2}{n_2}}}$

Step 4

$\alpha = 0.05$, two-tailed test

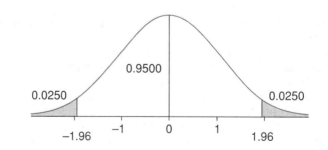

Decision rule: Reject H_0 if $z < -1.96$ or if $z > 1.96$.

Step 5

$z = 1.09$ (calculated by Excel)

Decision: Fail to reject H_0.

Conclusion: There is not sufficient sample evidence to reject the claim that the mean score for both styles is the same.

p-value

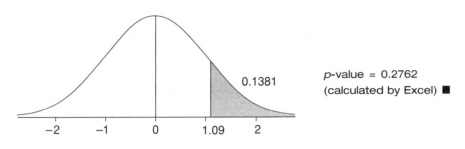

0.1381

p-value = 0.2762
(calculated by Excel) ∎

-2   -1   0   1.09   2

TI-83 | Hypothesis Test for Two Population Means
(Large Samples)

The TI-83 can help us perform the hypothesis test for two population means whether we have the two sets of sample data or two sets of sample statistics. We will step through an example of each type, beginning with an example using the sample data.

EXAMPLE 8.6

A mathematics instructor has changed his instruction style to include more group activities, both inside and outside of class. He wants to know whether the results of such an instruction style have any effect on students' performance on a standard department final exam. Here are the final exam scores of 30 students from the semester before the change in styles, and 32 students from the semester after the change. Scores are out of a maximum of 50 points. At the 0.05 level of significance, test the claim that the mean score for both styles is the same.

Before

34	21	36	28	46	31	45	40
48	42	27	41	39	29	37	37
29	46	34	38	43	46	35	34
32	45	27	40	47	41		

After

35	39	50	25	25	36	34	31
45	35	37	32	36	31	44	38
34	23	31	36	21	30	44	45
30	39	39	42	28	47	40	27

Enter the scores for the semester before the change in list L_1. Enter the scores for the semester after the change in list L_2. Before proceeding with the test, we must find both sample standard deviations, which we will use as estimates of the population standard deviations. We access the STAT CALC menu by pressing (STAT) and using the $\left(\rightarrow\right)$ key to move to the right. Highlight option **1:1-Var Stats** and press $\left(\text{ENTER}\right)$. When you are brought to the main screen, enter the list L_1 by pressing $\left(\text{2nd}\right)\left(1\right)$, and then press $\left(\text{ENTER}\right)$. Make note of the standard deviation, which is 7.10. Now repeat the process for list L_2. Its standard deviation is 7.25.

Now we access the STAT TESTS menu by pressing $\left(\text{STAT}\right)$ and using the $\left(\rightarrow\right)$ key to move to the right. Highlight option **3:2-SampZTest** and press $\left(\text{ENTER}\right)$.

Since we have the sample data, highlight **Data** next to **Inpt:**. Enter the first sample standard deviation (7.10) next to **μ1.** Enter the second sample standard deviation (7.25) next to **μ2.** Next to **List1:** enter L_1 by pressing $\left(\text{2nd}\right)\left(1\right)$. Next to **List2:** enter L_2 by pressing $\left(\text{2nd}\right)\left(2\right)$. Leave **Freq1:** and **Freq2:** as 1. Scroll down to see the alternate hypothesis. This test is a two-tailed test, so highlight ≠**μ2** and press $\left(\text{ENTER}\right)$. Finally, scroll down one more line to highlight **Calculate** and press $\left(\text{ENTER}\right)$. Here are the screens that you should see.

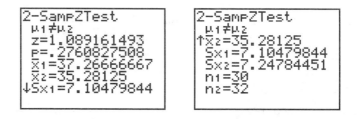

The first line of output gives us our test statistic, which rounds to 1.09. The next line gives us our *p*-value, which is 0.2761. The remaining information includes the mean, standard deviation, and size of each sample. If we repeated the exact same procedure but chose **Draw** instead of **Calculate,** we would see the following screen.

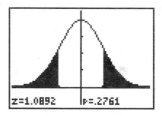

This screen also includes the value of the test statistic and *p*-value in addition to the graph. ∎

We can also use the TI-83 to help when we are given only the sample statistics instead of all of the data.

EXAMPLE 8.7

A mathematics instructor has changed his instruction style to include more group activities, both inside and outside of class. He wants to know whether the results of such an instruction style have any effect on students' performance on a standard department final exam. Thirty students from the semester before the change in styles had a mean score of 37.27 points, with a standard deviation of 7.10 points. Thirty-two students from the semester after the change had a mean score of 35.28 points, with a standard deviation of 7.25 points. Scores are out of a maximum of 50 points. At the 0.05 level of significance, test the claim that the mean score for both styles is the same.

Access the STAT TESTS menu by pressing $\boxed{\text{STAT}}$ and using the $\boxed{\rightarrow}$ key to move to the right. Highlight option **3:2-SampZTest** and press $\boxed{\text{ENTER}}$.

Since we only have the sample statistics, highlight **Stats** next to **Inpt:**. Enter the first sample standard deviation (7.10) next to **μ1.** Enter the second sample standard deviation (7.25) next to **μ2.** Enter the first sample mean (37.27) next to **x̄1:**, and the first sample size (30) next to **n1:**. Enter the second sample mean (35.28) next to **x̄2:,** and the second sample size (32) next to **n2:**. Scroll down to see the alternate hypothesis. This test is a two-tailed test, so highlight ≠**μ2** and press $\boxed{\text{ENTER}}$. Finally, scroll down one more line to highlight **Calculate** and press $\boxed{\text{ENTER}}$. Here are the screens that you should see.

```
2-SampZTest        2-SampZTest
μ1≠μ2              μ1≠μ2
z=1.091675822     ↑P=.274975692
P=.274975692       x̄1=37.27
x̄1=37.27           x̄2=35.28
x̄2=35.28           n1=30
↓n1=30             n2=32
```

Note that the results are slightly off when we compare them to the previous results. This is due to the rounding of the sample means and sample standard deviations. The first line of output gives us our test statistic, which rounds to 1.09. The next line gives us our p-value, which is 0.2750. The remaining information includes the mean and size of each sample.

Here is the hypothesis test for the first set of calculations.

Step 1

Population 1: previous teaching style; population 2: new teaching style
Claim in words: The mean score for both styles is the same.
Claim: $\mu_1 = \mu_2$
Complement: $\mu_1 \neq \mu_2$
H_0: $\mu_1 = \mu_2$
H_A: $\mu_1 \neq \mu_2$

Step 2

Level of significance: $\alpha = 0.05$

Step 3

Test statistic: $z = \dfrac{\bar{x}_1 - \bar{x}_2}{\sqrt{\dfrac{s_1^2}{n_1} + \dfrac{s_2^2}{n_2}}}$

Step 4

$\alpha = 0.05$, two-tailed test

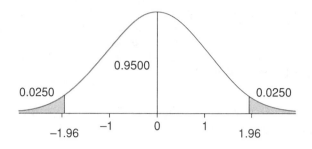

Decision rule: Reject H_0 if $z < -1.96$ or if $z > 1.96$.

Step 5

$z = 1.09$ (calculated by TI-83)

Decision: Fail to reject H_0.

Conclusion: There is not sufficient sample evidence to reject the claim that the mean score for both styles is the same.

_p_-value

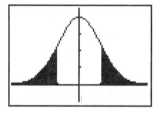

p-value = 0.2750 (calculated by TI-83) ∎

EXERCISES 8.1

Use the standard procedure for all hypothesis tests.

1. A college student randomly surveys 45 female students at her school and finds that their mean GPA is 3.22, with a standard deviation of 0.31. She also randomly surveys 45 male students and finds that their mean GPA is 3.12 with a standard deviation of 0.31.

   (a) At the 0.05 level of significance, test the claim that the mean GPA for female students at her school is greater than the mean GPA of male students.
   (b) Identify potential sources of bias for these samples.

2. A random sample of 88 male deaths in a county over a 1-year period had a mean age at death of 68.9 years, with a standard deviation of 16.71 years. A random sample of 84 female deaths in the same county over the same 1-year period had a mean age at death of 77.1 years, with a standard deviation of 15.65 years. At the 0.01 level of significance, test the claim that the mean age at death in this county is lower for males than it is for females.

3. At a college, 60 male students and 63 female students were randomly surveyed. The males were taking a mean of 12.37 course units, with a standard deviation of 3.740 units. The females were taking a mean of 12.47 units, with a standard deviation of 3.458 units. At the 0.05 level of significance, test the claim that the mean number of course units is the same for male and female students.

4. A random sample of 45 male college students and 50 female college students was taken, and each student was asked to write down their height. The male students had a mean height of 70.4 inches, with a standard deviation of 2.44 inches. The

female students had a mean height of 64.3 inches, with a standard deviation of 2.86 inches.

(a) At the 0.01 level, test the claim that male college students have a greater mean height than female college students.

(b) How can we eliminate the bias that may be present with students self-reporting their heights?

5. At a community college, students seem to perform better during the summer session than they do during the fall and spring semesters. Many instructors and administrators feel that this may be because summer classes have a different student makeup than the classes during the fall and spring semesters. A random sample of 100 students in the spring semester had a mean age of 23.1 years, with a standard deviation of 7.10 years. A random sample of 30 summer session students had a mean age of 26.2 years, with a standard deviation of 6.92 years. At the 0.01 level of significance, test the claim that the mean age of spring semester students is not the same as the mean age of summer session students.

6. A student conducted a study of 103 men and 104 women who had been married at least one time. The men had a mean age of 25.4 years at their first marriage, with a standard deviation of 5.19 years. The women had a mean age of 22.6 years at their first marriage, with a standard deviation of 4.63 years.

(a) At the 0.01 level of significance, test the claim that the mean age of men at their first marriage is higher than the mean age of women at their first marriage.

(b) If you were going to test the claim, explain how you would gather data to do so. Be sure to list potential biases that you are seeking to eliminate.

7. The PSAT Mathematics scores of 50 high school juniors and 35 high school sophomores were selected at random.

Juniors

24	36	72	48	28	50	44	46	38	72
57	69	48	64	39	48	59	51	59	48
42	37	52	72	47	55	38	58	40	60
60	59	38	42	49	40	31	61	42	47
56	37	62	59	59	44	69	41	60	33

Sophomores

29	52	42	68	38	60	58	49	45	55
30	60	52	37	42	49	43	36	38	52
58	61	31	61	46	58	49	36	59	49
48	47	37	43	31

At the 0.01 level of significance, test the claim that the mean score for juniors on this test is greater than the mean score of sophomores on this test.

8. Do students do as well on the PSAT verbal test as they do on the PSAT math test? PSAT verbal scores were obtained for a random sample of 40 high school juniors and PSAT math scores were obtained for a different sample of 40 high school juniors, as follows.

Verbal

28	41	49	41	40	41	48	47	48	37
56	41	36	49	40	47	44	72	42	50
64	47	56	54	58	71	38	63	40	50
42	53	68	43	38	39	30	70	46	60

Math

34	52	46	56	29	51	49	41	66	42
50	65	50	48	48	45	48	44	67	65
45	44	52	54	46	30	32	58	39	49
52	72	46	59	29	39	35	53	51	41

At the 0.01 level of significance, test the claim that the mean score for the two exams is the same.

9. It is a widely held belief that males have better mathematical skills than verbal skills. Here are the SAT math scores for 36 randomly selected male students and the SAT verbal scores for 30 randomly selected male students.

Math

370	670	460	530	690	640	740	540	610
470	610	660	410	450	720	740	430	480
460	520	580	580	560	510	320	640	650
580	540	580	440	310	260	420	320	410

Verbal

320	280	750	450	520	460	630	580	490
570	450	490	590	530	320	380	520	440
480	610	660	740	580	330	480	590	330
580	500	480						

At the 0.05 level of significance, test the claim that the mean SAT math score for males is greater than the mean SAT verbal score for males.

10. It is a widely held belief that females have better verbal skills than mathematical skills. Here are the SAT math scores for 42 randomly selected female students and the SAT verbal scores for 33 randomly selected female students.

Math

320	490	470	280	390	560	520	760	480
460	470	490	460	520	530	580	370	490
550	600	310	480	460	340	400	330	700
570	420	570	750	730	470	640	530	470
310	560	570	330	570	500			

Verbal

280	250	510	680	570	360	480	540	600
350	590	390	580	540	510	550	320	510
290	550	550	350	450	530	630	630	630
490	570	690	620	500	380			

At the 0.05 level of significance, test the claim that the mean SAT math score for females is less than the mean SAT verbal score for females.

11. A random sample of 56 departing flights at an airport over a 3-month period had a mean wait of 14.2 minutes between boarding and takeoff, with a standard deviation of 4.53 minutes. At the same airport, a random sample of 81 incoming flights over the same 3-month period had a mean wait of 17.5 minutes between the time that the plane arrived at the gate and the time that the baggage reached the baggage claim area, with a standard deviation of 9.87 minutes.

   (a) At the 0.05 level of significance, test the claim that at this airport the mean wait for takeoff is less than the mean wait for baggage.

   (b) Explain how you would gather data to test the claim that the mean wait for takeoff is less than the mean wait for baggage at U.S. airports.

12. A survey of 65 beer drinkers ages 21–29 had a mean of 22.7 servings of beer during the last month, with a standard deviation of 8.49 servings. A survey of 107 beer drinkers ages 30–39 had a mean of 19.8 servings of beer during the last month, with a standard deviation of 6.21 servings. At the 0.05 level of significance, test the claim that the mean monthly beer consumption of beer drinkers ages 21–29 is greater than the mean monthly beer consumption of beer drinkers ages 30–39. (Based on the results of a study by Maritz AmeriPoll.)

13. A professional bowler claims that his game is better suited to synthetic lanes than to natural lanes. Here are the scores of 42 randomly selected games on synthetic lanes and the scores of 42 randomly selected games on natural lanes.

    Synthetic

256	269	279	245	290	214	210	235	238
213	279	236	218	222	247	188	249	214
196	222	237	177	246	227	279	267	236
237	244	217	258	264	266	279	238	257
238	224	226	267	256	228			

    Natural

248	221	246	255	289	244	223	222	255
192	224	203	204	179	244	234	225	218
201	219	214	233	204	212	217	207	203
244	163	215	182	192	203	198	210	225
207	236	224	264	228	164			

    At the 0.05 level of significance, test the bowler's claim.

14. A math instructor is not sure that collecting homework from his class on a daily basis is beneficial to his students. He conducts an experiment by collecting homework from one algebra class of 37 students but not collecting homework from another algebra class of 32 students. Here are the test scores for both classes.

    Collected Homework

53	88	68	64	70	67	57	88	83	79
78	55	84	78	78	62	82	76	54	74
85	76	74	71	84	87	75	73	63	64
73	48	82	87	78	78	75			

    Did Not Collect Homework

60	79	65	74	76	62	71	72	76	77
62	73	66	80	74	74	81	68	61	68
76	76	78	70	79	88	68	82	63	68
76	87								

    At the 0.05 level of significance, test the claim that collecting homework produces the same mean score as not collecting homework.

15. Are American League baseball games higher scoring than National League baseball games? Here are the number of runs scored in 62 randomly selected American League games and 71 randomly selected National League games.

    American

5	4	12	16	11	10	11	6	12	4
20	11	7	6	8	13	23	16	3	9
4	13	9	17	8	16	4	12	8	9
4	14	7	14	4	9	15	7	14	7
8	9	11	4	8	13	9	11	10	6
6	7	9	17	19	25	17	12	9	6
11	1								

    National

11	11	6	1	15	19	9	11	12	13
6	27	6	12	11	6	14	13	6	6

(continues)

(continued)
National

10	13	9	16	3	14	9	22	11	9
6	4	6	9	13	13	11	3	8	12
3	13	9	14	5	5	6	13	7	12
3	7	9	22	10	3	10	6	6	6
7	6	11	19	13	10	5	4	10	14
4									

At the 0.01 level of significance, test the claim that the mean number of runs scored in American League games is greater than the mean number of runs scored in National League games.

16. Do American League baseball games have more hits than National League baseball games? Here are the number of hits in 62 randomly selected American League games and 71 randomly selected National League games.

American

13	16	23	23	23	16	14	14	23	9
32	27	17	12	14	27	31	24	9	17
11	19	18	26	13	28	14	18	18	17
14	21	15	27	16	13	24	19	22	16
20	20	17	14	18	22	20	18	20	18
14	18	21	31	28	33	27	20	16	19
17	7								

National

19	20	14	12	25	27	16	17	23	19
20	32	15	23	21	17	23	17	13	20
17	21	20	18	15	20	23	30	19	20
13	9	13	16	19	25	21	14	11	18
8	19	19	24	10	19	12	16	16	22
14	20	16	26	21	9	16	14	16	13
18	12	19	27	24	16	14	18	20	19
12									

At the 0.01 level of significance, test the claim that the mean number of hits in American League games is greater than the mean number of hits in National League games.

17. Do American League baseball games take longer than National League baseball games? Here are the lengths, in minutes, of 62 randomly selected American League games and 71 randomly selected National League games.

American

175	168	185	183	196	151	198	148	189
146	225	199	143	149	156	192	230	178
124	167	156	185	179	185	144	158	157
196	190	166	176	186	153	189	180	205
163	177	154	172	178	186	187	163	181
208	142	200	150	172	166	168	182	219
235	224	202	171	159	153	169	171	

National

170	187	169	129	159	209	181	183	197
170	157	214	154	166	191	178	198	178
161	198	179	150	265	136	143	152	190
213	158	176	146	172	137	200	158	181
202	152	128	161	153	164	169	194	108
146	157	167	169	190	160	177	171	198
159	160	146	188	149	151	172	143	189
204	215	181	141	152	176	180	158	

At the 0.05 level of significance, test the claim that the mean time to complete an American League game is longer than the mean time to complete a National League game.

18. Here are the prices for a gallon of unleaded regular gas at 48 randomly selected gas stations in the Los Angeles area and 39 randomly selected gas stations in the San Francisco Bay area from the week before Christmas in 1999.

Los Angeles

$ 0.949	$ 1.019	$ 0.989	$ 1.039	$ 0.969	$ 0.999
$ 0.979	$ 1.019	$ 1.039	$ 1.009	$ 0.939	$ 0.989
$ 1.059	$ 0.959	$ 1.049	$ 1.029	$ 1.039	$ 0.999
$ 1.099	$ 1.019	$ 1.019	$ 1.029	$ 1.069	$ 0.979
$ 1.039	$ 1.089	$ 1.049	$ 0.989	$ 1.039	$ 1.019
$ 1.019	$ 0.999	$ 1.009	$ 0.949	$ 1.019	$ 1.069
$ 1.069	$ 0.979	$ 0.999	$ 1.029	$ 1.079	$ 0.979
$ 0.999	$ 1.039	$ 0.969	$ 1.059	$ 1.049	$ 0.979

San Francisco

$ 1.049	$ 1.169	$ 1.029	$ 1.199	$ 1.059	$ 1.099
$ 1.139	$ 1.109	$ 1.129	$ 1.139	$ 1.139	$ 1.189
$ 1.089	$ 1.079	$ 1.149	$ 1.149	$ 1.059	$ 1.179
$ 1.169	$ 1.069	$ 1.139	$ 1.149	$ 1.109	$ 1.099
$ 1.099	$ 1.199	$ 1.129	$ 1.039	$ 1.109	$ 1.089
$ 1.159	$ 1.069	$ 1.089	$ 1.129	$ 1.239	$ 1.149
$ 1.169	$ 1.119	$ 1.169			

At the 0.01 level, test the claim that the mean price for a gallon of unleaded regular gas in the Los Angeles area is lower than the mean price for a gallon of unleaded regular gas in the San Francisco area.

19. A random sample of 140 women who first married in 1970 had a mean age of 20.8 years, with a standard deviation of 3.1 years. A similar sample of 108 women who first married in 1998 had a mean age of 25.0 years with a standard deviation of 4.1 years. At the 0.05 level of significance, test the claim that the mean age at which women get married has increased from 1970 to 1998. (Based on data from the U.S. Census Bureau.)

20. Do women have more doctor visits per year than men? A random sample of 111 men produced a mean of 3.8 visits per year, with a standard deviation of 2.1 visits. A random sample of 153 women produced a mean of 5.8 visits per year, with a standard deviation of 1.8 visits. At the 0.05 level of significance, test the claim that the mean number of doctor visits per year is higher for women than it is for men. (Based on the results of the March 2000 National Vital Statistics Report.)

21. Do households that have cable TV have higher incomes? A sample of 420 cable households had a mean annual income of $51,468 with a standard deviation of

$43,329. A sample of 317 households that do not have cable TV had a mean annual income of $36,735 with a standard deviation of $29,071. At the 0.05 level of significance, test the claim that the mean annual income of households with cable TV is higher than the mean annual income of households without cable TV. (Based on a study by Cablevision Advertising Bureau.)

22. A random sample of 163 men between the ages of 20 and 29 had a mean weight of 172.1 pounds with a standard deviation of 33.88 pounds. A random sample of 85 men between the ages of 50 and 59 had a mean weight of 189.2 pounds with a standard deviation of 32.99 pounds. At the 0.05 level of significance, test the claim that the mean weight of men who are between 20 and 29 years old is less than the weight of men who are between 50 and 59 years old. (Based on the results of a study by the Centers for Disease Control and Prevention.)

23. A child development student is interested in the amount of time that children watch television. A study of 45 children ages 2–5 at the college's day care center revealed that they watched 135.4 minutes of television per day, with a standard deviation of 82.1 minutes. A study of 120 elementary school children ages 6–12 revealed that they watched a mean of 170.6 minutes of television per day, with a standard deviation of 106.9 minutes.

    (a) Construct a 95% confidence interval for the difference between the mean amount of television watched per day by 2- to 5-year-olds and the mean amount of television watched per day by 6- to 12-year-olds.
    (b) At the 0.05 level of significance, test the claim that the mean amount of television watched per day by 2- to 5-year-olds is the same as the mean amount of television watched per day by 6- to 12-year-olds.
    (c) Does your hypothesis test from part (b) reinforce what your confidence interval from part (a) told you? Explain.

24. A plant produces various types of corn chips. Among the varieties of chips that this plant produces are blue corn tortilla chips and jalapeno tortilla chips, both of which are sold in 9-ounce bags. A random sample of 50 bags of blue corn tortilla chips had a mean weight of 9.06 ounces, with a standard deviation of 0.05 ounces. A random sample of 50 jalapeno tortilla chips had a mean weight of 9.01 ounces, with a standard deviation of 0.03 ounces.

    (a) Construct a 95% confidence interval for the difference between the mean fill of blue corn tortilla chip bags and the mean fill of jalapeno tortilla chip bags.
    (b) At the 0.05 level of significance, test the claim that the mean fill of blue corn tortilla chip bags is the same as the mean fill of jalapeno tortilla chip bags.
    (c) Does your hypothesis test from part (b) reinforce what your confidence interval from part (a) told you? Explain.

?   What Is Wrong with This Picture?

The following statistical project contains a major error. Potential errors could be the choice of the wrong hypothesis test or that the assumptions required for this test are not met. Find the error, and explain what could be done to correct it.

A student believes that there are more females at private universities than at public universities. She randomly selects 30 private universities and visits their Web sites, where she finds the percentage of females at each school. She repeats this for 30 public universities. Taking these numbers, she tests the claim that the mean number of females at private universities is higher than the mean number of females at public universities.

What is wrong with this picture?

Randomly sample at least 30 males and 30 females at your school, and ask them how many serious relationships they have had. Use these sample data to test the claim that college males and college females have had the same mean number of serious relationships at the 0.05 level of significance. In addition to your complete hypothesis test, include

- Your raw data
- A histogram for the number of serious relationships that the males have had and a second histogram for the number of serious relationships that the females have had

Finally, answer the following questions.

- How did you obtain your sample?
- Which type of sampling did you use?
- Was the sample truly random?
- What potential biases may show up in your sample?

MINI PROJECT

SECTION 8.2
Hypothesis Test for Two Population Means (Small Samples)

In this section we introduce a hypothesis test for comparing the means of two different populations, for which at least one of the samples has a size of less than 30. The two samples must be independent, and both must be drawn from populations that are normally distributed. If either of the samples are drawn from a population that is not normally distributed, we cannot use the methods presented in this section. There is a hypothesis test covered in Section 11.3 that may be used in this case.

Traditionally, there have been two different test statistics that could be used. The one that we will use does not assume the two samples come from populations that have equal variances. (There is a second test that does assume the two samples come from populations with the same variance, but we will leave that test as optional material in the "Extra" section that follows Section 9.1.)

The test statistic that we will be using is

$$t = \frac{\bar{x}_1 - \bar{x}_2}{\sqrt{\dfrac{s_1^2}{n_1} + \dfrac{s_2^2}{n_2}}}$$

where the degrees of freedom are equal to the smaller of $n_1 - 1$ and $n_2 - 1$. There is actually a formula that we could use to calculate the degrees of freedom, but our choice of the smaller of these two values is conservative, making it harder to reject a null hypothesis.

We begin with a few examples of the hypothesis test using this test statistic, and then we look at how to construct a confidence interval for the difference between two population means.

A female college student majoring in math has heard many times that males have higher SAT math scores than females. She believes that this is not true, but instead that the two groups have the same mean SAT math scores. She randomly samples 20 females at her school. Their mean score is 509.5, with a standard deviation of 110.62. She also randomly samples 25 males at her school. Their mean score is 536.4, with a standard deviation of 117.79. Use these data to test the claim that females and males have the same mean SAT math score at the 0.05 level of significance.

Since both samples have sizes of less than 30, the test statistic introduced in this section is the appropriate one. We begin by labeling the populations. We let females taking the SAT be population 1, and we let males taking the SAT be population 2.

Step 1

Population 1: females; population 2: males
Claim in words: Females and males have the same mean SAT math score.
Claim: $\mu_1 = \mu_2$
Complement: $\mu_1 \neq \mu_2$
H_0: $\mu_1 = \mu_2$
H_A: $\mu_1 \neq \mu_2$

Step 2

Level of significance: $\alpha = 0.05$

Step 3

Test statistic: $t = \dfrac{\bar{x}_1 - \bar{x}_2}{\sqrt{\dfrac{s_1^2}{n_1} + \dfrac{s_2^2}{n_2}}}$ d.f. = smaller of $(n_1 - 1)$ and $(n_2 - 1)$

Step 4

$\alpha = 0.05$, two-tailed test, with $20 - 1 = 19$ degrees of freedom
Decision rule: Reject H_0 if $t < -2.093$ or if $t > 2.093$.

Step 5

$$t = \frac{\bar{x}_1 - \bar{x}_2}{\sqrt{\dfrac{s_1^2}{n_1} + \dfrac{s_2^2}{n_2}}}$$

$$= \frac{509.5 - 536.4}{\sqrt{\dfrac{110.62^2}{20} + \dfrac{117.79^2}{25}}}$$

$$= -0.788$$

Decision: Fail to reject H_0.

Conclusion: There is not sufficient sample evidence to reject the claim that the mean SAT math scores of females and males are equal.

p-value

Using techniques developed in Section 7.2, we place the p-value between 0.20 and 0.50. A calculator or computer would tell us that the actual p-value of this test is 0.4354. ■

For the 1997–98 school year, the mean SAT math score for all males was 35 points higher than the mean SAT math score for all females—so why weren't we able to reject the null hypothesis that the two mean scores were equal? One potential problem could be the small sample sizes. There was a fairly wide gap between the two sample means (26.9 points), but the small sample sizes kept the value of the test statistic small. Similar results based on samples of 200 and 250, rather than 20 and 25, yield a test statistic of approximately –2.5.

Of course, bias could have affected the value of the test statistic as well. One possible source of bias may come from the students self-reporting their SAT scores. Do we know that the score that they reported was their actual score? Are there any other potential biases?

EXAMPLE 8.9

In tournaments of the Professional Bowler's Association, there are two rounds. The first round is a qualifying round where the bowlers' total scores determine who advances to the second round. The second round is a match play round, meaning that two bowlers compete against each other. In this round, the scores are not important, only the number of wins. It is believed that there is more pressure on the individual bowlers in the match play round, and that causes scores to drop during this round. Scores were randomly selected from 18 qualifying games and 24 match play games from the same tournament. The mean score for the qualifying games was 230.8, with a standard deviation of 33.83. The mean score for the match play games was 224.8, with a standard deviation of 27.76. Use these data to test the claim that the mean score for qualifying games is higher than the mean score for match play games, at the 0.05 level of significance.

Since both samples have sizes of less than 30, the test statistic introduced in this section is the appropriate one. We begin by labeling the populations. We let qualifying games be population 1, and we let match play games be population 2.

Step 1

Population 1: Qualifying; population 2: match play
Claim in words: The mean score for qualifying games is higher than the mean score for match play games.
Claim: $\mu_1 > \mu_2$
Complement: $\mu_1 \le \mu_2$
H_0: $\mu_1 \le \mu_2$
H_A: $\mu_1 > \mu_2$

Step 2

Level of significance: $\alpha = 0.05$

Step 3

Test statistic: $t = \dfrac{\bar{x}_1 - \bar{x}_2}{\sqrt{\dfrac{s_1^2}{n_1} + \dfrac{s_2^2}{n_2}}}$ d.f. = smaller of $(n_1 - 1)$ and $(n_2 - 1)$

Step 4

$\alpha = 0.05$, right-tailed test, with $18 - 1 = 17$ degrees of freedom

Decision rule: Reject H_0 if $t > 1.740$.

Step 5

$$t = \frac{\bar{x}_1 - \bar{x}_2}{\sqrt{\dfrac{s_1^2}{n_1} + \dfrac{s_2^2}{n_2}}}$$

$$= \frac{230.8 - 224.8}{\sqrt{\dfrac{33.82^2}{18} + \dfrac{27.76^2}{24}}}$$

$$= 0.613$$

Decision: Fail to reject H_0.

Conclusion: There is not sufficient sample evidence to support the claim that the mean score for qualifying games is higher than the mean score for match play games.

p-value

Using techniques developed in Section 7.2, the best that we can say is that the p-value is higher than 0.25, since the value of our test statistic is lower than the last entry on our table for 17 degrees of freedom. A calculator or computer would tell us that the actual p-value of this test is 0.2720. ∎

Many people believe that male racehorses run faster than female racehorses. To test this theory, the results of 55 randomly selected races at Santa Anita Park were selected, all at the distance of 1 mile. Thirty-four of the races were for male racehorses, and 21 of the races were restricted to fillies and mares (female racehorses). The time required by the winners of the male races to complete the 1-mile race had a mean of 96.9 seconds, with a standard deviation of 1.22 seconds. The time required by the winners of the races restricted to fillies and mares to complete the 1-mile race had a mean of 97.8 seconds, with a standard deviation of 1.55 seconds. At the 0.05 level of significance, test the claim that the mean time for a male racehorse to win a 1-mile race is less than the mean time for a female racehorse to win a 1-mile race.

Even though one of the samples (male races) has a size that is at least 30, the small sample test is appropriate for this two-mean test because the other sample has a size that is less than 30. We begin by labeling the populations. We let male races be population 1, and we let female races be population 2.

Step 1

Population 1: male races; population 2: female races
Claim in words: The mean time for a male racehorse to win a 1-mile race is less than the mean time for a female racehorse to win a 1-mile race.
Claim: $\mu_1 < \mu_2$
Complement: $\mu_1 \geq \mu_2$
H_0: $\mu_1 \geq \mu_2$
H_A: $\mu_1 < \mu_2$

Step 2

Level of significance: $\alpha = 0.05$

Step 3

Test statistic: $t = \dfrac{\bar{x}_1 - \bar{x}_2}{\sqrt{\dfrac{s_1^2}{n_1} + \dfrac{s_2^2}{n_2}}}$ d.f. = smaller of $(n_1 - 1)$ and $(n_2 - 1)$

Step 4

$\alpha = 0.05$, left-tailed test, with $21 - 1 = 20$ degrees of freedom

Decision rule: Reject H_0 if $t < -1.725$.

Step 5

$$t = \frac{\bar{x}_1 - \bar{x}_2}{\sqrt{\dfrac{s_1^2}{n_1} + \dfrac{s_2^2}{n_2}}}$$

$$= \frac{96.9 - 97.8}{\sqrt{\dfrac{1.22^2}{34} + \dfrac{1.55^2}{21}}}$$

$$= -2.263$$

Decision: Reject H_0.

Conclusion: There is sufficient sample evidence to support the claim that the mean time for a male racehorse to win a 1-mile race is less than the mean time for a female racehorse to win a 1-mile race.

p-value

Using techniques developed in Section 7.2, the best that we can say is that the p-value is between 0.025 and 0.01. A calculator or computer would tell us that the actual p-value of this test is 0.0150. ∎

Confidence Intervals

Recall that the idea of hypothesis testing is based on confidence intervals. There is a confidence interval that can be constructed for the difference between two population means, when at least one of the samples is small. We begin, as we did before, by calculating the margin of error. Here is the formula.

$$E = t_{\alpha/2} \sqrt{\frac{s_1^2}{n_1} + \frac{s_2^2}{n_2}}$$ degrees of freedom: the smaller of $n_1 - 1$ and $n_2 - 1$

We then subtract this margin of error from $\bar{x}_1 - \bar{x}_2$ to find the left endpoint of the confidence interval, and we add the margin of error to $\bar{x}_1 - \bar{x}_2$ to find the right endpoint of the confidence interval. (If you want to create a one-sided confidence interval, see the "Extra" material at the end of the Section 6.2 exercises.)

$$(\bar{x}_1 - \bar{x}_2) - E \leq (\mu_1 - \mu_2) \leq (\bar{x}_1 - \bar{x}_2) + E$$

EXAMPLE 8.11

A mathematics instructor holds a supplementary review session the day before each exam. Fourteen of the 24 students in his algebra class attended his review session before the exam. The mean score on this exam for these 14 students was 80.9 points out of 100, with a standard deviation of 9.62 points. The mean score for the 10 students who did not attend was 45.4 points out of 100, with a standard deviation of 16.36 points. Construct a 95% confidence interval for the difference between the mean scores of algebra students who attend a review session and the mean scores of algebra students who do not attend a review session. (Assume that the students in this class represent a random sample of all algebra students.)

We let population 1 be the algebra students who attend a review session, and population 2 be the algebra students who do not attend a review session. We begin by finding $t_{\alpha/2}$. We have 9 degrees of freedom, since the smaller sample had size 10. For 9 degrees of freedom, $t_{0.025} = 2.262$. Now we calculate the margin of error.

$$E = t_{\alpha/2} \sqrt{\frac{s_1^2}{n_1} + \frac{s_2^2}{n_2}}$$

$$= 2.262 \sqrt{\frac{9.62^2}{14} + \frac{16.36^2}{10}}$$

$$= 13.07$$

Next we construct the confidence interval.

$$(\bar{x}_1 - \bar{x}_2) - E \leq (\mu_1 - \mu_2) \leq (\bar{x}_1 - \bar{x}_2) + E$$
$$(80.9 - 45.4) - 13.07 \leq (\mu_1 - \mu_2) \leq (80.9 - 45.4) + 13.07$$
$$35.5 - 13.07 \leq (\mu_1 - \mu_2) \leq 35.5 + 13.07$$
$$22.43 \leq (\mu_1 - \mu_2) \leq 48.57$$

We are 95% confident that the mean score of algebra students who attend a review session will be between 22.43 and 48.57 points higher than the mean score of algebra students who do not attend a study session. ∎

MICROSOFT EXCEL   Hypothesis Test for Two Population Means (Small Samples)

Microsoft Excel has a built-in procedure for the hypothesis test for two population means using the *t*-distribution, assuming unequal variances. It also has a procedure for when we assume the populations have equal variance, so be careful to select the right one. We will step through an example showing how to do this using Excel, and then we will write the hypothesis test based on our results.

EXAMPLE
8.12

A business student wondered whether the stock market would be busier on Mondays or Fridays. She took a random sample of 23 Mondays and 16 Fridays and found the volume (number of shares sold) for the New York Stock Exchange. Here are the results, rounded to the nearest million.

Monday

731	542	559	592	548	629	597	692
542	543	530	564	560	620	564	714
610	690	610	592	774	689	531	

Friday

689	669	613	569	579	635	622	557
558	759	725	736	637	785	683	602

At the 0.05 level of significance, test the claim that the mean volume on Mondays is the same as the mean volume on Fridays.

In a new Excel worksheet, type Monday in cell **A1** and Friday in **B1.** Enter the volumes for Mondays in column **A,** from cell **A2** through **A24.** Enter the volumes for Fridays in column **B,** from cells **B2** through **B17.**

Select **Data Analysis** from the **Tools** menu. When the dialog box appears, select **t-Test: Two-Sample Assuming Unequal Variances** and click on **OK.** In the box labeled **Variable 1 Range,** enter A1:A24. In the box labeled **Variable 2 Range,** enter B1:B17. These two cell ranges include the labels *Monday* and *Friday,* so check the box labeled **Labels.** In the box labeled **Hypothesized Mean Difference,** enter 0. Put the level of significance, 0.05, in the box labeled **Alpha.** Click on **OK** to see the results, which will be on a new worksheet. Here is an example of what you should see.

t-Test: Two-Sample Assuming Unequal Variances

	Monday	*Friday*
Mean	609.6956522	651.125
Variance	5115.857708	5309.583333
Observations	23	16
Hypothesized Mean Difference	0	
df	32	
t Stat	-1.759723569	
P(T<=t) one-tail	0.044006409	
t Critical one-tail	1.693888407	
P(T<=t) two-tail	0.088012817	
t Critical two-tail	2.036931619	

We see the mean, variance, and size of each sample. The box labeled **df** contains the degrees of freedom. Excel uses the actual formula for degrees of freedom, rather than our conservative estimate. Here it is.

$$d.f. = \frac{\left(\dfrac{s_1^2}{n_1} + \dfrac{s_2^2}{n_2}\right)^2}{\dfrac{1}{n_1 - 1}\left(\dfrac{s_1^2}{n_1}\right)^2 + \dfrac{1}{n_2 - 1}\left(\dfrac{s_2^2}{n_2}\right)^2}$$

(Recall that we have been using 1 less than the smaller sample size as the number of degrees of freedom, which in this case is 15.) Here is the calculation that produces 32 degrees of freedom.

$$d.f. = \frac{\left(\dfrac{s_1^2}{n_1} + \dfrac{s_2^2}{n_2}\right)^2}{\dfrac{1}{n_1 - 1}\left(\dfrac{s_1^2}{n_1}\right)^2 + \dfrac{1}{n_2 - 1}\left(\dfrac{s_2^2}{n_2}\right)^2}$$

$$= \frac{\left(\dfrac{5115.9}{23} + \dfrac{5309.6}{16}\right)^2}{\dfrac{1}{22}\left(\dfrac{5115.9}{23}\right)^2 + \dfrac{1}{15}\left(\dfrac{5309.6}{16}\right)^2}$$

$$= 32.03$$

The test statistic is listed next, which is –1.760. For a two-tailed test, Excel lists the *p*-value as 0.0880, as well as giving us the critical value for the test (2.037).

We now have enough information to write the test.

Step 1

Population 1: Mondays; population 2: Fridays
Claim in words: The mean volume on Mondays is the same as the mean volume
  on Fridays.
Claim: $\mu_1 = \mu_2$
Complement: $\mu_1 \neq \mu_2$
H_0: $\mu_1 = \mu_2$
H_A: $\mu_1 \neq \mu_2$

Step 2

Level of significance: $\alpha = 0.05$

Step 3

Test statistic: $t = \dfrac{\bar{x}_1 - \bar{x}_2}{\sqrt{\dfrac{s_1^2}{n_1} + \dfrac{s_2^2}{n_2}}}$ $d.f. = \dfrac{\left(\dfrac{s_1^2}{n_1} + \dfrac{s_2^2}{n_2}\right)^2}{\dfrac{1}{n_1 - 1}\left(\dfrac{s_1^2}{n_1}\right)^2 + \dfrac{1}{n_2 - 1}\left(\dfrac{s_2^2}{n_2}\right)^2}$

Step 4

$\alpha = 0.05$, two-tailed test, with 32 degrees of freedom

Decision rule: Reject H_0 if $t < -2.037$ or if $t > 2.037$.
  (Note that if we had used our table to find the critical value, we would have
  obtained a critical value 2.042.)

Step 5

$t = -1.760$

Decision: Fail to reject H_0.

Conclusion: There is not sufficient sample evidence to reject the claim that the
  mean volume on Mondays is the same as the mean volume on Fridays.

***p*-value**

p-value = 0.0880 ∎

TI-83   Hypothesis Test for Two Population Means (Small Samples)

The TI-83 has a built-in test for the hypothesis test for two population means using the *t*-distribution, assuming unequal variances. The TI-83 will perform the test whether we have the data or only the sample statistics. We will step through an example showing how to do this using the TI-83 for both methods, and then we will write the hypothesis test based on our results.

Before we begin, note that the TI-83 uses the actual formula for degrees of freedom, rather than our conservative estimate. Here it is.

$$d.f. = \frac{\left(\dfrac{s_1^2}{n_1} + \dfrac{s_2^2}{n_2} \right)^2}{\dfrac{1}{n_1 - 1}\left(\dfrac{s_1^2}{n_1} \right)^2 + \dfrac{1}{n_2 - 1}\left(\dfrac{s_2^2}{n_2} \right)^2}$$

(Recall that we have been using 1 less than the smaller sample size as the number of degrees of freedom.)

EXAMPLE 8.13

A business student wondered whether the stock market would be busier on Mondays or Fridays. She took a random sample of 23 Mondays and 16 Fridays and found the volume (number of shares sold) for the New York Stock Exchange. Here are the results, rounded to the nearest million.

Monday

731	542	559	592	548	629	597	692
542	543	530	564	560	620	564	714
610	690	610	592	774	689	531	

Friday

689	669	613	569	579	635	622	557
558	759	725	736	637	785	683	602

At the 0.05 level of significance, test the claim that the mean volume on Mondays is the same as the mean volume on Fridays.

Enter the Monday volumes in list L_1. Enter the Friday volumes in list L_2. Access the STAT TESTS menu by pressing (STAT) and using the (→) key to move to the right. Highlight option **4:2-SampTTest** and press (ENTER).

Since we have the sample data, highlight **Data** next to **Inpt:**. Next to **List1:** enter L_1 by pressing (2nd) (1). Next to **List2:** enter L_2 by pressing (2nd) (2). Leave **Freq1:** and **Freq2:** as 1. Next we establish the alternate hypothesis. This test is a two-tailed test, so highlight ≠**μ2** and press (ENTER). Next to **Pooled:** highlight **No** and press (ENTER). (We would press **Yes** only if we are assuming that the two populations have equal variance. This situation will be discussed after Section 9.2.) Finally, highlight **Calculate** and press $\boxed{\text{ENTER}}$. You should see these screens.

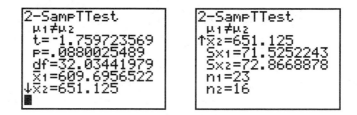

The first line of output gives us our test statistic, which rounds to –1.760. The next line gives us our *p*-value (0.0880), which is based on 32 degrees of freedom, rather than our conservative use of 15 (1 less than the smaller sample size). The remaining information includes the mean, standard deviation, and size of each sample.

If we repeated the exact same procedure but chose **Draw** instead of **Calculate,** we would see the following screen.

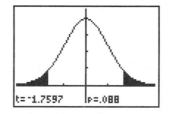

This screen also includes the value of the test statistic and *p*-value in addition to the graph. ■

We can also use the TI-83 to help when we are given only the sample statistics instead of all of the data.

EXAMPLE 8.14

A business student wondered whether the stock market would be busier on Mondays or Fridays. She took a random sample of 23 Mondays and 16 Fridays and found the volume (number of shares sold) for the New York Stock Exchange. The mean volume for the Monday volumes was 609.7 million shares, with a standard deviation of 71.53 million shares. The mean volume for the Friday volumes was 651.1 million shares, with a standard deviation of 72.87 million shares. At the 0.05 level of significance, test the claim that the mean volume on Mondays is the same as the mean volume on Fridays.

Access the STAT TESTS menu by pressing $\boxed{\text{STAT}}$ and using the $\boxed{\rightarrow}$ key to move to the right. Highlight option **4:2-SampTTest** and press $\boxed{\text{ENTER}}$.

Since we only have the sample statistics, highlight **Stats** next to **Inpt:**. Enter the first sample mean (609.7) next to **x̄1**. Enter the first sample standard deviation (71.53) next to **Sx1.** Enter the first sample size (23) next to **n1:.** Enter the second sample mean (651.1) next to **x̄2:**. Enter the second sample standard deviation (72.87) next to **Sx2**. Enter the second sample size (16) next to **n2:**. Scroll down to see the alternate hypothesis. This test is a two-tailed test, so highlight ≠**μ2** and press (ENTER). Scroll down one line, and next to **Pooled:** highlight **No** and press (ENTER). Finally, scroll down one more line to highlight **Calculate** and press (ENTER). Here are the screens that you should see.

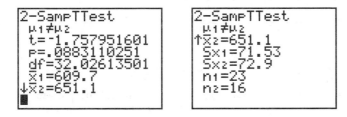

Note that the calculations are slightly off from our earlier results due to rounding. We now have enough information to write the hypothesis test.

Step 1

Population 1: Mondays; population 2: Fridays

Claim in words: The mean volume on Mondays is the same as the mean volume on Fridays.

Claim: $\mu_1 = \mu_2$

Complement: $\mu_1 \neq \mu_2$

H_0: $\mu_1 = \mu_2$

H_A: $\mu_1 \neq \mu_2$

Step 2

Level of significance: $\alpha = 0.05$

Step 3

Test statistic: $t = \dfrac{\bar{x}_1 - \bar{x}_2}{\sqrt{\dfrac{s_1^2}{n_1} + \dfrac{s_2^2}{n_2}}}$

$d.f. = \dfrac{\left(\dfrac{s_1^2}{n_1} + \dfrac{s_2^2}{n_2}\right)^2}{\dfrac{1}{n_1 - 1}\left(\dfrac{s_1^2}{n_1}\right)^2 + \dfrac{1}{n_2 - 1}\left(\dfrac{s_2^2}{n_2}\right)^2}$

Step 4

$\alpha = 0.05$, two-tailed test, with 32 degrees of freedom

Decision rule: Reject H_0 if $t < -2.042$ or if $t > 2.042$.

Step 5

$t = -1.758$

Decision: Fail to reject H_0.

Conclusion: There is not sufficient sample evidence to reject the claim that the mean volume on Mondays is the same as the mean volume on Fridays.

p-value

p-value = 0.0883 ∎

EXERCISES 8.2

Use the standard procedure for all hypothesis tests.

1. An instructor gives one version of a test to his 7:30 A.M. statistics class and another version to his 6 P.M. class. Here are their scores.

   Morning

85	73	83	100	86	70	98	80	97	66	41
83	89	57	64	76	91	77	68	81	63	72

   Evening

74	80	92	73	82	96	82	84	76	98	94	63
94	91	46	58	54	37	86	79	55	73	64	

   At the 0.05 level of significance, test the claim that the two versions are of equal difficulty; in other words, the two exams produce the same mean score.

2. A high school instructor is curious to see the effect that an open-notes policy would have on tests. He allows one of his classes to use their notes on their test, but has his other class take the test without them. Here are the scores.

   With Notes

86	95	97	98	53	84	91	64
97	97	97	84	64	94	73	

   Without Notes

70	92	97	50	81	97	84	61
98	98	58	23	69	84	91	78

   At the 0.05 level, test the claim that the use of notes produces a higher mean test score.

3. At a community college, evening classes tend to have more reentry students than day classes, because most reentry students work during the day. A random sample of 25 day students had a mean age of 21.5 years, with a standard deviation of 4.44 years. A random sample of 25 evening students had a mean age of 29.8 years, with a standard deviation of 9.88 years. At the 0.01 level of significance, test the claim that the mean age of day students is less than the mean age of evening students.

4. It is believed that most females have stopped growing taller by the time they graduate from high school. A random sample of 20 female college freshmen had a mean height of 66.2 inches, with a standard deviation of 3.03 inches. A random sample of 24 female high school freshmen had a mean height of 65.1 inches, with a standard deviation of 2.61 inches. At the 0.05 level of significance, test the claim that the mean height of female college freshmen is the same as the mean height of female high school freshmen.

5. A student, eager to apply the theories of small-sample hypothesis testing, chose a claim that he felt would certainly be supported using larger samples. He claimed that the mean height of male college freshmen is greater than the mean height of female college freshmen. He randomly selected 13 male college freshmen, who had a mean height of 67.7 inches, with a standard deviation of 4.03 inches. He randomly selected 11 female college freshmen, who had a mean height of 64.6 inches, with a standard deviation of 3.41 inches. At the 0.05 level of significance, test the student's claim.

6. In Exercises 4 and 5 we have two different samples of female college freshman heights. Here is the information.

	Exercise 4	*Exercise 5*
Mean	66.2 in.	64.6 in.
Standard Deviation	3.03 in.	3.41 in.
Size	20	11

Test the claim that the mean height of these two populations is equal, at the 0.05 level of significance. (We know that these two population means are equal because the samples are drawn from the same population. In fact, both samples were collected the same semester at the same school. If we reject the claim that the two means are equal, this suggests that one or both samples *may* be biased.)

7. A college student is interested in the GPAs of male and female students at her school. A random sample of 21 female students had a mean GPA of 3.01, with a standard deviation of 0.498. A random sample of 19 males had a mean GPA of 2.76, with a standard deviation of 0.536. At the 0.05 level of significance, test the claim that the mean GPA for female students at her school is greater than the mean GPA of male students.

8. The previous exercise contained a small sample of male GPAs, and the first exercise from Section 8.1 contained a large sample of male GPAs from the same school. Here is a summary of the information.

	Section 8.1 Exercise 1	Section 8.2 Exercise 7
Mean	3.12	2.76
Standard Deviation	0.31	0.536
Size	45	19

Test the claim that the mean GPA of these two populations are equal at the 0.05 level of significance.

9. Here are the serum cholesterol levels, in mg/dL, of 12 men and 26 women ages 60–69.

Men

176	196	189	184	230	169
159	180	284	221	196	173

Women

168	239	213	197	170	226	200
205	237	219	246	247	268	273
213	151	188	196	273	202	236
214	295	221	255	260		

At the 0.01 level of significance, test the claim that men and women ages 60–69 have the same mean serum cholesterol level.

10. A professional bowler has two bowling balls, and the one he uses depends on the lane conditions. In warm-ups at a tournament, he bowls 6 games with the first ball and 8 games with the second ball. Here are the scores.

First

257	248	192	246	245	237

Second

217	247	249	224	142	255	215	199

He then decides to use the first ball in the tournament. At the 0.05 level of significance, test the claim that the first ball will produce scores with a higher mean than the second ball.

11. A community college's administrators are investigating telephone registration systems. They have narrowed their choices down to two systems, both of which are supposed to make registering quicker and easier for students. They decide to visit two campuses that are using the two systems. At the first campus, a random sample of 20 registration phone calls took a mean of 7.1 minutes to complete, with a standard deviation of 1.89 minutes. Feeling that 20 calls may be too few, they decided to take a random sample of 40 registration phone calls at the second

campus. These 40 calls took a mean of 8.4 minutes to complete, with a standard deviation of 1.24 minutes. The college decided to buy the first system based on these results. At the 0.05 level of significance, test the claim that there is no difference in the mean length of registration phone calls for these two systems.

12. A student who works at a fast-food restaurant claims that customers who come to the drive-through window spend more money than customers who come in to the counter. A random sample of 38 customers who came in to the counter to order their food had a mean total of $5.19, with a standard deviation of $3.06. A random sample of 17 customers who used the drive-through window had a mean total of $5.94, with a standard deviation of $3.25. At the 0.05 level of significance, test the student's claim.

13. The Insurance Institute for Highway Safety performed a 5 mph crash test on 4 different large luxury cars and 4 different large family cars. The front of each car was driven into an angle barrier and then the repair costs were calculated. Here are the results.

Large Luxury

$559 $1072 $578 $1828

Large Family

$1101 $443 $674 $520

At the 0.05 level of significance, test the claim that the mean damage amount for this type of accident is greater for large luxury cars than it is for large family cars.

14. A golfer is considering buying a new driver. He hits 12 balls with his old driver, and then 15 balls with the new driver. Here are the distances, in yards.

New

| 238 | 261 | 263 | 236 | 269 | 268 |
| 270 | 250 | 262 | 242 | 251 | 251 |

Old

| 254 | 266 | 275 | 279 | 261 | 268 | 280 | 269 |
| 285 | 254 | 276 | 272 | 283 | 262 | 267 |

At the 0.05 level of significance, test the claim that the mean driving distance with his old driver is less than the mean driving distance with the new driver.

15. A study done at the South Alabama College of Medicine compared the laparoscopic appendectomy procedure (using a video scope) with the traditional open appendectomy procedure. Thirty-seven patients with acute appendicitis were randomly assigned to one of these two procedures. Nineteen patients had the laparoscopic appendectomy (LA), and 18 had the open appendectomy (OA).

   (a) The mean operating time for LA was 93 minutes, with a standard deviation of 26.7 minutes. The mean operating time for OA was 87 minutes, with a standard deviation of 17.3 minutes. At the 0.05 level of significance, test the claim that there is no difference between the mean operating times for these two procedures.

   (b) The mean postoperative hospital stay for LA was 57 hours, with a standard deviation of 26.7 hours. The mean postoperative hospital stay for OA was 66 hours, with a standard deviation of 21.6 hours. At the 0.05 level of significance, test the claim that there is no difference between the mean postoperative hospital stay for these two procedures.

   (c) The mean total hospital bill for LA was $4600, with a standard deviation of $355.83. The mean total hospital bill for OA was $1700, with a standard deviation of $151.52. At the 0.05 level of significance, test the claim that the mean total hospital bill is higher for LA than it is for OA.

16. A random sample of 40 tax returns of individuals who earned less than $40,000 showed a mean charitable contribution of $527 with a standard deviation of $167. Another random sample of 12 tax returns of individuals who earned between

$40,000 and $50,000 showed a mean charitable contribution of $951 with a standard deviation of $248. At the 0.05 level of significance, test the claim that individuals who earn between $40,000 and $50,000 have a higher mean charitable contribution than individuals who earn less than $40,000. (Data based on the results of a survey by Gallup for *Independent Sector.*)

17. A random sample of 25 women with Turner's syndrome had a mean height of 148.7 cm, with a standard deviation of 2.81 cm. A random sample of 155 women from the general population had a mean height of 164.2 cm, with a standard deviation of 5.35 cm. At the 0.01 level of significance, test the claim that the mean height of women with Turner's syndrome is less than that of the general female population. (*Source: European Journal of Pediatrics.*)

18. A random sample of 17 male college graduates required a mean of 6.28 years to earn their bachelor's degree with a standard deviation of 0.81 years. A similar sample of 21 female college graduates required a mean of 6.30 years to earn their bachelor's degree with a standard deviation of 0.76 years. At the 0.05 level of significance, test the claim that male and female college students require the same mean length of time to graduate. (Based on results of a study by the Census Bureau.)

19. A random sample of 14 college students majoring in mathematics or statistics required a mean of 5.28 years to earn their bachelor's degree with a standard deviation of 1.04 years. A similar sample of 86 college students majoring in psychology required a mean of 6.02 years to earn their bachelor's degree with a standard deviation of 0.93 years. At the 0.05 level of significance, test the claim that the mean time required by students majoring in mathematics or statistics to earn a bachelor's degree is less than the mean length of time required by college students majoring in psychology. (Based on results of a study by the Census Bureau.)

20. A random sample of 26 families that rent their home owned a mean of 1.29 cars with a standard deviation of 0.97 cars. A random sample of 100 families that own their own home owned a mean of 1.97 cars with a standard deviation of 0.96 cars. At the 0.05 level of significance, test the claim that the mean number of cars owned by families that rent their home is less than the mean number of cars owned by families that own their own home. (Based on the results of the Census Bureau American Housing Brief.)

21. A random sample of 26 women received a mean of 17.5 e-mails per day, with a standard deviation of 15.60 e-mails. A random sample of 49 men received a mean of 16.3 e-mails per day, with a standard deviation of 14.25 e-mails. At the 0.05 level of significance, test the claim that the mean number of e-mails received per day is the same for women and men. (Based on the results of a study by Solomon-Wolff Associates.)

22. A student believes that the younger a person is, the more likely he or she is to go to the movies. She asked 21 young adults between the ages of 16 and 21 how many times they attended the movies during the past month. They attended the movies a mean of 1.57 times with a standard deviation of 0.98 times. She also asked 19 young adults between the ages of 22 and 27, who had been to the movies a mean of 1.32 times with a standard deviation of 0.82 times. At the 0.05 level of significance, test the claim that the mean number of times that 16- to 21-year-olds attend the movies per month is higher than the mean number of times that 22- to 27-year-olds do.

23. A random sample of 10 cereals marketed to children had a mean of 13.2 grams of sugar per serving, with a standard deviation of 1.55 grams. A random sample of 10 cereals marketed to adults had a mean of 10.5 grams of sugar per serving, with a standard deviation of 5.76 grams. At the 0.05 level of significance, test the claim that the mean amount of sugar per serving in children's cereals is higher than the mean amount in adult cereals.

24. A random sample of 17 male smokers revealed that they began smoking at a mean age of 18.2 years old, with a standard deviation of 2.60 years. A random sample of 20 female smokers revealed that they began smoking at a mean age of 18.0 years old, with a standard deviation of 3.10 years.

   (a) Construct a 95% confidence interval for the difference between the mean age that men begin smoking and the mean age that women begin smoking.

(b) At the 0.05 level of significance, test the claim that the mean age that men begin smoking is the same as the mean age that women begin smoking.

25. Seven one-acre fields of cotton were planted in sandy soil at different locations on a ranch. The mean production per acre was 941.4 pounds, with a standard deviation of 195.5 pounds. Six one-acre fields of cotton were planted in soil that was high in clay at different locations on a ranch. The mean production per acre was 870.3 pounds, with a standard deviation of 133.5 pounds.

(a) Construct a 95% confidence interval for the difference between the mean production of cotton per acre in sandy soil and the mean production of cotton per acre in soil that is high in clay.

(b) At the 0.05 level of significance, test the claim that the mean production of cotton per acre is the same in both types of soil.

? What Is Wrong with This Picture?

The following statistical project contains a major error. Potential errors could be the choice of the wrong hypothesis test or that the assumptions required for this test are not met. Find the error, and explain what could be done to correct it.

A student records the number of cavities that the 7 members of his family have. He then goes on to record the number of cavities for each of the 5 members of his girlfriend's family. He uses these data to test the claim that the two families have the same mean number of cavities per person, using the hypothesis test developed in this section.

What is wrong with this picture?

MINI PROJECT

Randomly sample fewer than 30 males and fewer than 30 females at your school, and ask them to estimate how much money they have spent in the last week on food. Use these sample data to test the claim that college males and college females have had the same mean food expenditures at the 0.05 level of significance. In addition to your complete hypothesis test, include

- Your raw data
- A histogram for food expenditures of the males and a second histogram for the food expenditures of the females

Finally, answer the following questions.

- How did you obtain your sample?
- Which type of sampling did you use?
- Was the sample truly random?
- What potential biases may show up in your sample?

SECTION 8.3
Paired Difference Test for Dependent Samples

Before we begin, consider the following example.

A football coach claims that consistent weight training over a period of time improves a person's strength. Here are the number of bench press repetitions performed before and after a two-month training program for 18 randomly selected players.

Player	Before	After	Player	Before	After	Player	Before	After
1	4	11	7	2	2	13	0	3
2	16	21	8	2	4	14	9	16
3	14	19	9	11	19	15	4	7
4	4	6	10	15	17	16	4	10
5	12	14	11	2	9	17	0	1
6	10	16	12	11	10	18	6	9

Test the claim that the training program increases the number of repetitions a person can do at the 0.01 level of significance. ■

At first glance, this appears to be a small-sample, two-mean test. However, there is really only one sample, the 18 football players. If we consider the number of repetitions before the training program as one sample and the number of repetitions after the training program as a second sample, then our samples are **dependent samples.** Two samples are dependent if there is a relation between them. In the above example the samples are related because the number of repetitions before and after are for the same football player. These numbers are related, and depend in some degree on the overall strength of the player.

In a two-mean test the standard deviation of each sample plays a major role in determining whether the null hypothesis is rejected or not. The larger the standard deviations are, the smaller the test statistic will be. The smaller the test statistic is, the less likely it is that the null hypothesis is rejected. In the previous example, the standard deviation is influenced by the varying levels of strength among the players. The claim that is being tested does not involve the varying strength of the players. According to the claim, we only want to find out if the number of repetitions increases.

The appropriate test in this situation is the **paired difference** test, as long as the two samples are drawn from populations that are normally distributed. This will produce differences that are normally distributed as well. If the two populations are not both normally distributed, then we can apply the sign test from Section 11.1 or the Wilcoxon signed-ranks test from Section 11.2.

We begin by finding the difference between each pair of values, and essentially perform a one-mean test on these differences. (*Note:* Even though the number of pairs may be 30 or higher, we will use the *t*-distribution rather than the *z*-distribution for this test.) We will use the following test statistic:

$$t = \frac{\bar{d} - \mu_d}{s_d / \sqrt{n}}$$

where $\bar{d}$ = mean of sample differences
  μ_d = mean of population differences (claimed)
  s_d = standard deviation of sample differences
  n = number of pairs

More often than not, μ_d will be 0. This is the case when we are interested only in whether there was an increase, rather than an increase by 5, for example.

One tricky part of this test is trying to determine what the null and alternate hypotheses are. This all depends on which order we subtract in. When trying to determine which order to subtract in, keep the following two ideas in mind. First, try to subtract in such a way that a majority of the differences are positive. This helps to simplify calculating the mean and standard deviation of the differences. Second, try to subtract in a way that will let you easily convert the claim to symbols.

EXAMPLE 8.16

A football coach claims that consistent weight training over a period of time improves a person's strength. Here are the number of bench press repetitions performed before and after a two-month training program for 18 randomly selected players.

Player	Before	After	Player	Before	After	Player	Before	After
1	4	11	7	2	2	13	0	3
2	16	21	8	2	4	14	9	16
3	14	19	9	11	19	15	4	7
4	4	6	10	15	17	16	4	10
5	12	14	11	2	9	17	0	1
6	10	16	12	11	10	18	6	9

Test the claim that the training program increases the number of repetitions a person can do at the 0.01 level of significance.

We begin by determining which order to subtract in. Since the claim is that the number of repetitions increases, the "after" values should be higher than the "before" values. We will let d = after – before. If the claim is true, then these values should be positive in general, and the mean of all population differences should be positive as well ($\mu_d > 0$).

Step 1

d = after – before
Claim in words: The training program increases the number of repetitions a person can do.
Claim: $\mu_d > 0$
Complement: $\mu_d \leq 0$
H_0: $\mu_d \leq 0$
H_A: $\mu_d > 0$

Step 2

Level of significance: $\alpha = 0.01$

Step 3

Test statistic: $t = \dfrac{\bar{d} - \mu_d}{s_d/\sqrt{n}}$ $d.f. = n - 1$

Step 4

$\alpha = 0.01$, right-tailed test, with $18 - 1 = 17$ degrees of freedom
Decision rule: Reject H_0 if $t > 2.567$.

Step 5

Player	Before	After	d = After − Before
1	4	11	7
2	16	21	5
3	14	19	5
4	4	6	2
5	12	14	2
6	10	16	6
7	2	2	0
8	2	4	2
9	11	19	8
10	15	17	2
11	2	9	7
12	11	10	−1
13	0	3	3
14	9	16	7
15	4	7	3
16	4	10	6
17	0	1	1
18	6	9	3

$\bar{d} = 3.8, \quad s_d = 2.67$

$$t = \bar{d} - \frac{\mu_d}{s_d / \sqrt{n}}$$

$$= \frac{3.8 - 0}{2.67 / \sqrt{18}}$$

$$= 6.038$$

Decision: Reject H_0.

Conclusion: There is sufficient sample evidence to support the claim that the training program increases the number of repetitions a person can do.

p-value

Using techniques developed in Section 7.2, the best that we can say is that the p-value is less than 0.005. A calculator or computer would tell us that the actual p-value of this test is 0.00000665, which is very small. ∎

If the previous test had been done as a two-mean test, the test statistic would have been 1.956 and the critical value would have been 2.567. Thus we would have failed to reject the null hypothesis and failed to support the claim, with a p-value of 0.0295. Once again, the major difference between these two results is that in the paired difference test we disregard the deviation within each sample.

EXAMPLE 8.17

Many companies claim to raise students' SAT scores. When evaluating such a claim, we must consider something called the "testing effect." If a student has already taken the SAT exam once, there is a chance that the student's scores will improve the next time, due to familiarity with the test. Here are the scores of 12 randomly selected students on the SAT exam (combined math and English) on their first and second attempt at the exam. At the 0.05 level of significance, test the claim that scores improve from the first attempt to the second attempt.

Student	First Score	Second Score	Student	First Score	Second Score
1	680	650	7	710	740
2	1140	1140	8	830	870
3	910	910	9	1290	1270
4	1370	1420	10	1050	1130
5	1110	1110	11	1170	1160
6	1000	1090	12	1060	1110

We begin by determining which order to subtract in. Since the claim is that the scores improve, the second scores should be higher than the first scores. We will let d = second – first. If the claim is true, then these values should be positive in general, and the mean of all population differences should be positive as well ($\mu_d > 0$).

Step 1

d = second – first

Claim in words: SAT scores improve the second time that students take the test.

Claim: $\mu_d > 0$

Complement: $\mu_d \leq 0$

H_0: $\mu_d \leq 0$

H_A: $\mu_d > 0$

Step 2

Level of significance: $\alpha = 0.05$

Step 3

Test statistic: $t = \dfrac{\bar{d} - \mu_d}{s_d/\sqrt{n}}$ $d.f. = n - 1$

Step 4

$\alpha = 0.05$, right-tailed test, with $12 - 1 = 11$ degrees of freedom

Decision rule: Reject H_0 if $t > 1.796$.

Step 5

Student	First Score	Second Score	d = Second – First
1	680	650	–30
2	1140	1140	0
3	910	910	0
4	1370	1420	50
5	1110	1110	0
6	1000	1090	90
7	710	740	30
8	830	870	40
9	1290	1270	–20
10	1050	1130	80
11	1170	1160	–10
12	1060	1110	50

$\bar{d} = 23.3, \quad s_d = 39.16$

$$t = \bar{d} - \frac{\mu_d}{s_d/\sqrt{n}}$$

$$= \frac{23.3 - 0}{39.16 \, / \, \sqrt{12}}$$

$$= 2.061$$

Decision: Reject H_0.

Conclusion: There is sufficient sample evidence to support the claim that SAT scores improve the second time that students take the test.

p-value

Using techniques developed in Section 7.2, the best we can say is that the p-value is between 0.025 and 0.05. A calculator or computer would tell us that the actual p-value of this test is 0.0319. ∎

EXAMPLE 8.18

A doctor randomly selects 9 patients for an experiment. At the end of an examination, he tells the patients that they need to lose a few pounds to improve their health. He gives them a sugar-based placebo, and tells them that it is an appetite suppressant that they are to take three times daily. Here are their weights, in pounds, at the examination and at a follow-up appointment one month later. At the 0.05 level of significance, test the claim that this "appetite suppressant" produces no difference in the weight of the patients.

Patient	Weight at Initial Exam	Weight at Follow-Up
A	209	201
B	249	245
C	185	179
D	207	204
E	227	231
F	174	182
G	157	162
H	196	197
I	177	177

We begin by determining which order to subtract in. There is no suggestion that either set will be higher or lower, so we will arbitrarily subtract the initial weight minus the weight at the follow-up, d = initial weight – follow-up weight. If there is no difference in the weights, as the claim suggests, then these differences should be close to 0, and the mean of all population differences should equal $0(\mu_d = 0)$.

Step 1

d = initial weight – follow-up weight

Claim in words: This "appetite suppressant" produces no difference in the weight of the patients.

Claim: $\mu_d = 0$

Complement: $\mu_d \neq 0$

H_0: $\mu_d = 0$

H_A: $\mu_d \neq 0$

Step 2

Level of significance: $\alpha = 0.05$

Step 3

Test statistic: $t = \dfrac{\bar{d} - \mu_d}{s_d/\sqrt{n}}$ $d.f. = n - 1$

Step 4

$\alpha = 0.05$, two-tailed test, with $9 - 1 = 8$ degrees of freedom

Decision rule: Reject H_0 if $t > 2.306$ or if $t < -2.306$.

Step 5

Patient	Weight at Initial Exam	Weight at Follow-Up	d
A	209	201	8
B	249	245	4
C	185	179	6
D	207	204	3
E	227	231	-4
F	174	182	-8
G	157	162	-5
H	196	197	-1
I	177	177	0

$\bar{d} = -0.3, \quad s_d = 5.36$

$$t = \frac{\bar{d} - \mu_d}{s_d / \sqrt{n}}$$

$$= \frac{-0.3 - 0}{5.36 / \sqrt{9}}$$

$$= -0.168$$

Decision: Fail to reject H_0.

Conclusion: There is not sufficient sample evidence to reject the claim that this "appetite suppressant" produces no difference in the weight of the patients.

p-value

Using techniques developed in Section 7.2, the best we can say is that the p-value is higher than 0.5. A calculator or computer would tell us that the actual p-value of this test is 0.8708. ∎

Confidence Intervals

We can construct a confidence interval for the population mean of differences, μ_d. First, we begin with the margin of error E.

$$E = t_{\alpha/2} \cdot \frac{s_d}{\sqrt{n}} \qquad \text{degrees of freedom: } n - 1$$

Then we construct the interval by subtracting the margin of error from the sample mean to obtain the left endpoint of the interval. We find the right endpoint by adding the margin of error to the sample mean.

$$\bar{d} - E \leq \mu_d \leq \bar{d} + E$$

EXAMPLE
8.19

A scientist has developed a gasoline additive that she believes will increase the gas mileage of cars. Eight cars are given exactly 1 gallon of gasoline, and the distance that they can travel, in miles, is measured. The cars are then given another gallon of gasoline, in addition to the additive. The distance that the cars travel is measured again. Construct a 95% confidence interval for the mean of the differences between the additive-influenced mileage and the original mileage.

Car	Original Mileage	Mileage with Additive
1	10.3	11.7
2	14.5	15.7
3	15.8	18.9
4	10.9	12.4
5	16.5	18.2
6	17.1	19.8
7	17.8	20.7
8	14.6	16.5

There are 7 degrees of freedom, so the value for $t_{\alpha/2}$, or $t_{0.025}$, is 2.365. Before we can find the margin of error, we must calculate s_d. We begin by calculating the sample differences.

Car	Original Mileage	Mileage with Additive	d Additive – Original
1	10.3	11.7	1.4
2	14.5	15.7	1.2
3	15.8	18.9	3.1
4	10.9	12.4	1.5
5	16.5	18.2	1.7
6	17.1	19.8	2.7
7	17.8	20.7	2.9
8	14.6	16.5	1.9

$\bar{d} = 2.1, \quad s_d = 0.74$

Here is the margin of error E.

$$E = t_{\alpha/2} \cdot \frac{s_d}{\sqrt{n}}$$

$$= 2.365 \cdot \frac{0.74}{\sqrt{8}}$$

$$= 0.619$$

Now, the confidence interval.

$$\bar{d} - E \leq \mu_d \leq \bar{d} + E$$

$$2.1 - 0.62 \leq \mu_d \leq 2.1 + 0.62$$

$$1.48 \leq \mu_d \leq 2.72$$

We are 95% confident that the mean of the differences between the additive-influenced mileage and the original mileage is between 1.48 miles and 2.72 miles. Since all of the values in this interval are positive, this suggests that the additive increases a car's mileage. ∎

MICROSOFT EXCEL   Paired Difference Test

Microsoft Excel can assist us with the calculations required for the paired difference test. We can use Excel to quickly find the sample differences, as well as their mean and standard deviation. We can also use it to calculate the test statistic, critical value, and *p*-value of the test. We will step through an example that shows how to use Excel for these calculations, and then write the corresponding hypothesis test.

 EXAMPLE 8.20

An instructor claims that test scores drop from the Unit 4 test to the Unit 5 test. Here are the scores of 16 randomly selected students on the Unit 4 test and the Unit 5 test. At the 0.05 level of significance, test the instructor's claim.

				Student				
	1	2	3	4	5	6	7	8
Unit 4	67	96	91	100	100	96	81	93
Unit 5	41	92	95	96	100	82	89	96

				Student				
	9	10	11	12	13	14	15	16
Unit 4	65	89	90	88	100	100	87	84
Unit 5	69	74	87	72	93	86	94	80

In a new Excel worksheet, type Unit 4 in cell **A1,** and type Unit 5 in cell **B1.** Enter the scores for the Unit 4 test in column **A,** from cell **A2** through **A17.** Enter the scores for the Unit 5 test in column **B,** from cell **B2** through **B17.** According to the claim, the scores are decreasing, which means that the scores on the Unit 4 test should be higher than the scores on the Unit 5 test. We will let *d* = Unit 4 – Unit 5. Our claim will be that $\mu_d > 0$, so we have a right-tailed test.

Type d in cell **C1.** We will store the differences in column **C.** In cell **C2,** type

    =A2-B2

After you press the Enter key, the difference for the first student will appear in cell **C2.** Now copy that formula to cells **C3** through **C17;** first, click on cell **C2** and select **Copy** from the **Edit** menu. Now highlight cells **C3** through **C17** by clicking on cell **C3** and dragging the mouse down to cell **C17** before you release the mouse button. Once these cells have been highlighted, select **Paste** from the **Edit** menu. The differences should now appear in column **C.**

In cell **D1,** type d-bar. Type sd in cell **D2.** Type n in cell **D3.** Moving to the next column, we will calculate the sample statistics. Type the following in cell **E1.**

    =AVERAGE(C2:C17)

This calculates $\bar{d}$. To calculate s_d, type the following in cell **E2.**

    =STDEV(C2:C17)

Finally, type the sample size, 16, in cell **E3.**

We are now ready to calculate the test statistic

$$t = \frac{\bar{d} - \mu_d}{s_d / \sqrt{n}}$$

Since $\mu_d = 0$, we can leave it out of the calculation. Type Test Statistic in cell **D4.** To calculate the test statistic, type the following in cell **E4.**

    =E1/(E2/SQRT(E3))

The result should be 2.097, rounded to three decimal places.

Next we will calculate the critical value for this test. In cell **D5,** type Critical Value. We will use Excel's built-in TINV function, which requires a two-tailed level of significance, followed by the degrees of freedom. Since our test is a right-tailed test, we must double the level of significance for this function. (The critical value for a right-tailed test at the 0.05 level of significance is the same as the critical value for a two-tailed test at the 0.10 level of significance.) The degrees of freedom for this test are 1 less than the sample size, so we have 15 degrees of freedom. Type the following in cell **E5.**

    =TINV(0.10,15)

Excel returns the value 1.753, rounded to three decimal places. Since our test statistic is greater than our critical value, we will reject the null hypothesis.

Finally, we can use Excel to calculate the p-value of this test, using its built-in TDIST function. The format of this function is as follows.

    =TDIST(Test Statistic, Degrees of Freedom, Number of Tails)

There is one thing about the test statistic that we must be familiar with. If you have a negative test statistic, Excel requires that you input it as a positive value. If you have a left-tailed test and a negative test statistic, the p-value will still be accurate. However, if you have a right-tailed test and a negative p-value, Excel will give you the complement of the p-value. Simply subtract that value from 1 to find the p-value. This is not a concern in this example, as we have a right-tailed test with a positive test statistic. Type p-value in cell **D6,** and then type the following in cell **E6** to find the p-value.

    =TDIST(2.097,15,1)

Excel returns the value 0.0267, rounded to four decimal places. Here is an example of what your worksheet should look like.

Unit 4	Unit 5	d		
67	41	26	d-bar	5.0625
96	92	4	sd	9.657251
91	95	−4	n	16
100	96	4	test statistic	2.09687
100	100	0	Critical Value	1.753051
96	82	14	p-value	0.026678
81	89	−8		
93	96	−3		
65	69	−4		
89	74	15		
90	87	3		
88	72	16		
100	93	7		
100	86	14		
87	94	−7		
84	80	4		

We now can write the hypothesis test for this example.

Step 1

d = Unit 4 – Unit 5
Claim in words: Test scores drop from the Unit 4 test to the Unit 5 test.
Claim: $\mu_d > 0$
Complement: $\mu_d \leq 0$
H_0: $\mu_d \leq 0$
H_A: $\mu_d > 0$

Step 2

Level of significance: α = 0.05

Step 3

Test statistic: $t = \dfrac{\bar{d} - \mu_d}{s_d/\sqrt{n}}$ $d.f. = n - 1$

Step 4

α = 0.05, right-tailed test, with 12 – 1 = 11 degrees of freedom

Decision rule: Reject H_0 if $t > 1.796$ (calculated by Excel).

Step 5

$\bar{d}$ = 5.0265 (calculated by Excel)
t = 2.097 (calculated by Excel)

Decision: Reject H_0.

Conclusion: There is sufficient sample evidence to support the claim that test scores
   drop from the Unit 4 test to the Unit 5 test.

***p*-value**

p-value = 0.0267 (calculated by Excel) ■

TI-83 │ Paired Difference Test

We will use the TI-83's one-sample t test to help us perform the paired difference test.
We will use the TI-83 to store the differences between two lists in a third list, and then
perform a one-sample t test on that third list. We will step through an example that
shows how to use the TI-83 for these purposes, and then write the corresponding
hypothesis test.

EXAMPLE 8.21

An instructor claims that test scores drop from the Unit 4 test to
the Unit 5 test. Here are the scores of 16 randomly selected students on the Unit 4 test and the Unit 5 test. At the 0.05 level of
significance, test the instructor's claim.

				Student				
	1	*2*	*3*	*4*	*5*	*6*	*7*	*8*
Unit 4	67	96	91	100	100	96	81	93
Unit 5	41	92	95	96	100	82	89	96

				Student				
	9	*10*	*11*	*12*	*13*	*14*	*15*	*16*
Unit 4	65	89	90	88	100	100	87	84
Unit 5	69	74	87	72	93	86	94	80

Enter the scores on the Unit 4 test in list L_1, and the scores on the Unit 5 test in list L_2. According to the claim, the scores are decreasing, which means that the scores on the Unit 4 test should be higher than the scores on the Unit 5 test. We will let d = Unit 4 – Unit 5. Our claim will be that $\mu_d > 0$, so we have a right-tailed test. Move to list L_3 and use the ⟨↑⟩ key to move up and highlight the name L_3. Doing this allows us to define a formula for list L_3 in terms of other lists. We want to subtract the values in list L_2 from the values in list L_1, so press ⟨2nd⟩ ⟨1⟩ ⟨–⟩ ⟨2nd⟩ ⟨2⟩ and then ⟨ENTER⟩. This will put the differences in list L_3.

Now that we have the differences, we can begin the test. Access the STAT TESTS menu by pressing ⟨STAT⟩ and then using the ⟨→⟩ key to move to the right. Select option **2:T-Test.** Next to **Inpt:,** highlight **Data,** because we have the actual differences. On the next line, enter 0 next to **μ_0:.** This represents the claimed value of μ_d for our test. Enter L_3 next to **List:** by pressing ⟨2nd⟩ ⟨3⟩. Make sure that there is a 1 next to **Freq:.** On the next line, we select the form of the alternate hypothesis. Since our test is a right-tailed test, select the third option: **>μ_0.** Finally, highlight **Calculate** and press ⟨ENTER⟩. Here is the screen of output that you should see.

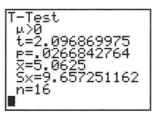

We see that the value of the test statistic is 2.097, rounded to three decimal places, and the p-value of this test is 0.0267. The last three lines contain $\bar{d}$, s_d, and n.

If, on the final step, we had highlighted **Draw** instead of **Calculate,** we would have seen the following output.

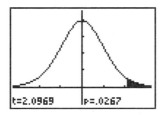

This display gives us the test statistic and p-value along with the graph.

We are now ready to write the hypothesis test.

Step 1

d = Unit 4 – Unit 5
Claim in words: Test scores drop from the Unit 4 test to the Unit 5 test.
Claim: $\mu_d > 0$
Complement: $\mu_d \leq 0$
H_0: $\mu_d \leq 0$
H_A: $\mu_d > 0$

Step 2

Level of significance: $\alpha = 0.05$

Step 3

Test statistic: $t = \dfrac{\bar{d} - \mu_d}{s_d/\sqrt{n}}$ $d.f. = n - 1$

Step 4

$\alpha = 0.05$, right-tailed test, with $12 - 1 = 11$ degrees of freedom

Decision rule: Reject H_0 if $t > 1.796$.

Step 5

$\bar{d} = 5.0265$ (calculated using TI-83)

$t = 2.097$ (calculated using TI-83)

Decision: Reject H_0.

Conclusion: There is sufficient sample evidence to support the claim that test scores drop from the Unit 4 test to the Unit 5 test.

p-value

p-value = 0.0267 (calculated using TI-83) ■

EXERCISES 8.3

1. Here are the prices of a dose of 10 medications in Canada and the United States (all given in U.S. dollars).

Drug	Canada	United States
Prilosec	$1.47	$3.31
Prozac	$1.07	$2.27
Lipitor	$1.34	$2.54
Prevacid	$1.34	$3.13
Epogen	$21.44	$23.40
Zocor	$1.47	$3.16
Zoloft	$1.07	$1.98
Zyprexa	$3.39	$5.27
Claritin	$1.11	$1.96
Paxil	$1.13	$2.22

(Source: USA Today.)

(a) At the 0.05 level of significance, test the claim that prescription medications cost less in Canada than in the United States.

(b) Explain, in your own words, why these samples are dependent and a test of paired differences is appropriate.

2. Here are the prices of medications that are prescribed to people and animals. The prices are for one human dose.

Drug	Humans	Animals
Medrol	$0.67	$0.13
Winstrol	$0.64	$0.18
Lodine	$1.20	$0.42
Robaxin	$0.51	$0.25
Vasotec	$0.87	$0.57
Cleocin	$0.74	$0.57
Robinul	$0.50	$0.49
Fulvicin U/F	$1.22	$1.28

(Source: House Committee on Government Reform, Minority Staff, USA Today.)

At the 0.01 level of significance, test the claim that the medication for humans is more expensive than the exact same medication for animals.

3. Some sports fans complain that referees show preferential treatment to the home team. Here are the number of free throws attempted by the home team and the away team in 13 randomly selected NBA games.

Home Team

28	27	34	26	29	30	34
27	37	16	46	48	29	

Away Team

26	31	27	14	21	22	18
23	27	30	53	43	28	

(a) At the 0.05 level of significance, test the claim that the home team shoots more free throws than the away team.

(b) Explain, in your own words, why these samples are dependent and a test of paired differences is appropriate.

4. A doctor believes that a sensible vegan diet and exercise program can be effective in lowering a person's serum cholesterol if it is high. He randomly selects 7 patients who are willing to follow his program. Here are their cholesterol levels (mg/dL) before the program, and again 1 month after beginning the program.

Before

255	243	264	249	275	280	259

After

241	245	251	248	268	269	246

At the 0.05 level of significance, test the claim that the doctor's program lowers the cholesterol levels of people with high cholesterol.

5. For a random sample of 20 dairy cows, here are the number of pounds of milk produced during their first lactation (after their first calf) and their second lactation (after their second calf).

Cow	First Lactation	Second Lactation	Cow	First Lactation	Second Lactation
1	22,792	29,655	11	15,058	8,013
2	31,693	23,817	12	16,681	23,301
3	18,367	35,360	13	21,002	32,529
4	17,440	21,848	14	17,987	23,620
5	29,798	29,828	15	15,968	26,437
6	28,540	33,898	16	23,580	26,227
7	23,661	22,444	17	14,334	25,529
8	39,242	23,648	18	24,030	27,368
9	21,574	25,303	19	20,392	29,527
10	34,542	34,812	20	24,347	25,592

(a) At the 0.05 level of significance, test the claim that cows produce more milk during their second lactation than during their first lactation.

(b) Explain, in your own words, why these samples are dependent and a test of paired differences is appropriate.

6. A random sample of 4 male college students had their pulse taken. They then walked briskly for 100 yards and had their pulse taken again.

Before Walk

66	60	64	72

After Walk

76	64	72	84

At the 0.05 level of significance, test the claim that a brisk 100-yard walk raises pulse rates.

7. Eight faculty members at a community college decided to get a flu shot for the first time. Here are the number of absences due to illness each person had during the flu season the year before the shot and also for the flu season following the shot.

Before Shot

| 1 | 4 | 4 | 4 | 6 | 3 | 3 | 2 |

After Shot

| 0 | 3 | 1 | 2 | 3 | 2 | 3 | 2 |

At the 0.05 level of significance, test the claim that the flu shot is effective in lowering the number of absences due to illness.

8. Here are the average prices for a gallon of unleaded regular gasoline in the Los Angeles area and the San Francisco Bay area for 8 randomly selected months.

| | *Month/Year* | | | | | | | |
	3/97	*7/97*	*9/97*	*1/98*	*10/98*	*12/98*	*4/99*	*6/99*
LA	$0.993	$0.965	$1.104	$0.930	$0.867	$0.874	$1.154	$0.994
SF	$1.061	$1.026	$1.108	$1.044	$0.912	$0.958	$1.303	$1.119

At the 0.01 level of significance, test the claim that gas prices in the Los Angeles area are cheaper than in the San Francisco Bay area.

9. Here are the results of 12 people who participated in a weight loss program. The values represent the pounds lost (negative values) or gained (positive values) over a 3-month period.

| −12 | −10 | 20 | 4 | 5 | −23 |
| −12 | −9 | −16 | 1 | −14 | −3 |

At the 0.05 level of significance, test the claim that the diet is effective. (Is this a paired difference test? Yes, the values that are given are the sample differences d.)

10. A new medication is being tested to see if it can lower a person's diastolic blood pressure. It is tested on 900 people. After 60 days, the average drop in diastolic blood pressure is −6.03 mmHg, with a standard deviation of 2.967 mm Hg. At the 0.01 level of significance, test the claim that this medication is effective in lowering blood pressure.

11. *USA Today* made a study of rental car companies (*USA Today*, May 22, 2000) to compare price quotes made on the company's Web site and again over the phone. Here are the daily base rates for a midsize or intermediate-size car at Dallas/Fort Worth International Airport for a 1-day rental.

Company	Web Site	Over the Phone
Budget	$73.99	$64.79
Hertz	$74.99	$74.99
Avis	$74.99	$74.99
National	$73.98	$71.98
Alamo	$62.99	$70.99
Enterprise	$43.95	$49.95
Dollar	$37.99	$41.99
Thrifty	$65.95	$65.95

At the 0.01 level of significance, test the claim that there is no difference between the daily base rates given out on Web sites and over the phone.

12. Do cows produce milk for a longer period after their second calf than after their first calf? Twenty cows were selected at random. Here are the lengths of their first and second lactation, in days.

Cow	First Lactation	Second Lactation	Cow	First Lactation	Second Lactation
1	263	328	11	329	315
2	279	354	12	338	352
3	284	414	13	266	367
4	249	316	14	286	302
5	335	525	15	229	335
6	329	331	16	387	474
7	399	367	17	387	361
8	283	285	18	317	305
9	285	377	19	294	380
10	293	388	20	315	372

At the 0.05 level of significance, test the claim that cows produce milk for more than 30 days longer during their second lactation than they do during their first lactation.

13. Here are the midterm exam scores of 20 algebra students, along with their scores on the final exam.

Student	Midterm	Final	Student	Midterm	Final
1	65	85	11	38	56
2	50	77	12	92	95
3	75	90	13	59	60
4	70	84	14	66	87
5	68	61	15	61	83
6	58	70	16	71	77
7	49	76	17	68	79
8	92	78	18	82	74
9	68	80	19	85	94
10	93	92	20	67	80

(a) Construct a 95% confidence interval for the mean of the differences between the final exam scores and the midterm scores.

(b) At the 0.025 level of significance, test the claim that student's scores improve from the midterm to the final.

14. Here are the predicted high temperatures (°F) for 15 U.S. cities on August 16, 1999 that appeared in the *New York Times,* and the actual high temperatures for that day. The temperatures were predicted 2 days in advance.

City	Predicted High	Actual High
Atlantic City	80	84
Austin	99	103
Billings	78	77
Birmingham	92	90
Colorado Springs	86	77
Columbus	79	79
Fairbanks	65	57
Jackson	93	90
Minn.-St. Paul	81	79
Norfolk	85	83
Pittsburgh	81	74
St. Louis	86	80
Sioux Falls	87	85
Tampa	90	90
Wichita	95	97

(a) Construct a 95% confidence interval for the mean of the differences between the predicted high temperatures and the actual high temperatures.

(b) At the 0.05 level of significance, test the claim that there is no difference between the predicted high temperatures and the actual high temperatures.

15. An SAT preparation company claims that their program improves the composite SAT scores (math + verbal) of students. Here is a random sample of the composite scores of 20 students before they took the program and again after they completed the program.

Before Program	After Program	Before Program	After Program
910	970	1250	1280
1300	1400	1200	1300
1030	1070	1090	1090
1010	1010	1150	1130
1090	1060	1010	1050
1150	1200	1060	1110
1110	1140	1000	1030
1240	1310	1010	1060
1290	1260	1120	1090
1100	1200	1150	1210

(a) Construct a 95% confidence interval for the mean of the differences between the scores after the program and the scores before the program.
(b At the 0.05 level of significance, test the claim that the program is effective in raising the scores of students.

16. A group of 18 concertgoers was selected at random. Before the concert they were given a hearing test, and then were given another one after the concert. (The volume varied during the test, and the person also had to state which ear the sound was in.) Here are the number of correctly identified sounds out of 10, both before and after the concert.

Before	After	Before	After
9	8	10	9
10	8	9	9
9	9	10	8
8	6	8	8
8	6	8	9
9	7	9	9
9	10	9	7
9	8	9	6
8	5	9	6

(a) Construct a 95% confidence interval for the mean of the differences between the scores before the concert and the scores after the concert.
(b) At the 0.05 level of significance, test whether a person's hearing is adversely affected by the noise of a concert.

? **What Is Wrong with This Picture?**

The following statistical project contains a major error. Potential errors could be the choice of the wrong hypothesis test or that the assumptions required for this test are not met. Find the error, and explain what could be done to correct it.

A student randomly selects 20 students, 10 males and 10 females. She numbers the 10 males from 1 through 10, and does the same for the 10 females. Each of the students is given an IQ test. She matches male #1 with female #1, male #2 with female #2, and so on. She uses the results of the IQ test to test the claim that males and females perform equally well on IQ tests, using the hypothesis test introduced in this section.

   What is wrong with this picture?

Randomly sample at least 15 students at your school, and ask them for their weight when they began attending your school and their present weight. This might be a difficult survey to conduct. An alternative would be to ask each student how much weight they have gained or lost since attending your school, which would provide you with the differences that you will eventually need for this paired difference test. Use these sample data to test the claim that college students gain weight while attending college at the 0.05 level of significance. In addition to your complete hypothesis test, include

- Your raw data
- A histogram for the amount of weight gained or lost

Finally, answer the following questions.

- How did you obtain your sample?
- Which type of sampling did you use?
- Was the sample truly random?
- What potential biases may show up in your sample?

SECTION 8.4
Hypothesis Test for Two Population Proportions
(Large Samples)

In this section, we develop a hypothesis test for the difference between two population proportions. In order to apply this test, the number of successes and failures in each sample must be at least 5 in keeping with the conditions to apply the normal approximation to the binomial distribution.

The test for two population proportions is similar to the previous two-sample tests in that we must identify one of the populations as population 1 and the other as population 2. The major difference is the test statistic:

$$z = \frac{p_1 - p_2}{\sqrt{\bar{p}(1-\bar{p})\left(\frac{n_1 + n_2}{n_1 \cdot n_2}\right)}}, \quad \text{where } \bar{p} = \frac{x_1 + x_2}{n_1 + n_2}$$

The numerator consists of the difference between the two sample proportions. In the denominator, $\bar{p}$ represents the proportion of the two samples when combined. In many texts the fraction

$$\frac{n_1 + n_2}{n_1 \cdot n_2}$$

is written as

$$\frac{1}{n_1} + \frac{1}{n_2}$$

The form that we will use in this text is easier to use with calculators.

As with the previous test statistics for two-sample hypothesis tests, the numerator represents how far apart the two sample statistics are. The denominator tells us how the distribution for the difference is distributed. We begin with an example.

EXAMPLE 8.22

Are male high school graduates equally likely to attend college the following fall as female high school graduates? A random sample of 1354 males who graduated high school in 1997 found that 860 of them were enrolled in college in October 1997. A sample of 1415 females who graduated high school in 1997 found that 995 of them were enrolled in college in October 1997. At the 0.05 level of significance, test the claim that the proportion of male graduates that go on to college is the same as the proportion of female graduates that go on to college. (Based on data from "College Enrollment and Work Activity of 1997 High School Graduates" by the U.S. Department of Labor, Bureau of Labor Statistics.)

Step 1

We begin, as usual, by identifying the claim that is to be tested.

Claim in words: The proportion of male graduates that go on to college is the same as the proportion of female graduates that go on to college.

Since males are mentioned first in the claim, it is a wise choice to label male high school graduates as population 1.

Population 1: male high school graduates; population 2: female high school graduates We will use π_1 to represent the first population proportion and π_2 to represent the second. Since the claim is that the two population proportions are the same (equal), our claim will be $\pi_1 = \pi_2$. To determine the complement, as well as the null and alternate hypotheses, we follow our work from past sections.

Claim: $\pi_1 = \pi_2$
Complement: $\pi_1 \neq \pi_2$
H_0: $\pi_1 = \pi_2$
H_A: $\pi_1 \neq \pi_2$

Step 2

Level of significance: $\alpha = 0.05$

Step 3

Test statistic: $z = \dfrac{p_1 - p_2}{\sqrt{\bar{p}(1 - \bar{p})\left(\dfrac{n_1 + n_2}{n_1 \cdot n_2}\right)}}$,   where $\bar{p} = \dfrac{x_1 + x_2}{n_1 + n_2}$

Step 4

Note that we are using the z-distribution, not the t-distribution.

$\alpha = 0.05$, two-tailed test

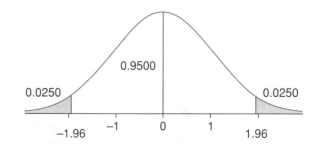

Decision rule: Reject H_0 if $z < -1.96$ or if $z > 1.96$.

Step 5

To begin the calculations, make note of the two sample proportions.

$$p_1 = \frac{860}{1354}, \quad p_2 = \frac{995}{1415}$$

Then we want to rewrite each of the three parts of the denominator ($\bar{p}$, $1 - \bar{p}$, $\frac{n_1 + n_2}{n_1 \cdot n_2}$) as single fractions.

$$\bar{p} = \frac{x_1 + x_2}{n_1 + n_2}$$

$$= \frac{860 + 995}{1354 + 1415}$$

$$= \frac{1855}{2769}$$

$$1 - \bar{p} = 1 - \frac{1855}{2769}$$

$$= \frac{914}{2769}$$

$$\frac{n_1 + n_2}{n_1 \cdot n_2} = \frac{1354 + 1415}{1354 \cdot 1415}$$

$$= \frac{2769}{1,915,910}$$

Now we put all of the pieces together.

$$z = \frac{p_1 - p_2}{\sqrt{\bar{p}(1 - \bar{p})\left(\dfrac{n_1 + n_2}{n_1 \cdot n_2}\right)}}$$

$$= \frac{\dfrac{860}{1354} - \dfrac{995}{1415}}{\sqrt{\dfrac{1855}{2769} \cdot \dfrac{914}{2769} \cdot \dfrac{2769}{1,915,910}}}$$

$$= -3.81$$

Decision: Reject H_0.

Conclusion: There is sufficient sample evidence to reject the claim that the proportion of male graduates that go on to college is the same as the proportion of female graduates that go on to college.

p-value

The closest z-value that we have to –3.81 on our chart is –3.50.

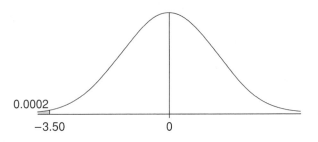

The area to the left of –3.50 is 0.0002. Since this is a two-tailed test, we double this area for a p-value of 0.0004. Technology could be used to show that the p-value of this test is actually 0.00014. ∎

We now continue with another example.

EXAMPLE 8.23

A high school health instructor is interested in teen smoking. In 1985, he randomly selected 50 students and found that 15 of them had smoked cigarettes during the previous year. In 1995, he randomly selected 60 students and found that 16 of them had smoked cigarettes during the previous year. Use this teacher's data to test the claim that the proportion of high school students that smoked in 1985 is higher than the proportion of high school students that smoked in 1995 at the 0.05 level of significance.

Step 1

Claim in words: The proportion of high school students that smoked in 1985 is higher than the proportion of high school students that smoked in 1995.

Since 1985 high school students are mentioned first in the claim, it is a wise choice to label 1985 high school students as population 1.

Population 1: 1985 high school students; population 2: 1995 high school students
Claim: $\pi_1 > \pi_2$
Complement: $\pi_1 \le \pi_2$
H_0: $\pi_1 \le \pi_2$
H_A: $\pi_1 > \pi_2$

Step 2

Level of significance: $\alpha = 0.05$

Step 3

Test statistic: $z = \dfrac{p_1 - p_2}{\sqrt{\overline{p}\left(1 - \overline{p}\right)\left(\dfrac{n_1 + n_2}{n_1 \cdot n_2}\right)}}$,   where $\overline{p} = \dfrac{x_1 + x_2}{n_1 + n_2}$

Step 4

$\alpha = 0.05$, right-tailed test

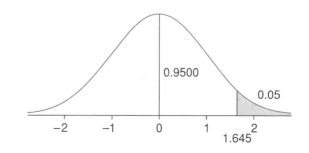

Decision rule: Reject H_0 if $z > 1.645$.

Step 5

$$p_1 = \frac{15}{50}, \qquad p_2 = \frac{16}{60}$$

$$\overline{p} = \frac{x_1 + x_2}{n_1 + n_2}$$

$$= \frac{15 + 16}{50 + 60}$$

$$= \frac{31}{110}$$

$$1 - \overline{p} = 1 - \frac{31}{110}$$

$$= \frac{79}{110}$$

$$\frac{n_1 + n_2}{n_1 \cdot n_2} = \frac{50 + 60}{50 \cdot 60}$$

$$= \frac{110}{3000}$$

$$z = \frac{p_1 - p_2}{\sqrt{\overline{p}(1 - \overline{p})\left(\dfrac{n_1 + n_2}{n_1 \cdot n_2}\right)}}$$

$$= \frac{\dfrac{15}{50} - \dfrac{16}{60}}{\sqrt{\dfrac{31}{110} \cdot \dfrac{79}{110} \cdot \dfrac{110}{3000}}}$$

$$= 0.39$$

Decision: Fail to reject H_0.

Conclusion: There is not sufficient sample evidence to support the claim that the proportion of high school students that smoked in 1985 is higher than the proportion of high school students that smoked in 1995.

p-value

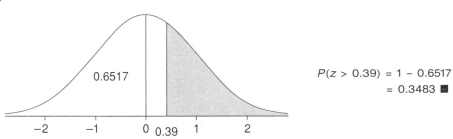

$$P(z > 0.39) = 1 - 0.6517$$
$$= 0.3483 \ \blacksquare$$

Confidence Intervals

We can construct confidence intervals for the difference between two population proportions ($\pi_1 - \pi_2$). Here is the formula for margin of error.

$$E = z_{\alpha/2} \cdot \sqrt{\bar{p}(1-\bar{p}) \cdot \frac{n_1 + n_2}{n_1 \cdot n_2}}$$

We then construct the confidence interval using the following.

$$(p_1 - p_2) - E \leq \pi_1 - \pi_2 \leq (p_1 - p_2) + E$$

Here is an example.

EXAMPLE 8.24   Many doctors believe that early prenatal care is very important to the health of a baby and its mother. Efforts have recently been focused on teen mothers. A random sample of 52 teenagers who gave birth revealed that 32 of them began prenatal care in the first trimester of their pregnancy. A random sample of 209 women in their twenties who gave birth revealed that 163 of them began prenatal care in the first trimester of their pregnancy. Construct a 95% confidence interval for the difference between the proportion of teen mothers who get early prenatal care and the proportion of mothers in their twenties who get early prenatal care.

First, we need a value for $z_{\alpha/2}$.

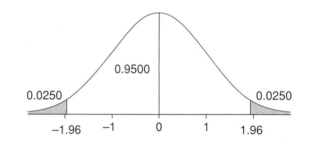

We will use 1.96. Now we calculate the margin of error, which we will round to four decimal places.

$$\bar{p} = \frac{x_1 + x_2}{n_1 + n_2}$$

$$= \frac{32 + 163}{52 + 209}$$

$$= \frac{195}{261}$$

$$1 - \bar{p} = 1 - \frac{195}{261}$$

$$= \frac{66}{261}$$

$$\frac{n_1 + n_2}{n_1 \cdot n_2} = \frac{52 + 209}{52 \cdot 209}$$

$$= \frac{261}{10,868}$$

$$E = z_{\alpha/2} \sqrt{\overline{p}\left(1 - \overline{p}\right) \cdot \frac{n_1 + n_2}{n_1 \cdot n_2}}$$

$$= 1.96 \sqrt{\frac{195}{261} \cdot \frac{66}{261} \cdot \frac{261}{10,868}}$$

$$= 0.1320$$

Now we construct the confidence interval.

$$(p_1 - p_2) - E \leq \pi_1 - \pi_2 \leq (p_1 - p_2) + E$$

$$\left(\frac{32}{52} - \frac{163}{209}\right) - 0.1320 \leq \pi_1 - \pi_2 \leq \left(\frac{32}{52} - \frac{163}{209}\right) + 0.1320$$

$$-0.2965 \leq \pi_1 - \pi_2 \leq -0.0325$$

We are 95% confident that between 3.25% and 29.65% fewer teen mothers get early prenatal care than mothers in their twenties. ∎

MICROSOFT EXCEL	Hypothesis Test for Two Population Proportions (Large Samples)

There is no built-in procedure in Excel to help us with this hypothesis test, so we will focus on how to use Excel to calculate the test statistic, and then how to determine the *p*-value. We will use the following example to walk through the calculations.

EXAMPLE 8.25

A study funded by the Medical Research Council and the Heart Foundation of New Zealand compared stroke patients to healthy people. Of the 521 stroke patients, 164 of them were active smokers. Of the 1851 healthy people, 255 were active smokers. Using these data, test the claim that people who suffer strokes are more likely to be smokers than are healthy people, at the 0.01 level of significance.

In a new Excel worksheet, type x1 in cell **A1** and type 164 in cell **B1.** Type n1 in cell **A2,** and type 521 in cell **B2.** Type x2 in cell **A3,** and type 255 in cell **B3.** Finally, type n2 in cell **A4,** and type 1851 in cell **B4.** These are all of the values that we need to calculate the test statistic.

In cell **A5,** type p1. Next to that, in cell **B5,** type =B1/B2. This is the first sample proportion. In cell **A6,** type p2. Next to that, in cell **B6,** type =B3/B4. This is the second sample proportion.

Next we will calculate the factors in the denominator, beginning with $\overline{p}$. In cell **A7,** type p-bar. In cell **B7,** type =(B1+B3)/(B2+B4). In cell **A8,** type 1 - p-bar. In cell **B8,** type =1-B7. The last piece of the puzzle is $n_1 + n_2/n_1 \cdot n_2$. In cell **A9,** type s. size fraction. In cell **B9,** type =(B2+B4)/(B2*B4).

We are now ready to calculate the test statistic. In cell **A10,** type test statistic. In cell **B10,** type =(B5-B6)/SQRT(B7*B8*B9). The result in this cell should be 9.36.

To calculate *p*-values using Excel, we use Excel's built-in function NORMSDIST. For this function we input a *z*-value, which returns the area under the standard normal curve to the left of that value. Since we have a right-tailed test, we want the complement of that value. In cell **A11,** type p-value. In cell **B11,** type =1-NORMSDIST(9.36). The output of this calculation is 0. (The actual p-value is not 0, but very close. Using the TI-83 calculator, we could see that the first 20 decimal places are zeros.)

Here is what your worksheet should look like.

x1	164
n1	521
x2	255
n2	1851
p1	0.314779
p2	0.137763
p-bar	0.176644
1 - p-bar	0.823356
s. size fraction	0.00246
test statistic	9.359089
p-value	0

Now we can write the hypothesis test.

Step 1

Claim in words: People who suffer strokes are more likely to be smokers than are healthy people.
Population 1: stroke patients; population 2: healthy people
Claim: $\pi_1 > \pi_2$
Complement: $\pi_1 \leq \pi_2$
H_0: $\pi_1 \leq \pi_2$
H_A: $\pi_1 > \pi_2$

Step 2

Level of significance: $\alpha = 0.01$

Step 3

Test statistic: $z = \dfrac{p_1 - p_2}{\sqrt{\bar{p}(1 - \bar{p})\left(\dfrac{n_1 + n_2}{n_1 \cdot n_2}\right)}}$, where $\bar{p} = \dfrac{x_1 + x_2}{n_1 + n_2}$

Step 4

$\alpha = 0.01$, right-tailed test

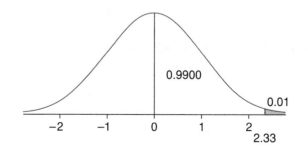

Decision rule: Reject H_0 if $z > 2.33$.

Step 5

$z = 9.36$ (calculated by Excel)

Decision: Reject H_0.

Conclusion: There is sufficient sample evidence to support the claim that people who suffer strokes are more likely to be smokers than are healthy people.

p-value

p-value = 0 (calculated by Excel) ∎

TI-83   Hypothesis Test for Two Population Proportions (Large Samples)

The TI-83 has a built-in procedure to help us with this hypothesis test. We will use the following example to walk through the calculations.

EXAMPLE 8.26   A study funded by the Medical Research Council and the Heart Foundation of New Zealand compared stroke patients to healthy people. Of the 521 stroke patients, 164 of them were active smokers. Of the 1851 healthy people, 255 were active smokers. Using these data, test the claim that people who suffer strokes are more likely to be smokers than healthy people are at the 0.01 level of significance.

Access the STAT TESTS menu by pressing $\boxed{\text{STAT}}$ and using the $\boxed{\rightarrow}$ key to move to the right. Select option **6:2-PropZTest.** The input for this test is easy. Treating the stroke patients as sample 1, enter 164 next to **x1:** and 521 next to **n1:.** The healthy people are sample 2; enter 255 next to **x2:** and 1851 next to **n2:.**

The next line allows us to choose the form of the alternate hypothesis. This test is a right-tailed test, so select the third option, **>p2.** Finally, highlight **Calculate** and press $\boxed{\text{ENTER}}$. Here are the results you should see.

```
2-PropZTest
 p1>p2
 z=9.359088553
 p=4.087574ᴇ-21
 p̂1=.3147792706
 p̂2=.1377633712
↓p̂=.1766441821
■
```

```
2-PropZTest
 p1>p2
↑p̂1=.3147792706
 p̂2=.1377633712
 p̂=.1766441821
 n1=521
 n2=1851
■
```

The first line holds the test statistic, which rounds to be 9.36 to two decimal places. The next line holds the p-value, which is approximately 4.1×10^{-21}. The next three lines are the two sample proportions and the combined proportion (which we label as p_1, p_2, and $\bar{p}$, respectively). The remaining two lines are the two sample sizes.

Now we can write the hypothesis test.

Step 1

Claim in words: People who suffer strokes are more likely to be smokers than are healthy people.
Population 1: stroke patients; population 2: healthy people
Claim: $\pi_1 > \pi_2$
Complement: $\pi_1 \leq \pi_2$
H_0: $\pi_1 \leq \pi_2$
H_A: $\pi_1 > \pi_2$

Step 2

Level of significance: $\alpha = 0.01$

Step 3

Test statistic: $z = \dfrac{p_1 - p_2}{\sqrt{\bar{p}(1 - \bar{p})\left(\dfrac{n_1 + n_2}{n_1 \cdot n_2}\right)}}$, where $\bar{p} = \dfrac{x_1 + x_2}{n_1 + n_2}$

Step 4

$\alpha = 0.01$, right-tailed test

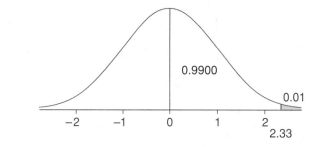

Decision rule: Reject H_0 if $z > 2.33$.

Step 5

$z = 9.36$ (calculated by TI-83)

Decision: Reject H_0.

Conclusion: There is sufficient sample evidence to support the claim that people who suffer strokes are more likely to be smokers than are healthy people.

p-value

p-value $= 4.1 \times 10^{-21}$ (calculated by TI-83) ∎

EXERCISES 8.4

Use the standard procedure for all hypothesis tests.

1. In a random sample of 60 adult health-care professionals, 6 were smokers. In a random sample of 65 adults who were not health-care professionals, 15 were smokers. At the 0.01 level of significance, test the claim that health-care professionals are less likely to be smokers than people who are not health-care professionals.

2. In a random sample of 70 college graduates, 9 were smokers. In a random sample of 144 adults with a high school diploma or less, 34 were smokers. At the 0.01 level of significance, test the claim that college graduates are less likely to be smokers than adults that did not attend college.

3. In a random sample of 250 adults, 33 of 114 males were smokers, while 27 of 136 females were smokers. At the 0.05 level of significance, test the claim that males are more likely to be smokers than females.

4. A student randomly selected 225 college students and asked them whether they ate breakfast that morning before coming to campus. There were 57 students who were at least 25 years old, of which 30 had breakfast that morning. Of the 168 students that were younger than 25, 82 had breakfast that morning. At the 0.05 level of significance, test the claim that there is no difference between the proportion of students age 25 or older that eat breakfast and the proportion of students younger than 25 that eat breakfast.

5. A random telephone survey asked 208 drivers if they run red lights. Of the 48 drivers who were 35 years old or younger, 32 admitted that they run red lights. Eighty-six of the 160 drivers that were older than 35 years old admitted to running red lights. At the 0.05 level of significance, test the claim that drivers who are 35 years old or younger are more likely to run red lights than drivers who are older than 35 years old. (Based on the results of a study by the Social Science Research Center, Department of Psychology, Old Dominion University.)

6. Exercise is an important part of weight loss. Federal recommendations call for at least $2\frac{1}{2}$ hours of exercise per week. A survey of 78 men who were trying to lose weight showed that only 33 were meeting this Federal exercise recommendation. A survey of 139 women who were trying to lose weight revealed that only 51 were meeting this Federal exercise recommendation. At the 0.05 level of significance, test the claim that men who are trying to lose weight are more likely to follow the Federal exercise recommendation than women who are trying to lose weight.

7. The results of 13 randomly selected NBA games were analyzed. The road team in those games attempted 364 free throws and made 260 of them. The home team in those games attempted 411 free throws and made 313 of them. At the 0.01 level of significance, test the claim that the proportion of free throws made by the home team is greater than the proportion made by the road team.

8. In recent years there has been a strong push for standards in education, to keep students from advancing to the next grade or graduating without demonstrating the acquisition of certain knowledge and skills. According to a poll from the Albert Shanker Institute, 759 out of 825 principals and 785 out of 1075 teachers are in favor of this standards movement. At the 0.01 level of significance, test the claim that the proportion of principals that support this movement is greater than the proportion of teachers that support it.

9. A random sample of 76 adults ages 18–24 showed that 11 had donated blood within the past year, while a random sample of 156 adults who were at least 25 years old had 18 people who had donated blood within the past year. At the 0.05 level of significance, test the claim that the proportion of blood donors is equal for these two age groups.

10. In a study of 1000 people who were 60 or older, New York City researchers found that 19 of the 459 women and 10 of the 541 men had lung cancer. At the 0.01 level of significance, test the claim that women over 60 are more likely to get lung cancer than men over 60. (*Source:* The Associated Press.)

11. Researchers at Old Dominion University conducted a survey of 880 adult drivers. Eight hundred sixty-two of the drivers acknowledged that running red lights is hazardous. Another sample of 240 adult drivers showed that 106 said they do not run red lights. At the 0.01 level of significance, test the claim that the proportion of drivers who label running red lights as hazardous is greater than the proportion of drivers who do not run red lights.

12. A random sample of 155 married couples with children and 42 single mothers asked if there was a computer in the home. Ninety-six of the married couples and 13 of the single mothers had a computer in the home. At the 0.01 level of significance, test the claim that married couples with children are more likely to have a computer at home than are single mothers. (Based on the results of a study by the Henry J. Kaiser Family Foundation.)

13. A random sample of 275 Italians showed that 100 of them owned a cell phone. A similar sample of 240 Americans showed that 59 of them owned a cell phone. At the 0.05 level of significance, test the claim that the proportion of Italians with a cell phone is greater than the proportion of Americans with a cell phone. (Based on the results of a study by Baskerville Communications.)

14. A survey of 415 kindergarten children who lived in the suburbs revealed that 24 had been diagnosed with asthma. A survey of 638 inner-city kindergarten children revealed that 69 had been diagnosed with asthma. At the 0.01 level of signi-ficance, test the claim that inner-city children are more likely to have asthma than children who live in the suburbs. (Published in *Annals of Allergy, Asthma and Immunology.*)

15. In a study of 60 randomly selected jury trials, the defendant won 31 of the trials. In a study of 80 randomly selected trials that were decided by a judge, the defendant won 30 of the trials. At the 0.05 level of significance, test the claim that the

proportion of jury trials won by the defendant is greater than the proportion of trials decided by a judge that are won by the defendant. (Based on the results of a study by the Bureau of Justice Statistics.)

16. A random sample of 400 Spanish adults revealed that 143 were smokers. A random sample of 300 American adults revealed that 70 were smokers. At the 0.05 level of significance, test the claim that the proportion of Spanish adults that smoke is greater than the proportion of American adults that smoke. (*Source: New York Times.*)

17. A random sample of 240 American men revealed that 68 held a bachelor's or graduate degree, while 43 of 151 randomly selected American women held a bachelor's or graduate degree. At the 0.05 level of significance, test the claim that the proportion of American men who have a bachelor's or graduate degree is equal to the proportion of American women that do. (Based on information from the Census Bureau.)

18. A random sample of 118 adult women revealed that 70 were satisfied with their weight, while a random sample of 62 adult males showed that 43 were satisfied with their weight. At the 0.05 level of significance, test the claim that the proportion of all adult men who are satisfied with their weight is greater than the proportion of all adult women who are satisfied with their weight. (Based on a survey conducted by Maritz Poll.)

19. Is maintaining a romantic relationship over the Internet while in an exclusive relationship in the real world cheating? Sixty-five percent of 184 randomly selected men answered yes, while 75% of 136 randomly selected women said yes. At the 0.05 level of significance, test the claim that the proportion of all men who think that this is cheating is equal to the proportion of all women who think it is. (Based on the results of a study by International Communications Research for Princess Cruises.)

20. Are parents in the dark about their teenagers' drug use? When 132 parents were asked if they thought that their teenager had tried marijuana, 24 said yes. When 144 teenagers were asked if they had tried marijuana, 59 said yes. At the 0.01 level of significance, test the claim that the proportion of parents who think their teenager has tried marijuana is lower than the proportion of teenagers who say that they have tried it. (Based on a study by the Partnership for a Drug Free America.)

21. A random sample of 436 workers by the Society of Financial Service showed that 44% of them felt that it was "seriously unethical" to monitor employee e-mail, while 33% of 121 senior-level bosses felt the same way. At the 0.05 level of significance, test the claim that the proportion of bosses who view this as "seriously unethical" is lower than the proportion of workers who do.

22. A random sample of 2000 American adults showed that 658 had a college degree. Another sample of 1500 Canadian adults revealed that 426 had a college degree. At the 0.01 level, test the claim that the proportion of adults that have a college degree is higher in the United States than it is in Canada. (Based on the results of a study by the Organization for Economic Co-operation and Development.)

23. A random sample of 114 men ages 30–34 showed that 33 of them had never been married. Another sample of 86 women ages 30–34 showed that 19 of them had never been married. At the 0.05 level of significance, test the claim that the proportion of men ages 30–34 that have never been married is greater than the proportion of all women ages 30–34 who have never been married. (Based on Census Bureau results.)

24. In a random sample of 703 male high school students, 146 said that they had rarely or never worn seat belts when riding in a car/truck driven by someone else. A random sample of 464 female high school students were asked the same question and 55 said that they had had rarely or never worn seat belts when riding in a car/truck driven by someone else. At the 0.05 level of significance, test the claim that the proportion of male high school students that had rarely or never worn seat belts when riding in a car/truck driven by someone else is higher than the proportion of female high school students who had rarely or never worn seat belts when riding in a car/truck driven by someone else. (Based on the Morbidity and Mortality Weekly Report by the Centers for Disease Control and Prevention.)

25. A survey of high school students was taken, and they were asked whether they had been treated by a doctor or nurse for injuries sustained while exercising, playing

sports, or being physically active during the 12 months preceding the survey. Five hundred forty-seven of 1288 high school males had been treated for such injuries, while 307 out of 939 high school females had been treated for such injuries. At the 0.01 level of significance, test the claim that high school males are more likely to be treated for injuries sustained while exercising, playing sports, or being physically active than are high school females. (Based on the Morbidity and Mortality Weekly Report by the Centers for Disease Control and Prevention.)

26. A sample of 237 twelfth-grade boys revealed that 74 of them had driven a car after drinking alcohol in the preceding 30 days. A similar sample of 282 twelfth-grade girls revealed that 41 of them had driven a car after drinking alcohol in the preceding 30 days. At the 0.05 level, test the claim that twelfth-grade girls are less likely to drive a car after drinking alcohol than are twelfth-grade boys. (Based on the Morbidity and Mortality Weekly Report by the Centers for Disease Control and Prevention.)

27. A random sample of 214 high school students in Washington, D.C. contained 42 students who said they felt too unsafe to go to school at least once in the preceding 30 days. Another sample of 160 high school students in New York City showed that 15 of them felt that way. At the 0.05 level of significance, test the claim that high school students in Washington, D.C. are more likely to feel too unsafe to go to school than are high school students in New York City. (Based on the Morbidity and Mortality Weekly Report by the Centers for Disease Control and Prevention.)

28. In a random sample of 1629 high school females there were 406 who had seriously considered attempting suicide in the past 12 months. In a similar survey, 243 of 1774 high school males had seriously considered attempting suicide in the past 12 months. At the 0.01 level, test the claim that high school females are more likely to have seriously considered attempting suicide than high school males. (Based on the Morbidity and Mortality Weekly Report by the Centers for Disease Control and Prevention.)

29. A random sample of 1434 male high school students showed that 82 had attempted suicide at least once in the 12 months preceding the survey, while 159 out of 1457 randomly selected high school females had done so. At the 0.05 level of significance, test the claim that male and female high school students are equally likely to attempt suicide. (Based on the Morbidity and Mortality Weekly Report by the Centers for Disease Control and Prevention.)

30. A random sample of 480 ninth-grade students showed that 132 had smoked at least once in the last 30 days, while another random sample showed that 133 of 311 twelfth-grade students had smoked at least once in the last 30 days. At the 0.01 level of significance, test the claim that ninth-grade students are less likely to have smoked in the last 30 days than twelfth-grade students. (Based on the Morbidity and Mortality Weekly Report by the Centers for Disease Control and Prevention.)

31. A random sample of 267 male high school students that are smokers contained 175 who were not asked to show proof of age the last time they purchased cigarettes. A sample of 112 high school females that smoke cigarettes revealed that 85 of them were not asked to show proof of age the last time they purchased cigarettes. At the 0.05 level of significance, test the claim that male high school students are less likely to be asked to provide proof of age when purchasing cigarettes than are female high school students. (Based on the Morbidity and Mortality Weekly Report by the Centers for Disease Control and Prevention.)

32. A random sample of 1197 male high school students showed that 418 of them had consumed at least 5 drinks at least once in the 30 days preceding the survey. Another sample showed that 495 of 1760 female high school students had consumed at least 5 drinks at least once in the 30 days preceding the survey. At the 0.05 level of significance, test the claim that male high school students are more likely to have consumed at least 5 drinks at least once in the past 30 days than are female high school students. (Based on the Morbidity and Mortality Weekly Report by the Centers for Disease Control and Prevention.)

33. A random of sample of 50 female inner-city high school students showed that 13 had used illicit drugs within the past year, while a random sample of 40 female high school students who lived in the suburbs showed that 18 had used illicit drugs within the past year.

(a) Construct a 95% confidence interval between the proportion of female inner-city high school students that have used illicit drugs within the past year and the proportion of female high school students from the suburbs that have used illicit drugs within the last year.

(b) At the 0.05 level of significance, test the claim that the proportion of female inner-city high school students that have used illicit drugs within the past year is the same as the proportion of female high school students from the suburbs that have used illicit drugs within the last year.

(c) Compare the conclusion of your hypothesis test to your confidence interval. Was this conclusion to be expected, based on your confidence interval? Explain your answer.

34. A random sample of 32 male inner-city high school students showed that 11 had used illicit drugs within the past year, while a random sample of 51 male high school students who lived in the suburbs showed that 30 had used illicit drugs within the past year.

(a) Construct a 95% confidence interval for the difference between the proportion of male inner-city high school students that have used illicit drugs within the past year and the proportion of male high school students from the suburbs that have used illicit drugs within the last year.

(b) At the 0.05 level of significance, test the claim that the proportion of male inner-city high school students that have used illicit drugs within the past year is the same as the proportion of male high school students from the suburbs that have used illicit drugs within the last year.

(c) Compare the conclusion of your hypothesis test to your confidence interval. Was this conclusion to be expected, based on your confidence interval? Explain your answer.

35. A study led by James Haddow of the Foundation for Blood Research in Scarborough, Maine, examined the affects of thyroid disorders in expecting mothers on their children. The researchers compared a group of 48 women with low thyroid levels that got no treatment to a control group of 25,168 women with normal levels. The researchers located the children when they were between 7 and 9 years old and gave them an IQ test. Nine of the children born to the mothers with low thyroid levels had an IQ that was below 85, while 1,258 of the other children had an IQ that was below 85.

(a) Construct a 99% confidence interval for the difference between the proportion of children in the low thyroid group that had an IQ below 85 and the proportion of children in the other group that had an IQ below 85.

(b) At the 0.01 level of significance, test the claim that the proportion of children in the low thyroid group that had an IQ below 85 is the same as the proportion of children in the other group that had an IQ below 85.

36. In a random sample of 2000 kids ages 12–17, 71% reported having an excellent or very good relationship with their mothers. Only 58% reported having an excellent or very good relationship with their fathers.

(a) Construct a 99% confidence interval for the difference between the proportion of all kids ages 12–17 that have an excellent or very good relationship with their mothers and the proportion of all kids ages 12–17 that have an excellent or very good relationship with their fathers.

(b) At the 0.01 level of significance, test the claim that the proportion of all kids ages 12–17 that have an excellent or very good relationship with their mothers is greater than the proportion of all kids ages 12–17 that have an excellent or very good relationship with their fathers.

[*Source:* 1999 Teen/Parent Drug Survey, sponsored by the National Center on Addiction and Substance Abuse (CASA) at Columbia University.]

? What Is Wrong with This Picture?

The following statistical project contains a major error. Potential errors could be the choice of the wrong hypothesis test or that the assumptions required for this test are not met. Find the error, and explain what could be done to correct it.

A student believes that more males than females take physics classes. He randomly selects five physics classes. In those classes there are 80 male students and 35 female students. He uses these results to test the claim that the proportion of males in physics classes is higher than the proportion of females in physics classes, using the hypothesis test introduced in this section.

What is wrong with this picture?

MINI PROJECT

Randomly sample at least 30 male and 30 female students at your school, and ask them if they approve of the job that the president of the United States is doing. Use these sample data to test the claim that the percentage of male students who approve of the job that the president is doing is the same as the percentage of female students that approve, at the 0.05 level of signifance. In addition to your complete hypothesis test, include

- Your raw data
- A pie chart showing the percentage of males who approve of the job that the president is doing, and a second pie chart showing the percentage of females who approve of the job that the president is doing

Finally, answer the following questions.

- How did you obtain your sample?
- Which type of sampling did you use?
- Was the sample truly random?
- What potential biases may show up in your sample?

Choosing the Appropriate Tool

If we wish to test a claim comparing two population means, we must first determine whether our samples are independent or dependent. If the samples are independent and both sample sizes are at least 30, then we use the z-distribution. If the two samples are independent, one or both sample sizes are below 30, and both samples appear to be drawn from populations that are normally distributed, then we use the t-distribution. If the samples are independent with at least one sample size below 30, but at least one of the populations appears to not be normally distributed, then we should use methods that will be introduced in Chapter 11.

If the two samples are dependent, we will apply the paired difference test as long as the two samples appear to be drawn from populations that are normally distributed. If the two populations are not both normally distributed, then we should use methods that will be introduced in Chapter 11.

Finally, if we wish to test a claim comparing two population proportions, we use the z-distribution if the number of successes and failures for both samples is at least 5.

(continues)

Choosing the Appropriate Tool (continued)

Test	Section	Test Statistic	Conditions
Two population means (large samples)	8.1	$z = \dfrac{\bar{x}_1 - \bar{x}_2}{\sqrt{\dfrac{s_1^2}{n_1} + \dfrac{s_2^2}{n_2}}}$	$n_1 \geq 30$ $n_2 \geq 30$
Two population means (small samples)	8.2	$t = \dfrac{\bar{x}_1 - \bar{x}_2}{\sqrt{\dfrac{s_1^2}{n_1} + \dfrac{s_2^2}{n_2}}}$ degrees of freedom = smaller of $(n_1 - 1)$ and $(n_2 - 1)$	• $n_1 < 30$ or $n_2 < 30$ • Both samples come from populations that are normally distributed
Paired difference	8.3	$t = \dfrac{\bar{d} - \mu_d}{s_d / \sqrt{n}}$ degrees of freedom = $n - 1$	• There are two dependent samples • Both samples must be taken from populations that appear to follow a normal distribution
Two population proportions	8.4	$z = \dfrac{p_1 - p_2}{\sqrt{\bar{p}(1 - \bar{p})\left(\dfrac{n_1 + n_2}{n_1 \cdot n_2}\right)}}$, where $\bar{p} = \dfrac{x_1 + x_2}{n_1 + n_2}$	• There are two independent samples • $n \cdot \pi \geq 5$ • $n(1 - \pi) \geq 5$

Other Hypothesis Tests

It may seem that tests in a chapter titled "other" hypothesis tests are not important, but they are. In Section 9.1, we find the test for two population variances. If we can show that two population variances are equal, there is an alternative to the test for two population means in Section 8.2 that can be applied. This test for two population variances introduces us to the *F*-distribution, and lays the groundwork for one of the more important procedures covered in this book—ANOVA.

Section 9.2 covers ANOVA, which stands for analysis of variance. We use ANOVA to test whether three or more population means are all equal. For example, we can test the claim that the mean yield of cotton per acre is the same for three different fertilizers.

We are introduced to the χ^2 (chi-square) distribution in Section 9.3, using the goodness of fit test. Here we measure how a population compares to a claimed categorical distribution. For instance, we could test a claim that 30% of all plain M&Ms are brown, 20% are red, 20% are yellow, 10% are orange, 10% are green, and 10% are blue. We take a sample and compare the proportion of each color to the claimed proportions.

Finally, Section 9.4 covers contingency tables and the tests for independence and homogeneity. We will be testing to determine whether two categorical variables (such as burger preference and gender) are related or not.

SECTION 9.1
Hypothesis Test for Two Population Variances

In this section we introduce a hypothesis test for two population variances. This test will use a new distribution—the F-distribution. We begin by examining this distribution, and seeing how it differs from the standard normal distribution (z) and Student's t-distribution.

To start with, although the F-distribution is continuous, it is not symmetric like the other two distributions; it is right skewed. Also, it is impossible for an F-value to be negative. Here is a graph of a particular F-distribution.

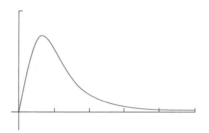

Another major difference is that the F-distribution has two different degrees of freedom associated with it, one for the numerator and one for the denominator. The final difference is in the decision rule. We will be using a chart that is set up for right-tailed decision rules only. There will be no left-tailed decision rules, and no two-tailed decision rules.

We will begin examining the hypothesis test, and will learn more about the chart and decision rules as we go. When we are performing a test that compares two population variances, we assume that the two populations are independent of one another and that they are both normally distributed. We begin by taking two random samples and calculating their standard deviations. The population that produces the sample with the *larger* sample standard deviation will be labeled as population 1, and the population that produces the *smaller* sample standard deviation will be labeled as population 2. This is a major difference from our two-sample tests from the previous chapter, where we were able to choose which population was going to be population 1 and which was going to be population 2.

Once we have established which population will be population 1, we can look at the test statistic for this hypothesis test.

$$F = \frac{s_1^2}{s_2^2}$$

Degrees of freedom for the numerator: $n_1 - 1$
Degrees of freedom for the denominator: $n_2 - 1$
s_1: Larger sample standard deviation
s_2: Smaller sample standard deviation
n_1: Size of sample 1
n_2: Size of sample 2

In the numerator we see the first sample's standard deviation squared, which is the first sample's variance. The denominator is the second sample's variance. If the two population variances are equal, then the sample standard deviations should be close to each other, which should produce a test statistic that is close to 1. However, if the population variances are not equal, then the numerator of the test statistic should be larger than the denominator. This will produce a test statistic that is larger than 1. Thus, a large value of the test statistic indicates that the two population variances are not equal.

By always putting the larger standard deviation in the numerator, we do not have to worry about small values of the test statistic (less than 1) and therefore do not have to worry about the left tail of the distribution. All of our decision rules will be written as right-tailed decision rules; that is, "reject H_0 if $F > \#$." This produces a bit of a dilemma when the null hypothesis is that the two population variances are equal ($\sigma_1^2 = \sigma_2^2$) and the alternate hypothesis is that they are not equal ($\sigma_1^2 \neq \sigma_2^2$). This is clearly a two-tailed test, and only half of the level of significance α belongs in the right tail where our decision rule is based. So we have to divide the level of significance by 2 to decide which of our F-distribution charts to use. In the back of the text there are two F-distribution charts (Table E), one for the 0.05 level of significance and another for the 0.01 level of significance. These correspond to two-tailed tests at the 0.10 and 0.02 levels of significance, respectively. We begin with a two-tailed test to explain this idea further.

EXAMPLE 9.1

A researcher is interested in the ages at which women and men first get married. A random sample of 18 women revealed that their mean age at their first marriage was 23.5 years, with a standard deviation of 8.52 years. A random sample of 17 men revealed that their mean age at their first marriage was 26.6 years, with a standard deviation of 14.57 years. At the 0.01 level of significance, test the claim that the variance of women's ages at their first marriage is the same as the variance of men's ages.

We begin by determining which population will be population 1 and which will be population 2. Since the standard deviation of the men's ages is greater than the standard deviation of the women's ages, we let men who have married be population 1.

Step 1

Population 1: men; population 2: women
Claim in words: The variance of women's ages at their first marriage is the same as the variance of men's ages. (The variance of men's ages at their first marriage is the same as the variance of women's ages.)
Claim: $\sigma_1^2 = \sigma_2^2$
Complement: $\sigma_1^2 \neq \sigma_2^2$

$H_0: \sigma_1^2 = \sigma_2^2$
$H_A: \sigma_1^2 \neq \sigma_2^2$

Step 2

Level of significance: $\alpha = 0.10$. Since this is a two-tailed test, we must divide α in half to find the area that belongs in the right tail and determine which chart to use. In this case we will use the $0.10/2 = 0.05$ chart.

Step 3

Test statistic: $F = \dfrac{s_1^2}{s_2^2}$, $d.f.$ numerator: $n_1 - 1$, $d.f.$ denominator: $n_2 - 1$

Step 4

To determine the decision rule, we must learn how to read the critical value from the F-distribution chart. Since there were 17 men in the first sample (remember that men are population 1), there are 16 degrees of freedom in the numerator. We look across the top of the chart for 16 in the area labeled "Degrees of Freedom for the Numerator." However, 16 is not listed in the chart, so our choices are 15 and 20 degrees of freedom. We will postpone this choice until after we look for the correct number of degrees of freedom for the denominator.

Since there were 18 women in the second sample, there are 17 degrees of freedom for the denominator. We look down the left side of the chart in the area labeled "Degrees of Freedom for the Denominator" until we find 17. We now scan to the right until we reach the columns for 15 and 20 degrees of freedom for the numerator. Our choice for the critical value is between 2.23 (20 degrees of freedom for the numerator) and 2.31 (15 degrees of freedom for the numerator). We choose the larger value, 2.31, for our critical value because it makes it harder for us to reject the null hypothesis. (A calculator or computer could be used to show that for 16 degrees of freedom for the numerator, the appropriate critical value is 2.29.)

In general, when you are unable to find the correct number of degrees of freedom for the numerator and/or denominator, choose the largest critical value among your nearest choices.

We are now able to state our decision rule.

Decision rule: Reject H_0 if $F > 2.31$.

Step 5

$$F = \frac{s_1^2}{s_2^2}$$

$$= \frac{14.57^2}{8.52^2}$$

$$= 2.92$$

Decision: Reject H_0.

Conclusion: There is sufficient sample evidence to reject the claim that the variance of women's ages at their first marriage is the same as the variance of men's ages.

p-value

Using the chart, the best we can say is that the p-value is less than 0.05 because we rejected the null hypothesis. However, a calculator or computer tells us that the p-value is 0.0373. ■

A note of caution: The information provided in the preceding example is exactly what we would see for a hypothesis test involving two population means. Be sure that you read each problem carefully so that you do not perform an inappropriate test.

Now we will look at a one-tailed test that shows the difficulty of setting up the claim that we are testing.

EXAMPLE 9.2

A math instructor claims that students' scores on test 1 are more consistent (have lower variance) than their scores on test 2. Thirty-two randomly selected students had a mean score of 75.91 with a standard deviation of 14.71 on test 1. Thirty-one randomly selected students had a mean score of 66.97 with a standard deviation of 21.43 on test 2. Test the claim that test 1 scores have less variance than test 2 scores at the 0.05 level of significance.

Population 1 will be test 2 scores, because the standard deviation is larger for the sample of test 2 scores.

Step 1

Population 1: test 2; population 2: test 1
Claim in words: Test 1 scores have less variance than test 2 scores.
   We need to "turn the claim around" so that test 2 comes first. We will use "Test 2 scores have more variance than test 1 scores" to formulate our claim in symbols.
Claim: $\sigma_1^2 > \sigma_2^2$
Complement: $\sigma_1^2 \leq \sigma_2^2$
H_0: $\sigma_1^2 \leq \sigma_2^2$
H_A: $\sigma_1^2 > \sigma_2^2$

Step 2

Level of significance: $\alpha = 0.05$. Since this is a one-tailed test, we will use the 0.05 chart.

Step 3

Test statistic: $F = \dfrac{s_1^2}{s_2^2}$, $d.f.$ numerator: $n_1 - 1$, $d.f.$ denominator: $n_2 - 1$

Step 4

There are 30 degrees of freedom for the numerator, and 31 degrees of freedom for the denominator. This time, we are missing the degrees of freedom for the denominator in our chart. Our choices are 30 and 40 degrees of freedom, which translates to a choice between 1.74 and 1.84 for our critical value. We choose the larger, which is 1.84. (Using a calculator or computer, the appropriate critical value is 1.83.)

Decision rule: Reject H_0 if $F > 1.84$.

Step 5

$$F = \frac{s_1^2}{s_2^2}$$

$$= \frac{21.43^2}{14.71^2}$$

$$= 2.12$$

Decision: Reject H_0.

Conclusion: There is sufficient sample evidence to support the claim that test 1 scores have less variance than test 2 scores.

p-value

Using the chart, the best we can say is that the *p*-value is less than 0.05 because we rejected the null hypothesis. However, a calculator or computer tells us that the *p*-value is 0.0205. ∎

Recall from Section 8.2 that there is another hypothesis test that can be used for two population means from small samples if it can be shown that the two populations' variances are equal. The hypothesis test learned in this section can be used to test whether the variances are equal. This alternate test for two means will be covered in the "Extra" section following the exercises.

MICROSOFT EXCEL   Hypothesis Test for Two Population Variances

We will learn how to use Microsoft Excel to perform a hypothesis test for two population variances. We will calculate the test statistic from the two sets of sample data, determine what the correct critical value is for our test, and find the *p*-value associated with our test.

Here are the scores of 24 randomly selected games from two different Professional Bowlers Association tournaments. The first is the Don Carter PBA Classic, held in January 1999. The second is the Greater Detroit Open, held in October 1999. Test the claim that the two bowling alleys where these tournaments were held produce scores with equal variances at the 0.10 level of significance.

Don Carter PBA Classic

248	289	238	181	255	187	244	214
215	227	188	244	233	195	198	238
227	203	225	235	248	194	258	234

Greater Detroit Open

217	224	192	160	300	269	269	257
246	224	209	279	248	237	252	278
258	247	257	236	235	257	268	186

Begin by typing the scores from the Don Carter PBA Classic in column **A,** in cells **A1** through **A24.** Do the same thing for the scores from the Greater Detroit Open in column **B,** in cells **B1** through **B24.** We now calculate the standard deviations for the two samples. In cell **C1,** type

=STDEV(A1:A24)

This will calculate the standard deviation of the scores from the Don Carter PBA Classic. To do the same for the scores from the Greater Detroit Open, type the following in cell **C2:**

    =STDEV(B1:B24)

To two decimal places we see that the standard deviation of the scores from the Don Carter PBA Classic is 26.77 pins and the standard deviation of the scores from the Greater Detroit Open is 32.37 pins. Since the standard deviation of the Greater Detroit Open scores is larger, the scores from the bowling alley where the Greater Detroit Open is contested will be population 1 and the bowling alley where the Don Carter PBA Classic is held will be population 2.

Now we can calculate the test statistic. In cell **D1,** type Test Statistic. For the calculation of the test statistic we need to divide the square of the larger standard deviation by the square of the smaller standard deviation. In cell **E1,** type the following:

    =(C2*C2)/(C1*C1)

The result should be 1.46 to two decimal places.

To find the critical value, we use the Excel function

    =FINV(alpha, df numerator, df denominator)

The first value in the parentheses, labeled *alpha,* is the right-tailed area. If a test is a two-tailed test, as this one is, we must divide the level of significance by 2 to know what value to enter here. For this example, type the following in any empty cell.

    FINV(0.05, 23, 23)

Excel tells us that the critical value is 2.01. Our decision rule will be to reject the null hypothesis if $F > 2.01$.

To find the *p*-value, we use the Excel function

    =FDIST(test statistic, df numerator, df denominator)

For this example the value of our test statistic was 1.46, so type the following in any empty cell to calculate the *p*-value.

    =FDIST(1.46, 23, 23)

Excel returns a value of 0.1854. This is the area in the right tail. Since this is a two-tailed test, we must multiply this value by 2 to account for the other tail. Thus, the *p*-value is 0.3708.

We now have everything that we need to write the hypothesis test.

Step 1

Population 1: Greater Detroit Open; population 2: Don Carter PBA Classic
Claim in words: The two bowling alleys produce scores with equal variances.
Claim: $\sigma_1^2 = \sigma_2^2$
Complement: $\sigma_1^2 \neq \sigma_2^2$
$H_0: \sigma_1^2 = \sigma_2^2$
$H_A: \sigma_1^2 \neq \sigma_2^2$

Step 2

Level of significance: $\alpha = 0.10$. Since this is a two-tailed test, we must divide α in half to find the area that belongs in the right tail, which in this case is 0.05.

Step 3

Test statistic: $F = \dfrac{s_1^2}{s_2^2}$, $d.f.$ numerator: $n_1 - 1$, $d.f.$ denominator: $n_2 - 1$

Step 4

Decision rule: Reject H_0 if $F > 2.01$ (calculated by Excel).

Step 5

$F = 1.46$ (calculated by Excel)

Decision: Fail to reject H_0.

Conclusion: There is not sufficient sample evidence to reject the claim that the two bowling alleys produce scores with equal variances.

p-value

p-value = 0.3708 (calculated by Excel) ∎

 Hypothesis Test for Two Population Variances

The TI-83 can be a valuable tool to help us perform a hypothesis test for two population variances. We will use an example to walk us through the procedure.

 Here are the scores of 24 randomly selected games from two different Professional Bowlers Association tournaments. The first is the Don Carter PBA Classic, held in January 1999. The second is the Greater Detroit Open, held in October 1999. Test the claim that the two bowling alleys where the tournaments are held produce scores with equal variances at the 0.10 level of significance.

Don Carter PBA Classic

248	289	238	181	255	187	244	214
215	227	188	244	233	195	198	238
227	203	225	235	248	194	258	234

Greater Detroit Open

217	224	192	160	300	269	269	257
246	224	209	279	248	237	252	278
258	247	257	236	235	257	268	186

Begin by entering the scores from the Don Carter PBA Classic in list L_1. Do the same thing for the scores from the Greater Detroit Open in list L_2. Before progressing, we need to determine which bowling alley will be population 1 and which will be population 2. This is done by determining which sample has the larger standard deviation. Recall that we use the STAT CALC menu to access **1-Var Stats,** which could give us the standard deviation of our two lists. Verify for yourself that the standard deviation for the Don Carter PBA Classic scores is 26.77 and the standard deviation for the Greater Detroit Open scores is 32.37. Population 1 will be the scores from the lanes that hosted the Greater Detroit Open (list L_2), and population 2 will be the alley that hosted the Don Carter PBA Classic (list L_1).

We can begin the test by going to the STAT TESTS menu by pressing the (STAT) key, and then moving to the right by using the (→) key. We want to use option **D: 2-SampFTest.** Highlight that option and press (ENTER).

When we are taken back to the main screen, be sure to highlight **Data** on the first line. Enter the list containing data from population 1 (L_2) next to **List1** and the other list (L_1) next to **List2.** Leave **Freq1** and **Freq2** as 1. On the next to last line we need to highlight the form of our alternate hypothesis. The test we are working on is a two-tailed test, so we highlight the first option ≠σ**2.** On the last line highlight **Calculate,** and then press (ENTER).

The output screen shows the value of the test statistic F (1.46), the p-value (0.3691), and sample information (mean, standard deviation, and size) for each sample. Here is what you should see, displayed over two screens.

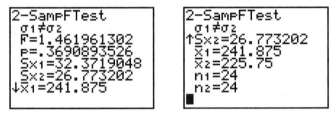

We can now write our hypothesis test.

Step 1

Population 1: Greater Detroit Open; population 2: Don Carter PBA Classic
Claim in words: The two bowling alleys produce scores with equal variances.
Claim: $\sigma_1^2 = \sigma_2^2$
Complement: $\sigma_1^2 \neq \sigma_2^2$
H_0: $\sigma_1^2 = \sigma_2^2$
H_A $\sigma_1^2 \neq \sigma_2^2$

Step 2

Level of significance: $\alpha = 0.10$. Since this is a two-tailed test, we must divide α in
half to find the area that belongs in the right tail, which in this case is 0.05.

Step 3

Test statistic: $F = \dfrac{s_1^2}{s_2^2}$, $d.f.$ numerator: $n_1 - 1$, $d.f.$ denominator: $n_2 - 1$

Step 4

Decision rule: Reject H_0 if $F > 2.05$ (from our chart).

Step 5

$F = 1.46$ (calculated by TI-83)

Decision: Fail to reject H_0.

Conclusion: There is not sufficient sample evidence to reject the claim that the two
bowling alleys produce scores with equal variances.

***p*-value**

p-value = 0.3691 (calculated by TI-83) ∎

EXERCISES 9.1

State the decision rule for the given circumstances.

1. Right-tailed test, $\alpha = 0.01$, d.f. numerator = 8, d.f. denominator = 24
2. Right-tailed test, $\alpha = 0.01$, d.f. numerator = 12, d.f. denominator = 11
3. Right-tailed test, $\alpha = 0.05$, d.f. numerator = 24, d.f. denominator = 21
4. Right-tailed test, $\alpha = 0.05$, d.f. numerator = 9, d.f. denominator = 30
5. Two-tailed test, $\alpha = 0.02$, d.f. numerator = 15, d.f. denominator = 19
6. Two-tailed test, $\alpha = 0.10$, d.f. numerator = 3, d.f. denominator = 17
7. Right-tailed test, $\alpha = 0.05$, d.f. numerator = 23, d.f. denominator = 23
8. Right-tailed test, $\alpha = 0.01$, d.f. numerator = 7, d.f. denominator = 28
9. Two-tailed test, $\alpha = 0.02$, d.f. numerator = 17, d.f. denominator = 45
10. Two-tailed test, $\alpha = 0.10$, d.f. numerator = 39, d.f. denominator = 175
11. A random sample of 88 male deaths in a county over a 1-year period had a mean age at death of 68.9 years, with a standard deviation of 16.71 years. A random sample of 84 female deaths in the same county over the same 1-year period had a mean age at death of 77.1 years, with a standard deviation of 15.65 years. At the 0.02 level of significance, test the claim that the age at death in this county for males has the same variance as the age at death for females in this county.
12. A random sample of 45 male college students and 50 female college students was taken, and each student was asked to write down their height. The male students had a mean height of 70.4 inches, with a standard deviation of 2.44 inches. The female students had a mean height of 64.3 inches, with a standard deviation of 2.86 inches. At the 0.02 level, test the claim that the heights of male college students have the same variance as the heights of female college students.
13. A student conducted a study of 103 men and 104 women that had been married at least one time. The men had a mean age of 25.4 years at their first marriage, with a standard deviation of 5.19 years. The women had a mean age of 22.36 years at their first marriage, with a standard deviation of 4.63 years. At the 0.10 level of significance, test the claim that the ages of men at their first marriage have the same variance as the ages of women at their first marriage.
14. A random sample of 56 departing flights at an airport over a 3-month period had a mean wait of 14.2 minutes between boarding and takeoff, with a standard deviation of 4.53 minutes. At the same airport, a random sample of 81 incoming flights over the same 3-month period had a mean wait of 17.5 minutes between the time that the plane arrived at the gate and the time that the baggage reached the baggage claim area, with a standard deviation of 9.87 minutes. At the 0.10 level of significance, test the claim that at this airport the variance of waiting times for takeoff is the same as the variance of waiting times for baggage.
15. A child development student is interested in the amount of time that children watch television. A study of 45 children ages 2–5 at the college's day care center revealed that they watched 135.4 minutes of television per day, with a standard deviation of 82.1 minutes. A study of 120 elementary school children ages 6–12 revealed that they watched a mean of 170.6 minutes of television per day, with a standard deviation of 106.9 minutes. At the 0.05 level of significance, test the claim that the variance in viewing times for children ages 2–5 is less than the variance in viewing times for children ages 6–12.
16. A national study on fourth graders was conducted. A sample of 400 fourth graders had a mean of 32.6 minutes of homework per night, with a standard deviation of 25.14 minutes. A later sample of 300 fourth graders watched a mean of 201.6 minutes of television per day, with a standard deviation of 125.63 minutes. At the 0.01 level of significance, test the claim that the variance of homework times is less than the variance of television viewing times.
17. An instructor believes that some students have a hard time with the material in Unit 2, while it comes easily to others. This produces a wide range of scores on a Unit 2 exam, as compared to the Unit 1 exam where few students have real difficulty. A random sample of 21 students produced a mean score on the Unit 1 exam of 78.3 points, with a standard deviation of 14.47 points. A random sample of 38

students produced a mean score on the Unit 1 exam of 67.4 points, with a standard deviation of 21.72 points. At the 0.05 level of significance, test the claim that Unit 2 exam scores have greater variance than Unit 1 exam scores.

18. A random sample of 25 day students at a community college had a mean age of 21.5 years, with a standard deviation of 4.44 years. A random sample of 25 evening students at the same community college had a mean age of 29.8 years, with a standard deviation of 9.88 years. At the 0.01 level of significance, test the claim that the variance in ages of day students is less than the variance in ages of evening students.

19. A random sample of 25 women with Turner's syndrome had a mean height of 148.7 cm, with a standard deviation of 2.81 cm. A random sample of 155 women from the general population had a mean height of 164.2 cm, with a standard deviation of 5.35 cm. At the 0.01 level of significance, test the claim that the variance of heights of women with Turner's syndrome is less than the variance of the heights of the general female population. (*Source: European Journal of Pediatrics.*)

20. A random sample of 48 football players had a mean weight of 246.6 pounds, with a standard deviation of 40.09 pounds. A random sample of 25 baseball players had a mean weight of 202.9 pounds, with a standard deviation of 15.81 pounds. At the 0.01 level of significance, test the claim that the weights of football players have greater variance than the weights of baseball players.

21. In PBA bowling tournaments, there are two different rounds—the qualifying round and the match play round. In the qualifying round, bowlers try for the highest score total in order to determine their rankings for later rounds. In the match play round, scores are not as important—only wins and losses are counted. Thirty-six qualifying scores and 48 match play scores were randomly selected, as follows.

Qualifying

248	207	235	239	221	239	257	213	157
234	235	184	235	246	216	279	248	211
269	228	247	239	226	248	249	289	206
259	221	299	259	280	248	255	200	242

Match Play

215	244	227	227	257	208	267	248	214
290	249	230	278	257	166	245	279	190
256	237	226	217	243	205	225	238	247
245	209	185	239	202	255	237	230	257
265	255	188	258	220	247	238	223	269
215	247	222						

At the 0.10 level of significance, test the claim that both formats produce scores with equal variance.

22. The PSAT mathematics scores of 50 high school juniors and 35 high school sophomores were selected at random:

Juniors

24	36	72	48	28	50	44	46	38	72
57	69	48	64	39	48	59	51	59	48
42	37	52	72	47	55	38	58	40	60
60	59	38	42	49	40	31	61	42	47
56	37	62	59	59	44	69	41	60	33

Sophomores

29	52	42	68	38	60	58	49	45	55
30	60	52	37	42	49	43	36	38	52
58	61	31	61	46	58	49	36	59	49
48	47	37	43	31					

At the 0.10 level of significance, test the claim that the variance of scores for juniors on this test is the same as the variance of scores for sophomores.

23. PSAT verbal scores were obtained for a random sample of 40 high school juniors and PSAT mathematics scores were obtained for a different sample of 40 high school juniors as follows:

Verbal

28	41	49	41	40	41	48	47	48	37
56	41	36	49	40	47	44	72	42	50
64	47	56	54	58	71	38	63	40	50
42	53	68	43	38	39	30	70	46	60

Mathematics

34	52	46	56	29	51	49	41	66	42
50	65	50	48	48	45	48	44	67	65
45	44	52	54	46	30	32	58	39	49
52	72	46	59	29	39	35	53	51	41

At the 0.02 level of significance, test the claim that the variance in scores for the two exams is the same.

24. A professional bowler claims that his game is better suited to synthetic lanes than to natural lanes. Here are the scores of 42 randomly selected games on synthetic lanes and the scores of 42 randomly selected games on natural lanes.

Synthetic

256	269	279	245	290	214	210	235	238
213	279	236	218	222	247	188	249	214
196	222	237	177	246	227	279	267	236
237	244	217	258	264	266	279	238	257
238	224	226	267	256	228			

Natural

248	221	246	255	289	244	223	222	255
192	224	203	204	179	244	234	225	218
201	219	214	233	204	212	217	207	203
244	163	215	182	192	203	198	210	225
207	236	224	264	228	164			

At the 0.02 level of significance, test the claim that the bowler's scores on the two types of lanes have different variances.

25. A math instructor is not sure that collecting homework from his class on a daily basis is beneficial to his students. He conducts an experiment by collecting homework from one algebra class of 37 students that he is teaching, but not collecting homework from another algebra class of 32 students. Here are the test scores for both classes.

Collected Homework

53	88	68	64	70	67	57	88	83	79
78	55	84	78	78	62	82	76	54	74
85	76	74	71	84	87	75	73	63	64
73	48	82	87	78	78	75			

Did Not Collect Homework

60	79	65	74	76	62	71	72	76	77
62	73	66	80	74	74	81	68	61	68
76	76	78	70	79	88	68	82	63	68
76	87								

At the 0.05 level of significance, test the claim that collecting homework produces scores with greater variance than not collecting homework.

26. Here are the number of runs scored in 62 randomly selected American League games and 71 randomly selected National League games.

American

5	4	12	16	11	10	11	6	12	4
20	11	7	6	8	13	23	16	3	9
4	13	9	17	8	16	4	12	8	9
4	14	7	14	4	9	15	7	14	7
8	9	11	4	8	13	9	11	10	6
6	7	9	17	19	25	17	12	9	6
11	1								

National

11	11	6	1	15	19	9	11	12	13
6	27	6	12	11	6	14	13	6	6
10	13	9	16	3	14	9	22	11	9
6	4	6	9	13	13	11	3	8	12
3	13	9	14	5	5	6	13	7	12
3	7	9	22	10	3	10	6	6	6
7	6	11	19	13	10	5	4	10	14
4									

At the 0.10 level of significance, test the claim that the variance of runs scored in American League games is the same as the variance of runs scored in National League games.

27. Here are the prices for a gallon of unleaded regular gas at 48 randomly selected gas stations in the Los Angeles area and 39 randomly selected gas stations in the San Francisco Bay area from the week before Christmas in 1999.

Los Angeles

$ 0.949	$ 1.019	$ 0.989	$ 1.039	$ 0.969	$ 0.999
$ 0.979	$ 1.019	$ 1.039	$ 1.009	$ 0.939	$ 0.989
$ 1.059	$ 0.959	$ 1.049	$ 1.029	$ 1.039	$ 0.999
$ 1.099	$ 1.019	$ 1.019	$ 1.029	$ 1.069	$ 0.979
$ 1.039	$ 1.089	$ 1.049	$ 0.989	$ 1.039	$ 1.019
$ 1.019	$ 0.999	$ 1.009	$ 0.949	$ 1.019	$ 1.069
$ 1.069	$ 0.979	$ 0.999	$ 1.029	$ 1.079	$ 0.979
$ 0.999	$ 1.039	$ 0.969	$ 1.059	$ 1.049	$ 0.979

San Francisco

$ 1.049	$ 1.169	$ 1.029	$ 1.199	$ 1.059	$ 1.099
$ 1.139	$ 1.109	$ 1.129	$ 1.139	$ 1.139	$ 1.189
$ 1.089	$ 1.079	$ 1.149	$ 1.149	$ 1.059	$ 1.179
$ 1.169	$ 1.069	$ 1.139	$ 1.149	$ 1.109	$ 1.099
$ 1.099	$ 1.199	$ 1.129	$ 1.039	$ 1.109	$ 1.089
$ 1.159	$ 1.069	$ 1.089	$ 1.129	$ 1.239	$ 1.149
$ 1.169	$ 1.119	$ 1.169			

At the 0.05 level of significance, test the claim that the variance of prices for a gallon of unleaded regular gas in the Los Angeles area is lower than the variance of prices for a gallon of unleaded regular gas in the San Francisco area.

28. An instructor gives one version of a test to his 7:30 A.M. statistics class and another version to his 6 P.M. class. Here are their scores.

Morning

85	73	83	100	86	70	98	80	97
66	41	72	83	89	57	64	76	91
77	68	81	63					

Evening

74	80	92	73	82	96	82	84	76
98	94	63	94	91	46	58	54	37
86	79	55	73	64				

At the 0.10 level of significance, test the claim that the two versions produce scores that have equal variances.

? What Is Wrong with This Picture?

The following statistical project contains a major error. Potential errors could be the choice of the wrong hypothesis test or that the assumptions required for this test are not met. Find the error, and explain what could be done to correct it.

A student reads that the variance of IQ test scores is 225 (the standard deviation is 15 points). She believes that the IQs of college students have greater variance than those of the general population. She randomly selects 25 college students and gives them an IQ test. She uses these results to test the claim that the IQ scores of college students have greater variance than the IQ scores of the general population, using the hypothesis test introduced in this section.

What is wrong with this picture?

MINI PROJECT

Randomly sample at least 50 students at your school, and ask them how many hours they study per week. Use these sample data to test the claim that the variance in the amount of time that males study is equal to the variance in the amount of time that females study, at the 0.10 level of significance. In addition to your hypothesis test, include

- Your raw data
- A histogram for both the male and female study times

Finally, answer the following questions.

- How did you obtain your sample?
- Which type of sampling did you use?
- Was your sample truly random?
- What potential biases may show up in your sample?

SECTION 9.1 EXTRA • Hypothesis Test for Two Population Means (Small Samples), Assuming Equal Population Variances

Recall from Section 8.2 that there is another test statistic that can be used to test two population means based on small samples if the variances of those populations are equal. To use this test, we must first test the claim that the two variances are equal. As a rule, we will perform this test at the 0.10 level of significance. If the null hypothesis is not rejected, we can then go on to perform this new test.

The test statistic and the number of degrees of freedom are the only differences between this test and the one covered in Section 8.2. Here is the test statistic.

$$t = \frac{\bar{x}_1 - \bar{x}_2}{\sqrt{\dfrac{(n_1 - 1) \cdot s_1^2 + (n_2 - 1) \cdot s_2^2}{n_1 + n_2 - 2} \cdot \dfrac{n_1 + n_2}{n_1 \cdot n_2}}}, \quad \text{degrees of freedom: } n_1 + n_2 - 2$$

The quantity

$$\frac{(n_1 - 1) \cdot s_1^2 + (n_2 - 1) \cdot s_2^2}{n_1 + n_2 - 2}$$

in the denominator is the pooled variance, and its square root is called the pooled standard deviation. It essentially is a weighted mean of the sample variances.

We will use an example that was originally done in Section 8.2.

EXAMPLE 9.5

Many people believe that male racehorses run faster than female racehorses. To test this theory, the results of 55 randomly selected races at Santa Anita Park were selected, all at the distance of 1 mile. Thirty-four of the races were for male racehorses, while 21 of the races were restricted to fillies and mares (female racehorses). The time required by the winners of the male races to complete the 1-mile race had a mean of 96.9 seconds, with a standard deviation of 1.22 seconds. The time required by the winners of the races restricted to fillies and mares to complete the 1-mile race had a mean of 97.8 seconds, with a standard deviation of 1.55 seconds. At the 0.05 level of significance, test the claim that the mean time for a male racehorse to win a 1-mile race is less than the mean time for a female racehorse to win a 1-mile race.

Even though one of the samples (male races) has a size that is at least 30, the small sample test is appropriate for this two-mean test because the other sample has a size that is less than 30. We now must test the equality of the two population variances.

Step 1

For this test, population 1 will be the times for the female races since they have the greater sample standard deviation.

Population 1: female races; population 2: male races
Claim in words: The variance of times for female racehorses to win a 1-mile race is the same as the variance of times for male racehorses to win a 1-mile race.
Claim: $\sigma_1^2 = \sigma_2^2$
Complement: $\sigma_1^2 \neq \sigma_2^2$
H_0: $\sigma_1^2 = \sigma_2^2$
H_A $\sigma_1^2 \neq \sigma_2^2$

Step 2

Level of significance: $\alpha = 0.10$. Since this is a two-tailed test, we must divide α in half to find the area that belongs in the right tail and determine which chart to use. In this case we will use the $0.10/2 = 0.05$ chart.

Step 3

Test statistic: $F = \dfrac{s_1^2}{s_2^2}$, $d.f.$ numerator: $n_1 - 1$, $d.f.$ denominator: $n_2 - 1$

Step 4

There are 20 degrees of freedom for the numerator (21 female races) and 33 degrees of freedom for the denominator (34 male races).

Decision rule: Reject H_0 if $F > 1.93$.

Step 5

$$F = \frac{s_1^2}{s_2^2}$$

$$= \frac{1.55^2}{1.22^2}$$

$$= 1.61$$

Decision: Fail to reject H_0.

Conclusion: There is not sufficient sample evidence to reject the claim that the variance of times for female racehorses to win a 1-mile race is the same as the variance of times for male racehorses to win a 1-mile race.

p-value

Using the chart, the best we can say is that the p-value is greater than 0.05 because we failed to reject the null hypothesis. However, a calculator or computer tells us that the p-value is 0.2196.

Since we were not able to reject the null hypothesis, we assume that the variances are equal for testing the population means. We will let the winning times of male races be population 1 because male racehorses are mentioned first in the claim. If this change from the variance test bothers you, it would be fine to let the winning times of female races be population 1.

Step 1

Population 1: male races; population 2: female races
Claim in words: The mean time for a male racehorse to win a 1-mile race is less than the mean time for a female racehorse to win a 1-mile race.
Claim: $\mu_1 < \mu_2$
Complement: $\mu_1 \geq \mu_2$
H_0: $\mu_1 \geq \mu_2$
H_A $\mu_1 < \mu_2$

Step 2

Level of significance: $\alpha = 0.05$

Step 3

Test statistic: $t = \dfrac{\bar{x}_1 - \bar{x}_2}{\sqrt{\dfrac{(n_1 - 1) \cdot s_1^2 + (n_2 - 1) \cdot s_2^2}{n_1 + n_2 - 2} \cdot \dfrac{n_1 + n_2}{n_1 \cdot n_2}}}$, $d.f. = n_1 + n_2 - 2$

Step 4

$\alpha = 0.05$, two-tailed test, with $21 + 34 - 2 = 53$ degrees of freedom

Decision rule: Reject H_0 if $t < -1.676$.

Step 5

We begin with the calculation of the pooled variance. We will work out the numerator and denominator separately without dividing, which will save accuracy for later calculations.

$$\frac{(n_1 - 1) \cdot s_1^2 + (n_2 - 1) \cdot s_2^2}{n_1 + n_2 - 2} = \frac{(34 - 1) \cdot 1.22^2 + (21 - 1) \cdot 1.55^2}{34 + 21 - 2}$$

$$= \frac{33 \cdot 1.22^2 + 20 \cdot 1.55^2}{53}$$

$$= \frac{97.1672}{53}$$

Next, we do the same for the fraction containing the sample sizes.

$$\frac{n_1 + n_2}{n_1 \cdot n_2} = \frac{34 + 21}{34 \cdot 21}$$

$$= \frac{55}{714}$$

Here is the final calculation of the test statistic.

$$t = \frac{\bar{x}_1 - \bar{x}_2}{\sqrt{\dfrac{(n_1 - 1) \cdot s_1^2 + (n_2 - 1) \cdot s_2^2}{n_1 + n_2 - 2} \cdot \dfrac{n_1 + n_2}{n_1 \cdot n_2}}}$$

$$= \frac{96.9 - 97.8}{\sqrt{\dfrac{97.1672}{53} \cdot \dfrac{55}{714}}}$$

$$= -2.395$$

Decision: Reject H_0.

Conclusion: There is sufficient sample evidence to support the claim that the mean time for a male racehorse to win a 1-mile race is less than the mean time for a female racehorse to win a 1-mile race.

p-value

Using techniques developed in Section 7.2, the best that we can say is that the p-value is between 0.025 and 0.01. A calculator or computer would tell us that the actual p-value of this test is 0.0101. ∎

EXTRA • EXERCISES 9.1

For the following hypothesis tests, first test that two populations have equal variances at the 0.10 level of significance before continuing to test that the two population means are equal. If you are unable to reject the null hypothesis that the two population variances are equal, use the test for two population means covered in this section. Otherwise, use the test from Section 8.2.

1. It is believed that most females have stopped growing taller by the time they graduate from high school. A random sample of 20 female college freshmen had a mean height of 66.2 inches, with a standard deviation of 3.03 inches. A random sample of 24 female high school freshmen had a mean height of 65.1 inches, with a standard deviation of 2.61 inches. At the 0.05 level of significance, test the claim that the mean height of female college freshmen is the same as the mean height of female high school freshmen.

2. Here are the serum cholesterol levels, in mg/dL, of 12 men and 26 women ages 60–69.

   Men

176	196	189	184	230	169
159	180	284	221	196	173

   Women

168	239	213	197	170	226	200
205	237	219	246	247	268	273
213	151	188	196	273	202	236
214	295	221	255	260		

   At the 0.01 level of significance, test the claim that men and women ages 60–69 have the same mean serum cholesterol level.

3. The Insurance Institute for Highway Safety performed a 5 mph crash test on 4 different large luxury cars and 4 different large family cars. The front of each car was driven into an angle barrier and then the repair costs were calculated. Here are the results.

   Large Luxury

$559	$1072	$578	$1828

   Large Family

$1101	$443	$674	$520

   At the 0.05 level of significance, test the claim that the mean damage amount for this type of accident is greater for large luxury cars than it is for large family cars.

4. A golfer is considering buying a new driver. He hits 12 balls with his old driver and then 15 balls with the new driver. Here are the distances, in yards.

   New

238	261	263	236	269	268
270	250	262	242	251	251

   Old

254	266	275	279	261	268	280	269
285	254	276	272	283	262	267	

   At the 0.05 level of significance, test the claim that the mean driving distance with his old driver is less than the mean driving distance with the new driver.

5. A random sample of 17 male smokers revealed that they began smoking at a mean age of 18.2 years old, with a standard deviation of 2.60 years. A random sample of 20 female smokers revealed that they began smoking at a mean age of 18.0 years old, with a standard deviation of 3.10 years. At the 0.05 level of significance, test the claim that the mean age that men begin smoking is the same as the mean age that women begin smoking.

SECTION 9.2
Analysis of Variance (ANOVA)

In this section we will develop a hypothesis test for the equality of three or more population means. The name of this test is ANOVA, which stands for analysis of variance. It may seem odd to see the word *variance* in a test for population *means,* but we will determine whether all of the means are equal by examining two estimates of the variance of the combined populations. The calculations are tedious, but we will use technology and another shortcut to help us out.

The first estimate is the *variance between the samples.* We take the mean of each sample and then look at how spread out those means are. If the sample means are far apart from each other, then this estimate will be quite large. The second estimate of the combined population variance is the *variance within the samples.* We will calculate the variance of each sample and then calculate the weighted mean of those variances. This estimate should be fairly close to the variance of the combined populations. Once these two estimates have been calculated, we proceed with a test that is similar to the test for two population variances that we saw in the last section.

Now we introduce some notation and terminology. We let k represent the number of samples that we have. This is also known as the number of **treatments**. A treatment is the characteristic that makes one population different from the others. Sometimes treatments are referred to as factors. The word treatment was originally used because the ANOVA technique was developed in agriculture, where different treatments were applied to the soil to see which produced the highest yield.

We denote the size of the first sample with n_1, the size of the second sample with n_2, and so on. We use the letter N to stand for the total of the sample sizes. We denote the first sample mean by $\bar{x}_1$, the second sample mean by $\bar{x}_2$, and so on. We use $\bar{x}$ to represent the weighted mean of the sample means.

$$\bar{x} = \frac{n_1\bar{x}_1 + n_2\bar{x}_2 + \cdots + n_k\bar{x}_k}{N}$$

The first estimate of the combined population variance is based on how far the individual sample means are from $\bar{x}$. Finally, we denote the first sample variance by s_1^2, the second population variance by s_2^2, and so on.

The first estimate of the combined population variance is based on the differences between the sample means. The estimate of this variance is called the **mean square for treatment**, or **MST**. We begin its calculation by first calculating the **treatment sum of squares, SST.**

$$SST = n_1(\bar{x}_1 - \bar{x})^2 + n_2(\bar{x}_2 - \bar{x})^2 + \cdots + n_k(\bar{x}_k - \bar{x})^2$$

Next, we divide the treatment sum of squares by $k - 1$ to obtain the mean square for treatment.

$$MST = \frac{SST}{k - 1}$$

The second estimate of the variance is based on the differences within the samples. Essentially, we will be taking a weighted mean of the sample variances. This variation is often referred to as variation due to error. We will call it the **mean square for error**, or **MSE.** Again, we begin its calculation by first calculating the **error sum of squares, SSE.**

$$SSE = (n_1 - 1)s_1^2 + (n_2 - 1)s_2^2 + \cdots + (n_k - 1)s_k^2$$

Next, we divide the error sum of squares by $N - k$ to obtain the mean square for error.

$$MSE = \frac{SSE}{N - k}$$

The test statistic for ANOVA is $F = MST/MSE$, which is simply a ratio of two variances, much like we saw in the last section. There are $k - 1$ degrees of freedom for the numerator and $N - k$ degrees of freedom for the denominator.

One final note: The k samples must be independent and normally distributed, and the variances of each population must be equal. This last assumption of equal variances is different from what we have seen before. If it appears that the samples are drawn from populations whose variances are not all equal, or that all of the populations are not normally distributed, then we need to apply the Kruskal–Wallis test from Section 11.4.

We begin with an example to demonstrate these calculations, as well as explain some of the differences between this hypothesis test and the tests that we covered previously.

Football players who play defense can be broken into three categories: linemen, linebackers, and defensive backs. The position requirements essentially dictate the size of the players. Linemen must be large and powerful to get by the offensive line, defensive backs must be quick and nimble to cover fleet wide receivers, and linebackers must be a combination of both. Here are the weights of 28 randomly selected football players—10 linemen, 7 linebackers, and 11 defensive backs. At the 0.05 level of significance, test the claim that the mean weight for all three positions is the same.

Linemen	275	310	260	295	302	
	268	280	305	276	260	
Linebackers	220	234	241	223		
	250	229	220			
Defensive Backs	200	195	215	195	197	190
	200	191	185	196	168	

There are three samples here, so $k = 3$. Also, $N = 28$, which is the total of the three sample sizes. We will let linemen be population 1, linebackers be population 2, and defensive backs be population 3. Other choices for the three populations would work as well.

Step 1

Population 1: linemen; population 2: linebackers; population 3: defensive backs
Claim in words: The mean weight for all three positions is the same.

How do we write this claim in symbols? Since we are claiming that all three population means are equal, we use $\mu_1 = \mu_2 = \mu_3$.

Claim: $\mu_1 = \mu_2 = \mu_3$

The complement of the claim is different from anything we have seen so far. There are several ways that all three means are not equal to each other. We cover all of the possibilities with the statement "At least one of the means is different from the others."

Complement: At least one of the means is different from the others.

The null hypothesis for this test will always be that all of the means are equal.

H_0: $\mu_1 = \mu_2 = \mu_3$
H_A: At least one of the means is different from the others.

Step 2

Level of significance: $\alpha = 0.05$

Step 3

Test statistic: $F = \dfrac{MST}{MSE}$, d.f. numerator = $k - 1$, d.f. denominator = $N - k$

Step 4

$\alpha = 0.05$

There are $3 - 1 = 2$ degrees of freedom for the numerator, and $28 - 3 = 25$ degrees of freedom for the denominator. Looking in our F chart (Table E), we see that it contains the correct number of degrees of freedom for the numerator and denominator. The appropriate critical value is 3.39. Keep in mind that we write the decision rules for tests using the F-distribution as right-tailed decision rules.

Decision rule: Reject H_0 if $F > 3.39$.

Step 5

We begin by calculating the mean and standard deviation for each individual sample. The results are summarized as follows. (Verify these calculations.)

	Linemen	Linebackers	Defensive Backs
Mean	283.1	231.0	193.8
Standard Deviation	18.63	11.40	11.44
Sample Size	10	7	11

(Looking at these sample means, it appears that the population means must be different.) Next, we need $\bar{x}$, the weighted mean of the sample means.

$$\bar{x} = \frac{n_1 \bar{x}_1 + n_2 \bar{x}_2 + n_3 \bar{x}_3}{N}$$

$$= \frac{10(283.1) + 7(231.0) + 11(193.8)}{28}$$

$$= \frac{2831 + 1617 + 2131.8}{28}$$

$$= \frac{6579.8}{28}$$

$$= 235.0$$

Now the calculation of the mean square for treatment (MST) begins with the calculation of the treatment sum of squares (SST).

$$SST = n_1(\bar{x}_1 - \bar{x})^2 + n_2(\bar{x}_2 - \bar{x})^2 + n_3(\bar{x}_3 - \bar{x})^2$$

$$= 10(283.1 - 235.0)^2 + 7(231.0 - 235.0)^2 + 11(193.8 - 235.0)^2$$

$$= 10(48.1)^2 + 7(-4.0)^2 + 11(-41.2)^2$$

$$= 10(2{,}313.61) + 7(16.0) + 11(1{,}697.44)$$

$$= 23{,}136.1 + 112.0 + 18{,}671.84$$

$$= 41{,}919.94$$

We then find MST.

$$MST = \frac{SST}{k-1}$$

$$= \frac{41{,}919.94}{2}$$

$$= 20{,}959.97$$

Next we begin work on the mean square for error (MSE), our second estimate of the combined population variance. First we calculate the error sum of squares (SSE).

$$SSE = (n_1 - 1)s_1^2 + (n_2 - 1)s_2^2 + (n_3 - 1)s_3^2$$

$$= (10 - 1) \cdot 18.63^2 + (7 - 1) \cdot 11.40^2 + (11 - 1) \cdot 11.44^2$$

$$= 9 \cdot 347.0769 + 6 \cdot 129.96 + 10 \cdot 130.8736$$

$$= 3123.6921 + 779.76 + 1308.736$$

$$= 5212.1881$$

This is now used to find the mean square for error.

$$MSE = \frac{SSE}{N-k}$$

$$= \frac{5212.1881}{28-3}$$

$$= \frac{5212.1881}{25}$$

$$= 208.4875$$

Finally, we calculate the test statistic.

$$F = \frac{MST}{MSE}$$

$$= \frac{20{,}959.97}{208.4875}$$

$$= 100.53$$

Decision: Reject H_0.

Conclusion: There is sufficient sample evidence to reject the claim that the mean weight for all three positions is the same.

p-value

All we know from our chart is that the p-value is less than 0.05. A calculator or computer tells us that the p-value is approximately 1.1×10^{-12}, or 0.0000000000011, which is a very small p-value. ∎

We will summarize the results of our calculations in an ANOVA table like the following.

Source of Variation	Sum of Squares (SS)	Degrees of Freedom	Mean Square (MS)	F
Treatment	SST	$k - 1$	$\frac{SST}{k-1}$	$\frac{MST}{MSE}$
Error	SSE	$N - k$	$\frac{SSE}{N-k}$	
Total	SST + SSE	$N - 1$		

Here is the ANOVA table for the previous example.

Source of Variation	Sum of Squares (SS)	Degrees of Freedom	Mean Square (MS)	F
Treatment	41,919.94	2	20,959.97	100.53
Error	5,212.1881	25	208.4875	
Total	47,132.1281	27		

Looking at these ANOVA tables, we see that if we were given the values of SST and SSE the remaining numbers would be fairly easy to complete. Through simple counting we can determine what k and N are, and therefore can easily fill in the column labeled Degrees of Freedom. Next we divide from left to right (sum of squares divided by degrees of freedom) to produce the mean squares. Finally, we divide the two mean squares to produce the test statistic.

 EXAMPLE 9.7   Do the exams on Units 1 through 4 produce the same mean score? An instructor randomly selects students' scores from all four exams. Here is the information.

Test Number	Mean	Standard Deviation	Sample Size
1	75.91	14.71	32
2	66.97	21.43	31
3	65.89	19.67	28
4	77.84	21.09	27

At the 0.05 level of significance, test the claim that the four population means are equal. (*Note:* SST = 3,213, SSE = 42,498.)

There are four samples here, so $k = 4$.

$N = 32 + 31 + 28 + 27$

$\quad = 118$

Step 1

Population 1: test 1; population 2: test 2; population 3: test 3; population 4: test 4
Claim in words: The mean score for all four tests is the same.
Claim: $\mu_1 = \mu_2 = \mu_3 = \mu_4$
Complement: At least one of the means is different from the others.
H_0: $\mu_1 = \mu_2 = \mu_3 = \mu_4$
H_A: At least one of the means is different from the others.

Step 2

Level of significance: $\alpha = 0.05$

Step 3

Test statistic: $F = \dfrac{MST}{MSE}$, d.f. numerator $= k - 1$, d.f. denominator $= N - k$

Step 4

$\alpha = 0.05$

There are $4 - 1 = 3$ degrees of freedom for the numerator, and $118 - 4 = 114$ degrees of freedom for the denominator. Looking in our F chart (Table E), we see that it contains the correct number of degrees of freedom for the numerator, but not for the denominator. Our choice for the denominator is between 60 degrees of freedom (critical value 2.76) and 120 degrees of freedom (critical value 2.68). Recall that we want to choose the larger critical value, which is 2.76, even though the actual value will be closer to 2.68 because the degrees of freedom are closer to 120 than 60. (Using technology, we could verify that the critical value actually is 2.68 to two decimal places.)

Decision rule: Reject H_0 if $F > 2.76$.

Step 5

We will complete an ANOVA table, since we already have values for SST (3,213) and SSE (42,498).

Source of Variation	Sum of Squares (SS)	Degrees of Freedom	Mean Square (MS)	F
Treatment	3,213			
Error	42,498			
Total				

We complete the first column by adding SST and SSE, and placing that result in the box containing the total sum of squares. Also, we place the appropriate degrees of freedom, 3 and 114, in the second column, and add them together to find the total degrees of freedom.

Source of Variation	Sum of Squares (SS)	Degrees of Freedom	Mean Square (MS)	F
Treatment	3,213	3		
Error	42,498	114		
Total	45,711	117		

To complete the Mean Square column, divide SST by $k - 1$ and place the result in the box for MST. Divide SSE by $N - k$ and place the result in the box for MSE.

Source of Variation	Sum of Squares (SS)	Degrees of Freedom	Mean Square (MS)	F
Treatment	3,213	3	1,071	
Error	42,498	114	372.79	
Total	45,711	117		

We complete the table by dividing MST by MSE and placing the result in the box labeled F. This is the value of our test statistic.

Source of Variation	Sum of Squares (SS)	Degrees of Freedom	Mean Square (MS)	F
Treatment	3,213	3	1,071	2.87
Error	42,498	114	372.79	
Total	45,711	117		

$F = 2.87$

Decision: Reject H_0.

Conclusion: There is sufficient sample evidence to reject the claim that the mean score for all four tests is the same.

p-value

All we know from our chart is that the p-value is less than 0.05. A calculator or computer tells us that the p-value is approximately 0.0395. ■

We will look at one more example.

EXAMPLE 9.8

A math department is considering new ways to teach its remedial math courses. They are trying to decide whether to incorporate computer-assisted instruction into their courses. They randomly assign classes to three instructional modes—traditional (no technology), programmed computer instruction, and a mixture of the two. At the end of the semester, the classes are given an achievement test. Test the claim that the mean scores for the three different methods of instruction are the same at the 0.01 level of significance. Here are the results.

Method of Instruction	Mean	Standard Deviation	Sample Size
Traditional	78.2	15.03	72
Computer	76.9	17.84	63
Mixture	82.8	11.31	87

(*Note: SST* = 1,491.02, *SSE* = 46,772.22.)

There are three samples here, so $k = 3$.

$$N = 72 + 63 + 87$$
$$= 222$$

Step 1

Population 1: traditional; population 2: computer; population 3: mixture
Claim in words: The mean score on the exam is the same for all three methods.
Claim: $\mu_1 = \mu_2 = \mu_3$
Complement: At least one of the means is different from the others.
H_0: $\mu_1 = \mu_2 = \mu_3$
H_A: At least one of the means is different from the others.

Step 2

Level of significance: $\alpha = 0.01$

Step 3

Test statistic: $F = \dfrac{MST}{MSE}$, d.f. numerator = $k - 1$, d.f. denominator = $N - k$

Step 4

$\alpha = 0.01$

There are $3 - 1 = 2$ degrees of freedom for the numerator, and $222 - 3 = 219$ degrees of freedom for the denominator. Looking in our F chart (Table E), we see that it contains the correct number of degrees of freedom for the numerator but not for the denominator. We will use the critical value associated with 120 degrees of freedom for the denominator.

Decision rule: Reject H_0 if $F > 4.79$.

(Using technology, we see that the actual critical value is 4.70.)

Step 5

We will complete an ANOVA table, since we already have values for *SST* (1,491.02) and *SSE* (46,772.22).

Source of Variation	Sum of Squares (SS)	Degrees of Freedom	Mean Square (MS)	F
Treatment	1,491.02	2	745.51	3.49
Error	46,772.22	219	213.57	
Total	48,263.24	221		

$F = 3.49$

Decision: Fail to reject H_0.

Conclusion: There is not sufficient sample evidence to reject the claim that the mean score on the exam is the same for all three methods.

p-value

All we know from our chart is that the *p*-value is higher than 0.01 because we failed to reject the null hypothesis. A calculator or computer tells us that the *p*-value is approximately 0.0322. ∎

One final note about ANOVA is that although it can tell us that all of the population means are not equal, it cannot be used to single out which population mean is significantly higher or lower than the others. Further tests must be used to do that.

MICROSOFT EXCEL   ANOVA

We can use Microsoft Excel to help us with the calculations required for ANOVA, including *p*-values and critical values. We will be using the Data Analysis ToolPak to aid us. Before we begin, be sure that the ToolPak has been added in. On the **Tools** menu, look for **Data Analysis.** If you do not see it, then we have to add it. From the **Tools** menu, select **Add-Ins.** When the dialog box appears, check the box next to **Analysis ToolPak,** and click on **OK.**

We will rework the first example of this section using Excel.

EXAMPLE 9.9   Football players who play defense can be broken into three categories: linemen, linebackers, and defensive backs. The position requirements essentially dictate the size of the players. Linemen must be large and powerful to get by the offensive line, defensive backs must be quick and nimble to cover fleet wide receivers, and linebackers must be a combination of both. Here are the weights of 28 randomly selected football players—10 linemen, 7 linebackers, and 7 defensive backs. At the 0.05 level of significance, test the claim that the mean weight for all three positions is the same.

Linemen	275	310	260	295	302	
	268	280	305	276	260	
Linebackers	220	234	241	223		
	250	229	220			
Defensive Backs	200	195	215	195	197	190
	200	191	185	196	168	

In cell **A1,** type Linemen, which is the name of our first population. Below that, in cells **A2** through **A11,** enter the weights of the linemen. In cell **B1,** type Linebackers, and then enter their weights in cells **B2** through **B8.** Finally, type Defensive Backs in cell **C1** and then enter their weights in cells **C2** through **C12.** Now that the data have been entered, we may proceed to the calculations.

From the **Tools** menu, select **Data Analysis.** When the dialog box appears, select **Anova: Single Factor** and click on **OK.** For **Input Range,** type A1:C12. For our data, cell **A1** is in the upper left-hand corner. We then use **C12** because column **C** is the column that is the end of our data and row **12** is the lowest row that contains any data. Be sure to check the box labeled **Labels in First Row,** because we have put the names of the different populations in the first row. Finally, enter the level of significance (0.05) in the box labeled **Alpha.** When you click on **OK,** you will be taken to another worksheet that contains the output. Here is what it should look like. (To see all of the results clearly, select **Columns** from the **Format** menu, and select **AutoFit Selection** while the results are still highlighted.)

Anova: Single Factor

SUMMARY

Groups	Count	Sum	Average	Variance
Linemen	10	2831	283.1	346.9888889
Linebackers	7	1617	231	130
Defensive Backs	11	2132	193.8181818	130.9636364

ANOVA

Source of Variation	SS	df	MS	F	P-value	F crit
Between Groups	41903.46364	2	20951.73182	100.4872214	1.11817E-12	3.385196123
Within Groups	5212.536364	25	208.5014545			
Total	47116	27				

The first set of values describes the samples including the mean (listed as Average) and the variance (s^2) of each sample. The second set of values is the ANOVA table. Under **Source of Variation, Between Groups** is what we have been calling Treatment. **Within Groups** is what we have been calling Error. The value of our test statistic is in the column under **F,** and it equals 100.49 to two decimal places. In the next two columns, Excel gives us the p-value for our test statistic and the critical value for our decision rule.

With all of this information, we can then write the hypothesis test.

Step 1

Population 1: linemen; population 2: linebackers; population 3: defensive backs
Claim in Words: The mean weight for all three positions is the same.
Claim: $\mu_1 = \mu_2 = \mu_3$
Complement: At least one of the means is different from the others.
H_0: $\mu_1 = \mu_2 = \mu_3$
H_A: At least one of the means is different from the others.

Step 2

Level of significance: $\alpha = 0.05$

Step 3

Test statistic: $F = \dfrac{MST}{MSE}$, d.f. numerator = $k - 1$, d.f. denominator = $N - k$

Step 4

$\alpha = 0.05$

Decision rule: Reject H_0 if $F > 3.39$ (Excel's value rounded to two places).

Step 5

$F = 100.49$

Decision: Reject H_0.

Conclusion: There is sufficient sample evidence to reject the claim that the mean weight for all three positions is the same.

***p*-value**

p-value $= 1.11817 \times 10^{-12}$, or 0.0000000000011817 (value given by Excel) ∎

 **ANOVA**

The TI-83 has a built-in ANOVA test that will calculate everything we need for this test. We will rework the first example of this section using the TI-83.

 Football players who play defense can be broken into three categories: linemen, linebackers, and defensive backs. The position requirements essentially dictate the size of the players. Linemen must be large and powerful to get by the offensive line, defensive backs must be quick and nimble to cover fleet wide receivers, and linebackers must be a combination of both. Here are the weights of 28 randomly selected football players—10 linemen, 7 linebackers, and 7 defensive backs. At the 0.05 level of significance, test the claim that the mean weight for all three positions is the same.

Linemen	275	310	260	295	302	
	268	280	305	276	260	
Linebackers	220	234	241	223		
	250	229	220			
Defensive Backs	200	195	215	195	197	190
	200	191	185	196	168	

In list L_1, enter the weights of the linemen. In list L_2, enter the weights of the linebackers. In list L_3, enter the weights of the defensive backs. When that has been completed, access the **Stat Tests** menu by pressing (STAT) and then moving to the right using the (→) key. The last option in that menu is option **F: ANOVA(.** Select that option and press (ENTER). When the TI-83 takes us back to the main screen we need to enter the lists to be used in ANOVA. Press (2nd)(1)(,) for list L_1, then press (2nd)(2)(,)(2nd)(3)()) to enter lists L_2 and L_3. Press the (ENTER) key, and the results of all the important calculations will appear. Here is what you should see, broken into two screens.

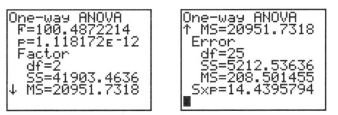

The first line of the display contains the value of the test statistic. The *p*-value is found on the second line. The TI-83 uses the word "Factor" for what we have been calling "Treatment." The next three lines contain the degrees of freedom for the numerator, *SST* and *MST*. The three lines after Error are the degrees of freedom for the denominator, *SSE* and *MSE*.

With all of this information, we can then write the hypothesis test.

Step 1

Population 1: linemen; population 2: linebackers; population 3: defensive backs
Claim in words: The mean weight for all three positions is the same.
Claim: $\mu_1 = \mu_2 = \mu_3$
Complement: At least one of the means is different from the others.
H_0: $\mu_1 = \mu_2 = \mu_3$
H_A: At least one of the means is different from the others.

Step 2

Level of significance: $\alpha = 0.05$

Step 3

Test statistic $F = \dfrac{MST}{MSE}$, d.f. numerator = $k - 1$, d.f. denominator = $N - k$

Step 4

$\alpha = 0.05$

Decision rule: Reject H_0 if $F > 3.39$.

Step 5

$F = 100.49$

Decision: Reject H_0.

Conclusion: There is sufficient sample evidence to reject the claim that the mean weight for all three positions is the same.

p-value

p-value 1.118172 $\times 10^{-12}$, or 0.00000000000118172 (calculated by TI-83) ∎

EXERCISES 9.2

Use the standard procedure for all hypothesis tests. Include an ANOVA table for each test.

1. Here are the waiting times, in minutes, at the airport check-in counters for four different airlines for randomly selected fliers.

Airline A	Airline B	Airline C	Airline D
3	19	8	7
11	13	11	16
19	11	17	15
11	13	7	9
7		12	
10			

At the 0.05 level of significance, test the claim that the mean waiting time at the checkout counters of the four airlines are equal.

2. A random sample of 25 beer drinkers was asked how many servings of beer they had in the past week. Here are the responses, broken down by age group.

	Ages	
21–29	30–39	40 and Above
5	4	2
7	5	4
5	5	4
6	5	3
6	5	4
6	5	3
5	6	
6	8	
6	5	
	5	

(a) At the 0.05 level of significance, test the claim that the mean number of servings is the same for the three age groups (21–29, 30–39, 40 and above).

(b) Identify any possible biases for this set of sample data.

3. A random sample of 16 adult readers was asked to report how many minutes they read per day. Here are their responses, broken down into four age groups: 18–25, 26–45, 46–65, and over 65.

	Ages		
18–25	26–45	46–65	Over 65
10	15	30	65
20	25	60	75
20	40	40	80
25	30	40	80

At the 0.01 level of significance, test the claim that the mean reading times for the four age groups are equal. (Based on the results of a study by the NPD Group.)

4. Here are the GPAs of five sorority members, six fraternity members, and eight students that do not belong to either.

Sorority	Fraternity	Neither
2.85	2.92	3.55
2.93	3.31	3.22
3.06	3.05	3.15
3.04	3.17	3.32
2.89	3.25	3.06
	3.18	3.01
		3.03
		3.30

At the 0.05 level of significance, test the claim that the mean GPAs of the three populations are equal.

5. A cotton farmer is investigating three fertilizers, trying to find the one that will produce the greatest yield of cotton. She randomly assigns a fertilizer to each 1-acre parcel of her 135 acre field in a way that each fertilizer is used on 45 acres. Here are the sample statistics for the yields, in pounds.

	Fertilizer A	Fertilizer B	Fertilizer C
Mean	950.0	961.1	884.4
Standard Deviation	206.24	212.55	151.81

At the 0.05 level of significance, test the claim that the three fertilizers produce equal cotton yields.

6. At a community college, students who place into a prealgebra course take a math competency exam at the end of the course. (The passing score is 39 out of 55.) There are three types of courses that a student could enroll in. One course meets 4 days per week for 17 weeks, the second course meets 5 days per week for 17 weeks, and the third is an intensive short-term class (3 hours/day, 5 days/week, 4 weeks). Here are the competency exam results for 355 students who met 4 days per week, 93 students who met 5 days per week, and 115 students who took the short-term class.

	4 Days/Week	5 Days/Week	Short-Term
Mean	41.1	37.5	44.4
Standard Deviation	4.94	5.23	5.13

At the 0.01 level of significance, test the claim that the three different courses produce the same mean scores on the math competency exam.

7. A random sample of 50 people was taken from each of the following age groups: under 5 years old, 5–17 years, 18–24 years, 25–44 years, 45–64 years, 65–74 years, and 75 years and older. They then determined how many times each of the 300 patients had contact with a physician during the past 12 months. Here are the results.

Age Group	Mean	Standard Deviation
Under 5	6.6	1.59
5–17	3.4	0.74
18–24	3.9	0.60
25–44	5.0	1.32
45–64	6.8	1.76
65–74	9.9	2.85
75 and older	13.1	3.42

At the 0.01 level of significance, test the claim that the mean number of physician contacts per year for these age groups are equal. (*Source:* Centers for Disease Control and Prevention, National Center for Health Statistics, Division of Health Interview Statistics. Data from the National Health Interview Survey.)

8. For a random sample of 250 people, it was determined how many contacts each person had with a physician during the past 12 months. Here are the results, summarized by family income.

Family Income	Mean	Standard Deviation	Sample Size
Under $10,000	8.2	1.89	22
$10,000–$19,999	6.8	1.61	37
$20,000–$34,999	5.6	1.56	107
$35,000 or more	5.5	1.58	84

At the 0.01 level of significance, test the claim that the mean number of physician contacts is the same for all four income levels. (*Source:* Centers for Disease Control and Prevention, National Center for Health Statistics, Division of Health Interview Statistics. Data from the National Health Interview Survey.)

9. Three statistics instructors gave the same exam to their classes. The first class had 19 students, the second class had 15 students, and the third class had 37 students.

Complete the following ANOVA table for these three samples, and then use it to test the claim that the mean score on this exam is the same for all three instructors at the 0.01 level.

Source of Variation	Sum of Squares (SS)	Degrees of Freedom	Mean Square (MS)	F
Treatment	470.80			
Error	26,971.59			
Total				

10. Twenty-five women were selected at random from each of the following six age groups: 20–34 years, 35–44 years, 45–54 years, 55–64 years, 65–74 years and 75 years and older. Researchers then found the serum cholesterol level of each of the 150 people. Complete the following ANOVA table for these six samples, and then use it to test the claim that the mean cholesterol level is the same for all six age groups at the 0.01 level.

Source of Variation	Sum of Squares (SS)	Degrees of Freedom	Mean Square (MS)	F
Treatment	73,246.16			
Error	236,197.69			
Total				

11. Sixty PSAT scores were randomly selected for each of the three parts of the test: verbal, mathematics, and writing. All scores were from high school juniors. Complete the following ANOVA table for the three samples, and then use it to test the claim that the mean scores for the three parts of the exam are equal at the 0.01 level of significance.

Source of Variation	Sum of Squares (SS)	Degrees of Freedom	Mean Square (MS)	F
Treatment	1,516.85			
Error	19,249.18			
Total				

12. Does the level of a parent's education have anything to do with their child's SAT scores? Composite SAT scores were found for a random sample of 100 students. There were 5 students with parents that did not finish high school, 34 students with a parent that had a high school diploma, 8 students with a parent who obtained an associate's degree, 28 students with a parent who obtained a bachelor's degree, and 25 students with a parent who obtained a graduate degree. Complete the following ANOVA table for the three samples, and then use it to test the claim that the mean scores for the three parts of the exam are equal at the 0.01 level of significance.

Source of Variation	Sum of Squares (SS)	Degrees of Freedom	Mean Square (MS)	F
Treatment	762,879.61			
Error	4,990,447.49			
Total				

13. Hole locations for professional golf courses are changed each day of a PGA tournament. Are any of the days set up to be more difficult or easier? Here are the scores of 74 golfers on the four days of a tournament.

	Round					Round			
Golfer	1	2	3	4	Golfer	1	2	3	4
1	63	65	68	65	38	71	68	70	71
2	66	67	68	67	39	70	68	75	68
3	66	67	70	67	40	68	68	76	69
4	70	65	66	69	41	70	70	71	70
5	67	68	69	67	42	69	71	71	70
6	68	67	70	67	43	71	65	74	71
7	68	67	67	70	44	71	67	72	71
8	68	69	65	70	45	71	69	70	71
9	68	68	72	67	46	72	67	71	71
10	67	69	70	69	47	66	67	73	75
11	70	69	67	69	48	71	68	74	69
12	69	67	68	71	49	68	71	73	70
13	67	67	69	72	50	67	70	73	72
14	67	67	73	69	51	70	69	71	72
15	72	63	72	69	52	69	70	70	73
16	70	68	69	69	53	69	70	74	70
17	67	69	70	70	54	68	71	72	72
18	68	68	70	70	55	71	68	71	73
19	66	68	75	68	56	71	68	71	73
20	70	70	68	69	57	71	69	70	73
21	67	71	69	70	58	69	71	76	68
22	70	70	72	66	59	66	71	76	71
23	68	66	75	69	60	70	69	74	71
24	66	70	71	71	61	69	71	72	72
25	68	69	69	72	62	70	68	72	74
26	70	68	74	67	63	72	68	70	74
27	68	70	74	67	64	70	70	74	71
28	72	67	71	69	65	72	68	74	71
29	70	70	69	70	66	69	71	73	72
30	68	67	72	72	67	68	72	75	71
31	67	68	72	72	68	69	71	73	73
32	69	70	68	72	69	70	67	76	74
33	69	69	68	73	70	71	68	76	73
34	68	69	74	69	71	66	71	73	78
35	69	70	72	69	72	70	70	78	72
36	69	70	72	69	73	69	70	72	79
37	72	66	71	71	74	68	72	71	81

At the 0.01 level of significance, test the claim that the mean scores produced by the four different rounds are equal.

14. Hole locations for professional golf courses are changed each day of an LPGA tournament. Are any of the days set up to be more difficult or easier? Here are the scores of 79 golfers on the four days of a tournament.

	Round					Round			
Golfer	1	2	3	4	Golfer	1	2	3	4
1	65	75	70	71	4	70	80	69	70
2	67	77	77	64	5	70	77	74	69
3	74	73	71	68	6	71	74	72	74

(continues)

(continued)

Golfer	Round 1	2	3	4	Golfer	Round 1	2	3	4
7	73	74	76	69	44	76	79	76	70
8	76	75	70	71	45	70	86	71	74
9	74	72	76	71	46	73	81	73	74
10	69	76	77	71	47	71	76	80	74
11	68	76	75	74	48	73	81	76	72
12	70	73	74	76	49	70	79	79	74
13	73	77	76	68	50	71	84	78	70
14	70	79	74	71	51	79	76	76	72
15	73	77	70	74	52	76	79	74	74
16	67	77	75	76	53	70	83	74	76
17	70	78	79	69	54	76	82	77	69
18	70	80	75	71	55	71	78	78	77
19	74	78	71	73	56	72	82	78	73
20	72	80	71	73	57	72	80	80	73
21	74	77	72	73	58	76	80	75	74
22	73	74	74	75	59	69	82	80	74
23	68	78	74	76	60	75	85	74	72
24	73	80	74	70	61	74	81	78	73
25	70	81	76	70	62	73	82	76	75
26	78	74	71	74	63	76	78	77	76
27	71	79	72	75	64	78	74	79	76
28	73	77	76	72	65	76	79	78	75
29	69	81	75	73	66	80	75	77	78
30	70	79	74	75	67	76	80	82	73
31	71	76	74	77	68	79	87	75	71
32	72	78	78	71	69	74	85	78	75
33	71	80	76	72	70	77	83	75	77
34	68	82	76	73	71	70	94	75	74
35	72	79	74	74	72	77	81	77	78
36	71	79	75	74	73	76	86	75	77
37	71	80	73	75	74	81	84	79	71
38	69	80	75	75	75	73	82	82	79
39	70	77	74	78	76	85	81	74	77
40	73	84	72	71	77	81	81	77	81
41	73	80	75	72	78	73	88	83	78
42	77	75	76	72	79	77	84	81	81
43	72	78	74	76					

At the 0.01 level of significance, test the claim that the mean scores produced by the four different rounds are equal.

15. A random sample of 30 Major League baseball players was divided into three groups: pitchers, infielders/catchers, and outfielders. Here are the salaries of those 30 players, broken down by position.

Pitchers	Infield/Catcher	Outfield
567,666	242,500	6,000,000
205,000	6,000,000	2,500,000
1,750,000	725,000	175,000
197,500	200,000	2,200,000
1,000,000	4,000,000	318,750
775,000	2,550,000	2,000,000
180,000	4,000,000	4,200,000
275,000	175,000	5,275,000
1,233,334	2,150,000	

(continues)

(continued) Pitchers	Infield/Catcher	Outfield
	240,000	
	500,000	
	3,000,000	
	300,000	

At the 0.05 level of significance, test the claim that the mean salary is the same for all three positions.

16. Here are the mean monthly prices of a gallon of unleaded regular gas for four areas of the country.

Northeast Urban	Midwest Urban	South Urban	West Urban
1.083	1.103	1.049	1.000
1.074	1.073	1.045	1.023
1.046	1.032	1.028	1.046
1.021	1.023	1.016	1.082
1.010	1.047	1.002	1.069
1.019	1.066	1.003	1.045
1.004	1.039	0.988	1.020
1.044	1.096	1.020	1.059
1.083	1.076	1.035	1.114
1.055	1.027	1.008	1.105
1.025	1.007	0.984	1.078
0.999	0.971	0.961	1.045
0.960	0.926	0.924	1.001
0.916	0.918	0.881	0.935
0.874	0.906	0.860	0.868
0.876	0.924	0.861	0.884
0.886	0.966	0.872	0.953
0.894	0.955	0.880	0.952
0.883	0.945	0.871	0.931
0.871	0.909	0.851	0.914
0.852	0.891	0.827	0.906
0.859	0.901	0.846	0.901
0.857	0.877	0.837	0.892
0.828	0.809	0.802	0.881
0.807	0.824	0.783	0.859
0.788	0.802	0.773	0.849
0.794	0.849	0.785	0.899
0.902	0.979	0.903	1.167
0.941	0.987	0.924	1.114
0.933	0.970	0.915	1.048
0.956	1.015	0.946	1.084
1.017	1.055	0.998	1.155
1.059	1.104	1.034	1.120
1.088	1.071	1.056	1.086
1.074	1.086	1.044	1.065
1.103	1.112	1.081	1.086

(Source: Bureau of Labor Statistics.)

At the 0.01 level of significance, test the claim that the mean price of 1 gallon of unleaded gasoline is the same for these four regions.

17. Here are the prices of a dose of 8 medications in the United States, Canada, Great Britain, Australia, and Mexico (all given in U.S. dollars).

Drug	United States	Canada	Great Britain	Australia	Mexico
Prilosec	3.31	1.47	1.67	1.29	0.99
Prozac	2.27	1.07	1.08	0.82	0.79
Lipitor	2.54	1.34	1.67	1.32	3.60
Prevacid	3.13	1.34	0.82	0.83	1.18
Zocor	3.16	1.47	1.73	1.75	3.66
Zoloft	1.98	1.07	0.95	0.84	1.96
Claritin	1.96	1.11	0.41	0.48	0.92
Paxil	2.22	1.13	1.70	0.82	1.83

(Source: USA Today.)

At the 0.01 level of significance, test the claim that the mean price of medications in these five countries is equal.

18. An onion farmer has three different locations, each planted by a different method. The farmer randomly selects 40 three-foot beds, and counts the number of plants in each bed. Here are the totals.

Farm A	Farm B	Farm C	Farm A	Farm B	Farm C
57	55	52	61	57	62
66	59	49	57	58	51
62	55	55	59	53	48
57	65	54	44	63	68
68	66	70	59	60	54
52	52	59	53	61	74
61	52	67	56	58	60
54	71	64	58	50	64
63	59	61	60	67	64
56	57	58	50	58	56
59	61	67	60	53	66
62	78	64	57	42	65
71	63	57	59	69	62
69	42	66	55	50	75
58	55	66	52	66	68
61	62	58	62	61	67
62	51	70	69	68	66
47	51	64	54	57	66
64	59	60	59	61	55
63	54	63	61	64	60

At the 0.05 level of significance, test the claim that the three farms have the same mean number of plants per 3-foot bed.

19. A study conducted at the University of Maryland at Baltimore examined the effect of cigarette smoking status on 6-minute walk distance. A sample of 37 nonsmokers walked a mean of 413 meters in 6 minutes with a standard deviation of 43.4 meters. A sample of 196 former smokers walked a mean of 370 meters with a standard deviation of 50.0 meters. A sample of 182 current smokers walked a mean of 352 meters with a standard deviation of 48.2 meters. At the 0.05 level of significance, test the claim that the mean length walked is the same for all three groups. (Based on "The effect of cigarette smoking status on six-minute walk distance in patients with intermittent claudication" by M. A. Cahan, P. Montgomery, R. B. Otis, R. Clancy, W. Flinn, and A. Gardner.)

20. A study conducted at the University of Maryland at Baltimore examined the effect of cigarette smoking status on 6-minute walk distance. A sample of 37 nonsmokers walked a mean of 665 steps in 6 minutes with a standard deviation of 43.4 steps. A sample of 196 former smokers walked a mean of 600 steps with a standard deviation of 57.1 steps. A sample of 182 current smokers walked a mean of 563 steps with a standard deviation of 61.9 steps. At the 0.05 level of significance,

test the claim that the mean number of steps is the same for all three groups. (Based on "The effect of cigarette smoking status on six-minute walk distance in patients with intermittent claudication" by M. A. Cahan, P. Montgomery, R. B. Otis, R. Clancy, W. Flinn, and A. Gardner.)

21. In testing the effect of a new zinc spray on the common cold, 81 people were randomly selected and given a spray to take home. Subjects were randomly assigned to one of three groups: placebo—reduced-dosage zinc spray, and full-dosage zinc spray—in such a way that there were 27 people in each group. At the first sign of a cold they were to begin using the spray. Here is a summary of the duration of each subject's first cold.

Group	Mean Duration	Standard Deviation
Placebo	10 days	4.1 days
Reduced Dosage	3.5 days	2.8 days
Full Strength	1.5 days	2.1 days

At the 0.05 level of significance, test the claim that the mean length of a cold being treated by this spray is equal for these three groups. (Based on the results of a study funded by Gel-Tech.)

? What Is Wrong with This Picture?

The following statistical project contains a major error. Potential errors could be the choice of the wrong hypothesis test or that the assumptions required for this test are not met. Find the error, and explain what could be done to correct it.

An agriculture student wants to compare the effect of three different fertilizers. Each fertilizer is given to 10 different tomato plants, and their yields are measured. Here are the sample results.

Fertilizer	Mean (lbs./plant)	Standard Deviation (lbs./plant)
A	181.9	17.74
B	190.6	19.16
C	208.3	84.75

He uses these results to test the claim that the mean yield per plant is the same for all three fertilizers, using the hypothesis test introduced in this section.
What is wrong with this picture?

MINI PROJECT

Visit a local grocery store. Randomly select several brands of soda, beer, and juice. Record the number of calories in one serving. Test the claim that all three types of drinks have the same mean calorie content. In addition to your hypothesis test, include

- Your raw data
- A histogram for the caloric content for the sodas, beers, and juices

Finally, answer the following questions.

- How did you obtain your sample?
- Which type of sampling did you use?
- Was the sample truly random?
- What potential biases may show up in your sample?

SECTION 9.3
The Goodness of Fit Test

In this section we develop a hypothesis test to determine whether a population follows a claimed percentage distribution by category. For example, a counselor may claim that 20% of all students who take the SAT score at least 1200, 25% score between 1000 and 1190, and the remaining 55% score below 1000. To test this claim, we will take a random sample and see whether our sample proportions are close enough to the claimed population proportions. This test will introduce a new distribution—the **chi-square (χ^2) distribution.**

The chi-square distribution is similar to the *F*-distribution in that it is positively skewed and its values are always nonnegative. Here is a graph of a chi-square distribution.

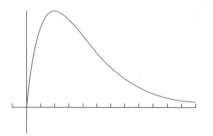

Another similarity to the *F*-distribution is that our decision rules will be right-tailed. The chi-square distribution depends on one type of degrees of freedom, unlike the *F*-distribution that depends on two types (for the numerator and denominator).

We will use the following example to introduce the notation and terminology associated with the test for goodness of fit.

EXAMPLE 9.10 | A counselor claims that 20% of all students who take the SAT score at least 1200, 25% score between 1000 and 1190, and the remaining 55% score below 1000. A random sample of 120 students revealed that 20 students had a score of at least 1200. Thirty-five of the students had a score between 1000 and 1190, while the remaining 65 students scored below 1000. Test the counselor's claim at the 0.05 level of significance.

We use *k* to denote the number of categories that we are working with. In this example *k* = 3 because there are three categories:

- 1200 and above
- 1000 through 1190
- below 1000

We use *n* to denote the sample size or the number of trials. In this example, *n* = 120.

We will be dealing with two different frequencies. The first are the **observed frequencies,** denoted f_o. These are the frequencies for each category in the sample. For our example, the observed frequencies are 20 for the first category (1200 and above), 35 for the second category (1000 through 1190), and 65 for the third category (below 1000). The other frequencies that we will use are **expected frequencies,** denoted f_e. These are the frequencies that we expect to see in the sample if the claimed proportions are true. We calculate the expected frequency for each category by multiplying the sample size n by the proportion, π, claimed for that category.

$$f_e = n \cdot \pi$$

In order to perform this test, it is necessary that each expected frequency (f_e) must be at least 5. If one or more categories has an expected frequency that is less than 5, we can combine categories together until each expected frequency is at least 5. Note that there is no such requirement on the observed frequencies for each class.

If the claim that 20% of all students score 1200 or higher were true, then we would expect 20% of the 120 students to score 1200 or higher, or $0.2 \cdot 120 = 24$ students. We will be examining the differences between the observed and expected frequencies for each category to determine whether the claimed proportions seem reasonable.

Once the expected frequencies have been calculated we are ready to calculate the test statistic.

$$\chi^2 = \Sigma \frac{(f_o - f_e)^2}{f_e}, \qquad \text{degrees of freedom} = k - 1$$

As we learn how to calculate this test statistic, you may notice that the calculation is similar to calculating variance by hand. This makes sense, as we are measuring how the observed frequencies vary from the expected frequencies.

To determine the decision rule we need to know the level of significance and how many degrees of freedom there are. We go to our chart, locate the column for our level of significance, and scan down until we reach the row for our number of degrees of freedom. That value is the critical value. Keep in mind that this decision rule will be written as a right-tailed decision rule. Also remember that if we cannot find our number of degrees of freedom on the chart, we want to choose the adjacent critical value that is larger. This makes it harder to reject the null hypothesis, so our mistake will be on the safe side. ∎

Now we will work our first example completely.

EXAMPLE 9.11

A counselor claims that 20% of all students who take the SAT score at least 1200, 25% score between 1000 and 1190, and the remaining 55% score below 1000. A random sample of 120 students revealed that 20 students had a score of at least 1200. Thirty-five of the students had a score between 1000 and 1190, while the remaining 65 students scored below 1000. Test the counselor's claim at the 0.05 level of significance.

We begin by identifying k and n. Since the counselor is dividing students into three categories, $k = 3$. Because there are 120 students in our sample, $n = 120$.

Step 1

Claim in words: Twenty percent of all students who take the SAT score at least 1200, 25% score between 1000 and 1190, and the remaining 55% score below 1000.

Claim: $\pi_1 = 0.2$, $\pi_2 = 0.25$, $\pi_3 = 0.55$

Complement: At least one of the proportions is different than claimed. Note that the complement of our claim takes a similar form to the complement of a claim using ANOVA.

H_0: $\pi_1 = 0.2$, $\pi_2 = 0.25$, $\pi_3 = 0.55$

H_A: At least one of the proportions is different than claimed.

Step 2

Level of significance: $\alpha = 0.05$

Step 3

Test statistic: $\chi^2 = \Sigma \dfrac{(f_o - f_e)^2}{f_e}$, $d.f. = k - 1$

Step 4

There are $3 - 1 = 2$ degrees of freedom. We look for the column that has 0.05 as its level of significance. The critical value is 5.991.

Decision rule: Reject H_0 if $\chi^2 > 5.991$.

Step 5

We will use a column approach to calculate this test statistic. We need five columns, labeled as follows.

f_o	$f_e = n \cdot \pi$	$f_o - f_e$	$(f_o - f_e)^2$	$\dfrac{(f_o - f_e)^2}{f_e}$

The observed (sample) frequencies go in the first column, and we calculate the expected frequencies in the second column. We find the difference between the observed and expected frequencies and put that difference in the third column. We square those values for the fourth column. Finally, we divide those squares by the expected frequency for that category and total that column. This total is our test statistic.

f_o	$f_e = n \cdot \pi$	$f_o - f_e$	$(f_o - f_e)^2$	$\dfrac{(f_o - f_e)^2}{f_e}$
20	$0.20 \cdot 120 = 24$	-4	16	16/24
35	$0.25 \cdot 120 = 30$	5	25	25/30
65	$0.55 \cdot 120 = 66$	-1	1	1/66

$$\chi^2 = 1.515$$

Decision: Fail to reject H_0.

Conclusion: There is not sufficient sample evidence to reject the claim that 20% of all students who take the SAT score at least 1200, 25% score between 1000 and 1190, and the remaining 55% score below 1000.

p-value

We know that the p-value must be greater than 0.05 because we failed to reject the null hypothesis. Using the chi-square distribution chart, the value of our test statistic is less than the critical value associated with a level of significance of 0.10 for 2 degrees of freedom, so the p-value is greater than 0.10. The use of technology would show us that the actual p-value is 0.4688. ∎

Before proceeding to the next example, let's go over what we have done here. First, we must be able to identify when a goodness of fit test is appropriate. We need to look for a population divided into different categories, with a claim about what proportion of the population fits into each category. The null hypothesis will always be that the claimed proportions are true, and the alternate hypothesis will always be that at least one of the proportions is different than claimed. The decision rule for this test will always be right-tailed. Finally, we begin the calculations by first determining the expected frequency (f_e) for each category. We do this by multiplying the sample size n by the claimed proportion for that category. We complete the calculation of the test statistic using a column approach.

A favorite example for the goodness of fit test involves the colors of M&M candies, because samples are easy to obtain. Mars Inc. claims that plain M&Ms are made in the following percentages.

Color	Brown	Yellow	Red	Orange	Green	Blue
Percentage	30%	20%	20%	10%	10%	10%

A random sample of 55 plain M&M candies had the following breakdown by colors.

Color	Brown	Yellow	Red	Orange	Green	Blue
Number	20	15	3	4	5	8

At the 0.05 level of significance, test the claim that Mars Inc.'s claimed percentage breakdown of colors is correct.

We begin by identifying k and n. Since there are six colors of plain M&M candies, $k = 6$. Our sample size is 55, so $n = 55$.

Step 1

Claim in words: Thirty percent of plain M&M candies are brown, 20% are yellow, 20% are red, 10% are orange, 10% are green, and 10% are blue.
Claim: $\pi_{BR} = 0.3$, $\pi_Y = 0.2$, $\pi_R = 0.2$, $\pi_O = 0.1$, $\pi_G = 0.1$, $\pi_{BL} = 0.1$
   (The subscripts used are more descriptive than simply using the numbers 1 through 6.)
Complement: At least one of the proportions is different than claimed.
H_0: $\pi_{BR} = 0.3$, $\pi_Y = 0.2$, $\pi_R = 0.2$, $\pi_O = 0.1$, $\pi_G = 0.1$, $\pi_{BL} = 0.1$
H_A: At least one of the proportions is different than claimed.

Step 2

Level of significance: $\alpha = 0.05$

Step 3

Test statistic: $\chi^2 = \Sigma \dfrac{(f_o - f_e)^2}{f_e}$, $d.f. = k - 1$

Step 4

There are $6 - 1 = 5$ degrees of freedom. The critical value is 11.070.

Decision rule: Reject H_0 if $\chi^2 > 11.070$.

Step 5

	f_o	$f_e = n \cdot \pi$	$f_o - f_e$	$(f_o - f_e)^2$	$\dfrac{(f_o - f_e)^2}{f_e}$
Brown	20	0.3(55) = 16.5	3.5	12.25	12.25/16.5
Yellow	15	0.2(55) = 11	4	16	16/11
Red	3	0.2(55) = 11	−8	64	64/11
Orange	4	0.1(55) = 5.5	−1.5	2.25	2.25/5.5
Green	5	0.1(55) = 5.5	−0.5	0.25	0.25/5.5
Blue	8	0.1(55) = 5.5	2.5	6.25	6.25/5.5

$$\chi^2 = 9.606$$

Decision: Fail to reject H_0.

Conclusion: There is not sufficient sample evidence to reject the claim that 30% of plain M&M candies are brown, 20% are yellow, 20% are red, 10% are orange, 10% are green, and 10% are blue.

p-value

We know that the p-value must be greater than 0.05 because we failed to reject the null hypothesis. Using the chi-square distribution chart (Table F), the value of our test statistic is between the critical values associated with a level of significance of 0.05 and 0.10 for 5 degrees of freedom, so the p-value is between 0.05 and 0.10. The use of technology would show us that the actual p-value is 0.0872. ∎

Although we have seen percentages in the claim for the previous two examples, there is another way that a claim could be stated. It could be claimed that the proportions for each category are equal. The only difference we face here is that we will have to determine the proportion for each group, which is simply 1 divided by the number of categories.

 EXAMPLE 9.13

A detergent maker has decided to change the appearance of the box. They have come up with four potential replacements, which we will call A, B, C, and D. They show the four designs to 400 randomly selected consumers, and ask them which design they like best. Here are the results.

Design	A	B	C	D
Consumers	107	105	122	66

At the 0.05 level of significance, test the claim that all consumers equally like the four designs.

We begin by identifying k and n. Since there are four designs to choose from, $k = 4$. Our sample size is 400, so $n = 400$.

Step 1

Claim in words: All consumers equally like the four designs.
Claim: $\pi_A = \pi_B = \pi_C = \pi_D = \frac{1}{4}$
Complement: At least one of the proportions is different than claimed.
H_0: $\pi_A = \pi_B = \pi_C = \pi_D = \frac{1}{4}$
H_A: At least one of the proportions is different than claimed.

Step 2

Level of significance: $\alpha = 0.05$

Step 3

Test statistic: $\chi^2 = \Sigma \dfrac{(f_o - f_e)^2}{f_e}$, $d.f. = k - 1$

Step 4

There are $4 - 1 = 3$ degrees of freedom. The critical value is 7.815.

Decision rule: Reject H_0 if $\chi^2 > 7.815$.

Step 5

	f_o	$f_e = n \cdot \pi$	$f_o - f_e$	$(f_o - f_e)^2$	$\dfrac{(f_o - f_e)^2}{f_e}$
A	107	0.25(400) = 100	7	49	49/100
B	105	0.25(400) = 100	5	25	25/100
C	122	0.25(400) = 100	22	484	484/100
D	66	0.25(400) = 100	−34	1156	1156/100

$$\chi^2 = 17.14$$

Decision: Reject H_0.

Conclusion: There is sufficient sample evidence to reject the claim that all consumers equally like the four designs.

***p*-value**

We know that the *p*-value must be less than 0.05 because we rejected the null hypothesis. Using the chi-square distribution chart, the value of our test statistic is higher than the critical values associated with a level of significance of 0.01 for 3 degrees of freedom, so the *p*-value is less than 0.01. The use of technology would show us that the actual *p*-value is 0.0007. ∎

If we are trying to determine which category or categories are "off" when we reject the null hypothesis, we can look at the value in the last column to see which category contributed the most to the test statistic. Design D contributed 11.56 to the test statistic, and design C contributed 4.84. It appears that the proportion of consumers who prefer design D may be significantly lower than its claimed value. It also appears that the proportion of consumers who prefer design C may be higher than its claimed value. However, these would require further investigation with a new sample.

We can use the goodness of fit test to determine whether a set of data follows a certain distribution, such as the binomial, Poisson, or normal distribution. Recall that many of our previous hypothesis tests required that our samples were selected from a population that followed a normal distribution. The following example will show how to test a set of data for **normality**. This test uses the same test statistic with $(k - 1) - 2$ degrees of freedom.

EXAMPLE 9.14

A fair coin is tossed 20 times, and the number of heads are recorded. This experiment is then repeated for a total of 100 times. Here are the results.

5	10	5	10	10	10	9	7	11	14
14	8	6	8	12	10	5	12	12	9
11	12	11	14	12	11	12	11	7	11
8	9	11	9	9	12	8	11	9	11
14	11	9	10	13	12	10	9	9	11
13	9	11	7	13	13	10	9	10	9
8	12	8	7	9	9	9	13	9	9
9	7	10	13	7	13	9	10	11	9
7	10	16	8	6	9	12	11	9	8
10	10	10	9	10	9	10	9	11	12

At the 0.05 level of significance, test the claim that the data follow a normal distribution.

It is a good idea to start by looking at a histogram of the data. If the histogram is skewed, or if outliers are present, then there is a good chance that the data do not follow a normal distribution.

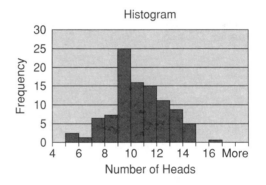

This histogram is roughly symmetric and looks close to "bell-shaped," so we now proceed. The sample mean is 9.95, and the sample standard deviation is 2.13. We will use a frequency distribution to group the data, and compare these frequencies to the frequencies that we would expect from a normal distribution with a mean of 9.95 and a standard deviation of 2.13.

Number of Heads	Frequency
6 or fewer	5
7 or 8	15
9 or 10	42
11 or 12	26
13 or 14	11
15 or more	1

We find the probability of counting 6 or fewer as follows. (Look back to Sections 5.1 and 5.2 if you need a refresher on finding normal probabilities.)

$$P(x \leq 6) \overset{cc}{=} P(x \leq 6.5)$$

$$= P\left(z \leq \frac{6.5 - 9.95}{2.13}\right)$$

$$= P(z \leq -1.62)$$

$$= 0.0526$$

You can go on to verify the following probabilities.

$$P(7 \leq x \leq 8) = 0.1957$$

$$P(9 \leq x \leq 10) = 0.3543$$

$$P(11 \leq x \leq 12) = 0.2823$$

$$P(13 \leq x \leq 14) = 0.0989$$

$$P(x \geq 15) = 0.0162$$

We can multiply these probabilities by the sample size (100) to determine the expected frequencies. Here are the results, listed next to the observed frequencies.

Number of Heads	Observed Frequency	Expected Frequency
6 or fewer	5	5.26
7 or 8	15	19.57
9 or 10	42	35.43
11 or 12	26	28.23
13 or 14	11	9.89
15 or more	1	1.62

Recall that each expected frequency has to be at least 5 to perform this test, and the expected frequency for the last group (15 or higher) is only 1.62. We can combine the last two classes to create a new class (13 or more) that has an expected frequency of 11.51. Here are the new observed and expected frequencies.

Number of Heads	Observed Frequency	Expected Frequency
6 or fewer	5	5.26
7 or 8	15	19.57
9 or 10	42	35.43
11 or 12	26	28.23
13 or more	12	11.51

Now we can perform the normality test.

Step 1

Claim in words: The data follow a normal distribution with a mean of 9.95 and a standard deviation of 2.13.
(#1: 6 or fewer, #2: 7 or 8, #3: 9 or 10, #4: 11 or 12, #5: 13 or more)
Claim: $\pi_1 = 0.0526$, $\pi_2 = 0.1957$, $\pi_3 = 0.3543$, $\pi_4 = 0.2823$, $\pi_5 = 0.1151$
Complement: At least one of the proportions is different than claimed.
H_0: $\pi_1 = 0.0526$, $\pi_2 = 0.1957$, $\pi_3 = 0.3543$, $\pi_4 = 0.2823$, $\pi_5 = 0.1151$
H_A: At least one of the proportions is different than claimed.

Step 2

Level of significance: $\alpha = 0.05$

Step 3

Test statistic: $\chi^2 = \sum \dfrac{(f_o - f_e)^2}{f_e}$, $d.f. = (k - 1) - 2$

Step 4

There are $(5 - 1) - 2 = 2$ degrees of freedom. The critical value is 5.991.
Decision rule: Reject H_0 if $\chi^2 > 5.991$.

Step 5

	f_o	$f_e = n \cdot \pi$	$f_o - f_e$	$(f_o - f_e)^2$	$\dfrac{(f_o - f_e)^2}{f_e}$
6 or fewer	5	5.26	−0.26	0.0676	0.0676/5.26
7 or 8	15	19.57	−4.57	20.8849	20.8849/19.57
9 or 10	42	35.43	6.57	43.1649	43.1649/35.43
11 or 12	26	28.23	−2.23	4.9729	4.9729/28.23
13 or more	12	11.51	0.49	0.2401	0.2401/11.51

$$\chi^2 = 2.495$$

Decision: Fail to reject H_0.

Conclusion: There is not sufficient sample evidence to reject the claim that these data follow a normal distribution.

p-value

The p-value is greater than 0.05. The use of technology would show us that the actual p-value is 0.2872. This is strong evidence that the data do follow a normal distribution. ■

MICROSOFT EXCEL   Goodness of Fit

Microsoft Excel does not provide a built-in procedure for goodness of fit, but we can use its spreadsheet functions to calculate the test statistic. We will also learn how to find the critical value for a goodness of fit test, as well as the p-value for its test statistic.

 EXAMPLE 9.15 A favorite example for the goodness of fit test involves the colors of M&M candies, because samples are easy to obtain. Mars Inc. claims that plain M&Ms are made in the following percentages.

Color	Brown	Yellow	Red	Orange	Green	Blue
Percentage	30%	20%	20%	10%	10%	10%

A random sample of 55 plain M&M candies had the following breakdown by colors.

Color	Brown	Yellow	Red	Orange	Green	Blue
Number	20	15	3	4	5	8

At the 0.05 level of significance, test the claim that Mars Inc.'s claimed percentage breakdown of colors is correct.

In cell **A2,** type Brown, and then put the remaining colors Yellow, Red, Orange, Green, and Blue in cells **A3** through **A7,** respectively. In cell **B1** type Observed, and then put the appropriate observed frequencies from the sample (20, 15, 3, 4, 5, 8) in cells **B2** through **B7.** In cell **C1** type Expected, and then put the appropriate expected frequencies in cells **C2** through **C7.** Once all of these values have been entered, we can begin our calculations.

In cell **D2,** type the following.

=(B2-C2)*(B2-C2)/C2

This formula represents $(f_o - f_e)^2/f_e$ for the first category, where **B2** is the observed frequency f_o and **C2** is the expected frequency for the first category. Repeat the same in cells **D3** through **D7,** being sure to change your formula to match the row that you are in. For example, **D3** should contain the following.

=(B3-C3)*(B3-C3)/C3

Once we have made our way through **D7,** we need to total these results to obtain our test statistic. Typing the following in cell **D8** will do the trick.

=SUM(D2:D7)

Here is what your worksheet should look like.

	Observed	Expected			
Brown	20	16.5	0.742424		
Yellow	15	11	1.454545		
Red	3	11	5.818182		
Orange	4	5.5	0.409091		
Green	5	5.5	0.045455		
Blue	8	5.5	1.136364		
			9.606061	←	Test Statistic

To find the critical value for a goodness of fit test, we use the following built-in Excel function.

=CHIINV(alpha, degrees of freedom)

The level of significance is denoted by alpha. For the above example, type the following in any cell to find the critical value.

=CHIINV(0.05, 5)

To three decimal places, Excel tells us that the critical value is 11.070.

Excel can help us find the p-value at the end of the test. We will use the built-in function that follows.

=CHIDIST(test statistic, degrees of freedom)

For our example, the test statistic was 9.606 and there were 5 degrees of freedom. To find the p-value, type the following in any empty cell.

=CHIDIST(9.606,5)

The critical value that Excel gives us is 0.0872. ■

TI-83 Goodness of Fit

The TI-83 does not provide a built-in procedure for goodness of fit, but we can use its lists to calculate the test statistic. We will also learn how to find the p-value for its test statistic.

EXAMPLE 9.16

A favorite example for the goodness of fit test involves the colors of M&M candies, because samples are easy to obtain. Mars Inc. claims that plain M&Ms are made in the following percentages.

Color	Brown	Yellow	Red	Orange	Green	Blue
Percentage	30%	20%	20%	10%	10%	10%

A random sample of 55 plain M&M candies had the following breakdown by colors.

Color	Brown	Yellow	Red	Orange	Green	Blue
Number	20	15	3	4	5	8

At the 0.05 level of significance, test the claim that Mars Inc.'s claimed percentage breakdown of colors is correct.

In list L_1 we enter the six claimed proportions (0.3, 0.2, 0.2, 0.1, 0.1, 0.1). In list L_2 we enter the observed frequencies (20, 15, 3, 4, 5, 8). In list L_3 we will calculate the expected frequencies. While L_3 is active, use the ⟨↑⟩ key to enter a formula for L_3. We need to multiply the proportions in list L_1 by 55. Enter this by typing ⟨2nd⟩ ⟨1⟩ (to access L_1) ⟨×⟩ 55 ⟨ENTER⟩.

In list L_4 we will calculate $(f_o - f_e)^2/f_e$ for each category. While L_4 is active, use the ⟨↑⟩ key to enter a formula for L_4. We need to find the difference between lists L_2 and L_3, square these differences, and divide each by list L_3. Enter this by typing ⟨(⟩ ⟨2nd⟩ ⟨2⟩ (to access list L_2) ⟨−⟩ ⟨2nd⟩ ⟨3⟩ (to access list L_3) ⟨)⟩ ⟨x^2⟩ ⟨÷⟩ ⟨2nd⟩ ⟨3⟩ (again, list L_3) ⟨ENTER⟩. The screen should look like this prior to pressing ⟨ENTER⟩, and again after pressing ⟨ENTER⟩.

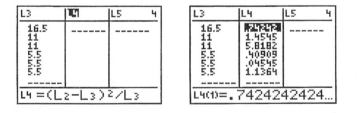

Finally, to calculate the test statistic, we need to total list L_4. Press ⟨STAT⟩ to access the STAT menu, press ⟨→⟩ to move to the CALC menu, and select option **1: 1-Var Stats.** When you return to the main screen, press ⟨2nd⟩ ⟨4⟩ (to access list L_4) and ⟨ENTER⟩. The test statistic will be found next to Σ**x.**

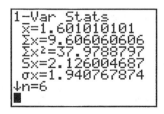

The test statistic is 9.606.

The TI-83 can help us find the p-value at the end of the test. We will use the built-in function χ^2**cdf(** that can be found in the DISTR menu. This function requires three arguments. The first is the calculated value of the test statistic (lower bound), the second is 1E99 (upper bound), and the third is the number of degrees of freedom.

Here are the steps to calculate the p-value for this example. Press (2nd)(VARS) to access the DISTR menu. Select option **7:** χ^2**cdf(.** When you return to the main screen, press 9.606 (,) 1 (2nd)(,) (to access EE) 99 (,) 5 ( ))(ENTER). Recall that 9.606 is the test statistic and there are 5 degrees of freedom. Here are the results.

```
X²cdf(9.606,1E99
,5)
          .0872005405
```

The p-value is 0.0872. ∎

EXERCISES 9.3

Use the standard procedure for all hypothesis tests.

1. A package of bell pepper seeds contained seeds for green, red, yellow, and orange peppers. The seed company claims that 25% of the seeds that are put in these packages are green, 25% are red, 25% are yellow, and 25% are orange. After planting, there were 31 green pepper plants, 26 red pepper plants, 21 yellow pepper plants, and 22 orange pepper plants. At the 0.05 level of significance, test the company's claim that each color makes up 25% of the seeds.
2. To a check a die for fairness, it is rolled 300 times. If the die is fair, then all of the outcomes should be equally likely. Here are the results of the rolls.

Number on die	1	2	3	4	5	6
Number of rolls	57	43	44	53	46	57

   At the 0.05 level of significance, test the claim that the die is fair; in other words, the outcomes 1, 2, 3, 4, 5, and 6 are all equally likely.
3. In the game of roulette:
   - the probability that the outcome is red is 0.474
   - the probability that the outcome is black is 0.474
   - the probability that the outcome is green is 0.052

   In 150 spins of the roulette wheel, the outcome was red 76 times, black 62 times, and green 12 times. At the 0.05 level of significance, test that the claimed probabilities are true for this roulette wheel.
4. Three candidates are running for one open seat on the city council. A random sample of 100 likely voters shows that 29 prefer candidate A, 40 prefer candidate B, and 31 prefer candidate C. At the 0.05 level of significance, test the claim that the three candidates are equally liked. Based on your results, can we state that the voters prefer a candidate?

5. A study reported that 28% of the primary vehicles owned by U.S. adults are less than 3 years old, 34% are between 3 and 7 years old, and 38% are 8 years old or older. (*Source:* Bruskin-Golding for Goodyear/Gemini Automotive Care.) A student did not believe that cars in her area followed that breakdown, and she gathered data on a random sample of 80 cars. Sixteen of the primary vehicles were less than 3 years old, 34 were between 3 and 7 years old, and the remaining 30 were 8 years old or older. At the 0.05 level of significance, test the claim that the cars in the student's area follow the percentage breakdown reported by the study.

6. A study of K-12 teachers asked them how prepared they were to integrate technology into instruction. Thirteen percent were very well prepared, 26% were well prepared, 51% were somewhat prepared, and 10% were not prepared at all. (*Source:* Market Data Retrieval.) A random sample of 60 small town K-12 teachers showed the following results.

Level of Preparedness	Number of Respondents
Very well	5
Well	11
Somewhat	33
Not at all	11

At the 0.05 level of significance, test the claim that teachers in this small town follow the distribution reported by the study.

7. A pair of dice are rolled 200 times. Here are the probabilities for possible sums, if the dice are fair.

Sum	Probability	Sum	Probability
2	1/36	8	5/36
3	2/36	9	4/36
4	3/36	10	3/36
5	4/36	11	2/36
6	5/36	12	1/36
7	6/36		

Here are the sums produced by the 200 rolls.

Sum	2	3	4	5	6	7	8	9	10	11	12
Number of Rolls	4	16	17	16	27	43	25	18	15	16	3

At the 0.01 level of significance, test the claim that the sums produced by this pair of dice follow the given probability distribution.

8. A computer program claims to generate random numbers. Here are 100 numbers between 1 and 10 that were supposed to be randomly generated.

1	1	2	1	9	4	8	6	5	3
8	6	3	7	7	3	9	6	4	6
5	1	1	9	2	10	4	6	6	2
8	7	5	10	8	8	8	7	2	4
2	3	2	5	7	4	9	10	2	9
7	8	6	9	8	4	8	7	9	9
6	9	5	3	3	3	8	1	3	7
3	6	6	6	2	7	9	5	9	5
5	8	3	10	3	8	9	3	6	6
3	4	9	2	7	8	4	8	1	8

At the 0.05 level of significance, test the claim that these numbers are randomly generated. (If the numbers were randomly generated, then any of the numbers between 1 and 10 should be equally likely to be chosen.)

9. Forty-five percent of Americans have type O blood, 40% have type A, 11% have type B, and 4% have type AB. A biology class tests the blood of 250 students. Of these students, 124 have type O, 108 have type A, 14 have type B, and 4 have type AB. At the 0.05 level of significance, test the claim that the claimed blood type percentages are correct.

10. The maker of a nondairy "ice cream" holds a taste test. Subjects taste the nondairy "ice cream," as well as three popular brands of ice cream. Of 120 ice cream eaters, 21 selected the nondairy "ice cream" as their favorite, while 42 selected ice cream A, 28 selected ice cream B, and 29 selected ice cream C. At the 0.01 level of significance, test the claim that ice cream eaters equally prefer the four different products.

11. Are all seven days of the week at restaurants equally busy? Here are the number of customers that ate at a certain restaurant last month, broken down by the days of the week.

Sunday	Monday	Tuesday	Wednesday	Thursday	Friday	Saturday
1139	901	954	1017	989	1443	1536

At the 0.01 level of significance, test the claim that the proportion of the week's restaurant business is equal for each day of the week. (Based on the results of a survey by the National Restaurant Association.)

12. During a 4-week period, a particular hospital had 172 births. Here is the breakdown of deliveries by day of the week.

Sunday	Monday	Tuesday	Wednesday	Thursday	Friday	Saturday
18	25	28	27	27	27	20

At the 0.05 level of significance, test the claim that the proportion of births is equal for each day of the week. (*Source:* Centers for Disease Control and Prevention, Monthly Vital Statistics Report.)

13. During the last calendar year, a particular hospital had 2209 births. Here is the breakdown by season.

Winter (Dec., Jan., Feb.)	Spring (Mar.–May)	Summer (June–Aug.)	Fall (Sep.–Nov.)
540	536	562	571

At the 0.05 level of significance, test the claim that the proportion of all babies born is equal for each season. (*Source:* Centers for Disease Control and Prevention, Monthly Vital Statistics Report.)

14. An algebra instructor tells his students on the first day of class that 40% will pass, 30% will fail, and 30% will withdraw that semester if historical patterns hold true. The class began with 38 students. Seventeen of the students passed the class, 6 failed, and the rest dropped. At the 0.05 level of significance, test the claim that the instructor made on the first day of class.

15. In 1990, 24.5% of Americans 25 years old and over had not graduated from high school, 48.9% held a high school diploma, and 26.6% held at least an associate's degree. (*Source:* U.S. Department of Commerce, Bureau of the Census. *Note:* Percentages altered to total to 100%.) A recent study of 440 Americans 25 years old and over showed that 89 had not graduated from high school, 208 were high school graduates, and 143 held at least an associate's degree. At the 0.05 level, test the claim that the 1990 proportions no longer hold true.

16. A 1970 study showed that of American married-couple families, 42.9% had no children, 18.3% had one child, 18.0% had two children, and 20.8% had three or more children. (*Source:* U.S. Department of Commerce, Bureau of the Census.) A recent survey of 500 American married-couple, families revealed that 267 had no children, 87 had one child, 96 had two children, and 50 had three or more children. At the 0.01 level of significance, test the claim that the 1970 proportions are no longer valid.

17. In 1987, teachers were asked to rate the availability and responsiveness of parents when the teachers needed to contact them. Here is the percentage breakdown for the responses to that question.

Excellent	Good	Fair	Poor
14%	38%	37%	11%

(*Source:* Metropolitan Life/Louis Harris Associates, Inc.)

Recently, a random sample of 529 teachers were asked to rate the availability and responsiveness of parents when they needed to contact them. Here are the responses.

Excellent	Good	Fair	Poor
85	249	143	52

At the 0.05 level of significance, test the claim that the proportions from 1987 are valid today.

18. A random sample of 80 moviegoers was asked which of the following movie genres was their favorite: action/adventure, comedy, horror, and romance. Here are the results.

Genre	Number Who Preferred
Action/Adventure	17
Comedy	31
Horror	20
Romance	12

At the 0.01 level of significance, test the claim that the four genres are equally likely to be listed as a moviegoer's favorite.

19. Here are the number of marriages that took place in a county last year by season.

Season	Number
Jan.–Mar.	51
Apr.–June	86
July–Sept.	99
Oct.–Dec.	64

At the 0.05 level of significance, test the claim that the proportion of marriages is the same for all four seasons. (Based on a study by *Bride's* magazine.)

At the 0.05 level of significance, test the claim that the following sets of data follow a normal distribution.

20. SAT math scores of 60 students

280	610	660	530	550	480	640	500	640
580	370	640	520	540	390	510	410	470
320	560	530	230	500	490	390	600	680
540	510	530	550	750	440	660	570	710
600	510	590	570	500	790	500	320	580
480	650	610	700	430	520	750	690	480
510	750	510	530	400	720			

Use the following classes: 390 and below, 400 to 490, 500 to 590, 600 to 690, 700 and above.

21. Number of pull-ups performed by 40 fourth graders

0	8	9	6	6	4	9	5	9	7
1	9	5	6	1	5	2	4	0	7
6	0	5	4	1	8	9	6	5	6
6	10	3	9	7	10	8	5	8	7

Use the following classes: 2 or fewer, 3 to 4, 5 to 6, 7 to 8, 9 and above.

22. Number of rejects produced per day on an assembly line, for a 56-day period

4	3	2	0	0	1	1	3
2	4	2	4	3	2	2	2
1	3	2	0	3	3	1	1
1	0	1	0	0	1	1	0
3	0	1	0	1	3	1	1
2	0	0	1	1	0	2	2
1	1	0	0	3	1	3	2

23. Number of tornadoes that struck Utah for each year from 1950 through 1995

Year	Number	Year	Number	Year	Number	Year	Number	Year	Number
1950	0	1960	0	1970	5	1980	0	1990	4
1951	0	1961	1	1971	1	1981	2	1991	5
1952	0	1962	1	1972	0	1982	3	1992	4
1953	2	1963	1	1973	0	1983	0	1993	6
1954	1	1964	1	1974	0	1984	6	1994	0
1955	2	1965	5	1975	0	1985	0	1995	2
1956	0	1966	2	1976	0	1986	3		
1957	1	1967	2	1977	0	1987	3		
1958	0	1968	4	1978	0	1988	1		
1959	0	1969	3	1979	0	1989	6		

Use the following classes: less than 1 (i.e., 0), 1, 2, 3, 4 or more. (Data from Utah Disaster Center Web site.)

? What Is Wrong with This Picture?

The following statistical project contains a major error. Potential errors could be the choice of the wrong hypothesis test or that the assumptions required for this test are not met. Find the error, and explain what could be done to correct it.

An instructor tells a statistics class that the following is the historical breakdown of grades on test 1, and that this class should expect the same.

A	B	C	D	F
5%	15%	35%	25%	20%

A student believes that this is merely a motivating ploy to encourage students to study harder. Fifty students took the first exam, producing the following results.

A	B	C	D	F
9	11	19	6	5

The student uses these results to test the instructor's claim about the grade breakdown on test 1, using the hypothesis test introduced in this section.
What is wrong with this picture?

MINI PROJECT

Do some research on eye color to determine what percentage of people have brown eyes, what percentage of people have blue eyes, and what percentage of people have another color eyes. Randomly sample a minimum of 100 students at your school and record their eye color. Use this information to test the eye color percentage breakdown at the 0.05 level of significance. In addition to your hypothesis test, include

- Your raw data
- A pie chart showing the percentage breakdown of eye color

Finally, answer the following questions.

- How did you obtain your sample?
- Which type of sampling did you use?
- Was the sample truly random?
- What potential biases may show up in your sample?

SECTION 9.4
Contingency Tables and the Hypothesis Test for Independence

In this section we will learn a hypothesis test to determine whether two categorical variables are related or not. If the two variables are not related, then the variables are said to be **independent.** In our first example, we will test to see whether a person's burger preference is independent of the person's gender. If we find that the two variables are not independent, we say that they are **dependent.** This does not necessarily suggest that one of the variables affects the other, just that there is a relationship between the two.

For this test we will make use of a **contingency table.** The frequencies in the contingency table correspond to the two variables. One variable categorizes the rows and the other categorizes the columns. Here is an example of a contingency table.

	Burger King	McDonald's	In-N-Out	Other
Male	15	18	35	15
Female	35	12	10	10

The **row variable** is the gender of the person, and the **column variable** is the burger preference of the person.

As was the case in the last section, we will use f_o to denote an observed frequency. The observed frequencies are the values that are found in the cells of the contingency table. We will use f_e to denote an expected frequency. This will be the number that we expect in each cell if the two variables are indeed independent. If the observed and expected frequencies are similar, this is evidence that the two variables are independent. If the observed and expected frequencies are not similar, this is evidence that the two variables are not independent. As with the goodness of fit test, it is necessary that each of the expected frequencies is at least 5.

Recall from Chapter 3 that two events A and B are independent if $P(A \cap B) = P(A) \cdot P(B)$. We will use this formula to help us find the expected frequencies for a contingency table. Consider this contingency table.

	Burger King	McDonald's	In-N-Out	Other
Male	15	18	35	15
Female	35	12	10	10

First we should total each row and column.

	Burger King	McDonald's	In-N-Out	Other	Total
Male	15	18	35	15	83
Female	35	12	10	10	67
Total	50	30	45	25	150

The probability that a person selected from this sample is male is 83/150. The probability that a person selected from this sample prefers Burger King is 50/150. If these two variables are independent, then the probability that a person selected from this sample is male *and* prefers Burger King can be found by multiplying the previous two probabilities, (83/150) • (50/150). As in the last section, if we know the probability that an individual observation ends up in a particular cell, then we can multiply that probability by the sample size to find the expected frequency of that cell. In this case, we expect that 150 • (83/150) • (50/150) or 27.7 males prefer Burger King. This calculation could be simplified to be (83 • 50)/150 by canceling the sample size of 150 with one of the denominators. Note that 83 is the total of the row and 50 is the total of the column that the cell is in, while 150 is the sample size. This translates into the following formula for the expected frequency of any cell.

$$f_e = \frac{(\text{row total})(\text{column total})}{\text{sample size}}$$

How many males should we have expected to prefer McDonald's? The total of the row is 83, and the total of the column is 30.

$$f_e = \frac{83 \cdot 30}{150}$$

$$= 16.6$$

The expected number of males that prefer McDonald's is 16.6. How many males should we have expected to prefer In-N-Out?

$$f_e = \frac{83 \cdot 45}{150}$$

$$= 24.9$$

The expected frequency for that cell is 24.9. We could multiply to find the last expected frequency in the top row, but there is another property that we could take advantage of. The total of expected frequencies for any one row or column is equal to the total of observed frequencies for that row or column. Since there are 83 individuals in the top row and we have already accounted for 69.2 individuals (27.7 + 16.6 + 24.9) in the first three cells, the expected frequency for the last cell is 83 − 69.2 = 13.8.

We could use the same addition property to find the expected frequencies for the second row. The expected frequencies for the first column (Burger King) are supposed to add up to equal 50. Since we have 27.7 in the top row, there must be 50 − 27.7 = 22.3 in the second row. After all of the expected frequencies have been calculated, we will write them in their respective cells inside parentheses as follows.

	Burger King	McDonald's	In-N-Out	Other	Total
Male	15 (27.7)	18 (16.6)	35 (24.9)	15 (13.8)	83
Female	35 (22.3)	12 (13.4)	10 (20.1)	10 (11.2)	67
Total	50	30	45	25	150

The null hypothesis in the test for independence is always that the two variables are independent, and the alternate hypothesis is always that the two variables are dependent. The test statistic for the test for independence is exactly the same as the test statistic for the test for goodness of fit, except that the number of degrees of freedom is different.

$$\chi^2 = \Sigma \frac{(f_o - f_e)^2}{f_e}, \qquad \text{degrees of freedom} = (r-1)(c-1)$$

When calculating the number of degrees of freedom, r is the number of rows in the table and c is the number of columns in the table.

EXAMPLE 9.17

One hundred fifty students at the College of the Sequoias were asked their preference of burgers. Their responses were categorized as Burger King, McDonald's, In-N-Out, and Other. Here are the results.

	Burger King	McDonald's	In-N-Out	Other
Male	15	18	35	15
Female	35	12	10	10

At the 0.05 level of significance, test the claim that burger preference and gender are independent.

Step 1

Claim: Burger preference and gender are independent.
Complement: Burger preference and gender are dependent.
H_0: Burger preference and gender are independent.
H_A: Burger preference and gender are dependent.

Step 2

Level of significance: $\alpha = 0.05$

Step 3

Test statistic: $\chi^2 = \Sigma \frac{(f_o - f_e)^2}{f_e}, \qquad d.f. = (r-1)(c-1)$

Step 4

There are $(2-1)(4-1) = 3$ degrees of freedom. The critical value is 7.815.
Decision rule: Reject H_0 if $\chi^2 > 7.815$.

Step 5

We begin by calculating the expected frequencies, which was already done earlier in the section. Here are the expected frequencies in parentheses, along with their observed frequencies.

	Burger King	*McDonald's*	*In-N-Out*	*Other*	*Total*
Male	15 (27.7)	18 (16.6)	35 (24.9)	15 (13.8)	83
Female	35 (22.3)	12 (13.4)	10 (20.1)	10 (11.2)	67
Total	50	30	45	25	150

When putting the observed and expected frequencies for each cell in columns to calculate the test statistic, be sure to keep them next to each other. Here is the calculation of the test statistic, which is exactly the same as in the last section.

f_o	f_e	$f_o - f_e$	$(f_o - f_e)^2$	$\dfrac{(f_o - f_e)^2}{f_e}$
15	27.7	−12.7	161.29	161.29/22.7
18	16.6	1.4	1.96	1.96/16.6
35	24.9	10.1	102.01	102.01/24.9
15	13.8	1.2	1.44	1.44/13.8
35	22.3	12.7	161.29	161.29/22.3
12	13.4	−1.4	1.96	1.96/13.4
10	20.1	−10.1	102.01	102.01/20.1
10	11.2	−1.2	1.44	1.44/11.2

$$\chi^2 = 24.007$$

Decision: Reject H_0.

Conclusion: There is sufficient sample evidence to reject the claim that burger preference and gender are independent.

p-value

We know that the *p*-value must be less than 0.05 because we rejected the null hypothesis. Using the chi-square distribution chart, we can tell that the *p*-value is also less than 0.01. The use of technology would show us that the actual *p*-value is 0.000025. ∎

We see a lot that is familiar for this hypothesis test; the major difference is how the expected frequencies are found. We will take a look at another example before moving on to a variation of this hypothesis test.

EXAMPLE 9.18

A random sample of 200 students (K–12) were asked to list their favorite sport out of the following four choices: football, basketball, baseball, and soccer. Here are the results, broken down by the grade level of the student.

	Football	*Basketball*	*Baseball*	*Soccer*
Elementary	16	11	23	38
Junior High	14	17	6	8
High School	21	18	15	13

At the 0.01 level of significance, test the claim that sport preference and type of school are independent.

Step 1

Claim: Sport preference and type of school are independent.
Complement: Sport preference and type of school are dependent.
H_0: Sport preference and type of school are independent.
H_A: Sport preference and type of school are dependent.

Step 2

Level of significance: $\alpha = 0.01$

Step 3

Test statistic: $\chi^2 = \Sigma \dfrac{(f_o - f_e)^2}{f_e}$, $d.f. = (r-1)(c-1)$

Step 4

There are $(3-1)(4-1) = 6$ degrees of freedom. The critical value is 16.812.
Decision rule: Reject H_0 if $\chi^2 > 16.812$.

Step 5

To calculate the expected frequencies, we begin by totaling the rows and columns.

	Football	*Basketball*	*Baseball*	*Soccer*	*Total*
Elementary	16	11	23	38	88
Junior High	14	17	6	8	45
High School	21	18	15	13	67
Total	51	46	44	59	200

Here are the calculations and expected frequencies in parentheses, along with their observed frequencies.

	Football	*Basketball*	*Baseball*	*Soccer*	*Total*
Elementary	16 $\left(\dfrac{88 \cdot 51}{200} = 22.4\right)$	11 $\left(\dfrac{88 \cdot 46}{200} = 20.2\right)$	23 $\left(\dfrac{88 \cdot 44}{200} = 19.4\right)$	38 $\left(\dfrac{88 \cdot 59}{200} = 26.0\right)$	88
Junior High	14 $\left(\dfrac{45 \cdot 51}{200} = 11.5\right)$	17 $\left(\dfrac{45 \cdot 46}{200} = 10.4\right)$	6 $\left(\dfrac{45 \cdot 44}{200} = 9.9\right)$	8 $\left(\dfrac{45 \cdot 59}{200} = 13.3\right)$	45
High School	21 $\left(\dfrac{67 \cdot 51}{200} = 17.1\right)$	18 $\left(\dfrac{67 \cdot 46}{200} = 15.4\right)$	15 $\left(\dfrac{67 \cdot 44}{200} = 14.7\right)$	13 $\left(\dfrac{67 \cdot 59}{200} = 19.8\right)$	67
Total	51	46	44	59	200

(*Note:* Some expected frequencies do not total to observed frequencies due to rounding.)

Here is the calculation of the test statistic.

f_o	f_e	$f_o - f_e$	$(f_o - f_e)^2$	$\dfrac{(f_o - f_e)^2}{f_e}$
16	22.4	−6.4	40.96	40.96 / 22.4
11	20.2	−9.2	84.64	84.64 / 20.2
23	19.4	3.6	12.96	12.96 / 19.4
38	26.0	12.0	144.0	144.0 / 26.0
14	11.5	2.5	6.25	6.25 / 11.5
17	10.4	6.6	43.56	43.56 / 10.4
6	9.9	−3.9	15.21	15.21 / 9.9
8	13.3	−5.3	28.09	28.09 / 13.3
21	17.1	3.9	15.21	15.21 / 17.1
18	15.4	2.6	6.76	6.76 / 15.4
15	14.7	0.3	0.09	0.09 / 14.7
13	19.8	−6.8	46.24	46.24 / 19.8

$$\chi^2 = 24.275$$

Decision: Reject H_0.

Conclusion: There is sufficient sample evidence to reject the claim that sport preference and type of school are independent.

p-value

We know that the p-value must be less than 0.01 because we rejected the null hypothesis. The use of technology would show us that the actual p-value is 0.000465. ∎

This procedure that we have been using to test for independence is also used to test for **homogeneity.** In such a test, we are testing to determine whether two or more populations follow the same distribution. The major difference between these two tests is how the sample is gathered. In the test for independence, we sample from one population, and then divide up that population into categories for the two variables. In the test for homogeneity, we divide one of the variables into categories before we sample. Then we gather samples from these individual populations. In this type of sampling we are able to control the total number of values for each category of that one variable. In the previous example, if we decided ahead of time to sample 120 students from elementary school, 40 students from junior high school, and 80 students from high school instead of sampling from the combined population of these three schools, then this would be a test of homogeneity.

As far as we are concerned, there is no other difference between these two tests. Our results will still end up in a contingency table, and the way that we calculate the expected frequencies for each cell, as well as the test statistic and number of degrees of freedom, is exactly the same. Here is an example.

EXAMPLE 9.19

A researcher wants to study the lengths of time required to graduate for college students. A random sample of 195 males who graduated after the spring term showed that 73 had taken 4 years or less to earn their degree, 44 had taken 5 years, 21 had taken 6 years, and 57 had taken 7 years or longer. A random sample of 174 females who graduated after the spring term showed that 83 had taken 4 years or less to earn their degree, 38 had taken 5 years, 14 had taken 6 years, and 39 had taken 7 years or longer. The results are collected in the following table.

	4 Years	5 Years	6 Years	7 Years +
Males	73	44	21	57
Females	83	38	14	39

At the 0.05 level of significance, test the claim that both males and females follow the same distribution of time required to graduate.

Step 1

Claim: Males and females follow the same distribution of time required to graduate.
Complement: At least one of the proportions is different for males and females.
H_0: Males and females follow the same distribution of time required to graduate.
H_A: At least one of the proportions is different for males and females.

Step 2

Level of significance: $\alpha = 0.05$

Step 3

Test statistic: $\chi^2 = \Sigma \dfrac{(f_o - f_e)^2}{f_e}$, $d.f. = (r - 1)(c - 1)$

Step 4

There are $(2 - 1)(4 - 1) = 3$ degrees of freedom. The critical value is 7.815.
Decision rule: Reject H_0 if $\chi^2 > 7.815$.

Step 5

Here is the table with the rows and columns totaled, and expected values already calculated.

	4 Years	5 Years	6 Years	7 Years +	Total
Males	73	44	21	57	195
	(82.4)	(43.3)	(18.5)	(50.7)	
Females	83	38	14	39	174
	(73.6)	(38.7)	(16.5)	(45.3)	
Total	156	82	35	96	369

Here is the calculation of the test statistic.

f_o	$f_e = n \cdot \pi$	$f_o - f_e$	$(f_o - f_e)^2$	$\dfrac{(f_o - f_e)^2}{f_e}$
73	82.4	−9.4	88.36	88.36 / 82.4
44	43.3	0.7	0.49	0.49 / 43.3
21	18.5	2.5	6.25	6.25 / 18.5
57	50.7	6.3	39.69	39.69 / 50.7
83	73.6	9.4	88.36	88.36 / 73.6
38	38.7	−0.7	0.49	0.49 / 38.7
14	16.5	−2.5	6.25	6.25 / 16.5
39	45.3	−6.3	39.69	39.69 / 45.3

$$\chi^2 = 4.672$$

Decision: Fail to reject H_0.

Conclusion: There is not sufficient sample evidence to reject the claim that males and females follow the same distribution of time required to graduate.

p-value

We know that the p-value must be greater than 0.05 because we failed to reject the null hypothesis. The use of technology would show us that the actual p-value is 0.1975. ∎

MICROSOFT EXCEL Test for Independence

Microsoft Excel can assist us with calculating the test statistic for the chi-square test for independence. We will also use Excel to find the critical value and p-value for this test. Here is the example we will use.

EXAMPLE 9.20

One hundred fifty students at the College of the Sequoias were asked their preference of burgers. Their responses were categorized as Burger King, McDonald's, In-N-Out, and Other. Here are the results.

	Burger King	*McDonald's*	*In-N-Out*	*Other*
Male	15	18	35	15
Female	35	12	10	10

At the 0.05 level of significance, test the claim that burger preference and gender are independent.

In a new Excel worksheet, type Observed in cell **A1.** In cells **A2** through **D3,** type the observed values from the contingency table. Here is what your worksheet should look like so far.

Observed			
15	18	35	15
35	12	10	10

The next step is to calculate the expected frequencies. In cell **A4,** type Expected. In cell **A5,** we will calculate the expected frequency for the first cell in the top row (males that prefer Burger King). Type the following in cell **A5.**

=SUM($A2:$D2)*SUM(A$2:A$3)/SUM($A$2:$D$3)

This will always be the format for the expected frequency of the first cell in the top row. In the above formula, A2 represents the observed frequency at the top left of the table, A3 is the bottom left, D2 is the top right, and D3 is the bottom right. If you work with a table that is of a different size than the table in this example, be sure to adjust this formula accordingly. Also, be sure to put the dollar signs ($) in the same location. To finish calculating the expected frequencies, copy cell **A5** and paste it into cells **A5** through **D6.** Here is what your worksheet should look like at this point.

Observed			
15	18	35	15
35	12	10	10
Expected			
27.6667	16.6	24.9	13.8333
22.3333	13.4	20.1	11.1667

The next step is to calculate the test statistic. We will begin by calculating $(f_o - f_e)^2/f_e$ for each cell. In cell **A7,** type Calculations. Type the following in cell **A8:**

=(A2-A5)^2/A5

This calculates $(f_o - f_e)^2/f_e$ for the first cell in the top row. Copy cell **A8** and paste the contents into cells **A8** through **D9.** Your worksheet should now look like this.

Observed			
15	18	35	15
35	12	10	10
Expected			
27.6667	16.6	24.9	13.8333
22.3333	13.4	20.1	11.1667
Calculations			
5.7992	0.1181	4.0968	0.0984
7.1841	0.1463	5.0751	0.1219

Finally, we must total these values to find the test statistic. Type Test Statistic in cell **A10,** and type =SUM(A8:D9) in cell **A11.** This test statistic calculates to be 22.640 when rounded to three decimal places. (Earlier in the section, we calculated the test statistic to be 22.725. The error is due to the fact that we rounded all expected frequencies to one decimal place, whereas Excel did not round them at all.)

To find the critical value for a test for independence, we use the following built-in Excel function.

=CHIINV(alpha, degrees of freedom)

The level of significance is denoted by alpha. For the above example, there are 3 degrees of freedom. Type the following in any cell to find the critical value.

=CHIINV(0.05, 3)

To three decimal places, Excel tells us that the critical value is 7.815.

Excel can help us find the p-value at the end of the test. We will use the built-in function that follows.

=CHIDIST(test statistic, degrees of freedom)

For our example, the test statistic was 22.640 and there were 3 degrees of freedom. To find the p-value, type the following in any empty cell.

=CHIDIST(22.640,3)

The critical value that Excel gives us is 0.000048, which is less than 0.0001. ∎

 Test for Independence

The TI-83 has a built-in chi-square test for independence. We will also use the TI-83 to find the critical value and p-value for this test. Here is the example we will use.

 One hundred fifty students at the College of the Sequoias were asked their preference of burgers. Their responses were categorized as Burger King, McDonald's, In-N-Out, and Other. Here are the results.

	Burger King	McDonald's	In-N-Out	Other
Male	15	18	35	15
Female	35	12	10	10

At the 0.05 level of significance, test the claim that burger preference and gender are independent.

To begin, we must store the observed frequencies in a matrix labeled **A.** To do this, press $(2nd)$ (x^{-1}) to access the MATRIX menu, and press $(\rightarrow)$ $(\rightarrow)$ to reach the EDIT menu. Select option **1** to access matrix **A.** The size of the matrix (our table) is listed at the top right corner of the window. Type 2 before the "x" for the 2 rows, and 4 after the "x" for the 4 columns. Fill in the matrix with the values from the table, using the $(ENTER)$ key to move from one cell to another. Here is what the screen should look like after all of the values have been entered. (Note that you can only see three of the columns at one time.)

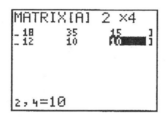

Press $(STAT)$ $(\rightarrow)$ $(\rightarrow)$ to access the STAT TESTS menu, and select option **C: χ^2 - Test.** Following **Observed,** we should see **[A]** (matrix A). This matrix contains our observed frequencies. Next to **Expected,** we should see **[B].** Matrix **B** is the matrix that will hold our expected frequencies, should we want to take a look at them. Press **Calculate** to complete the test. The output window contains the calculated test statistic and the p-value, as well as the number of degrees of freedom. Here is the output for our example.

```
χ²-Test
 χ²=22.63981298
 P=4.8001169E-5
 df=3

■
```

The test statistic is 22.640, and the p-value is 0.000048. ■

EXERCISES 9.4

Use the standard procedure for all hypothesis tests.

1. A random sample of 400 Americans were asked whether they had health insurance. Here are the results, broken down by race.

	Asian	Black	Hispanic	White, Non-Hispanic
Insurance	16	52	40	207
No Insurance	5	16	22	42

At the 0.01 level of significance, test the claim that race is independent of health insurance status. (Based on results from the Current Population Survey by the Census Bureau.)

2. A random sample of 747 people who were about to take a flight were asked to rate how safe from terrorists they felt. Here are the results, broken down by gender.

	Very Safe	Fairly Safe	Slightly Uneasy	Very Uneasy
Men	215	126	24	11
Women	163	183	17	8

At the 0.05 level of significance, test the claim that gender is independent of the safety rating. (Based on the results of a CNN/*USA Today*/Gallup poll.)

3. A random sample of 1000 was asked whether they voted in the last general election. Here are the results, broken down by age group.

	18–24 Years	25–39 Years	40–59 Years	60 Years or Older
Voted	46	159	178	159
Did Not Vote	90	179	118	71

At the 0.01 level of significance, test the claim that voting status and age are independent. (Based on the results of a poll for the American Association of Health Plans.)

4. In January 2000, 500 college students were asked to classify a Democratic Presidential candidate as liberal or moderate. Here are their responses, broken down by the political affiliation of the respondent.

	Liberal	Moderate
Democrats	44	111
Independents	95	99
Republicans	95	56

At the 0.05 level of significance, test the claim that the political affiliation of the respondent and the response are independent.

5. A biology class performed 2500 blood tests on randomly selected students. Here are the blood types and Rh factors of these 2500 students.

	O	A	B	AB
Positive	955	839	209	88
Negative	161	170	49	29

At the 0.05 level of significance, test the claim that blood type and Rh factor are independent.

6. An exit poll of 500 voters in the 1998 Minnesota gubernatorial election shows the number of voters who voted for Humphrey (D), Coleman (R), or Ventura (I) by age group.

	Humphrey (D)	Coleman (R)	Ventura (I)
18–29	15	33	37
30–44	35	54	76
45–59	50	49	46
60 or older	43	40	22

At the 0.05 level of significance, test the claim that choice of candidate and age are independent. (Based on the results of a CNN exit poll.)

7. A new osteoporosis medication is being tested to see if it can lower the chance of a woman suffering a spinal fracture. A sample of 2250 women were randomly assigned to one of three groups: standard dosage, reduced dosage, placebo. The women were monitored for a period of three years. Here are the results, showing the number of women who did/did not suffer a spinal fracture, broken down by the medication groups.

	High Dosage	Reduced Dosage	Placebo
Fracture	37	51	81
No Fracture	713	699	669

At the 0.05 level of significance, test the claim that the proportion of women who suffer a spinal fracture is the same for all three medication treatments. (Based on the results of a study led by Dr. Bruce Ettinger of the Kaiser Permanente Medical Care Program, published in the *Journal of the American Medical Association*.)

8. A random sample of 200 Americans were asked how concerned they were about the Y2K bug in December 1998, one year before potential problems. A second random sample of 300 Americans were polled in September 1999, three months prior to the potential problems. Here are the results.

	Very Concerned	Somewhat Concerned	Not Very Concerned	Not Concerned At All
Dec. '98	31	83	60	26
Sept. '99	19	94	133	54

At the 0.05 level of significance, test the claim that the feelings of Americans remained the same from December '98 to September '99. (The proportional distribution for December '98 is the same as the proportional distribution for September '99.) (Based on the results of a poll by *USA Today* and the National Science Foundation.)

9. A researcher divided up American drivers into five age groups (18–25, 26–35, 36–45, 46–55, and 56 and older), and took random samples of 60 drivers from each group. Each driver was asked whether they run red lights. Here are the results.

Age Group	Yes	No
18–25	46	14
26–35	43	17
36–45	38	22
46–55	34	26
56 and older	20	40

At the 0.01 level of significance, test the claim that the proportion of drivers who run red lights is the same for all five age groups. (Based on a study by the Social Science Research Center, Department of Psychology, Old Dominion University.)

10. A random sample of 440 students were asked whether they had a computer at home. Here are the results, by family income.

	Computer	No Computer
Less than $10,000	8	29
$10,000–$24,999	67	41
$25,000–$34,999	34	40
$35,000–$49,999	58	42
$50,000 and over	97	24

At the 0.01 level of significance, test the claim that the proportion of students with a computer at home is the same for each income level.

11. A blood bank researcher contacted 125 Americans ages 25–34 and another 350 Americans that were at least 35 years old, and inquired about the last time the person donated blood. Here are the results.

	18–34 Year Olds	35 Years and Older
Within the last year	18	44
Within the last 5 years	31	49
More than 5 years ago	25	86
Never	51	171

At the 0.05 level of significance, test the claim that the proportional distribution is the same for both age groups.

12. A random sample of 601 parents of elementary school children were asked about their own education level and whether they volunteered at their child's school. Here are the results.

Education Level	Volunteered at School	Did Not Volunteer at School
Less than high school	27	122
High school graduate	56	124
Some postsecondary	45	67
College graduate	61	55
Graduate/professional	25	19

At the 0.01 level of significance, test the claim that educational attainment and volunteering at school are independent. (Based on the results of the National Household Education Survey by the U.S. Department of Education, National Center for Education Statistics.)

13. Here are the results of a random sample of 535 adults.

Education Level	Smoker	Nonsmoker
Less than high school	47	96
High school graduate	56	117
Some postsecondary	27	81
College graduate	17	94

At the 0.01 level of significance, test the claim that educational attainment and smoking status are independent.

14. A random sample of 483 workers were asked how they saved things: in piles, in files, or some combination of both. Here are the results, broken down by gender.

	Piles	Files	Both
Male	78	138	69
Female	30	110	58

At the 0.05 level, test the claim that the technique used is independent of gender. (Based on the results of a study by Taylor Nelson Sofres Intersearch.)

15. A sample of 300 men and women were asked "How long should couples date before getting married?" Here are the results.

	Less Than 1 Year	1 Year	1–2 Years	2–3 Years	Longer Than 3 Years
Men	31	45	48	16	10
Women	29	43	51	16	11

At the 0.05 level of significance, test the claim that a person's response to this question is independent of the person's gender. (Based on "The Virginia Slims Opinion Poll 2000" by Roper Starch Worldwide.)

16. A random sample of 1977 high school students were asked whether they felt too unsafe to go to school at least once in the preceding 30 days. Here are the results, broken down by race.

	White	Black	Hispanic
Yes	33	45	42
No	819	707	331

At the 0.05 level of significance, test the claim that whether or not a high school student has felt too unsafe to go to school is independent of his or her race. (Based on the Morbidity and Mortality Weekly Report, June 9, 2000, by the Centers for Disease Control and Prevention.)

17. A random sample of 2278 high school females were asked whether they had seriously considered attempting suicide in the 12 months prior to the survey. Here are the results, broken down by grade.

	Grade 9	Grade 10	Grade 11	Grade 12
Yes	114	253	103	111
No	352	588	344	413

At the 0.05 level of significance, test the claim that whether or not a female high school student has seriously considered suicide is independent of her grade. (Based on the Morbidity and Mortality Weekly Report by the Centers for Disease Control and Prevention.)

18. A random sample of 600 people were asked whether they go to the movies at least once a month. Here are the results, broken down by age.

	Age					
	18–24	25–34	35–44	45–54	55–64	65 and up
Yes	83	54	43	37	27	20
No	17	46	57	63	73	80

At the 0.05 level of significance, test the claim that whether a person attends the movies regularly is dependent on a person's age. (Based on the results of a study by TELENATION/Market Facts.)

19. A sample of 605 parents of high school students were asked "Are the public high schools in your state doing an excellent, good, fair, or poor job?" (Sixty-five parents said they did not know enough to say.) Here are the results, broken down by race.

	White Parents	African American Parents	Hispanic Parents
Excellent	22	12	34
Good	80	69	55
Fair	60	75	61
Poor	24	24	24

At the 0.05 level of significance, test the claim that a person's response is independent of his or her race. (*Source:* "Great Expectations: How the Public and Parents—White, African American and Hispanic—View Higher Education," by Public Agenda.)

20. A sample of 605 parents of high school students were asked "When it comes to your own child, do you think a college education is something absolutely necessary to get, something helpful but not necessary, or not that important?" (Sixty-five

parents said they did not know enough to say.) Here are the results, broken down by race.

	White Parents	African American Parents	Hispanic Parents
Absolutely necessary	114	136	159
Helpful but not necessary	79	63	39
Not that important	8	3	4

At the 0.05 level of significance, test the claim that a person's response is independent of his or her race. (*Source*: "Great Expectations: How the Public and Parents—White, African American and Hispanic—View Higher Education," by Public Agenda.)

? What Is Wrong with This Picture?

The following statistical project contains a major error. Potential errors could be the choice of the wrong hypothesis test or that the assumptions required for this test are not met. Find the error, and explain what could be done to correct it.

A student reads a news report about the U.S. Senate. Each senator was asked to classify himself or herself as a liberal, moderate, or conservative. Here are the results, broken down by political party.

	Liberal	Moderate	Conservative
Democrat	11	33	1
Republican	0	37	18

The student uses these results to test the claim that a U.S. senator's political ideology is independent of the senator's party, using the hypothesis test introduced in this section.

What is wrong with this picture?

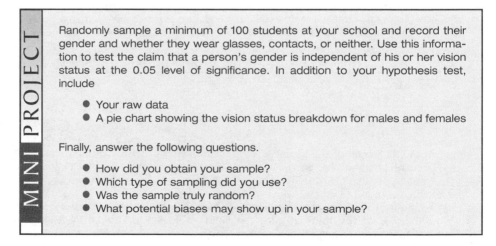

MINI PROJECT

Randomly sample a minimum of 100 students at your school and record their gender and whether they wear glasses, contacts, or neither. Use this information to test the claim that a person's gender is independent of his or her vision status at the 0.05 level of significance. In addition to your hypothesis test, include

- Your raw data
- A pie chart showing the vision status breakdown for males and females

Finally, answer the following questions.

- How did you obtain your sample?
- Which type of sampling did you use?
- Was the sample truly random?
- What potential biases may show up in your sample?

Choosing the Appropriate Tool

The four hypothesis tests in this section are used in four very different situations, although there are some similarities in the actual hypothesis test. If we want to compare the dispersion of two populations, we do so by using the hypothesis test for two population variances. This test uses the *F*-distribution, as long as the two samples are independent and appear to come from populations that are normally distributed.

We use ANOVA to test claims regarding the equality of three or more population means. This test also uses the *F*-distribution. We may use ANOVA if the *k* different samples are independent and appear to come from populations that are all normally distributed. In addition, the variance of these *k* populations must be equal. If any of these conditions fail to be true, then we should use methods that will be introduced in Chapter 11.

To test a claimed proportion breakdown for categories in a population, we use the goodness of fit test, which uses the χ^2-distribution. To use this test, each category's expected frequency must be at least 5. The goodness of fit is also used to test whether a set of data comes from a population that is normally distributed. This is an important test as the assumption of normality is present in several hypothesis tests (one mean—small sample, two means—small samples, paired difference, two variances, and ANOVA).

To test the claim that two categorical variables are independent, we apply the χ^2 test for independence. Again, the expected frequency for each cell must be at least 5.

Test	Section	Test Statistic	Conditions
Two Population Variances	9.1	$F = \dfrac{s_1^2}{s_2^2}$ d.f. numerator $= n_1 - 1$ d.f. denominator $= n_2 - 1$	Both samples come from populations that are normally distributed.
ANOVA	9.2	$F = \dfrac{MST}{MSE}$ d.f. numerator $= k - 1$ d.f. denominator $= N - k$	• Each sample comes from populations that are normally distributed. • The variances of each population must all be equal.
Goodness of Fit	9.3	$\chi^2 = \sum \dfrac{(f_o - f_e)^2}{f_e}$ d.f. $= k - 1$	The expected frequencies for each category are all at least 5.
Independence	9.4	$\chi^2 = \sum \dfrac{(f_o - f_e)^2}{f_e}$ d.f. $= (r - 1)(c - 1)$	The expected frequencies for each category are all at least 5.

Overview

This unit applies our hypothesis test procedure to new types of problems. The major differences from the hypothesis tests from Chapter 7 are the test statistics used and the way to find the decision rule. This unit introduces two new distributions: the F-distribution and the χ^2-distribution.

Chapter 8 deals exclusively with two-sample tests. There are two tests that compare the means of two populations based on independent samples. If the sample sizes are both at least 30, we use the z-distribution. This test is introduced in Section 8.1. If at least one of the samples has a size below 30, then we use the t-distribution as long as both samples appear to be drawn from populations that are approximately normally distributed. If the samples do not meet this condition, we should apply a nonparametric test from Section 11.3.

For a two-mean test based on dependent samples, the paired difference test is introduced in Section 8.3. This test also has nonparametric alternatives in Sections 11.1 and 11.2, which should be used if one or both samples appear to be drawn from a population that is not approximately normally distributed. Finally, a test for comparing two population proportions is introduced in Section 8.4.

Chapter 9 contains two tests that use the F-distribution. The first, in Section 9.1, compares two population variances. The second, in Section 9.2, covers analysis of variance (ANOVA). This test is used to test for the equality of three or more means. For ANOVA, all samples must be drawn from populations that are normally distributed with the same variance. If these conditions are not met, there is a nonparametric alternative in Section 11.4.

Chapter 9 also introduces the χ^2-distribution and two different hypothesis tests that use it. The first is a test for goodness of fit in Section 9.3. This test is used to determine whether a population follows a particular percentage breakdown. This test could also be used to test for normality, which is required to apply the hypothesis tests of Sections 8.2, 8.3, and 9.2. The other test introduced is the test for independence in Section 9.4. This test is used to determine whether two categorical variables are related.

Review Exercises

1. A random sample of 80 elementary school children who attend a public school revealed that 33 of them use a computer at home. A similar sample of 40 elementary school children who attend a private school revealed that 26 of them use a computer at home. At the 0.01 level of significance, test the claim that the proportion of all elementary school children that attend private schools and use a computer at home is higher than the proportion of all elementary school children that attend public schools who use a computer at home.

2. A 1998 exit poll asked voters "During the next year, do you think the nation's economy will be better, worse, or about the same?" Here are the results, broken down by political affiliation for the two major political parties.

	Better	Worse	About the Same
Democrat	57	42	99
Republican	15	57	121

At the 0.05 level of significance, test the claim that a person's view of the economy is independent of his or her political affiliation.

3. Students at a community college can get into an elementary algebra course either by passing a prealgebra course or by taking a placement test. The math department faculty believes that students who pass the placement test are not as prepared for elementary

algebra as students who pass the prealgebra course. Four elementary algebra courses are randomly selected, and the students in the class are given an algebra readiness exam. There were 85 students who had passed the prealgebra course, and their mean score was 61.2 points with a standard deviation of 12.91 points. There were 51 students who had passed the placement test, and their mean score was 54.9 points with a standard deviation of 14.26 points. At the 0.01 level of significance, test the claim that the mean score of students who pass the prealgebra course is higher than the mean score of students who passed the placement test.

4. A random sample of 94 full-time professors at public four-year universities produced a mean age of 49.5 years with a standard deviation of 9.25 years. A similar sample of 110 instructors at public two-year colleges produced a mean age of 48.7 years with a standard deviation of 8.89 years. At the 0.10 level of significance, test the claim that the ages of professors at public four-year universities have the same variance as the ages of instructors at two-year public colleges.

5. If a fair coin is flipped 4 times, here are the probabilities of getting 0 through 4 heads.

Number of Heads	Probability
0	0.0625
1	0.25
2	0.375
3	0.25
4	0.0625

A class of 100 students passed around a coin, flipping it 4 times each. Here are the results.

Number of Heads	Number of Times
0	3
1	10
2	42
3	30
4	15

At the 0.05 level of significance, test the claim that the coin is a fair coin—in other words, that the results agree with the given probabilities for a fair coin.

6. A mini-mart is considering hiring one of two security alarm system monitoring companies. To test their response times, they randomly select 12 stores that use each of the companies, set off the alarm, and measure the time before the company phones to check the situation. The response times of the first company had a mean of 39.6 seconds, with a standard deviation of 7.2 seconds. The response times of the second company had a mean of 45.1 seconds, with a standard deviation of 6.0 seconds. At the 0.01 level of significance, test the claim that the mean response time for the two companies is the same.

7. The owner of a flower shop chain is looking to buy a fleet of delivery vans. She is concerned about the mileage of the vans, and wants to compare four different models of vans. She obtains the following chart that shows the average mileage (miles per gallon) for various different vans of the four different models under test conditions.

Model A	Model B	Model C	Model D
11.4	13.1	9.3	17.7
8.3	9.9	13.5	12.1
13.1	15.0	12.5	14.3
16.3	8.5	12.8	12.2
16.0	8.9	12.4	11.5
14.7	9.4	12.3	11.9
11.6	8.4	10.5	16.4
	9.3	14.5	9.3
	11.3	15.1	14.7
	12.2	11.5	
	10.8		

Use this chart to test the claim that the four different models have the same mean mileage at the 0.05 level of significance.

8. A swimsuit company claims that their new bodysuits lower the race times for competitive swimmers. A random sample of 10 college swimmers (5 male and 5 female) swam two 100-meter freestyle races, once with the new bodysuit and again with a standard swimsuit. The two races are held three days apart to eliminate any concerns about swimmers being tired for the second races. Here are the times for the swimmers.

Swimmer	Bodysuit	Swimsuit	Swimmer	Bodysuit	Swimsuit
1 (M)	53.58	53.55	6 (F)	57.27	57.19
2 (M)	52.70	52.52	7 (F)	56.40	56.33
3 (F)	57.56	57.71	8 (M)	52.22	52.18
4 (M)	51.98	52.09	9 (M)	51.40	51.52
5 (F)	57.56	57.71	10 (F)	57.04	56.97

At the 0.05 level of significance, test the claim that the use of the new bodysuit lowers a swimmer's race times.

Formulas

Two-mean hypothesis test (large samples):

$$z = \frac{\bar{x}_1 - \bar{x}_2}{\sqrt{\dfrac{s_1^2}{n_1} + \dfrac{s_2^2}{n_2}}}$$

Two-mean hypothesis test (small samples):

$$t = \frac{\bar{x}_1 - \bar{x}_2}{\sqrt{\dfrac{s_1^2}{n_1} + \dfrac{s_2^2}{n_2}}}, \qquad d.f. = \text{smaller of } (n_1 - 1) \text{ and } (n_2 - 1)$$

Paired difference hypothesis test:

$$t = \frac{\bar{d} - \mu_d}{s_d / \sqrt{n}}$$

Two-proportion hypothesis test:

$$z = \frac{p_1 - p_2}{\sqrt{\bar{p}(1 - \bar{p})\left(\dfrac{n_1 + n_2}{n_1 \cdot n_2}\right)}}, \qquad \text{where } \bar{p} = \frac{x_1 + x_2}{n_1 + n_2}$$

Two-variance hypothesis test:

$$F = \frac{s_1^2}{s_2^2}, \qquad d.f. \text{ numerator} = n_1 - 1, \; d.f. \text{ denominator} = n_2 - 1$$

Analysis of variance:

$$F = \frac{MST}{MSE}, \qquad d.f. \text{ numerator} = k - 1, \; d.f. \text{ denominator} = N - k$$

Goodness of fit:

$$\chi^2 = \sum \frac{(f_o - f_e)^2}{f_e}, \qquad d.f. = k - 1, f_e = n \cdot \pi$$

Hypothesis test for independence:

$$\chi^2 = \sum \frac{(f_o - f_e)^2}{f_e}, \qquad d.f. = (r - 1)(c - 1), f_e = \frac{(\text{row total})(\text{column total})}{\text{sample size}}$$

unit six

6

Linear Correlation and Regression

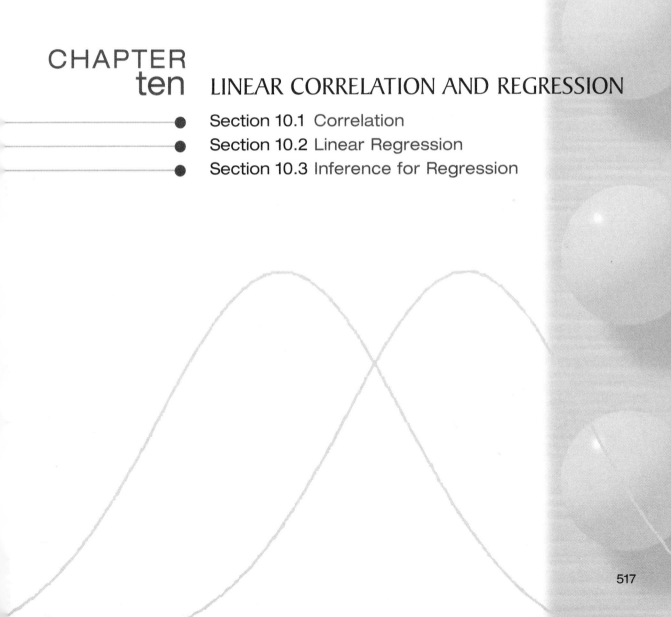

Linear Correlation and Regression

I n this chapter we examine the relation between two numerical variables. We begin with correlation in Section 10.1. The correlation coefficient measures the strength of the relation between paired data from the two variables. We will go on to make inferences about the population correlation coefficient using the sample correlation coefficient.

Section 10.2 covers linear regression. We will use the graph of a line to describe the relation between the two variables. In this section we will learn how to find the equation of the line that best fits a set of paired data, and how to use the equation to predict values for one variable for a particular value of the other variable. Inferences concerning the regression equation and its predictions are covered in Section 10.3.

SECTION 10.1
Correlation

In this section, we examine the relationship between two numerical variables. For instance, we will examine the relation between a person's height and weight, determine if there is indeed a relation, and determine how strong that relation is.

Our investigation begins by determining which of the two variables affects the other. Does a person's height affect their weight, or does a person's weight affect their height? A person's height affects their weight, and you could say that a person's weight *depends* on their height. In this case we refer to the weight as the **dependent variable**, and we refer to the height as the **independent variable.** Recall from algebra that in an equation giving y in terms of x, x was labeled the independent variable and y the dependent variable. We will use the same notation here.

Suppose we were trying to determine if there was a relationship between a student's score on a final exam and the number of hours the student studied for that exam. Which of the two variables is the independent variable (x), and which is the dependent variable (y)? Since the amount of study time affects a person's score, not vice versa, then the number of hours studied is the independent variable (x) and the score on the exam is the dependent variable (y).

A graphical tool that is very helpful in determining whether there is a relationship between two numerical variables is the **scatterplot.** If we plotted each ordered pair (x, y) of the sample data on the Cartesian plane, putting a dot on each point, then we would have a scatterplot. Here is an example.

 EXAMPLE 10.1

Here are the number of hours that 10 students spent studying for a final exam, and their scores on that exam. Create a scatterplot for these data.

Hours	7	8	4	9	13
Score	70	76	57	77	91
Hours	5	9	6	16	3
Score	66	82	64	96	50

We must first identify which variable is the independent variable and which is the dependent variable. Again, since the amount of time spent studying affects a student's score, the independent variable x is the number of hours spent studying and the dependent variable y is the score on the final exam. We must make sure that our x-axis extends to at least 16, which is the largest number of hours spent studying. We also must make sure that our y-axis extends to at least 96, which is the highest exam score given. The first ordered pair that needs to be put on the scatterplot is (7, 70). Here is what the scatterplot looks like after the first point is placed on it.

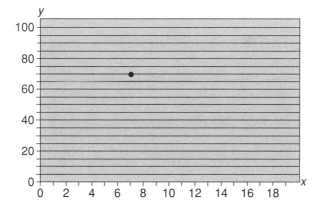

Here is the scatterplot after all 10 points are placed on it.

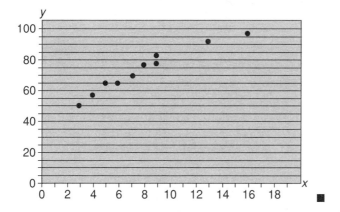

As you look at a scatterplot, it is important to complete the following statement:

As *x* increases, *y* appears to . . . (increase, decrease, be unaffected)

In the previous example, *y* appears to be increasing as *x* increases. In general, the longer a person studies, the higher that person's score will be. (Hmm! Better make a note of that!) This is an example of **positive correlation.** We have positive correlation when an increase in the independent variable corresponds to an increase in the dependent variable as well. **Negative correlation** describes the situation when *y* decreases as *x* increases. Here is an example of negative correlation.

EXAMPLE 10.2 Here are the number of hours that 10 students spent working at their jobs the week before a final exam, and their score on that exam. Create a scatterplot for these data, and determine whether the correlation is positive or negative.

Hours	20	15	0	30	20
Score	87	71	89	60	86
Hours	15	0	30	0	0
Score	89	91	65	94	92

The number of hours that a student spends working at his or her job affects their score on an exam, so the number of hours worked is the independent variable and the score on the exam is the dependent variable. Here is a scatterplot created using Microsoft Excel.

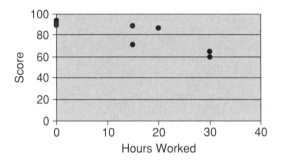

Hours Worked/Score on Final Exam

Note the general downward trend from left to right. This suggests a negative correlation. In general, the more time a student spends at work, the lower that student's exam score will be. ■

If the values of y follow no general pattern of increase/decrease as x increases, then the two variables may have no correlation. (Actually, when we discuss correlation we are talking in terms of *linear* relationships. The paired data may follow some other relationship, but there will be no linear correlation.)

To measure the strength of the correlation, we could calculate the **correlation coefficient r.** The correlation coefficient will always be a value between –1 and 1. If the correlation coefficient is positive, then the correlation for the two variables is positive as well. If the correlation coefficient is negative, then the correlation for the two variables is negative as well. If $r = 1$, then we have **perfect positive correlation,** which means that all of the points on the scatterplot lie on a straight line with positive slope. If $r = -1$, then we have **perfect negative correlation,** which means that all of the points on the scatterplot lie on a straight line with negative slope. The closer the value of r is to 1 or –1, the stronger the correlation is. The closer the value of r is to 0, the weaker the correlation is. If $r = 0$, then there is no correlation.

So, how do we find the value for the correlation coefficient r? It is a wise idea to use technology (calculator or software package) to calculate r. It is far more important that you can interpret and use the value of r than be able to calculate it. However, we will walk through the calculation of r a couple of times in order to cover all bases. Here is the formula:

$$r = \frac{\sum xy - \frac{(\sum x)(\sum y)}{n}}{\sqrt{\left(\sum x^2 - \frac{(\sum x)^2}{n}\right)\left(\sum y^2 - \frac{(\sum y)^2}{n}\right)}}$$

The formula appears tedious, and it is. We will use a column approach with five columns. The first column is for the x-values and the second column is for the y-values. In the third column, we multiply the x- and y-values. The fourth column will hold the squares of the x-values and the fifth column will hold the squares of the y-values. We need to total all of these columns to plug into the formula. Let's take a look at an example.

EXAMPLE 10.3

Here are the number of hours that 10 students spent studying for a final exam (x), and their score on that exam (y). Calculate the correlation coefficient r for these data.

Hours	7	8	4	9	13
Score	70	76	57	77	91
Hours	5	9	6	16	3
Score	66	82	64	96	50

We begin by filling in the five columns, and totaling those columns. Again, after listing the x- and y-values, we proceed to multiply them together to produce the third column. The squares of the x-values go in the fourth column and the squares of the y-values go in the fifth column.

x	y	xy	x^2	y^2
7	70	490	49	4,900
8	76	608	64	5,776
4	57	228	16	3,249
9	77	693	81	5,929
13	91	1,183	169	8,281
5	66	330	25	4,356
9	82	738	81	6,724
6	64	384	36	4,096
16	96	1,536	256	9,216
3	50	150	9	2,500
80	729	6,340	786	55,027

For our formula, $\Sigma x = 80$, $\Sigma y = 729$, $\Sigma xy = 6340$, $\Sigma x^2 = 786$, and $\Sigma y^2 = 55,027$. By the way, n is the number of pairs, which in this case is 10. Now we can plug into the formula to calculate r.

$$r = \frac{\Sigma xy - \dfrac{(\Sigma x)(\Sigma y)}{n}}{\sqrt{\left(\Sigma x^2 - \dfrac{(\Sigma x)^2}{n}\right)\left(\Sigma y^2 - \dfrac{(\Sigma y)^2}{n}\right)}}$$

$$= \frac{6340 - \dfrac{80 \cdot 729}{10}}{\sqrt{\left(786 - \dfrac{80^2}{10}\right)\left(55,027 - \dfrac{729^2}{10}\right)}}$$

$$= \frac{6340 - \dfrac{58,320}{10}}{\sqrt{\left(786 - \dfrac{6400}{10}\right)\left(55,027 - \dfrac{531,441}{10}\right)}}$$

$$= 0.969$$

We see that this value of r is very close to 1, so the correlation is positive and very strong. If we reexamine the scatterplot, we will see that the points virtually fall on a straight line.

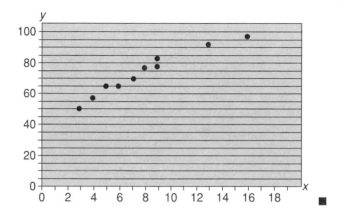

EXAMPLE 10.4	Here are the number of hours that 10 students spent working at their jobs the week before a final exam, and their scores on that exam. Calculate the correlation coefficient r for these data.

Hours	20	15	0	30	20
Score	87	71	89	60	86

Hours	15	0	30	0	0
Score	89	91	65	94	92

We begin by filling in the five columns, and totaling those columns.

x	y	xy	x^2	y^2
20	87	1,740	400	7,569
15	71	1,065	225	5,041
0	89	0	0	7,921
30	60	1,800	900	3,600
20	86	1,720	400	7,396
15	89	1,335	225	7,921
0	91	0	0	8,281
30	65	1,950	900	4,225
0	94	0	0	8,836
0	92	0	0	8,464
130	824	9,610	3,050	69,254

For our formula, $\Sigma x = 130$, $\Sigma y = 824$, $\Sigma xy = 9610$, $\Sigma x^2 = 3050$, and $\Sigma y^2 = 69{,}254$. Again, $n = 10$. Now we can plug into the formula to calculate r.

$$r = \frac{\Sigma xy - \dfrac{(\Sigma x)(\Sigma y)}{n}}{\sqrt{\left(\Sigma x^2 - \dfrac{(\Sigma x)^2}{n}\right)\left(\Sigma y^2 - \dfrac{(\Sigma y)^2}{n}\right)}}$$

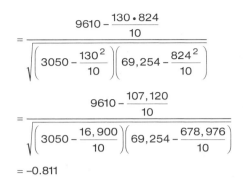

$$= \frac{9610 - \dfrac{130 \cdot 824}{10}}{\sqrt{\left(3050 - \dfrac{130^2}{10}\right)\left(69,254 - \dfrac{824^2}{10}\right)}}$$

$$= \frac{9610 - \dfrac{107,120}{10}}{\sqrt{\left(3050 - \dfrac{16,900}{10}\right)\left(69,254 - \dfrac{678,976}{10}\right)}}$$

$$= -0.811$$

The correlation coefficient is –0.811, so we have negative correlation. Since this value is close to –1, the correlation appears to be strong. ∎

It is a good idea to create a scatterplot before calculating the correlation coefficient. First, you will get an idea about what type of value you should get from your calculation. Also, there may be a point or two that seem not to fit with the other points. In other words, there may be outliers. As with any outlier, we should carefully examine it to make sure that it is a valid piece of data.

Another important measure to calculate is the **coefficient of determination r^2**. This is calculated by squaring the correlation coefficient r. It tells us what proportion of the variation in the dependent variable y can be explained by variation in the independent variable x. For the first example in this section (hours studying/score on exam) the correlation coefficient r was 0.969. Here is the coefficient of determination.

$$r^2 = 0.969^2$$
$$= 0.939$$

This tells us that 93.9% of the variation in exam scores is explained by the variation in study times. In the second example in this section (hours working/score on exam) the correlation coefficient r was –0.811. You can verify that the coefficient of determination for this example is 0.658. This tells us that 65.8% of the variation in exam scores can be explained by the variation in the hours worked by the students.

EXAMPLE 10.5

Here are the heights (in inches) and weights (in pounds) of 9 randomly selected major league baseball players. Construct a scatterplot, and calculate the correlation coefficient and coefficient of determination.

Height	73	69	72	70	72
Weight	201	170	180	200	190

Height	66	72	72	74
Weight	175	205	185	186

The independent variable is height and the dependent variable is weight. Here is a scatterplot created using Microsoft Excel.

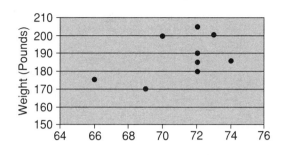

Height/Weight of Major League Baseball Players

The correlation appears to be positive. Here are the calculations.

x	y	xy	x^2	y^2
73	201	14,673	5,329	40,401
69	170	11,730	4,761	28,900
72	180	12,960	5,184	32,400
70	200	14,000	4,900	40,000
72	190	13,680	5,184	36,100
66	175	11,550	4,356	30,625
72	205	14,760	5,184	42,025
72	185	13,320	5,184	34,225
74	186	13,764	5,476	34,596
640	1,692	120,437	45,558	319,272

For our formula, $\Sigma x = 640$, $\Sigma y = 1692$, $\Sigma xy = 120{,}437$, $\Sigma x^2 = 45{,}558$, and $\Sigma y^2 = 319{,}272$. Also, $n = 9$. Now we can plug into the formula to calculate r.

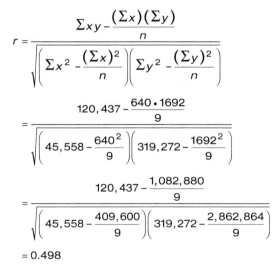

$$r = \frac{\Sigma xy - \dfrac{(\Sigma x)(\Sigma y)}{n}}{\sqrt{\left(\Sigma x^2 - \dfrac{(\Sigma x)^2}{n}\right)\left(\Sigma y^2 - \dfrac{(\Sigma y)^2}{n}\right)}}$$

$$= \frac{120{,}437 - \dfrac{640 \cdot 1692}{9}}{\sqrt{\left(45{,}558 - \dfrac{640^2}{9}\right)\left(319{,}272 - \dfrac{1692^2}{9}\right)}}$$

$$= \frac{120{,}437 - \dfrac{1{,}082{,}880}{9}}{\sqrt{\left(45{,}558 - \dfrac{409{,}600}{9}\right)\left(319{,}272 - \dfrac{2{,}862{,}864}{9}\right)}}$$

$$= 0.498$$

The correlation coefficient is 0.498. The correlation is positive, but we cannot say that it is strong. We could describe the strength of the correlation as moderate. The coefficient of determination is 0.248. This tells us that only 24.8% of the variation in weight can be explained by the variation in height. ∎

Before moving on, there is one common misconception that should be addressed here. Correlation between two variables does not imply that there is a cause-and-effect relationship; it simply says that there is a linear relation or association between them. That determination is up to experts in the field, not statisticians. An experiment could be set up to eliminate other possible factors to determine whether there is a cause-and-effect relationship between the two variables. Also, rarely is a dependent variable influenced by a single variable, but instead is influenced by several variables. Just because you have found two variables that have a relation, do not hesitate to search for other independent variables that may play a role as well.

In general, the paired data that we are dealing with represent a sample. The correlation coefficient r is a sample statistic. The population correlation coefficient is ρ (rho). There is a hypothesis test that we can use with regard to the population correlation coefficient ρ. We can test to determine whether there is a significant correlation between the two variables ($\rho \neq 0$), as well as test whether the correlation is positive ($\rho > 0$) or negative ($\rho < 0$). The test statistic for this test is

$$t = \frac{r}{\sqrt{\dfrac{1 - r^2}{n - 2}}}, \qquad \text{degrees of freedom: } n - 2$$

Here is an example of a hypothesis test.

EXAMPLE 10.6

Is there a correlation between the height of a tree and its circumference? Here are the heights and circumferences, both in feet, of 12 giant sequoia trees. At the 0.05 level of significance, test the claim that there is a significant correlation between the height and circumference of giant sequoia trees.

Tree	Height (ft.)	Circumference (ft.)
1	274.9	102.6
2	246.1	101.1
3	267.4	107.6
4	240.9	93.0
5	255.8	98.3
6	243.0	109.0
7	257.5	85.3
8	268.8	113.0
9	223.8	94.8
10	270.3	104.2
11	247.8	91.3
12	254.7	88.3

It is difficult to assign one of these variables as the independent variable. Does a tree's height affect its circumference? Does a tree's circumference affect its height? In such a case, keep in mind that we can choose either variable to be the independent variable. If we switch x and y, we will get the exact same value for r. Here we arbitrarily choose the height to be x, and let y represent the circumference.

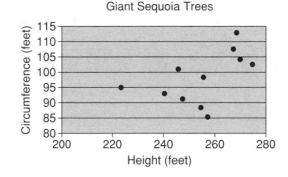

It seems to make sense here that there should be a positive correlation between these two variables. As a tree gets older, it grows taller and its circumference increases. The sample correlation coefficient r calculates to be 0.389 (verify this yourself).

Step 1

Claim in words: There is a significant correlation between the height and circumference of giant sequoia trees.

Claim: $\rho \neq 0$

Complement: $\rho = 0$

H_0: $\rho = 0$

H_A: $\rho \neq 0$

Step 2

Level of significance: $\alpha = 0.05$

Step 3

Test statistic: $t = \dfrac{r}{\sqrt{\dfrac{1 - r^2}{n - 2}}}$, $d.f. = n - 2$

Step 4

$\alpha = 0.05$, two-tailed test, $12 - 2 = 10$ degrees of freedom

Decision rule: Reject H_0 if $t > 2.228$ or if $t < -2.228$.

Step 5

$$t = \frac{r}{\sqrt{\dfrac{1 - r^2}{n - 2}}}$$

$$= \frac{0.389}{\sqrt{\dfrac{1 - 0.389^2}{12 - 2}}}$$

$$= 1.335$$

Decision: Fail to reject H_0.

Conclusion: There is not sufficient sample evidence to support the claim that there is a significant correlation between the height and circumference of giant sequoia trees.

p-value

Using techniques developed in Section 7.2, the best that we can say is that the p-value is between 0.2 and 0.5. A calculator or computer would tell us that the actual p-value of this test is 0.2115. ∎

MICROSOFT EXCEL | Correlation

Microsoft Excel can help us to create a scatterplot for a set of paired data and also calculate the correlation coefficient. We will rework the example involving the giant sequoia trees, using Excel to construct a scatterplot and calculate the correlation coefficient r.

Here are the heights and circumferences, both in feet, of 12 giant sequoia trees. Construct a scatterplot for these two variables, treating height as x. Also, calculate the correlation coefficient r.

Tree	Height (ft.)	Circumference (ft.)
1	274.9	102.6
2	246.1	101.1
3	267.4	107.6
4	240.9	93.0
5	255.8	98.3
6	243.0	109.0
7	257.5	85.3
8	268.8	113.0
9	223.8	94.8
10	270.3	104.2
11	247.8	91.3
12	254.7	88.3

In a new Excel worksheet, type the heights in column **A,** from cell **A1** down to **A12.** In the next column, **B,** type the circumferences in cells **B1** through **B12.** When creating a scatterplot, it is crucial that you type the values of x to the left of the values of y. Also be sure that the data are paired after you type them in; in other words, each x value should be next to its corresponding y value.

Highlight the data by clicking on cell **A1** and dragging the mouse to cell **B12** before releasing the button. Now, from the **Insert** menu select **Chart.** This starts the Chart Wizard. In Step 1, select **XY (Scatter)** under **Chart Type.** Click on **Next** to advance to Step 2. There is nothing that we need to do in Step 2, so click on **Next** to advance to Step 3.

In Step 3, we can change the appearance of our scatterplot. Click on the **Titles** tab to add a title to our graph, as well as labeling the x-axis (Height) and y-axis (Circumference). If you click on the **Legend** tab you can get rid of the box to the right of the graph. Click on the box labeled **Show Legend** so the check disappears. Click on **Next** to advance to Step 4. In Step 4 you can decide whether you want your scatterplot to appear on the same worksheet or have its own page. After you make this decision, click on **Finish.** Here is an example of what you should see.

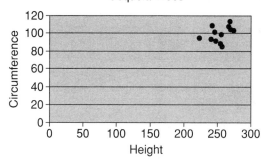

If you are unhappy with all of the points being in the upper right-hand corner, you can fix that by adjusting the x- and/or y-axis. Note that all of the heights are between 200 and 300 feet. Right-click with your mouse on the values on the x-axis, and select **Format Axis.** Click on the **Scale** tab, and change the minimum value to 200 instead of 0. You can adjust the maximum value also if you wish to. Click **OK** for the changes to take effect. Here is an example of what you should see after changing the scale.

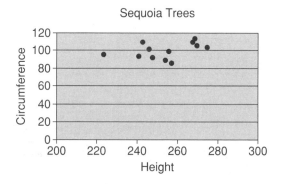

Note that all of the points are at the top of the graph, with circumferences between 80 feet and 120 feet. We can change the scale on the y-axis in the same fashion. Here is what you should see after changing the minimum value to 80 instead of 0.

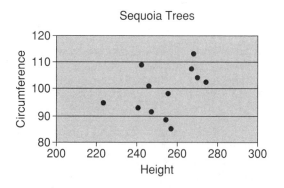

The strength of the correlation seems to change as we change the scale. The scatterplot helps us to get an idea of what type of correlation we have, but we must calculate the correlation coefficient *r* to be sure. Excel has a built-in function to calculate the correlation coefficient.

=CORREL(x cell range, y cell range)

For our example, type the following in any cell.

=CORREL(A1:A12,B1:B12)

The result that Excel gives us, rounded to four decimal places, is 0.389. ■

 Correlation

The TI-83 can help us to create a scatterplot for a set of paired data. With a minor adjustment, it can also help us calculate the correlation coefficient through one of its built-in functions. We will rework the example involving the giant sequoia trees, using the TI-83 to construct a scatterplot and calculate the correlation coefficient *r*.

EXAMPLE 10.8

Here are the heights and circumferences, both in feet, of 12 giant sequoia trees. Construct a scatterplot for these two variables, treating height as *x*. Also, calculate the correlation coefficient *r*.

Tree	Height (ft.)	Circumference (ft.)
1	274.9	102.6
2	246.1	101.1
3	267.4	107.6
4	240.9	93.0
5	255.8	98.3
6	243.0	109.0
7	257.5	85.3
8	268.8	113.0
9	223.8	94.8
10	270.3	104.2
11	247.8	91.3
12	254.7	88.3

In list L_1 enter the heights of the trees and in list L_2 enter the circumferences. Be sure that the data are paired after you type them in; in other words, each *x* value should be next to its corresponding *y* value. To construct the scatterplot, press (2nd) (Y=) to access the **Stat Plot** menu, which looks like the following.

```
STAT PLOTS
1:Plot1…Off
   L1   L2
2:Plot2…Off
   L1   L2
3:Plot3…Off
   L1   L2
4↓PlotsOff
```

Highlight number 1 and press the (ENTER) key. Be sure that **On** is highlighted on the first line of the display. Next to type, we want to highlight the first option to construct a scatterplot, which looks like (⌊∴). Next to **Xlist** enter L_1, and enter L_2 next to **Ylist.** Next to **Mark,** you have the choice of three different ways to display the points on the scatterplot. Pick the one you want by highlighting it and press (ENTER).

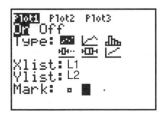

Now press the (GRAPH) key and your scatterplot should appear looking like this.

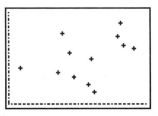

Your calculator may need a minor adjustment in order to calculate the correlation coefficient *r*. Press the (2nd) key, followed by (0) to access the **Catalog** menu. Scroll down until you find the choice DiagnosticOn. When the cursor is next to it, as shown here,

```
CATALOG            ▣
  DependAsk
  DependAuto
  det(
  DiagnosticOff
▶ DiagnosticOn
  dim(
  Disp
```

press the (ENTER) key twice.

Now to calculate the correlation coefficient, we will use a tool that will be explained in full in the next section. Press the (STAT) key, move to the **Calc** menu, and select option **8: LinReg (a+bx).** When you are brought back to the main screen, press (( )(2nd)(1)(,)(2nd)(2)( )). The screen should look like this.

```
LinReg(a+bx) (L1
,L2)
```

Press the (ENTER) key, and you will see r, as well as r^2. Ignore the other values until the next section.

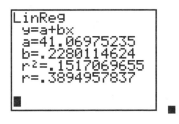

For Exercises 1–4, create a scatterplot for the given data. Use your scatterplot to determine whether there is a positive correlation, negative correlation, or no correlation between these two variables. (Do *not* calculate r.) If there is a correlation, do you feel that it is weak or strong? Explain your responses in your own words.

1. Here are the shoe sizes of 10 randomly selected men, along with their heights in inches. (Treat shoe size as the independent variable x.)

Shoe Size	4	7	7.5	9	9.5
Height (in.)	62	64	66	68	68

Shoe Size	10.5	9	10	11	11.5
Height (in.)	69	69	70	71	72

2. Here are the waist and hip measurements, in inches, of twelve 4-year-old dance students. (Treat the waist measurements as the independent variable x.)

Waist	24	26	22	23	23	25
Hips	29	31	28	28	28	31

Waist	24	27	23	23	25	31
Hips	30	34	28	28	28	37

(*Source:* Dancer's Edge dance studio.)

3. Here are the midterm exam scores of 20 algebra students, along with their scores on the final exam. (Treat the midterm scores as the independent variable x.)

Student	Midterm	Final	Student	Midterm	Final
1	65	85	11	38	56
2	50	77	12	92	95
3	75	90	13	59	60
4	70	84	14	66	87
5	68	61	15	61	83
6	58	70	16	71	77
7	49	76	17	68	79
8	92	78	18	82	74
9	68	80	19	85	94
10	93	92	20	67	80

4. Here are the scores of 16 randomly selected statistics students on a test on Unit 3 and on the final exam. (Treat the Unit 3 scores as the independent variable x.)

Student	Unit 3	Final	Student	Unit 3	Final
1	68	59	9	84	81
2	93	91	10	74	85
3	90	75	11	90	84
4	97	100	12	54	78
5	97	98	13	78	95
6	48	59	14	86	94
7	89	89	15	56	86
8	96	97	16	79	57

For Exercises 5–8, calculate the correlation coefficient r for the given data and compare your result to your responses from Exercises 1–4.

5. Here are the shoe sizes of 10 randomly selected men, along with their heights in inches. (Treat shoe size as the independent variable x.)

Shoe Size	4	7	7.5	9	9.5
Height (in.)	62	64	66	68	68

Shoe Size	10.5	9	10	11	11.5
Height (in.)	69	69	70	71	72

6. Here are the waist and hip measurements, in inches, of twelve 4-year-old dance students. (Treat the waist measurements as the independent variable x.)

Waist	24	26	22	23	23	25
Hips	29	31	28	28	28	31

Waist	24	27	23	23	25	31
Hips	30	34	28	28	28	37

(*Source:* Dancer's Edge dance studio.)

7. Here are the midterm exam scores of 20 algebra students, along with their scores on the final exam. (Treat the midterm scores as the independent variable x.)

Student	Midterm	Final	Student	Midterm	Final
1	65	85	11	38	56
2	50	77	12	92	95
3	75	90	13	59	60
4	70	84	14	66	87
5	68	61	15	61	83
6	58	70	16	71	77
7	49	76	17	68	79
8	92	78	18	82	74
9	68	80	19	85	94
10	93	92	20	67	80

8. Here are the scores of 16 randomly selected statistics students on a test on Unit 3 and on the final exam. (Treat the Unit 3 scores as the independent variable x.)

Student	Unit 3	Final	Student	Unit 3	Final
1	68	59	9	84	81
2	93	91	10	74	85
3	90	75	11	90	84
4	97	100	12	54	78
5	97	98	13	78	95
6	48	59	14	86	94
7	89	89	15	56	86
8	96	97	16	79	57

9. Here are the gross values, in millions of dollars, of milk produced and cattle raised in Tulare County, California, for the ten years from 1989 through 1998.

Year	Milk ($ millions)	Cattle ($ millions)
1989	287	177
1990	363	214
1991	413	212
1992	411	237
1993	455	238
1994	477	223
1995	547	223
1996	569	229
1997	712	252
1998	718	271

(Source: Visalia Times Delta.)

Treating the value of milk production as the independent variable, calculate the correlation coefficient r and the coefficient of determination r^2. Explain what the coefficient of determination tells us for this problem.

10. A tutorial lab on campus offers free tutoring for any student on campus. Here are the GPAs for 12 randomly selected students, and the number of tutoring appointments that those students missed.

GPA	2.66	2.05	2.07	2.62	1.30	3.00	3.25	2.58	2.36	2.81	3.11	2.56
Missed Appointments	3	1	2	0	7	0	2	0	3	1	1	2

Is there a relation between a student's GPA and the number of tutoring appointments that the student missed? Treating the student GPAs as the independent variable, calculate the correlation coefficient r and the coefficient of determination r^2. Explain what the coefficient of determination tells us for this problem.

11. Wilt Chamberlain played 14 NBA seasons in Philadelphia, San Francisco, and Los Angeles. Here are his point totals for those 14 seasons.

Season	Team	Points	Rebounds
1959–60	Philadelphia	2707	1941
1960–61	Philadelphia	3033	2149
1961–62	Philadelphia	4029	2052
1962–63	San Francisco	3586	1946
1963–64	San Francisco	2948	1787
1964–65	San Francisco	2534	1673
1965–66	Philadelphia	2649	1943
1966–67	Philadelphia	1956	1957
1967–68	Philadelphia	1992	1952
1968–69	Los Angeles	1664	1712
1969–70	Los Angeles	328	221
1970–71	Los Angeles	1696	1493
1971–72	Los Angeles	1213	1572
1972–73	Los Angeles	1084	1526

(a) Construct a scatterplot for these data, and calculate the correlation coefficient. (Since the data represent a population, this is ρ.)

(b) Note that the point on the scatterplot associated with the 1969–70 season is far removed from the rest of the points. (Wilt was injured for most of the season.) If we disregard that point, we will be able to adjust the scale on our axes to get a better view of the other points and how they are related. Construct a scatterplot with this point removed.

(c) Calculate the correlation coefficient without including data from the 1969–70 season. Does this radically change the coefficient that was calculated in part (a)?

12. It seems that there should be a strong relation between one season's NBA ticket prices and the previous season's prices. Here are the average ticket prices for NBA arenas for the 1998–99 and 1999–2000 seasons. Calculate the correlation coefficient for these two variables. (Since the data represent a population, this is ρ.)

Team	1998–99	1999–2000	Team	1998–99	1999–2000
New York	79.34	86.82	Sacramento	34.11	44.68
L.A. Lakers	51.11	81.89	Philadelphia	41.96	44.26
Seattle	63.47	64.60	Orlando	44.46	44.18
Houston	58.18	62.63	L.A. Clippers	31.75	43.89
Washington	61.40	59.65	Toronto	26.17	42.76
New Jersey	49.24	59.22	Dallas	34.84	40.76
Utah	43.47	54.60	Detroit	33.32	40.04
Chicago	53.17	52.84	Cleveland	39.75	39.75
Portland	52.28	52.28	Minnesota	38.61	39.08
Boston	49.79	49.79	San Antonio	38.01	38.92
Indiana	43.36	48.39	Denver	30.53	38.34
Golden State	36.79	48.10	Vancouver	31.90	34.71
Miami	36.55	46.57	Charlotte	28.12	32.04
Atlanta	36.79	45.75	Milwaukee	29.06	30.83
Phoenix	48.84	45.39			

(Source: Team Marketing Report, USA Today.)

13. Barry Sanders was an NFL running back with the Detroit Lions for 10 NFL seasons. Here are his yearly statistics for the 1989 through 1998 seasons. Included are the number of rushing attempts, number of rushing touchdowns, and the number of rushing yards and receiving yards. Calculate the correlation coefficient between the number of rushing attempts and the number of rushing yards.

Year	Attempts	Rush TD	Rushing Yards	Receiving Yards
1989	280	14	1470	282
1990	255	13	1304	480
1991	342	16	1548	307
1992	312	9	1352	225
1993	243	3	1115	205
1994	331	7	1883	283
1995	314	11	1500	398
1996	307	11	1553	147
1997	335	11	2053	305
1998	343	4	1491	289

14. Jerry Rice began his NFL career as a wide receiver with the San Francisco Forty-Niners in 1985. Here are the number of receptions, receiving yards, and touchdown receptions for the 1985–1996 seasons. Calculate the correlation coefficient between the number of receptions and receiving yards.

Year	Receptions	Yards	Touchdowns
1985	49	927	3
1986	86	1570	15
1987	65	1078	22
1988	64	1306	9
1989	82	1483	17
1990	100	1502	13
1991	80	1206	14
1992	84	1201	10
1993	98	1503	15
1994	112	1499	13
1995	122	1848	15
1996	108	1254	8

15. Is there a relation between family income in a city and the price of homes in that city? Here are the median family incomes for 16 California cities, and the median sales price for homes in those cities. Calculate the correlation coefficient *r*. Then, at the 0.05 level of significance, test the claim that there is a significant correlation between these two variables.

City	Median Income ($ thousands)	Median Sales Price ($ thousands)
Bakersfield	38.7	94
Riverside	47.2	133
Modesto	43.1	128
Visalia	34.3	98
Sacramento	51.9	158
Redding	37.5	113
Fresno	37.2	109
Merced	36.9	122
Stockton	44.3	154
Ventura	65.3	235
Los Angeles	51.3	189
Santa Barbara	52.1	210
San Diego	52.5	208
San Luis Obispo	48.0	192
San Jose	82.6	355
San Francisco	72.4	407

(*Source: National Association of Home Builders,* Fresno Bee.)

16. For a random sample of 20 dairy cows, here are the number of pounds of milk produced during their first lactation (after their first calf) and their second lactation (after their second calf). Is there a relation between these two variables? Calculate the correlation coefficient *r* for these data. Then, at the 0.05 level of significance, test the claim that there is a significant correlation between the amount of milk produced during the first lactation and the second lactation.

Cow	First Lactation	Second Lactation	Cow	First Lactation	Second Lactation
1	22,792	29,655	11	15,058	8,013
2	31,693	23,817	12	16,681	23,301
3	18,367	35,360	13	21,002	32,529
4	17,440	21,848	14	17,987	23,620
5	29,798	29,828	15	15,968	26,437
6	28,540	33,898	16	23,580	26,227
7	23,661	22,444	17	14,334	25,529
8	39,242	23,648	18	24,030	27,368
9	21,574	25,303	19	20,392	29,527
10	34,542	34,812	20	24,347	25,592

17. If a company has a quick Web server, does that mean that it is reliable as well? Here is a listing of 10 major electronic commerce Web sites. For each site, the average length of time for the site to come up on a user's computer and the percentage of time the site is available are shown. Calculate the correlation coefficient *r*. Then, at the 0.05 level of significance, test the claim that there is a significant correlation between these two variables.

Site	Seconds	Availability (%)
Amazon.com	17.66	90.1
Barnesandnoble.com	15.85	94.2

(*continues*)

(continued)

Site	Seconds	Availability (%)
CDnow	15.52	89.2
eBay	17.25	78.4
eToys.com	20.11	84.0
Gateway	19.01	86.1
Landsend.com	13.49	94.2
Macys.com	31.42	93.0
Wal Mart Online	22.24	92.0
Wine.com	20.58	93.6

(Source: Keynote Systems.)

18. Many politicians and citizen groups complain about the cost of prescription medications in the United States. Here are the prices of a dose of 10 medications in Canada and the United States (all in U.S. dollars).

Drug	Canada	United States
Prilosec	1.47	3.31
Prozac	1.07	2.27
Lipitor	1.34	2.54
Prevacid	1.34	3.13
Epogen	21.44	23.40
Zocor	1.47	3.16
Zoloft	1.07	1.98
Zyprexa	3.39	5.27
Claritin	1.11	1.96
Paxil	1.13	2.22

(Source: USA Today.)

(a) Calculate the correlation coefficient r. Then, at the 0.01 level of significance, test the claim that there is a significant correlation between these two variables.
(b) Construct a scatterplot for these data.
(c) Note that the point on the scatterplot associated with Epogen is far removed from the rest of the points. If we disregard that point, we will be able to adjust the scale on our axes to get a better view of the other points and how they are related. Construct a scatterplot with this point removed.
(d) Calculate the correlation coefficient without including the data for Epogen. Does this radically change the coefficient that was calculated in part (a)?

19. For eight NHL goalies in the season's second month, here are the number of minutes played by the goalie, the number of goals given up by the goalie, and the number of shots attempted against the goalie.

Minutes	788	661	783	831	476	643	767	608
Goals	22	20	28	34	24	30	39	32
Shots	335	258	355	413	243	313	339	295

(a) Calculate the correlation coefficient between the number of minutes played and the number of goals given up.
(b) Calculate the correlation coefficient between the number of shots attempted and the number of goals given up.
(c) Based on your results, would the number of minutes played or the number of shots attempted be a better predictor of the number of goals given up?

20. Cal Ripken set a major league record by playing in 2424 consecutive games for the Baltimore Orioles between 1982 and 1998. Here are his at-bats (AB), runs (R), hits (H), home runs (HR), runs batted in (RBI), walks (BB), and strike-outs (SO).

Year	AB	R	H	HR	RBI	BB	SO
1982	598	90	158	28	93	46	95
1983	663	121	211	27	102	58	97
1984	641	103	195	27	86	71	89
1985	642	116	181	26	110	67	68
1986	627	98	177	25	81	70	60
1987	624	97	157	27	98	81	77
1988	575	87	152	23	81	102	69
1989	646	80	166	21	93	57	72
1990	600	78	150	21	84	82	66
1991	650	99	210	34	114	53	46
1992	637	73	160	14	72	64	50
1993	641	87	165	24	90	65	58
1994	444	71	140	13	75	32	41
1995	550	71	144	17	88	52	59
1996	640	94	178	26	102	59	78
1997	615	79	166	17	84	56	73
1998	601	65	163	14	61	51	68

(a) Calculate the correlation coefficient between the number of at-bats and the number of runs batted in.
(b) Calculate the correlation coefficient between the number of hits and the number of runs batted in.
(c) Calculate the correlation coefficient between the number of home runs and the number of runs batted in.
(d) Based on your results, would the number of at-bats, the number of hits, or the number of home runs be a better predictor of the number of runs batted in?

Exercises 21–25 use the following data. For the 50 states and Washington, D.C., here are the 1999

- average ACT composite scores (ACT)
- percentage of high school graduates that took the ACT (ACT %)
- average SAT verbal score (SAT V)
- average SAT math score (SAT M)
- percentage of high school graduates that took the SAT (SAT %)

State	ACT	ACT %	SAT V	SAT M	SAT %
AL	20.2	65	561	555	9
AK	21.1	35	516	514	50
AZ	21.4	28	524	524	34
AR	20.3	69	563	556	6
CA	21.3	12	497	514	49
CO	21.5	62	536	540	32
CT	21.6	3	510	509	80
DE	20.5	3	503	497	67
DC	18.6	13	494	478	77
FL	20.6	39	499	498	53
GA	20.0	16	487	482	63
HI	21.6	18	482	513	52
ID	21.4	60	542	540	16
IL	21.4	67	569	585	12
IN	21.2	19	496	498	60
IA	22.0	66	594	598	5
KS	21.5	75	578	576	9
KY	20.1	68	547	547	12
LA	19.6	76	561	558	8

(continues)

(continued)

State	ACT	ACT %	SAT V	SAT M	SAT %
ME	22.1	4	507	503	68
MD	20.9	10	507	507	65
MA	22.0	6	511	511	78
MI	21.3	69	557	565	11
MN	22.1	64	586	598	9
MS	18.7	82	563	548	4
MO	21.6	67	572	572	8
MT	21.8	54	545	546	21
NE	21.7	41	568	571	8
NV	21.5	5	512	517	34
NH	22.2	5	520	518	72
NJ	20.7	4	498	510	80
NM	20.1	64	549	542	12
NY	22.0	14	495	502	76
NC	19.4	12	493	493	61
ND	21.4	79	594	605	5
OH	21.4	59	534	538	25
OK	20.6	69	567	560	8
OR	22.6	11	525	525	53
PA	21.4	7	498	495	70
RI	22.7	3	504	499	70
SC	19.1	18	479	475	61
SD	21.2	70	585	588	4
TN	19.9	77	559	553	13
TX	20.3	31	494	499	50
UT	21.4	68	570	565	5
VT	21.9	9	514	506	70
VA	20.6	7	508	499	65
WA	22.6	18	525	526	52
WV	20.2	58	527	512	18
WI	22.3	67	584	595	7
WY	21.4	66	546	551	10

(Source: College Entrance Examination Board, American College Testing Program.)

21. Calculate the correlation coefficient for SAT verbal scores and SAT math scores.
22. Calculate the correlation coefficient for ACT scores and the composite SAT scores (SAT verbal score + SAT math score).
23. A state superintendent of schools, when asked about her state's low scores, claims that the low scores are due to the high percentage of high school graduates that take the test. "There is a significant negative correlation between scores and the percentage of high school graduates that take the SAT."

    (a) Calculate the correlation coefficient for the percentage of graduates who took the SAT and the composite SAT scores (SAT verbal score + SAT math score).
    (b) Based on your results, does the superintendent's statement appear to be valid?

24. Calculate the correlation coefficient for the percentage of graduates who took the ACT and the ACT scores. How does this coefficient compare to the comparable coefficient for the SAT calculated in the previous exercise?
25. Calculate the correlation coefficient for the percentage of graduates who took the ACT and the percentage of graduates who took the SAT. Does this correlation make sense? Explain in your own words, and include a scatterplot to support your argument.
26. The results of 47 horse races at Santa Anita Park were selected at random. Here are the number of horses in the race, and the price that the winning horse in that race paid to win.

Race	Number of Horses	Winning Price	Race	Number of Horses	Winning Price
1	7	9.20	25	7	5.80
2	9	26.60	26	10	8.80
3	4	4.40	27	10	64.00
4	9	19.20	28	7	8.40
5	10	23.40	29	6	6.20
6	8	27.40	30	12	14.80
7	8	14.20	31	12	4.00
8	9	15.20	32	7	8.40
9	6	14.80	33	6	6.20
10	9	5.60	34	12	14.80
11	8	7.80	35	12	4.00
12	9	17.00	36	6	3.00
13	7	8.80	37	10	8.80
14	10	15.40	38	7	5.80
15	10	17.40	39	10	64.00
16	11	22.80	40	7	8.80
17	11	8.00	41	7	10.80
18	12	21.00	42	9	8.60
19	5	21.80	43	8	7.00
20	11	7.40	44	11	12.80
21	12	33.20	45	9	8.20
22	10	5.00	46	11	62.80
23	10	8.80	47	12	10.80
24	6	3.00			

(a) Calculate the correlation coefficient for the number of horses in the race and the winning price.

(b) At the 0.05 level of significance, test the claim that there is a significant correlation between these two variables.

27. For 100 randomly selected community college students, here are the number of units they are enrolled in and the number of hours that they study per week.

Units	Study Hours	Units	Study Hours	Units	Study Hours	Units	Study Hours
12	4	15	9	6	9	16	5
9	3	18	7	15	7	6	2
16	6	12	3	16	9	18	7
12	4	21	9	21	11	16	5
21	8	12	3	6	3	12	3
15	3	9	3	9	3	21	9
12	0	16	10	12	6	4	1
12	2	9	3	12	4	9	2
15	4	12	5	15	6	12	5
12	5	6	1	12	4	12	6
16	7	15	4	21	9	15	4
9	3	16	4	16	9	12	3
16	5	16	3	12	6	14	7
15	6	12	4	9	2	9	2
21	9	21	9	12	3	16	5
18	9	12	8	16	6	12	4
16	7	15	8	9	2	9	3
12	3	14	7	12	4	12	4
12	5	9	2	12	9	16	5

(continues)

(continued)

Units	Study Hours	Units	Study Hours	Units	Study Hours	Units	Study Hours
15	4	16	5	12	4	12	3
16	5	12	4	15	5	16	5
12	4	15	7	16	6	9	2
9	2	9	3	12	5	15	7
6	2	12	3	12	3	12	3
12	4	12	5	9	2	12	6

(a) Calculate the correlation coefficient for the number of units and the number of study hours per week, using the data from only the first 25 students (first column).

(b) Use the result of part (a) to test the claim that there is a significant correlation between these two variables at the 0.01 level of significance.

(c) Calculate the correlation coefficient between the number of students and the number of study hours per week, using the data from the complete sample of 100 students.

(d) Use the result of part (c) to test the claim that there is a significant correlation between these two variables at the 0.01 level of significance.

(e) Note that the test statistic is higher for the sample of 100 students than it is for the 25 students, even though the correlation coefficient was actually greater for the 25 students. Explain why this happened in your own words.

28. For 71 randomly selected 1999 National League baseball games, here are the times required to complete the games (in minutes), the total number of hits in the game, and the combined number of runs in the game.

Time	Hits	Runs	Time	Hits	Runs	Time	Hits	Runs
170	19	11	152	20	14	160	14	3
187	20	11	190	23	9	177	20	7
169	14	6	213	30	22	171	16	9
129	12	1	158	19	11	198	26	22
159	25	15	176	20	9	159	21	10
209	27	19	146	13	6	160	9	3
181	16	9	172	9	4	146	16	10
183	17	11	137	13	6	188	14	6
197	23	12	200	16	9	149	16	6
170	19	13	158	19	13	151	13	6
157	20	6	181	25	13	172	18	7
214	32	27	202	21	11	143	12	6
154	15	6	152	14	3	189	19	11
166	23	12	128	11	8	204	27	19
191	21	11	161	18	12	215	24	13
178	17	6	153	8	3	181	16	10
198	23	14	164	19	13	141	14	5
178	17	13	169	19	9	152	18	4
161	13	6	194	24	14	176	20	10
198	20	6	108	10	5	180	19	14
179	17	10	146	19	5	158	12	4
150	21	13	157	12	6			
265	20	9	167	16	13			
136	18	16	169	16	7			
143	15	3	190	22	12			

(a) Calculate the correlation coefficient between the number of hits and the time required to complete the game.

(b) At the 0.01 level of significance, test the claim that there is a significant correlation between the number of hits and the time required to complete the game.

    (c) Calculate the correlation coefficient between the number of runs and the time required to complete the game.

    (d) At the 0.01 level of significance, test the claim that there is a significant correlation between the number of runs and the time required to complete the game.

    (e) Based on your results, would the number of hits or the number of runs be better for predicting the time required to complete a game?

SECTION 10.2
Linear Regression

If we can determine that there is a linear correlation between two variables, then the behavior of those two variables can be described graphically by a line. In this section we learn how to find the equation of the line that best fits a set of data. We will go on to use that equation to predict the value of one of the variables for a particular value of the other variable.

A **regression line** is a line that best fits a set of data. First, we will learn how to find the equation of this line, and then later we will see what it means to best fit a set of data. The general formula of a regression line is $\hat{y} = a + bx$. In the equation $\hat{y}$, which is read "y-hat," is the predicted value of y for a given value of x. The slope of the line is b, and we calculate it first. We then use the value of b to help calculate a, which is the y-intercept of the line. Here are the formulas to calculate b and a.

$$b = \frac{n \cdot \Sigma xy - (\Sigma x) \cdot (\Sigma y)}{n \Sigma x^2 - (\Sigma x)^2}$$

$$a = \bar{y} - b\bar{x}$$

You may notice that the calculation of the slope b involves many of the same pieces as the formula for calculating the correlation coefficient r. Again, it would be best to use technology to find b and a, but we will use the formula for demonstration purposes in this section.

EXAMPLE 10.9

Here are the scores of five randomly selected students on test 1 and test 2 in a math class. Find the equation of the regression line, treating the score on test 1 as x and the score on test 2 as y.

Student	Test 1 Score	Test 2 Score
1	83	82
2	86	84
3	76	63
4	92	83
5	71	55

We begin by creating a scatterplot, to make sure that the data seem to have a linear relationship.

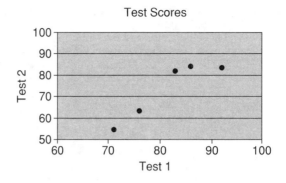

There seems to be a linear association, so we continue with the calculation of b, the slope of the regression line.

x	y	xy	x^2
83	82	6,806	6,889
86	84	7,224	7,396
76	63	4,788	5,776
92	83	7,636	8,464
71	55	3,905	5,041
408	367	30,359	33,566

So $\Sigma x = 408$, $\Sigma y = 367$, $\Sigma xy = 30{,}359$, $\Sigma x^2 = 33{,}566$. In the formula, n is the number of pairs, which is 5.

$$b = \frac{n \cdot \Sigma xy - (\Sigma x) \cdot (\Sigma y)}{n \Sigma x^2 - (\Sigma x)^2}$$

$$= \frac{5 \cdot 30{,}359 - 408 \cdot 367}{5 \cdot 33{,}566 - 408^2}$$

$$= \frac{151{,}795 - 149{,}736}{167{,}830 - 166{,}464}$$

$$= \frac{2059}{1366}$$

$$= 1.5073$$

The slope of this regression line is 1.5073. This tells us that for each point increase on test 1, the score on test 2 will increase by 1.5073 points. Now we calculate the y-intercept a.

$$\bar{x} = \frac{\Sigma x}{n} \qquad \bar{y} = \frac{\Sigma y}{n}$$

$$= \frac{408}{5} \qquad = \frac{367}{5}$$

$$= 81.6 \qquad = 73.4$$

$$a = \bar{y} - b\bar{x}$$

$$= 73.4 - (1.5073) \cdot 81.6$$

$$= -49.5957$$

The regression equation is $\hat{y} = -49.5957 + 1.5073x$. ∎

Now let's examine why this line best fits the set of data. Here is the graph of the regression line on the scatterplot of the data.

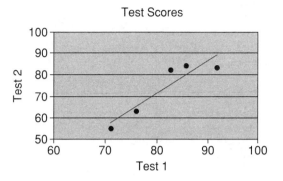

Note that the line does not go through any of the data points. The vertical distances between each point and the regression line are called the **residuals.**

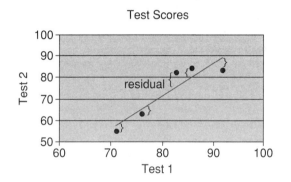

The equations for the slope and y-intercept are designed to minimize the sum of the squares of the residuals. In other words, if we took each residual and squared it, and then we totaled the results, this line would have the lowest possible total for this set of data. For this reason, the regression line is often called the **least-squares line.**

We can use the regression line to predict a y-value for a given x-value. Recall that the equation of the regression line for the previous example was $\hat{y} = -49.5957 + 1.5073x$, where x represents a student's score on test 1 and $\hat{y}$ is the predicted score for that student on test 2. Suppose that a student got a score of 95 on test 1. What score can we expect from that student on test 2? All we need to do is plug in 95 for x in the regression equation.

$$\hat{y} = -49.5957 + 1.5073x$$
$$= -49.5957 + 1.5073(95)$$
$$= -49.5957 + 143.1935$$
$$= 93.5978$$

The predicted score from the regression equation is 93.5978. We could round this to produce a predicted score of 94. This score seems to fit the pattern of the data.

How about predicting the score of a student who scored 50 on test 1? Plug in 50 for x in the regression equation.

$$\hat{y} = -49.5957 + 1.5073x$$
$$= -49.5957 + 1.5073(50)$$
$$= -49.5957 + 75.365$$
$$= 25.7693$$

The predicted score from the regression equation is 25.7693, which rounds to a score of 26. Does this value seem reasonable? It is hard to say, because we do not have a student in our sample data with a score on test 1 that is close to 50. When we plug in a value for x that is not close to the values of x on which the regression equation is based, this is called **extrapolation.** When we predict values using extrapolation, the results are not reliable. We should use a regression equation for values of the independent variable that are close to the values on which the equation is based.

As a further example of extrapolation, consider the y-intercept of the regression equation $\hat{y} = -49.5957 + 1.5073x$. The y-intercept of an equation is the y-value that corresponds to an x-value of 0. So, if a student had a score of 0 on test 1 we would predict a score of -49.5957 (or approximately -50) on test 2. Of course, this value makes no sense. Note that none of the test 1 scores on which the regression equation is based were close to 0. In fact, none of the scores were below 71. The y-intercept of some regression equations can be meaningful if there are x-values close to 0, representing some sort of initial condition.

Let's revisit an example from the previous section.

 EXAMPLE 10.10   Here are the number of hours that 10 students spent studying for a final exam, and their score on that exam. Find the equation of the regression line that best fits the data, and use it to predict the score of students who study for 0 hours, 5 hours, 10 hours, 15 hours, 20 hours, and 40 hours.

Hours	7	8	4	9	13
Score	70	76	57	77	91
Hours	5	9	6	16	3
Score	66	82	64	96	50

We begin with a scatterplot.

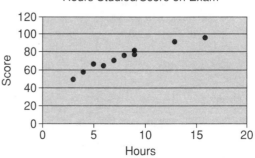

Hours Studied/Score on Exam

There appears to be a linear relationship to the data, so we continue with the calculations.

x	y	xy	x^2
7	70	490	49
8	76	608	64
4	57	228	16
9	77	693	81
13	91	1183	169
5	66	330	25
9	82	738	81
6	64	384	36
16	96	1536	256
3	50	150	9
80	729	6340	786

Thus, $\Sigma x = 80$, $\Sigma y = 729$, $\Sigma xy = 6340$, $\Sigma x^2 = 786$, and $n = 10$.

$$b = \frac{n \cdot \Sigma xy - (\Sigma x) \cdot (\Sigma y)}{n\Sigma x^2 - (\Sigma x)^2}$$

$$= \frac{10 \cdot 6340 - 80 \cdot 729}{10 \cdot 786 - 80^2}$$

$$= \frac{63,400 - 58,320}{7860 - 6400}$$

$$= \frac{5080}{1460}$$

$$= 3.4795$$

The slope of this regression line is 3.4795. This tells us that for each additional hour that a student studies, his or her score on the exam will increase by 3.4795 points. Now we calculate the y-intercept a.

$$\bar{x} = \frac{\Sigma x}{n} \qquad \bar{y} = \frac{\Sigma y}{n}$$

$$= \frac{80}{10} \qquad = \frac{729}{10}$$

$$= 8 \qquad = 72.9$$

$$a = \bar{y} - b\bar{x}$$

$$= 72.9 - (3.4795) \cdot 8$$

$$= 45.064$$

The regression equation is $\hat{y} = 45.064 + 3.4795x$. Here is the graph of the regression line. Note how well it fits. This is due to the fact that there is a strong correlation between the two variables. Recall from the last section that for these two variables $r = 0.969$, which is nearly perfect correlation.

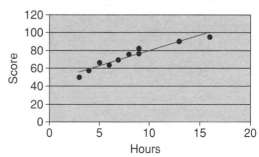

Hours Studied/Score on Exam

Now, for the predictions.

$x = 0$	$x = 5$	$x = 10$
$\hat{y} = 45.064 + 3.4795x$	$\hat{y} = 45.064 + 3.4795x$	$\hat{y} = 45.064 + 3.4795x$
$= 45.064 + 3.4795(0)$	$= 45.064 + 3.4795(5)$	$= 45.064 + 3.4795(10)$
$= 45.064$	$= 62.4615$	$= 79.859$

$x = 15$	$x = 20$	$x = 40$
$\hat{y} = 45.064 + 3.4795x$	$\hat{y} = 45.064 + 3.4795x$	$\hat{y} = 45.064 + 3.4795x$
$= 45.064 + 3.4795(15)$	$= 45.064 + 3.4795(20)$	$= 45.064 + 3.4795(40)$
$= 97.2565$	$= 114.654$	$= 184.244$

Here's a summary of these results in a table, with the scores rounded to the nearest whole number.

Hours	0	5	10	15	20	40
Score	45	62	80	97	115	184

Let's begin by investigating the y-intercept. The lowest number of hours studied was 3 hours by a student who had a score of 50, so a score of 45 for a student who did not study seems to fit. We can think of the equation as telling us that each student "begins" with a score of 45.064, and then gains 3.4795 points for each hour studied. The scores for 5 hours, 10 hours, and 15 hours seem to fit as well. There is a problem with the prediction for 20 hours of study time. The highest number of hours studied in the sample data was 16 hours, the next highest was 3 hours lower at 13 hours, and the rest were below 10 hours. So this extrapolation has some problems, especially considering that the highest score possible is 100. One problem with regression lines is that in any small "window" a set of data may appear to be linear, but if we expand outward the data may actually follow some sort of curve. Finally, the extrapolation for 40 hours of study time is obviously no good. ■

Here are the gross values, in millions of dollars, of nectarines and peaches grown in Tulare County, California, for the ten years from 1989 through 1998. First, find the correlation coefficient for these two variables. Then, find the equation of the regression line for these two variables. Also, predict the value of peaches produced for a year in which the value of nectarines produced is $60 million.

Year	Nectarines ($ millions)	Peaches ($ millions)
1989	47	32
1990	53	47
1991	52	57
1992	68	43
1993	51	53
1994	90	63
1995	74	76
1996	89	64
1997	83	66
1998	56	56

(Source: Visalia Times Delta.)

It is hard to determine which of these variables affects the other, but since we will be trying to predict the value of peaches for a certain value of nectarines, we will let *x* be the gross value of nectarines and *y* be the gross value of peaches.

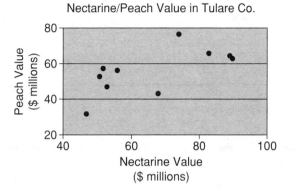

Nectarine/Peach Value in Tulare Co.

There seems to be a general positive trend. Since the correlation coefficient and regression equation involve many of the same sums, we will calculate them all first.

x	y	xy	x^2	y^2
47	32	1,504	2,209	1,024
53	47	2,491	2,809	2,209
52	57	2,964	2,704	3,249
68	43	2,924	4,624	1,849
51	53	2,703	2,601	2,809
90	63	5,670	8,100	3,969
74	76	5,624	5,476	5,776
89	64	5,696	7,921	4,096
83	66	5,478	6,889	4,356
56	56	3,136	3,136	3,136
663	557	38,190	46,469	32,473

For our formula, $\Sigma x = 663$, $\Sigma y = 557$, $\Sigma xy = 38,190$, $\Sigma x^2 = 46,469$, $\Sigma y^2 = 32,473$, and $n = 10$. Now we can plug into the formula to calculate r.

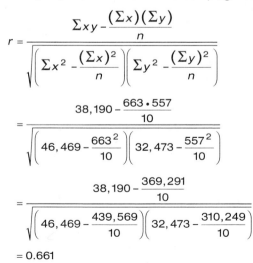

$$r = \frac{\Sigma xy - \dfrac{(\Sigma x)(\Sigma y)}{n}}{\sqrt{\left(\Sigma x^2 - \dfrac{(\Sigma x)^2}{n}\right)\left(\Sigma y^2 - \dfrac{(\Sigma y)^2}{n}\right)}}$$

$$= \frac{38,190 - \dfrac{663 \cdot 557}{10}}{\sqrt{\left(46,469 - \dfrac{663^2}{10}\right)\left(32,473 - \dfrac{557^2}{10}\right)}}$$

$$= \frac{38,190 - \dfrac{369,291}{10}}{\sqrt{\left(46,469 - \dfrac{439,569}{10}\right)\left(32,473 - \dfrac{310,249}{10}\right)}}$$

$$= 0.661$$

The correlation coefficient is 0.661, so we do have positive correlation. Now for the calculation of the regression equation.

$$b = \frac{n \cdot \Sigma xy - (\Sigma x) \cdot (\Sigma y)}{n \Sigma x^2 - (\Sigma x)^2}$$

$$= \frac{10 \cdot 38,190 - 663 \cdot 557}{10 \cdot 46,469 - 663^2}$$

$$= \frac{381,900 - 369,291}{464,690 - 439,569}$$

$$= \frac{12,609}{25,121}$$

$$= 0.5019$$

The slope of this regression line is 0.5019. This tells us that for each additional million dollars in nectarine production, the value of peaches produced will increase by 0.5019 million dollars, or $501,900. Now we calculate the y-intercept a.

$$\bar{x} = \frac{\Sigma x}{n} \qquad \bar{y} = \frac{\Sigma y}{n}$$

$$= \frac{663}{10} \qquad = \frac{557}{10}$$

$$= 66.3 \qquad = 55.7$$

$$a = \bar{y} - b\bar{x}$$

$$= 55.7 - (0.5019) \cdot 66.3$$

$$= 22.4240$$

The regression equation is $\hat{y} = 22.4240 + 0.5019x$. Here is the graph of the regression line.

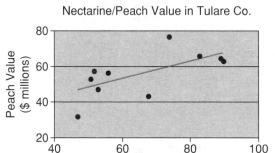

Nectarine/Peach Value in Tulare Co.

What will be the value of peaches produced in a year in which the value of nectarines produced is $60 million? Simply plug in 60 for x in the regression equation.

$$\hat{y} = 22.4240 + 0.5019x$$

$$= 22.4240 + 0.5019(60)$$

$$= 52.538$$

We predict the production of $52.538 million of peaches. ■

MICROSOFT EXCEL   Linear Regression

We can use Microsoft Excel to quickly find the regression equation, and even to draw the graph of our regression line over our scatterplot. We begin with the regression equation, reworking the example involving nectarines and peaches.

> **EXAMPLE 10.12**
>
> Here are the gross values, in millions of dollars, of nectarines and peaches grown in Tulare County, California, for the ten years from 1989 through 1998. Find the equation of the regression line for these two variables, letting x represent the value of nectarines produced and y the value of peaches produced.

Year	Nectarines ($ millions)	Peaches ($ millions)
1989	47	32
1990	53	47
1991	52	57
1992	68	43
1993	51	53
1994	90	63
1995	74	76
1996	89	64
1997	83	66
1998	56	56

(*Source:* Visalia Times Delta.)

In a new Excel worksheet, type the values of the nectarines in column **A,** from cell **A1** through **A10.** In the next column, type the values of the peaches beginning in cell **B1** and continuing through cell **B10.**

We need to use the Excel's Data Analysis ToolPak, so we must be sure that it has been added to the **Tools** menu. Click on the **Tools** menu; if you see **Data Analysis** then you may skip to the next paragraph. If you do not see **Data Analysis,** then click on **Add-Ins.** When a dialog box opens, check the box next to **Analysis ToolPak,** and then click **OK.**

From the **Tools** menu, select **Data Analysis.** When the dialog box appears, scroll down to select **Regression** and click **OK.** When the dialog box appears, type B1:B10 next to **Input Y Range.** Next to **Input X Range,** type A1:A10. Click on **OK,** and Excel will give you a great deal of information in a new worksheet. Most of this information will not be used until the next section. While all of the regression information is still highlighted, from the **Format** menu select **Column.** Then choose **AutoFit Selection.** This will make the information easier to see.

Look down the first column of regression information until you find **Intercept** and **X Variable 1.** Under the column labeled **Coefficients** you will find the value of a directly next to **Intercept,** and you will find the value of b directly next to **X Variable 1.** Here is a picture of what the output looks like.

SUMMARY OUTPUT

Regression Statistics	
Multiple R	0.661093529
R Square	0.437044654
Adjusted R Square	0.366675236
Standard Error	10.0946498
Observations	10

ANOVA

	df	SS	MS	F	Significance F
Regression	1	632.8843637	632.8843637	6.210718593	0.037397804
Residual	8	815.2156363	101.9019545		
Total	9	1448.1			

a

	Coefficients	Standard Error	t Stat	P-value	Lower 95%	Upper 95%
Intercept	22.42199753	13.72949169	1.63312656	0.141085013	–9.23828756	54.08228262
X Variable 1	0.501930656	0.201406181	2.492131335	0.037397804	0.037486868	0.966374443

b

The regression equation is $\hat{y} = 22.4240 + 0.5019x$ after rounding both coefficients to four decimal places.

To draw the regression line over the scatterplot, first create the scatterplot as we did in the last section. Once the scatterplot is finished, highlight the graph by clicking on it. It should look like the following, with eight boxes around the border.

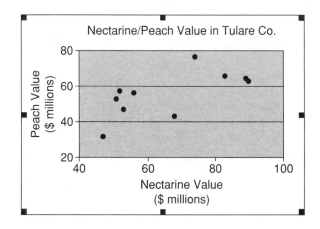

Click on the **Chart** menu, and select **Add Trendline.** The **Chart** menu only appears if the chart (graph) is highlighted. Click on the **Options** tab, and check the boxes next to **Display Equation on chart** and **Display R-squared value on chart,** and then click on **OK.** This will produce the following graph. If the equation or the R-squared value are in your way, just click on them and drag them to a more convenient location.

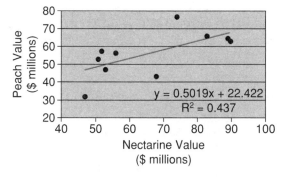

The value labeled R^2 is useful because we can use it to calculate r without using another function. Simply take the square root, and then determine whether r is positive or negative. The graph should tell you, but the slope of the regression line has the same sign as r. ■

TI-83   Linear Regression

The TI-83 can help us to find a linear regression equation quite easily, and then we can graph the regression line over the scatterplot to see just how well it fits the data. We will rework the example involving nectarines and peaches.

EXAMPLE 10.13

Here are the gross values, in millions of dollars, of nectarines and peaches grown in Tulare County, California for the ten years from 1989 through 1998. Find the equation of the regression line for these two variables, letting x represent the value of nectarines produced and y the value of peaches produced.

Year	Nectarines ($ millions)	Peaches ($ millions)
1989	47	32
1990	53	47
1991	52	57
1992	68	43
1993	51	53
1994	90	63
1995	74	76
1996	89	64
1997	83	66
1998	56	56

(*Source:* Visalia Times Delta.)

Enter the values of nectarines produced in list L_1, and enter the values of peaches produced in list L_2. Press (STAT), and then move to the right to select the **Calc** menu. Select option **8: LinReg (a+bx).** You will be taken to the main screen. Press ((2nd) (1) , (2nd) (2)) and then (ENTER). Here is the output that you should see.

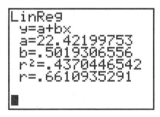

```
LinReg
 y=a+bx
 a=22.42199753
 b=.5019306556
 r²=.4370446542
 r=.6610935291
```

When we round the regression coefficients to four decimal places, we get the regression equation $\hat{y} = 22.4240 + 0.5019x$.

To draw the regression line over the scatterplot, we must create the scatterplot as we did in the last section. Now press the (Y=) key, and next to Y_1 type the regression equation. In this case that would be $22.4220 + 0.5019x$. To type the x, use the (X,T,q,n) key. Here is what the display should look like.

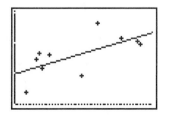

To clear the line from the graph window, press the (Y=) key, move the cursor to the equation whose graph you want removed, and press (CLEAR). To clear the scatterplot from the graph window, simply turn off the plot that created it. ∎

EXERCISES 10.2

Many of the data sets in these exercises were contained in the exercise set for Section 10.1. For problems that you are calculating by hand, the sums from the last section will come in handy. (Don't reinvent the wheel.)

1. Here are the shoe sizes of 10 randomly selected men, along with their heights in inches.

Shoe Size	4	7	7.5	9	9.5	10.5	9	10	11	11.5
Height (in.)	62	64	66	68	68	69	69	70	71	72

   (a) Find the regression equation that predicts height using shoe size.
   (b) Predict the height of a man whose shoe size is 8.
   (c) Predict the height of a man whose shoe size is 6.
   (d) Predict the height of NBA great Bob Lanier, whose shoe size is 22.
   (e) Explain, in your own words, why it is inappropriate to use the regression equation to predict the height of a man whose shoe size is 22.

2. Here are the waist and hip measurements, in inches, of twelve 4-year-old dance students.

Waist	24	26	22	23	23	25	24	27	23	23	25	31
Hips	29	31	28	28	28	31	30	34	28	28	28	37

   (*Source:* Dancer's Edge dance studio.)

   (a) Find the regression equation that predicts the size of a dancer's hips using her waist size.
   (b) Predict the size of a dancer's hips if her waist size is 24 inches.
   (c) Predict the size of a dancer's hips if her waist size is 28 inches.
   (d) Predict the size of a dancer's hips if her waist size is 38 inches.
   (e) Is there anything wrong with using the regression equation to predict the size of a dancer's hips in parts (b) through (d)? Explain why or why not for each prediction.

3. Here are the midterm exam scores of 20 algebra students, along with their scores on the final exam.

Student	Midterm	Final	Student	Midterm	Final
1	65	85	11	38	56
2	50	77	12	92	95
3	75	90	13	59	60
4	70	84	14	66	87
5	68	61	15	61	83
6	58	70	16	71	77
7	49	76	17	68	79
8	92	78	18	82	74
9	68	80	19	85	94
10	93	92	20	67	80

   (a) Find the regression equation that predicts a student's score on the final exam based on the student's score on the midterm.
   (b) Predict the final exam score of a student who scored 80 on the midterm.
   (c) Predict the final exam score of a student who scored 55 on the midterm.
   (d) What is the lowest score on the midterm that would produce a predicted score of 70 on the final exam?

4. Here are the scores of 16 randomly selected statistics students on a test on Unit 3 and on the final exam. (Treat the Unit 3 scores as the independent variable x.)

Student	Unit 3	Final	Student	Unit 3	Final
1	68	59	9	84	81
2	93	91	10	74	85
3	90	75	11	90	84
4	97	100	12	54	78
5	97	98	13	78	95
6	48	59	14	86	94
7	89	89	15	56	86
8	96	97	16	79	57

(a) Find the regression equation that predicts a student's score on the final exam based on the student's score on the Unit 3 exam.
(b) Predict the final exam score of a student who scored 75 on the Unit 3 exam.
(c) Predict the final exam score of a student who scored 20 on the Unit 3 exam.
(d) What is the lowest score on the Unit 3 exam that would produce a predicted score of 90 on the final exam?

5. Here are the gross values, in millions of dollars, of milk produced and cattle raised in Tulare County, California for the ten years from 1989 through 1998.

Year	Milk ($ millions)	Cattle ($ millions)
1989	287	177
1990	363	214
1991	413	212
1992	411	237
1993	455	238
1994	477	223
1995	547	223
1996	569	229
1997	712	252
1998	718	271

(Source: Visalia Times Delta.)

(a) Find the regression equation that predicts the value of cattle produced in a year based on the value of milk produced that year.
(b) Predict the value of cattle produced in a year when $600 million worth of milk is produced.

6. A tutorial lab on campus offers free tutoring for any student on campus. Here are the GPAs for 12 randomly selected students, and the number of tutoring appointments that those students missed.

GPA	2.66	2.05	2.07	2.62	1.30	3.00	3.25	2.58	2.36	2.81	3.11	2.56
Missed Appointments	3	1	2	0	7	0	2	0	3	1	1	2

(a) Find the regression equation that predicts the number of tutoring appointments that a student misses based on the student's GPA.
(b) Predict the number of tutoring appointments that a student misses if the student's GPA is 2.50.

7. Wilt Chamberlain played 14 NBA seasons in Philadelphia, San Francisco, and Los Angeles. Here are his point totals for those 14 seasons.

Season	Team	Points	Rebounds
1959–60	Philadelphia	2707	1941
1960–61	Philadelphia	3033	2149
1961–62	Philadelphia	4029	2052
1962–63	San Francisco	3586	1946
1963–64	San Francisco	2948	1787
1964–65	San Francisco	2534	1673
1965–66	Philadelphia	2649	1943
1966–67	Philadelphia	1956	1957
1967–68	Philadelphia	1992	1952
1968–69	Los Angeles	1664	1712
1969–70	Los Angeles	328	221
1970–71	Los Angeles	1696	1493
1971–72	Los Angeles	1213	1572
1972–73	Los Angeles	1084	1526

(a) Find the regression equation that predicts the number of points that Wilt scored based on the number of rebounds that he had.

(b) Note that the totals associated with the 1969–70 season are far removed from the rest of the totals. (Wilt was injured for most of the season.) If we disregard that season, find the regression equation that predicts the number of points that Wilt scored based on the number of rebounds that he had.

(c) Does leaving that season out radically change the regression equation? Why or why not? Explain your answer in your own words.

8. It seems that there should be a strong relation between one season's NBA ticket prices and the previous season's prices. Here are the average ticket prices for NBA arenas for the 1998–99 and 1999–2000 seasons.

Team	1998–99	1999–2000	Team	1998–99	1999–2000
New York	79.34	86.82	Sacramento	34.11	44.68
L.A. Lakers	51.11	81.89	Philadelphia	41.96	44.26
Seattle	63.47	64.60	Orlando	44.46	44.18
Houston	58.18	62.63	L.A. Clippers	31.75	43.89
Washington	61.40	59.65	Toronto	26.17	42.76
New Jersey	49.24	59.22	Dallas	34.84	40.76
Utah	43.47	54.60	Detroit	33.32	40.04
Chicago	53.17	52.84	Cleveland	39.75	39.75
Portland	52.28	52.28	Minnesota	38.61	39.08
Boston	49.79	49.79	San Antonio	38.01	38.92
Indiana	43.36	48.39	Denver	30.53	38.34
Golden State	36.79	48.10	Vancouver	31.90	34.71
Miami	36.55	46.57	Charlotte	28.12	32.04
Atlanta	36.79	45.75	Milwaukee	29.06	30.83
Phoenix	48.84	45.39			

(Source: Team Marketing Report, USA Today.)

(a) Find the regression equation for 1999–2000 average ticket prices based on the 1998–99 ticket prices.

(b) Using a scatterplot and the graph of the regression line, does there seem to be one point that does not fit? When you identify the point, explain why it does not seem to fit with the rest of the points.

(c) Find the regression equation for 1999–2000 average ticket prices based on the 1998–99 ticket prices, leaving the outlier out of your calculations. Does this significantly change your regression equation from part (a)?

9. Is there a relation between family income in a city and the price of homes in that city? Here are the median family incomes for 16 California cities, and the median sales price for homes in those cities.

City	Median Income ($ thousands)	Median Sales Price ($ thousands)
Bakersfield	38.7	94
Riverside	47.2	133
Modesto	43.1	128
Visalia	34.3	98
Sacramento	51.9	158
Redding	37.5	113
Fresno	37.2	109
Merced	36.9	122
Stockton	44.3	154
Ventura	65.3	235
Los Angeles	51.3	189
Santa Barbara	52.1	210
San Diego	52.5	208
San Luis Obispo	48.0	192
San Jose	82.6	355
San Francisco	72.4	407

(Source: National Association of Home Builders, Fresno Bee.)

(a) Find the regression equation for the median home sales price in a city based on the median income of that city.
(b) Use the regression equation to predict the median sales price in Ventura, and interpret the residual.
(c) Use the regression equation to predict the median sales price in San Francisco, and interpret the residual.

10. For a random sample of 20 dairy cows, here are the number of pounds of milk produced during their first lactation (after their first calf) and their second lactation (after their second calf).

Cow	First Lactation	Second Lactation	Cow	First Lactation	Second Lactation
1	22,792	29,655	11	15,058	8,013
2	31,693	23,817	12	16,681	23,301
3	18,367	35,360	13	21,002	32,529
4	17,440	21,848	14	17,987	23,620
5	29,798	29,828	15	15,968	26,437
6	28,540	33,898	16	23,580	26,227
7	23,661	22,444	17	14,334	25,529
8	39,242	23,648	18	24,030	27,368
9	21,574	25,303	19	20,392	29,527
10	34,542	34,812	20	24,347	25,592

(a) Find the regression equation for the amount of milk produced during a cow's second lactation based on the amount of milk produced during the cow's first lactation.
(b) Predict the amount of milk produced during a cow's second lactation, if that cow produced 20,000 pounds of milk during her first lactation.
(c) Explain, in your own words, the meaning of the slope in this regression equation. Be specific, using the actual value of the slope.
(d) Explain, in your own words, why the y-intercept of this regression equation lacks a practical interpretation.

11. If a company has a quick Web server, does that mean that it is reliable as well? Here is a listing of 10 major electronic commerce sites. For each site here is the average length of time for the site to come up on a user's computer, and the percentage of time the site is available.

Site	Seconds	Availability (%)
Amazon.com	17.66	90.1
Barnesandnoble.com	15.85	94.2
CDnow	15.52	89.2
eBay	17.25	78.4
eToys.com	20.11	84.0
Gateway	19.01	86.1
Landsend.com	13.49	94.2
Macys.com	31.42	93.0
Wal Mart Online	22.24	92.0
Wine.com	20.58	93.6

(Source: Keynote Systems.)

(a) Calculate the regression equation for the percentage of time that a Web site is available based on the time that it takes for the site to come up on a user's computer.

(b) If it takes a site 12.5 seconds to come up on a user's computer, predict the percentage of time that the Web site will be available.

(c) Explain, in your own words, the meaning of the slope in this regression equation. Be specific, using the actual value of the slope.

12. Many politicians and citizen groups complain about the cost of prescription medications in the United States. Here are the prices of a dose of 10 medications in Canada and the United States, in U.S. dollars.

Drug	Canada	United States
Prilosec	1.47	3.31
Prozac	1.07	2.27
Lipitor	1.34	2.54
Prevacid	1.34	3.13
Epogen	21.44	23.40
Zocor	1.47	3.16
Zoloft	1.07	1.98
Zyprexa	3.39	5.27
Claritin	1.11	1.96
Paxil	1.13	2.22

(Source: USA Today.)

(a) Find the regression equation that predicts the cost for a dose of a prescription medication in the United States, based on its cost in Canada.

(b) Note that the prices for a dose of Epogen are far removed from the rest of the prices. If we disregard Epogen, find the regression equation that predicts the cost for a dose of a prescription medication in the United States, based on its cost in Canada.

(c) Does leaving that medication out radically change the regression equation? Why or why not? Explain your answer in your own words.

(d) Interpret, in your own words, what the slope of this new regression equation means in a practical sense.

(e) Interpret, in your own words, what the y-intercept of this new regression equation means in a practical sense.

13. The results of 19 horse races for fillies and mares (female racehorses) at Santa Anita Park were selected at random. Here are the number of horses in the race, and the price that the winning horse in that race paid to win.

Race	Number of Horses	Winning Price	Race	Number of Horses	Winning Price
1	7	7.40	11	12	7.40
2	6	3.20	12	6	5.60
3	7	12.20	13	7	19.40
4	6	3.80	14	5	2.40
5	12	8.00	15	10	5.40
6	11	6.80	16	8	5.20
7	3	4.00	17	9	47.40
8	5	2.60	18	4	5.60
9	10	49.40	19	9	9.00
10	12	70.00			

(a) Find the regression equation for the winning price in a race for fillies and mares based on the number of horses in the race.
(b) Predict the winning price for a race with 6 horses in it.
(c) In your own words, interpret the practical meaning of this regression equation's slope.

14. Here are the number of teachers at public schools and the number of students in those schools for 14 states in 1997.

State	Teachers	Students
AL	45,793	749,187
CA	268,581	5,803,734
FL	124,473	2,294,077
IL	118,734	1,998,289
KY	40,488	669,322
MA	67,170	949,006
MO	60,869	910,654
NH	12,931	201,629
NC	77,785	1,236,083
OR	26,935	541,346
SD	9,282	142,443
UT	21,115	482,957
WA	75,524	1,110,815
WI	57,227	881,780

(Source: U.S. Department of Education.)

(a) Find the regression equation for the number of public school students in a state based on the number of teachers in those schools.
(b) In 1997, the state of Rhode Island had 10,598 public school teachers. Use the regression equation to predict how many students were in Rhode Island public schools.
(c) In 1997, there were 153,321 students in Rhode Island public schools. Does your prediction in part (b) seem like a good prediction?
(d) In 1997, the state of Connecticut had 37,658 public school teachers. Use the regression equation to predict how many students were in Connecticut public schools.
(e) In 1997, there were 535,164 students in Connecticut public schools. Does your prediction in part (d) seem like a good prediction?

15. For 1997, here are the number of nursing home beds in each state, as well as the number of complaints received that year.

State	Beds	Complaints	State	Beds	Complaints
AL	24,444	998	MT	7,617	1,174
AK	767	114	NE	17,195	1,453
AZ	28,110	1,625	NV	4,250	7,158
AR	24,325	621	NH	8,146	593
CA	134,312	17,764	NJ	49,541	2,192
CO	19,876	7,669	NM	7,376	9,171
CT	32,012	231	NY	115,000	5,483
DE	5,697	1,043	NC	46,467	2,457
DC	3,733	5,199	ND	8,755	546
FL	80,055	5,029	OH	128,181	4,016
GA	41,544	3,274	OK	34,624	2,164
HI	3,952	118	OR	14,470	2,950
ID	6,638	804	PA	97,593	4,641
IL	110,817	4,731	RI	10,365	686
IN	58,355	1,792	SC	20,060	1,896
IA	35,790	254	SD	2,171	541
KS	26,877	4,552	TN	38,786	1,332
KY	28,956	5,314	TX	126,476	9,495
LA	38,253	1,992	UT	8,954	1,958
ME	9,255	499	VT	3,844	330
MD	30,546	2,260	VA	30,060	406
MA	57,934	10,438	WA	27,662	1,670
MI	53,314	2,674	WV	11,529	1,043
MN	44,669	2,953	WI	49,740	3,314
MS	19,905	661	WY	3,116	733
MO	56,719	7,324			

(Source: U.S. Department of Health and Human Services.)

(a) Find the regression equation for the number of nursing home beds in a state based on the number of complaints received.
(b) Use the regression equation to predict how many beds are in Nevada, which received 7,158 complaints.
(c) Why is the predicted number of beds in part (b) so far off? Which other states seem to have the same problem? (Either examine the data or graph the regression line over the scatterplot of points.)
(d) Use the regression equation to predict how many beds are in Ohio, which received 4,016 complaints.
(e) Why is the predicted number of beds in part (d) so far off? Which other states seem to have the same situation?

16. For eight NHL goalies in the season's second month, here are the number of minutes played by the goalie, the number of goals given up by the goalie, and the number of shots attempted against the goalie.

Minutes	788	661	783	831	476	643	767	608
Goals	22	20	28	34	24	30	39	32
Shots	335	258	355	413	243	313	339	295

(a) Find the regression equation for the number of goals given up, based on the number of minutes played.
(b) Find the regression equation for the number of goals given up, based on the number of shots attempted.
(c) Use the two regression equations to predict the number of goals given up by goalie who has played 692 minutes and faced 291 attempted shots.
(d) If the goalie in part (c) had actually allowed 24 goals, can we say which equation is a better predictor of goals allowed?

Exercises 17–20 use the following data. For the 50 states and Washington, D.C., here are the 1999

● average ACT composite scores (ACT)

- percentage of high school graduates that took the ACT (ACT %)
- average SAT verbal score (SAT V)
- average SAT math score (SAT M)
- percentage of high school graduates that took the SAT (SAT %)

State	ACT	ACT %	SAT V	SAT M	SAT %
AL	20.2	65	561	555	9
AK	21.1	35	516	514	50
AZ	21.4	28	524	524	34
AR	20.3	69	563	556	6
CA	21.3	12	497	514	49
CO	21.5	62	536	540	32
CT	21.6	3	510	509	80
DE	20.5	3	503	497	67
DC	18.6	13	494	478	77
FL	20.6	39	499	498	53
GA	20.0	16	487	482	63
HI	21.6	18	482	513	52
ID	21.4	60	542	540	16
IL	21.4	67	569	585	12
IN	21.2	19	496	498	60
IA	22.0	66	594	598	5
KS	21.5	75	578	576	9
KY	20.1	68	547	547	12
LA	19.6	76	561	558	8
ME	22.1	4	507	503	68
MD	20.9	10	507	507	65
MA	22.0	6	511	511	78
MI	21.3	69	557	565	11
MN	22.1	64	586	598	9
MS	18.7	82	563	548	4
MO	21.6	67	572	572	8
MT	21.8	54	545	546	21
NE	21.7	41	568	571	8
NV	21.5	5	512	517	34
NH	22.2	5	520	518	72
NJ	20.7	4	498	510	80
NM	20.1	64	549	542	12
NY	22.0	14	495	502	76
NC	19.4	12	493	493	61
ND	21.4	79	594	605	5
OH	21.4	59	534	538	25
OK	20.6	69	567	560	8
OR	22.6	11	525	525	53
PA	21.4	7	498	495	70
RI	22.7	3	504	499	70
SC	19.1	18	479	475	61
SD	21.2	70	585	588	4
TN	19.9	77	559	553	13
TX	20.3	31	494	499	50
UT	21.4	68	570	565	5
VT	21.9	9	514	506	70
VA	20.6	7	508	499	65
WA	22.6	18	525	526	52
WV	20.2	58	527	512	18
WI	22.3	67	584	595	7
WY	21.4	66	546	551	10

(Source: College Entrance Examination Board, American College Testing Program.)

17. Find the regression equation for SAT verbal scores based on SAT math scores.
18. Find the regression equation for ACT scores based on composite SAT scores (SAT verbal score + SAT math score).
19. A state superintendent of schools, when asked about her state's low scores, claims that the low scores are due to the high percentage of high school graduates that take the test. "There is a significant negative correlation between scores and the percentage of high school graduates that take the SAT."

    (a) Find the regression equation for composite SAT scores (SAT verbal score + SAT math score) based on the percentage of graduates who took the SAT.
    (b) If 49% of the graduates in her state took the SAT and their average composite score was 1011, are her state's students performing above or below the regression line's prediction?

20. Find the regression equation for the percentage of graduates who took the SAT based on the percentage of graduates who took the ACT.
21. For the 1997 major league baseball season, here are the salaries for the 28 teams (in $ millions), and the number of wins for each team.

| American League | | | National League | | |
Team	Payroll	Wins	Team	Payroll	Wins
Anaheim	39.2	84	Atlanta	52.2	101
Baltimore	63.0	98	Chicago	30.8	68
Boston	40.0	78	Cincinnati	36.9	76
Chicago	41.9	80	Colorado	41.6	83
Cleveland	58.5	86	Florida	53.5	92
Detroit	16.2	79	Houston	34.2	84
Kansas City	35.2	67	Los Angeles	48.4	88
Milwaukee	26.7	78	Montreal	18.5	78
Minnesota	31.8	67	New York	39.2	88
New York	65.0	96	Philadelphia	30.7	68
Oakland	13.6	65	Pittsburgh	12.2	79
Seattle	45.6	90	San Diego	31.7	76
Texas	42.9	77	San Francisco	44.1	90
Toronto	43.8	76	St. Louis	48.3	73

    (a) For these 28 teams, find the regression equation for the number of wins in terms of their payroll.
    (b) Many people were complaining after the 1997 season that teams with deep pockets could buy a championship. Does your regression equation support this idea; in other words, does greater spending produce more wins?

22. For the 1999 major league baseball season, here are the salaries for the 30 teams (in $ millions), and the number of wins for each team.

| American League | | | National League | | |
Team	Payroll	Wins	Team	Payroll	Wins
Anaheim	51.3	70	Arizona	70.0	100
Baltimore	75.4	78	Atlanta	79.3	103
Boston	72.3	94	Chicago	55.4	67
Chicago	24.5	75	Cincinnati	38.0	96
Cleveland	73.5	97	Colorado	54.4	72
Detroit	37.0	69	Florida	14.7	64
Kansas City	16.6	64	Houston	56.4	97
Minnesota	15.8	63	Los Angeles	76.6	77
New York	92.0	98	Milwaukee	43.0	74
Oakland	25.2	87	Montreal	15.0	68
Seattle	45.4	79	New York	71.5	97
Tampa Bay	37.9	69	Philadelphia	30.4	77
Texas	80.8	95	Pittsburgh	23.7	78
Toronto	48.8	84	San Diego	46.5	74
			San Francisco	46.0	86
			St. Louis	46.3	75

(a) For these 30 teams, find the regression equation for the number of wins in terms of their payroll.

(b) Does your regression equation support the idea that greater spending produces more wins? Were there any teams that do not seem to fit with this idea? (Check a scatterplot.)

23. For 71 randomly selected 1999 National League baseball games, here are the times required to complete the games (in minutes), the total number of hits in the game, and the combined number of runs in the game.

Time	Hits	Runs	Time	Hits	Runs	Time	Hits	Runs
170	19	11	152	20	14	160	14	3
187	20	11	190	23	9	177	20	7
169	14	6	213	30	22	171	16	9
129	12	1	158	19	11	198	26	22
159	25	15	176	20	9	159	21	10
209	27	19	146	13	6	160	9	3
181	16	9	172	9	4	146	16	10
183	17	11	137	13	6	188	14	6
197	23	12	200	16	9	149	16	6
170	19	13	158	19	13	151	13	6
157	20	6	181	25	13	172	18	7
214	32	27	202	21	11	143	12	6
154	15	6	152	14	3	189	19	11
166	23	12	128	11	8	204	27	19
191	21	11	161	18	12	215	24	13
178	17	6	153	8	3	181	16	10
198	23	14	164	19	13	141	14	5
178	17	13	169	19	9	152	18	4
161	13	6	194	24	14	176	20	10
198	20	6	108	10	5	180	19	14
179	17	10	146	19	5	158	12	4
150	21	13	157	12	6			
265	20	9	167	16	13			
136	18	16	169	16	7			
143	15	3	190	22	12			

(a) Find the regression equation for the time required to complete the game based on the number of hits.

(b) Interpret the slope of this regression equation.

(c) Find the regression equation for the time required to complete the game based on the number of runs.

(d) Interpret the slope of this regression equation.

24. For 62 randomly selected 1999 American League baseball games, here are the times required to complete the games (in minutes), the total number of hits in the game, and the combined number of runs in the game.

Time	Hits	Runs	Time	Hits	Runs	Time	Hits	Runs
175	13	5	158	28	16	166	14	6
168	16	4	157	14	4	168	18	7
185	23	12	196	18	12	182	21	9
183	23	16	190	18	8	219	31	17
196	23	11	166	17	9	235	28	19
151	16	10	176	14	4	224	33	25
198	14	11	186	21	14	202	27	17
148	14	6	153	15	7	171	20	12
189	23	12	189	27	14	159	16	9

(continues)

(continued)

Time	Hits	Runs	Time	Hits	Runs	Time	Hits	Runs
146	9	4	180	16	4	153	19	6
225	32	20	205	13	9	169	17	11
199	27	11	163	24	15	171	7	1
143	17	7	177	19	7			
149	12	6	154	22	14			
156	14	8	172	16	7			
192	27	13	178	20	8			
230	31	23	186	20	9			
178	24	16	187	17	11			
124	9	3	163	14	4			
167	17	9	181	18	8			
156	11	4	208	22	13			
185	19	13	142	20	9			
179	18	9	200	18	11			
185	26	17	150	20	10			
144	13	8	172	18	6			

(a) Find the regression equation for the time required to complete the game based on the number of hits.

(b) Interpret the slope of this regression equation.

(c) Is this equation substantially different from the equation found in part (a) of the last problem? Explain your response.

(d) Find the regression equation for the time required to complete the game based on the number of runs.

(e) Interpret the slope of this regression equation.

(f) Is this equation substantially different from the equation found in part (c) of the last problem? Explain your response.

MINI PROJECT

Randomly sample at least 30 male and 30 female students at your school, and record their shoe size and their height. Calculate the correlation coefficient for the males and the correlation coefficient for the females. Next determine the regression equation for predicting height based on a person's shoe size. In addition to these calculations, include

- Your raw data
- A scatterplot for the males, with the regression line drawn on it
- A scatterplot for the females, with the regression line drawn on it

Finally, answer the following questions.

- How did you obtain your sample?
- What type of sample did you use?
- Was the sample truly random?
- What potential biases may show up in your sample?
- Why is it inappropriate to combine the male data and the female data into one sample?

SECTION 10.3
Inference for Regression

In this section, we explore inference for regression. The regression equation that was introduced in the last section, $\hat{y} = a + bx$, is based on sample data. We can consider *a* and *b* as sample regression coefficients that we can use to estimate the *y*-intercept and slope for the population. We denote the population regression coefficients α (alpha) and β (beta), which represent the *y*-intercept and slope for the population, respectively. The population regression line has the equation $y = \alpha + \beta x$. We will examine a hypothesis test involving the slope β. We will then go on to construct a prediction interval for *y* for a given value of *x*, instead of predicting a single value as we did in the last section.

Before going on, there are some assumptions that are being made. We assume that for each value of *x*, the values of *y* are normally distributed. The mean of each of these distributions lies on the population regression line. Finally, every one of these distributions has the same standard deviation.

We begin by introducing the **standard error of estimate**, s_e, which will be used in later calculations.

$$s_e = \sqrt{\frac{\Sigma(y - \hat{y})^2}{n - 2}}$$

Recall that the residual for a sample data point is the vertical distance between that data point and the regression line. In other words, the residual is the difference between the data point's *y*-value and the predicted value $\hat{y}$ associated with its *x*-value. In the standard error of estimate the residuals are squared and then totaled. This sum is divided by $n - 2$, and we finish by taking the square root. The standard error of estimate is a measure of how the data points deviate from the regression line.

EXAMPLE 10.14   Here are the scores of five randomly selected students on test 1 and test 2 in a math class. Treating the score on test 1 as *x* and the score on test 2 as *y*, the regression equation was found to be $\hat{y} = -49.5957 + 1.5073x$ in the last section. Calculate the standard error of estimate.

Student	Test 1 Score	Test 2 Score
1	83	82
2	86	84
3	76	63
4	92	83
5	71	55

Once again, we will use a column approach. We will calculate the predicted value $\hat{y}$ for each data point using the regression equation. The results are shown in the table that follows, rounded to four decimal places (verify these predicted values). Then we find the residual $y - \hat{y}$, and place it in the next column. Next we square these values and place them in the next column.

x	*y*	$\hat{y}$	$y - \hat{y}$	$(y - \hat{y})^2$
83	82	75.5102	6.4898	42.1175
86	84	80.0322	3.9678	15.7434
76	63	64.9590	−1.9590	3.8377
92	83	89.0761	−6.0761	36.9190
71	55	57.4224	−2.4224	5.8680

| | | | | 104.4856 |

Now we are ready to plug into the formula.

$$s_e = \sqrt{\frac{\Sigma(y - \hat{y})^2}{n - 2}}$$

$$= \sqrt{\frac{104.4856}{5 - 2}}$$

$$= \sqrt{\frac{104.4856}{3}}$$

$$= 5.902$$

The standard error of estimate s_e is 5.902. ■

This formula gets tedious as the number of data points increases, because we have to find $\hat{y}$ for each data point. There is another formula that you may find easier to work with because it involves values that we have already calculated on the way to calculating the correlation coefficient and finding the regression equation. Here's the formula.

$$s_e = \sqrt{\frac{\Sigma y^2 - a\Sigma y - b\Sigma(xy)}{n - 2}}$$

where a and b are the y-intercept and slope of the regression line. We will rework the previous example with this new formula.

 EXAMPLE 10.15

Here are the scores of five randomly selected students on test 1 and test 2 in a math class. Treating the score on test 1 as x and the score on test 2 as y, calculate the standard error of estimate.

Student	Test 1 Score	Test 2 Score
1	83	82
2	86	84
3	76	63
4	92	83
5	71	55

We begin by assembling all of the sums that are necessary for calculating the correlation coefficient and the regression equation.

x	y	xy	x^2	y^2
83	82	6,806	6,889	6,724
86	84	7,224	7,396	7,056
76	63	4,788	5,776	3,969
92	83	7,636	8,464	6,889
71	55	3,905	5,041	3,025
408	367	30,359	33,566	27,663

The sums that we need for the standard error of estimate are $\Sigma y^2 = 27{,}663$, $\Sigma y = 367$, and $\Sigma(xy) = 30{,}359$. We also need to find a and b, and we begin with b.

$$b = \frac{n \cdot \Sigma xy - (\Sigma x) \cdot (\Sigma y)}{n\Sigma x^2 - (\Sigma x)^2}$$

$$= \frac{5 \cdot 30,359 - 408 \cdot 367}{5 \cdot 33,566 - 408^2}$$

$$= \frac{151,795 - 149,736}{167,830 - 166,464}$$

$$= \frac{2059}{1366}$$

$$= 1.5073$$

Now we calculate a.

$$\bar{x} = \frac{\Sigma x}{n} \qquad \bar{y} = \frac{\Sigma y}{n}$$

$$= \frac{408}{5} \qquad = \frac{367}{5}$$

$$= 81.6 \qquad = 73.4$$

$$a = \bar{y} - b\bar{x}$$

$$= 73.4 - (1.5073) \cdot 81.6$$

$$= -49.5957$$

Now we have all of the pieces to calculate the standard error of estimate.

$$s_e = \sqrt{\frac{\Sigma y^2 - a\Sigma y - b\Sigma(xy)}{n-2}}$$

$$= \sqrt{\frac{27,663 - (-49.5957)(367) - 1.5073(30,359)}{5-2}}$$

$$= \sqrt{\frac{27,663 + 18,201.6219 - 45,760.1207}{3}}$$

$$= \sqrt{\frac{104.5012}{3}}$$

$$= 5.902 \qquad \blacksquare$$

Inferences about the Slope

If we are unable to reject the claim that the population regression line's slope β is equal to 0, then we are saying that the population regression line may be horizontal. The equation would simply be of the form $y = \alpha$, so in other words, y does not depend on x at all. In such a case, x does not help us when trying to predict values of y. The test statistic for all claims comparing the slope β to 0 follows.

$$t = \frac{b}{s_e} \sqrt{\Sigma x^2 - \frac{(\Sigma x)^2}{n}}, \qquad \text{degrees of freedom} = n - 2$$

Note that we are using the t-distribution. We will be following our standard hypothesis testing procedure. Here is an example.

EXAMPLE 10.16

Here are the gross values, in millions of dollars, of nectarines and peaches grown in Tulare County, California, for the ten years from 1989 through 1998. Let *x* represent the gross value of nectarines produced and *y* the gross value of peaches produced. At the 0.05 level of significance, test the claim that the value of nectarines can be used to predict the value of peaches; in other words, the slope of the population's regression line is not 0.

Year	Nectarines ($ millions)	Peaches ($ millions)
1989	47	32
1990	53	47
1991	52	57
1992	68	43
1993	51	53
1994	90	63
1995	74	76
1996	89	64
1997	83	66
1998	56	56

(Source: Visalia Times Delta.)

We need to calculate Σy^2, Σy, and Σxy to calculate the standard error of estimate s_e. We will also need to calculate the regression coefficients a and b for the calculation of s_e. Finally, we need the sums Σx^2 and Σx to complete the calculation of the test statistic. Fortunately, all of these are calculated along the way to finding the correlation coefficient and regression equation.

Recall, from the last section, that we calculated the following sums:

$$\Sigma x = 663, \quad \Sigma y = 557, \quad \Sigma xy = 38{,}190, \quad \Sigma x^2 = 46{,}469, \quad \Sigma y^2 = 32{,}473$$

We went on to calculate that the equation of the regression line was $\hat{y} = 22.4240 + 0.5019x$, so $a = 22.4240$ and $b = 0.5019$. We can now go on to calculate the standard error of estimate. We will be using the second formula that was provided, so that we do not have to plug 10 different values of *x* into the regression equation in order to calculate the residuals.

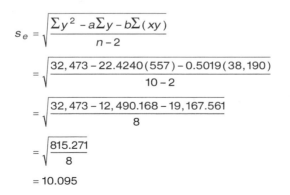

$$s_e = \sqrt{\frac{\Sigma y^2 - a\Sigma y - b\Sigma(xy)}{n-2}}$$

$$= \sqrt{\frac{32{,}473 - 22.4240(557) - 0.5019(38{,}190)}{10-2}}$$

$$= \sqrt{\frac{32{,}473 - 12{,}490.168 - 19{,}167.561}{8}}$$

$$= \sqrt{\frac{815.271}{8}}$$

$$= 10.095$$

Now we have everything we need for calculating our test statistic, so we will begin our test.

Step 1

Claim in words: The value of nectarines can be used to predict the value of peaches; in other words, the slope of the population's regression line is not 0.

Claim: $\beta \neq 0$

Complement: $\beta = 0$

H_0: $\beta = 0$

H_A: $\beta \neq 0$

Step 2

Level of significance: $\alpha = 0.05$

Step 3

Test statistic: $t = \dfrac{b}{s_e}\sqrt{\Sigma x^2 - \dfrac{(\Sigma x)^2}{n}}$, $d.f. = n - 2$

Step 4

$\alpha = 0.05$, two-tailed test, $10 - 2 = 8$ degrees of freedom

Decision rule: Reject H_0 if $t > 2.306$ or if $t < -2.306$.

Step 5

$$t = \frac{b}{s_e}\sqrt{\Sigma x^2 - \frac{(\Sigma x)^2}{n}}$$

$$= \frac{0.5019}{10.095}\sqrt{46,469 - \frac{663^2}{10}}$$

$$= 2.492$$

Decision: Reject H_0.

Conclusion: There is sufficient sample evidence to support the claim that the value of nectarines can be used to predict the value of peaches, or in other words, that the slope of the population's regression line is not 0.

p-value

Using techniques developed in Section 7.2, the best that we can say is that the p-value is between 0.02 and 0.05. A calculator or computer would tell us that the actual p-value of this test is 0.0374. ∎

EXAMPLE 10.17   Here are the 1998 average SAT scores (math score plus verbal score divided by 2) and average ACT scores for 7 randomly selected states. At the 0.05 level of significance, test the claim that a state's average ACT score can be used to predict a state's average SAT score; in other words, the slope of the population's regression line is not equal to 0.

State	Average SAT	Average ACT
CT	508	21.6
FL	499	20.6
HI	498	21.6
NM	550	20.1
NC	489	19.4
OH	536	21.4
OR	525	22.6

Since we are asked to determine whether a state's average ACT score can predict that state's average SAT score, we will let x be the ACT score and y be the SAT score. First, let's look at the scatterplot for these data points.

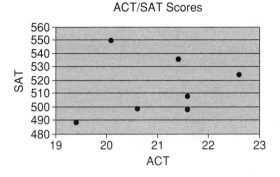

ACT/SAT Scores

The pattern does not appear to be linear for this entire set of data. We begin our calculations by constructing the appropriate sums.

x	y	xy	x^2	y^2
21.6	508	10,972.8	466.56	258,064
20.6	499	10,279.4	424.36	249,001
21.6	498	10,756.8	466.56	248,004
20.1	550	11,055.0	404.01	302,500
19.4	489	9,486.6	376.36	239,121
21.4	536	11,470.4	457.96	287,296
22.6	525	11,865.0	510.76	275,625
147.3	3,605	75,886.0	3,106.57	1,859,611

Thus, $\Sigma x = 147.3$, $\Sigma y = 3605$, $\Sigma xy = 75,886$, $\Sigma x^2 = 3106.57$, and $\Sigma y^2 = 1,859,611$. Now we need to find the equation of the regression equation.

$$b = \frac{n \cdot \Sigma xy - (\Sigma x) \cdot (\Sigma y)}{n\Sigma x^2 - (\Sigma x)^2}$$

$$= \frac{7 \cdot 75,886 - 147.3 \cdot 3605}{7 \cdot 3106.57 - 147.3^2}$$

$$= \frac{531,202 - 531,016.5}{21,745.99 - 21,697.29}$$

$$= \frac{185.5}{48.7}$$

$$= 3.8090$$

$$\bar{x} = \frac{\Sigma x}{n} \qquad \bar{y} = \frac{\Sigma y}{n}$$

$$= \frac{147.3}{7} \qquad = \frac{3605}{7}$$

$$= 21.0 \qquad = 515$$

$$a = \bar{y} - b\bar{x}$$

$$= 515 - (3.8090) \cdot 21.0$$

$$= 435.011$$

Next we can calculate the standard error of estimate s_e.

$$s_e = \sqrt{\frac{\Sigma y^2 - a\Sigma y - b\Sigma(xy)}{n-2}}$$

$$= \sqrt{\frac{1,859,611 - 435.011(3605) - 3.8090(75,886)}{7-2}}$$

$$= \sqrt{\frac{1,859,611 - 1,568,214.655 - 289,049.774}{5}}$$

$$= \sqrt{\frac{2346.571}{5}}$$

$$= 21.664$$

Now we can begin our test.

Step 1

Claim in words: A state's average ACT score can be used to predict a state's average SAT score; in other words, the slope of the population's regression line is not equal to 0.

Claim: $\beta \neq 0$

Complement: $\beta = 0$

H_0: $\beta = 0$

H_A: $\beta \neq 0$

Step 2

Level of significance: $\alpha = 0.05$

Step 3

Test statistic: $t = \dfrac{b}{s_e}\sqrt{\Sigma x^2 - \dfrac{(\Sigma x)^2}{n}}$, $d.f. = n - 2$

Step 4

$\alpha = 0.05$, two-tailed test, $7 - 2 = 5$ degrees of freedom

Decision rule: Reject H_0 if $t > 2.571$ or if $t < -2.571$.

Step 5

$$t = \frac{b}{s_e}\sqrt{\Sigma x^2 - \frac{(\Sigma x)^2}{n}}$$

$$= \frac{3.8090}{21.664}\sqrt{3106.57 - \frac{147.3^2}{7}}$$

$$= 0.464$$

Decision: Fail to reject H_0.

Conclusion: There is not sufficient sample evidence to support the claim that a state's average ACT score can be used to predict a state's average SAT score, or in other words, that the slope of the population's regression line is not 0. Therefore, it would not be wise to use this regression line for predicting SAT scores based on ACT scores.

p-value

Using techniques developed in Section 7.2, the best that we can say is that the *p*-value is greater than 0.5. A calculator or computer would tell us that the actual *p*-value of this test is 0.6621. ∎

Confidence Intervals for β

We can construct confidence intervals for β as well. We begin by finding the margin of error.

$$E = t_{\alpha/2} \cdot \frac{s_e}{\sqrt{\Sigma x^2 - \frac{(\Sigma x)^2}{n}}}$$

We complete the interval by finding the left and right endpoints using the following.

$$b - E \leq \beta \leq b + E$$

This uses all the calculations that go into calculating the previous statistic. Here is a brief example, using the data from the previous example.

 EXAMPLE 10.18

Here are the 1998 average SAT scores (math score plus verbal score divided by 2) and average ACT scores for 7 randomly selected states. Let x represent the ACT scores. Construct a 95% confidence interval for the slope, β, of the population's regression line.

State	Average SAT	Average ACT
CT	508	21.6
FL	499	20.6
HI	498	21.6
NM	550	20.1
NC	489	19.4
OH	536	21.4
OR	525	22.6

First, calculate the margin of error. The value $t_{0.025}$ with 5 degrees of freedom is 2.571.

$$E = t_{\alpha/2} \cdot \frac{s_e}{\sqrt{\Sigma x^2 - \frac{(\Sigma x)^2}{n}}}$$

$$= 2.571 \cdot \frac{21.664}{\sqrt{3106.57 - \frac{147.3^2}{7}}}$$

$$= 21.1167$$

$$b - E \leq \beta \leq b + E$$

$$3.8090 - 21.1167 \leq \beta \leq 3.8090 + 21.1167$$

$$-17.3077 \leq \beta \leq 24.9257$$

We are 95% confident that the population regression slope is somewhere between –17.3077 and 24.9257. With such a wide range of values that includes 0, we can clearly see now why we did not reject the null hypothesis in the previous example. There is a strong chance that the slope could actually be 0, and this was seen in the p-value of the previous example. ∎

Prediction Intervals for y for a Given Value of x

In the previous section we were able to predict a value of y for a given value of x by plugging that value of x into the regression equation $\hat{y} = a + bx$. Instead of giving a "point estimate" of the predicted value, we will generate an interval that will contain the value y associated with a given value of x with a certain level of confidence. By being able to give a range of values that we believe y will lie in, we can see just how accurate our prediction is.

As with any confidence interval, we begin with a margin of error.

$$E = t_{\alpha/2} \cdot s_e \cdot \sqrt{1 + \frac{1}{n} + \frac{n(x - \bar{x})^2}{n\left(\Sigma x^2\right) - \left(\Sigma x\right)^2}}, \qquad \text{degrees of freedom} = n - 2$$

Again, the calculation is made up of values that have already been determined in the calculation of the regression equation. The variable x represents the given value of x for which we are trying to construct the prediction interval. We complete the interval as follows.

$$\hat{y} - E \leq y \leq \hat{y} + E$$

In this formula, $\hat{y}$ represents the predicted value for the given value of x. Let's examine a complete example.

EXAMPLE 10.19

Here are the number of people executed in the state of Texas and in the United States for the 9 years from 1990 through 1998. First, calculate the regression equation treating the number of executions in Texas as x. Then predict how many executions would occur in the United States in a year in which there were 30 executions in Texas. Finally, construct a 95% prediction interval for the number of executions that would occur in a year in which there were 30 executions in Texas.

Year	1990	1991	1992	1993	1994	1995	1996	1997	1998
Texas	4	5	12	17	14	19	3	37	20
U.S.	23	14	31	38	31	56	45	74	68

We begin by constructing all of the sums that are necessary for calculating the regression coefficients a and b, and the standard error of estimate s_e.

x	y	xy	x^2	y^2
4	23	92	16	529
5	14	70	25	196
12	31	372	144	961
17	38	646	289	1,444
14	31	434	196	961
19	56	1,064	361	3,136
3	45	135	9	2,025
37	74	2,738	1,369	5,476
20	68	1,360	400	4,624
131	380	6,911	2,809	19,352

Thus, $\Sigma x = 131$, $\Sigma y = 380$, $\Sigma xy = 6911$, $\Sigma x^2 = 2809$, and $\Sigma y^2 = 19{,}352$. Now we can find the equation of the regression equation.

$$b = \frac{n \cdot \Sigma xy - (\Sigma x) \cdot (\Sigma y)}{n \Sigma x^2 - (\Sigma x)^2}$$

$$= \frac{9 \cdot 6911 - 131 \cdot 380}{9 \cdot 2809 - 131^2}$$

$$= \frac{62{,}199 - 49{,}780}{25{,}281 - 17{,}161}$$

$$= \frac{12{,}419}{8120}$$

$$= 1.5294$$

$$\bar{x} = \frac{\Sigma x}{n} \qquad\qquad \bar{y} = \frac{\Sigma y}{n}$$

$$= \frac{131}{9} \qquad\qquad\quad = \frac{380}{9}$$

$$= 14.5556 \qquad\quad = 42.2222$$

$$a = \bar{y} - b\bar{x}$$

$$= 42.2222 - (1.5294) \cdot 14.5556$$

$$= 19.9609$$

The regression equation is $\hat{y} = 19.9609 + 1.5294x$. To predict how many executions would occur in the United States in a year in which there were 30 executions in Texas, we plug in 30 for x in the regression equation.

$$\hat{y} = 19.9609 + 1.5294x$$

$$= 19.9609 + 1.5294(30)$$

$$= 65.8429$$

The regression line predicts 65.8429, or approximately 66 executions.

To begin our work with the 95% prediction interval, we need to calculate the standard error of estimate s_e.

$$s_e = \sqrt{\frac{\Sigma y^2 - a\Sigma y - b\Sigma(xy)}{n-2}}$$

$$= \sqrt{\frac{19{,}352 - 19.9609(380) - 1.5294(6911)}{9-2}}$$

$$= \sqrt{\frac{19{,}352 - 7585.142 - 10{,}569.6834}{7}}$$

$$= \sqrt{\frac{1197.1746}{7}}$$

$$= 13.078$$

Because there are nine pairs of data, we have 7 degrees of freedom. Our value for $t_{0.025}$ is 2.365. Now we can calculate the margin of error.

$$E = t_{\alpha/2} \cdot s_e \cdot \sqrt{1 + \frac{1}{n} + \frac{n(x - \bar{x})^2}{n(\Sigma x^2) - (\Sigma x)^2}}$$

$$= 2.365 \cdot 13.078 \cdot \sqrt{1 + \frac{1}{9} + \frac{9(30 - 14.5556)^2}{9(2809) - 131^2}}$$

$$= 2.365 \cdot 13.078 \cdot \sqrt{1 + \frac{1}{9} + \frac{9(15.4444)^2}{25,281 - 17,161}}$$

$$= 2.365 \cdot 13.078 \cdot \sqrt{1 + \frac{1}{9} + \frac{2146.765422}{8120}}$$

$$= 36.2745$$

To complete the interval, we need to use the predicted value ($\hat{y}$) of y for 30 Texas executions. This was calculated above as 65.8429.

$$\hat{y} - E \leq y \leq \hat{y} + E$$

$$65.8429 - 36.2745 \leq y \leq 65.8429 + 36.2745$$

$$29.5684 \leq y \leq 102.1174$$

If we round these endpoints to whole numbers, we would be 95% sure that the number of executions in the United States in a year in which there were 30 executions in Texas would be between 30 and 102. (Naturally if there were 30 executions in Texas there would have to be at least 30 executions in the United States.) Note the wide range of values. Be careful not to be too confident with the predictions that were made in the last section without checking the prediction interval first. ■

MICROSOFT EXCEL   Calculations for Inference for Regression

Microsoft Excel can help with some of the calculations discussed in this section. Specifically, we will learn to use Excel to calculate the standard error of estimate s_e, the test statistic for the hypothesis test comparing the slope β of the population regression line, and the p-value of that test statistic. Recall this example from earlier in this section.

EXAMPLE 10.20   Here are the 1998 average SAT scores (math score plus verbal score divided by 2) and average ACT scores for 7 randomly selected states. At the 0.05 level of significance, test the claim that a state's average ACT score can be used to predict a state's average SAT score; in other words, the slope of the population's regression line is not equal to 0.

State	Average SAT	Average ACT
CT	508	21.6
FL	499	20.6
HI	498	21.6
NM	550	20.1
NC	489	19.4
OH	536	21.4
OR	525	22.6

We will follow the same procedure that we did in the last section. All of the information is contained in the same sheet of output. If we were going to create a scatterplot, we would need to type the ACT scores (*x*) in the first column and the SAT scores (*y*) next to them. We will do it that way in case a scatterplot is needed. (Recall that a scatterplot should be the first step throughout this chapter.) In column **A,** type the ACT scores in cells **A1** through **A7.** In column **B,** type the corresponding SAT scores in cells **B1** through **B7.**

From the **Tools** menu, select **Data Analysis.** When the dialog box appears, select **Regression** under **Analysis Tools** and click on **OK.** Type B1:B7 in the box labeled **Input Y Range,** since column **B** contains our *y* values (SAT scores). Type A1:A7 in the box labeled **Input X Range,** since column **A** contains our *x* values (ACT scores). The other options in that box do not directly concern us here, so click on **OK.** Here is what your output should look like after you select **Columns** from the **Format** menu and click on **AutoFit Selection.**

SUMMARY OUTPUT

Regression Statistics	
Multiple R	0.182338984
R Square	0.033247505
Adjusted R Square	–0.160102994
Standard Error	24.2283329
Observations	7

ANOVA

	df	SS	MS	F	Significance F
Regression	1	100.9394251	100.9394251	0.171954586	0.695571816
Residual	5	2935.060575	587.012115		
Total	6	3036			

	Coefficients	Standard Error	t Stat	P-value	Lower 95%
Intercept	434.8470026	193.5083116	2.24717491	0.074540566	–62.58111541
X Variable 1	3.809034908	9.185611466	0.414674072	0.695571816	–19.80329249

In the first set of values, under **Regression Statistics,** we find the standard error of estimate labeled as **Standard Error.** The value, to three decimal places, is 24.228. Note that in the same set of values we see the coefficient of determination r^2, which for this example is only 0.0332. Such a low value suggests that there is little to no correlation between these two variables.

If you skip to the last set of values where the regression coefficients can be found, this is where you can find the value of the test statistic and its *p*-value. In the bottom row, labeled **X Variable 1,** look to the right until you find the column labeled **t Stat.** Here is where you will find the value of the test statistic for this test. Our value, rounded to three decimal places, is 0.415. The *p*-value can be found immediately to the right of the test statistic, under the column labeled **P-value.** The *p*-value for this test was 0.6956, which is very high, suggesting that there is a strong chance that the slope of the regression line actually is 0.

We use Excel in this section to ease the calculations. However, we still need to perform the hypothesis test by hand. ∎

EXERCISES 10.3

Many of the data sets in these exercises were contained in the exercise sets for Sections 10.1 and 10.2. For problems that you are calculating by hand, the sums from the last two sections will come in handy. (Don't reinvent the wheel.)

1. Here are the shoe sizes of 10 randomly selected men, along with their heights in inches.

Shoe Size	4	7	7.5	9	9.5	10.5	9	10	11	11.5
Height (in.)	62	64	66	68	68	69	69	70	71	72

   (a) Find the regression equation that predicts height using shoe size.
   (b) Calculate the standard error of measurement s_e.

2. Here are the waist and hip measurements, in inches, of twevle 4-year-old dance students.

Waist	24	26	22	23	23	25	24	27	23	23	25	31
Hips	29	31	28	28	28	31	30	34	28	28	28	37

   (Source: Dancer's Edge dance studio.)

   (a) Find the regression equation that predicts the size of a dancer's hips using her waist size.
   (b) Calculate the standard error of measurement s_e.

3. Here are the midterm exam scores of 20 algebra students, along with their scores on the final exam.

Student	Midterm	Final	Student	Midterm	Final
1	65	85	11	38	56
2	50	77	12	92	95
3	75	90	13	59	60
4	70	84	14	66	87
5	68	61	15	61	83
6	58	70	16	71	77
7	49	76	17	68	79
8	92	78	18	82	74
9	68	80	19	85	94
10	93	92	20	67	80

   (a) Find the regression equation that predicts a student's score on the final exam based on the student's score on the midterm.
   (b Calculate the standard error of measurement s_e.

4. Here are the scores of 16 randomly selected statistics students on a test on Unit 3 and on the final exam. (Treat the Unit 3 scores as the independent variable x.)

Student	Unit 3	Final	Student	Unit 3	Final
1	68	59	9	84	81
2	93	91	10	74	85
3	90	75	11	90	84
4	97	100	12	54	78
5	97	98	13	78	95
6	48	59	14	86	94
7	89	89	15	56	86
8	96	97	16	79	57

   (a) Find the regression equation that predicts a student's score on the final exam based on the student's score on the Unit 3 exam.
   (b) Calculate the standard error of measurement s_e.

5. Here are the gross values, in millions of dollars, of milk produced and cattle raised in Tulare County, California, for the ten years from 1989 through 1998.

Year	Milk ($ millions)	Cattle ($ millions)
1989	287	177
1990	363	214
1991	413	212
1992	411	237
1993	455	238
1994	477	223
1995	547	223
1996	569	229
1997	712	252
1998	718	271

(Source: Visalia Times Delta.)

(a) Find the regression equation that predicts the value of cattle produced in a year based on the value of milk produced that year.
(b) Calculate the standard error of measurement s_e.
(c) At the 0.05 level of significance, test the claim that the value of milk produced can be used to predict the value of cattle produced; in other words, the slope of the population's regression line is not equal to 0.

6. A tutorial lab on campus offers free tutoring for any student on campus. Here are the GPAs for 12 randomly selected students, and the number of tutoring appointments that those students missed.

GPA	2.66	2.05	2.07	2.62	1.30	3.00	3.25	2.58	2.36	2.81	3.11	2.56
Missed Appointments	3	1	2	0	7	0	2	0	3	1	1	2

(a) Find the regression equation that predicts the number of tutoring appointments that a student misses based on the student's GPA.
(b) Calculate the standard error of measurement s_e.
(c) At the 0.05 level of significance, test the claim that a student's GPA can be used to predict the number of tutoring appointments that a student misses; in other words, the slope of the population's regression line is not equal to 0.

7. Here are the median family incomes for 16 California cities, and the median sales price for homes in those cities.

City	Median Income ($ thousands)	Median Sales Price ($ thousands)
Bakersfield	38.7	94
Riverside	47.2	133
Modesto	43.1	128
Visalia	34.3	98
Sacramento	51.9	158
Redding	37.5	113
Fresno	37.2	109
Merced	36.9	122
Stockton	44.3	154
Ventura	65.3	235
Los Angeles	51.3	189
Santa Barbara	52.1	210
San Diego	52.5	208
San Luis Obispo	48.0	192
San Jose	82.6	355
San Francisco	72.4	407

(Source: National Association of Home Builders, Fresno Bee.)

(a) Find the regression equation for the median home sales price in a city based on the median income of that city.
(b) Calculate the standard error of measurement s_e.
(c) At the 0.05 level of significance, test the claim that the median income of a city can be used to predict the median home sales price in that city; in other words, the slope of the population's regression line is not equal to 0.
(d) Construct a 95% confidence interval for the slope β of the population's regression line, and explain what it means.

8. For a random sample of 20 dairy cows, here are the number of pounds of milk produced during their first lactation (after their first calf) and their second lactation (after their second calf).

Cow	First Lactation	Second Lactation	Cow	First Lactation	Second Lactation
1	22,792	29,655	11	15,058	8,013
2	31,693	23,817	12	16,681	23,301
3	18,367	35,360	13	21,002	32,529
4	17,440	21,848	14	17,987	23,620
5	29,798	29,828	15	15,968	26,437
6	28,540	33,898	16	23,580	26,227
7	23,661	22,444	17	14,334	25,529
8	39,242	23,648	18	24,030	27,368
9	21,574	25,303	19	20,392	29,527
10	34,542	34,812	20	24,347	25,592

(a) Find the regression equation for the amount of milk produced during a cow's second lactation based on the amount of milk produced during the cow's first lactation.
(b) Calculate the standard error of measurement s_e.
(c) At the 0.05 level of significance, test the claim that the amount of milk produced during a cow's first lactation can be used to predict the amount of milk produced during that cow's second lactation; in other words, the slope of the population's regression line is not equal to 0.
(d) Construct a 95% confidence interval for the slope β of the population's regression line, and explain what it means.

9. If a company has a quick Web server, does that mean that it is reliable as well? Here is a listing of 10 major electronic commerce sites. Here is the average length of time for the Web site to come up on a user's computer, and the percentage of time the site is available.

Site	Seconds	Availability (%)
Amazon.com	17.66	90.1
Barnesandnoble.com	15.85	94.2
CDnow	15.52	89.2
eBay	17.25	78.4
eToys.com	20.11	84.0
Gateway	19.01	86.1
Landsend.com	13.49	94.2
Macys.com	31.42	93.0
Wal Mart Online	22.24	92.0
Wine.com	20.58	93.6

(Source: Keynote Systems)

(a) Find the regression equation for the percentage of time that a Web site is available based on the time that it takes for the site to come up on a user's computer.

(b) Calculate the standard error of measurement s_e.

(c) At the 0.05 level of significance, test the claim that the time it takes for the site to come up on a user's computer can be used to predict the percentage of time that a Web site is available; in other words, the slope of the population's regression line is not equal to 0.

(d) Construct a 95% confidence interval for the slope β of the population's regression line, and explain what it means.

10. Many politicians and citizen groups complain about the cost of prescription medications in the United States. Here are the prices of a dose of 10 medications in Canada and the United States, all in U.S. dollars.

Drug	Canada	United States
Prilosec	1.47	3.31
Prozac	1.07	2.27
Lipitor	1.34	2.54
Prevacid	1.34	3.13
Epogen	21.44	23.40
Zocor	1.47	3.16
Zoloft	1.07	1.98
Zyprexa	3.39	5.27
Claritin	1.11	1.96
Paxil	1.13	2.22

(Source: USA Today.)

(a) Find the regression equation that predicts the cost for a dose of a prescription medication in the United States, based on its cost in Canada.

(b) Calculate the standard error of measurement s_e.

(c) At the 0.05 level of significance, test the claim that the cost for a dose of a prescription medication in Canada can be used to predict its cost in the United States; in other words, the slope of the population's regression line is not equal to 0.

(d) Construct a 95% confidence interval for the slope β of the population's regression line, and explain what it means.

11. The results of 19 horse races for fillies and mares (female racehorses) at Santa Anita Park were selected at random. Here are the number of horses in the race, and the price that the winning horse in that race paid to win.

Race	Number of Horses	Winning Price	Race	Number of Horses	Winning Price
1	7	7.40	11	12	7.40
2	6	3.20	12	6	5.60
3	7	12.20	13	7	19.40
4	6	3.80	14	5	2.40
5	12	8.00	15	10	5.40
6	11	6.80	16	8	5.20
7	3	4.00	17	9	47.40
8	5	2.60	18	4	5.60
9	10	49.40	19	9	9.00
10	12	70.00			

(a) Find the regression equation for the winning price in a race for fillies and mares based on the number of horses in the race.

(b) Predict the winning price for a race with 6 horses in it.

(c) Construct a 95% prediction interval for the winning price for a race with 6 horses in it.

12. Here are the number of teachers at public schools and the number of students in those schools for 14 states in 1997.

State	Teachers	Students
AL	45,793	749,187
CA	268,581	5,803,734
FL	124,473	2,294,077
IL	118,734	1,998,289
KY	40,488	669,322
MA	67,170	949,006
MO	60,869	910,654
NH	12,931	201,629
NC	77,785	1,236,083
OR	26,935	541,346
SD	9,282	142,443
UT	21,115	482,957
WA	75,524	1,110,815
WI	57,227	881,780

(Source: U.S. Department of Education.)

(a) Find the regression equation for the number of public school students in a state based on the number of teachers in those schools.
(b) In 1997, the state of Rhode Island had 10,598 public school teachers. Use the regression equation to predict how many students were in Rhode Island public schools.
(c) Construct a 95% prediction interval for the number of students that were in Rhode Island public schools.
(d) In 1997, there were 153,321 students in Rhode Island public schools. Does your prediction interval in part (c) contain this value?
(e) In 1997, the state of Connecticut had 37,658 public school teachers. Use the regression equation to predict how many students were in Connecticut public schools.
(f) Construct a 95% prediction interval for the number of students that were in Connecticut public schools.
(g) In 1997, there were 535,164 students in Connecticut public schools. Does your prediction interval in part (f) contain this value?

13. For 1997, here are the number of nursing home beds in each state, as well as the number of complaints received that year.

State	Beds	Complaints	State	Beds	Complaints
AL	24,444	998	IA	35,790	254
AK	767	114	KS	26,877	4,552
AZ	28,110	1,625	KY	28,956	5,314
AR	24,325	621	LA	38,253	1,992
CA	134,312	17,764	ME	9,255	499
CO	19,876	7,669	MD	30,546	2,260
CT	32,012	231	MA	57,934	10,438
DE	5,697	1,043	MI	53,314	2,674
DC	3,733	5,199	MN	44,669	2,953
FL	80,055	5,029	MS	19,905	661
GA	41,544	3,274	MO	56,719	7,324
HI	3,952	118	MT	7,617	1,174
ID	6,638	804	NE	17,195	1,453
IL	110,817	4,731	NV	4,250	7,158
IN	58,355	1,792	NH	8,146	593

(continues)

(continued)

State	Beds	Complaints	State	Beds	Complaints
NJ	49,541	2,192	SD	2,171	541
NM	7,376	9,171	TN	38,786	1,332
NY	115,000	5,483	TX	126,476	9,495
NC	46,467	2,457	UT	8,954	1,958
ND	8,755	546	VT	3,844	330
OH	128,181	4,016	VA	30,060	406
OK	34,624	2,164	WA	27,662	1,670
OR	14,470	2,950	WV	11,529	1,043
PA	97,593	4,641	WI	49,740	3,314
RI	10,365	686	WY	3,116	733
SC	20,060	1,896			

(Source: U.S. Department of Health and Human Services.)

(a) Find the regression equation for the number of nursing home beds in a state based on the number of complaints received.

(b) Use the regression equation to predict how many beds are in Nevada, which received 7,158 complaints.

(c) Construct a 99% prediction interval for the number of beds in Nevada. Does your interval contain the actual number of beds in Nevada?

(d) Use the regression equation to predict how many beds are in Ohio, which received 4,016 complaints.

(e) Construct a 99% prediction interval for the number of beds in Ohio. Does your interval contain the actual number of beds in Ohio?

14. For eight NHL goalies in the season's second month, here are the number of minutes played by the goalie, the number of goals given up by the goalie, and the number of shots attempted against the goalie.

Minutes	788	661	783	831	476	643	767	608
Goals	22	20	28	34	24	30	39	32
Shots	335	258	355	413	243	313	339	295

(a) Find the regression equation for the number of goals given up, based on the number of minutes played.

(b) At the 0.05 level of significance, test the claim that the number of minutes played can be used to predict the number of goals given up; in other words, the slope of the population's regression line is not equal to 0.

(c) Predict the number of goals given up by a goalie who has played 692 minutes, and then construct a 95% prediction interval.

(d) Find the regression equation for the number of goals given up, based on the number of shots attempted.

(e) At the 0.05 level of significance, test the claim that the number of shots attempted can be used to predict the number of goals given up; in other words, the slope of the population's regression line is not equal to 0.

(f) Predict the number of goals given up by a goalie who has faced 291 attempted shots, and construct a 95% prediction interval.

Use the following data for Exercises 15–16. For 71 randomly selected 1999 National League baseball games, here are the times required to complete the games (in minutes), the total number of hits in the game, and the combined number of runs in the game.

Time	Hits	Runs	Time	Hits	Runs	Time	Hits	Runs
170	19	11	152	20	14	160	14	3
187	20	11	190	23	9	177	20	7
169	14	6	213	30	22	171	16	9
129	12	1	158	19	11	198	26	22

(continues)

(continued)

Time	Hits	Runs	Time	Hits	Runs	Time	Hits	Runs
159	25	15	176	20	9	159	21	10
209	27	19	146	13	6	160	9	3
181	16	9	172	9	4	146	16	10
183	17	11	137	13	6	188	14	6
197	23	12	200	16	9	149	16	6
170	19	13	158	19	13	151	13	6
157	20	6	181	25	13	172	18	7
214	32	27	202	21	11	143	12	6
154	15	6	152	14	3	189	19	11
166	23	12	128	11	8	204	27	19
191	21	11	161	18	12	215	24	13
178	17	6	153	8	3	181	16	10
198	23	14	164	19	13	141	14	5
178	17	13	169	19	9	152	18	4
161	13	6	194	24	14	176	20	10
198	20	6	108	10	5	180	19	14
179	17	10	146	19	5	158	12	4
150	21	13	157	12	6			
265	20	9	167	16	13			
136	18	16	169	16	7			
143	15	3	190	22	12			

15. (a) Find the regression equation for the time required to complete the game based on the number of hits.
    (b) Calculate the standard error of estimate s_e.
    (c) At the 0.05 level of significance, test the claim that the number of hits in a game can be used to predict the time required to complete the game; in other words, the slope of the population's regression line is not equal to 0.
    (d) Construct a 95% confidence interval for the slope β of the population regression line.
    (e) Construct a 95% prediction interval for the length of a game with 12 hits.

16. (a) Find the regression equation for the time required to complete the game based on the number of runs.
    (b) Calculate the standard error of estimate s_e.
    (c) At the 0.05 level of significance, test the claim that the number of hits in a game can be used to predict the time required to complete the game; in other words, the slope of the population's regression line is not equal to 0.
    (d) Construct a 95% confidence interval for the slope β of the population regression line.
    (e) Construct a 95% prediction interval for the length of a game with 8 runs.

Use the following data for Exercises 17 and 18. For 62 randomly selected 1999 American League baseball games, here are the times required to complete the games (in minutes), the total number of hits in the game, and the combined number of runs in the game.

Time	Hits	Runs	Time	Hits	Runs	Time	Hits	Runs
175	13	5	158	28	16	166	14	6
168	16	4	157	14	4	168	18	7
185	23	12	196	18	12	182	21	9
183	23	16	190	18	8	219	31	17
196	23	11	166	17	9	235	28	19
151	16	10	176	14	4	224	33	25
198	14	11	186	21	14	202	27	17
148	14	6	153	15	7	171	20	12
189	23	12	189	27	14	159	16	9

(continues)

Time	Hits	Runs	Time	Hits	Runs	Time	Hits	Runs
146	9	4	180	16	4	153	19	6
225	32	20	205	13	9	169	17	11
199	27	11	163	24	15	171	7	1
143	17	7	177	19	7			
149	12	6	154	22	14			
156	14	8	172	16	7			
192	27	13	178	20	8			
230	31	23	186	20	9			
178	24	16	187	17	11			
124	9	3	163	14	4			
167	17	9	181	18	8			
156	11	4	208	22	13			
185	19	13	142	20	9			
179	18	9	200	18	11			
185	26	17	150	20	10			
144	13	8	172	18	6			

17. (a) Find the regression equation for the time required to complete the game based on the number of hits.
    (b) Calculate the standard error of estimate s_e.
    (c) At the 0.05 level of significance, test the claim that the number of hits in a game can be used to predict the time required to complete the game; in other words, the slope of the population's regression line is not equal to 0.
    (d) Construct a 95% confidence interval for the slope β of the population regression line.
    (e) Construct a 95% prediction interval for the length of a game with 12 hits.

18. (a) Find the regression equation for the time required to complete the game based on the number of runs.
    (b) Calculate the standard error of estimate s_e.
    (c) At the 0.05 level of significance, test the claim that the number of hits in a game can be used to predict the time required to complete the game; in other words, the slope of the population's regression line is not equal to 0.
    (d) Construct a 95% confidence interval for the slope β of the population regression line.
    (e) Construct a 95% prediction interval for the length of a game with 8 runs.

Overview

This unit deals with the (linear) relation between two numerical values. The best way to examine this relationship is to begin with a **scatterplot** to view the relation graphically, and follow up by calculating the **correlation coefficient r.** A hypothesis test is introduced in Section 10.1 to determine whether the correlation coefficient is not equal to 0.

If there is a significant linear correlation, we may proceed to find the equation of the **regression line,** which puts the relationship between the two variables into the form of a formula. The equation of the regression line can be used to make predictions about one variable based on the value of the other variable.

Finally, the last section involves inference for linear regression. We can test that the slope is not equal to 0, which tells us that y depends in some way on x. This also allows us to construct confidence intervals for the slope of the regression line for the population. We also learned how to create prediction intervals for y based on a particular value of x.

Review Exercises

1. Here are the number of passes thrown and passing yards gained for eight randomly selected NFL quarterbacks.

Passes Thrown	Passing Yards
34	141
40	212
33	154
35	254
33	245
44	168
30	287
27	261

   (a) Construct a scatterplot for these data, treating the number of passes thrown as the independent variable (x).
   (b) A TV commentator states "It's quite simple. The more passes you throw, the more yards you gain." Use the scatterplot to determine whether the commentator's comment seems valid.
   (c) Calculate the correlation coefficient r.
   (d) At the 0.05 level of significance, test the claim that there is a significant correlation between the number of passes thrown and the number of passing yards gained.

2. Thirteen fast foods were selected at random, and the following chart lists the number of grams of fat found in each, as well as the number of calories in each.

Item	Fat (g)	Calories	Item	Fat (g)	Calories
1	18	368	8	14	410
2	13	388	9	3	222
3	39	640	10	10	398
4	7	290	11	15	340
5	28	510	12	36	640
6	27	560	13	18	285
7	30	530			

(a) Find the regression equation that predicts the number of calories in an item based on the number of grams of fat in an item.

(b) If you know that a sandwich has 12 grams of fat, predict the number of calories in the sandwich.

3. Use the same set of data as Exercise 2.

(a) Calculate the standard error of measurement s_e.

(b) At the 0.05 level of significance, test the claim that the number of grams of fat can be used to predict the number of calories in an item; in other words, the slope of the population's regression line does not equal 0.

(c) Construct a 95% confidence interval for the slope β of the population's regression line.

(d) Construct a 95% prediction interval for the number of calories in an item with 15 grams of fat.

Formulas

Correlation coefficient:

$$r = \frac{\Sigma xy - \dfrac{(\Sigma x)(\Sigma y)}{n}}{\sqrt{\left(\Sigma x^2 - \dfrac{(\Sigma x)^2}{n}\right)\left(\Sigma y^2 - \dfrac{(\Sigma y)^2}{n}\right)}}$$

Linear regression coefficients:

$$b = \frac{n \cdot \Sigma xy - (\Sigma x) \cdot (\Sigma y)}{n\Sigma x^2 - (\Sigma x)^2} \qquad a = \bar{y} - b\bar{x}$$

Standard error of estimate:

$$s_e = \sqrt{\frac{\Sigma(y - \hat{y})^2}{n - 2}} \quad \text{or} \quad s_e = \sqrt{\frac{\Sigma y^2 - a\Sigma y - b\Sigma(xy)}{n - 2}}$$

Hypothesis test for slope:

$$t = \frac{b}{s_e} \cdot \sqrt{\Sigma x^2 - \frac{(\Sigma x)^2}{n}}, \quad d.f. = n - 2$$

Confidence intervals for slope:

$$E = t_{\alpha/2} \cdot \frac{s_e}{\sqrt{\Sigma x^2 - \dfrac{(\Sigma x)^2}{n}}}, \quad d.f. = n - 2, \quad b - E \leq \beta \leq b + E$$

Prediction intervals for y for a given value of x:

$$E = t_{\alpha/2} \cdot s_e \cdot \sqrt{1 + \frac{1}{n} + \frac{n(x - \bar{x})^2}{n(\Sigma x^2) - (\Sigma x)^2}}, \quad d.f. = n - 2, \quad \hat{y} - E \leq y \leq \hat{y} + E$$

unit
seven
7

Nonparametric Tests

CHAPTER
eleven

NONPARAMETRIC TESTS

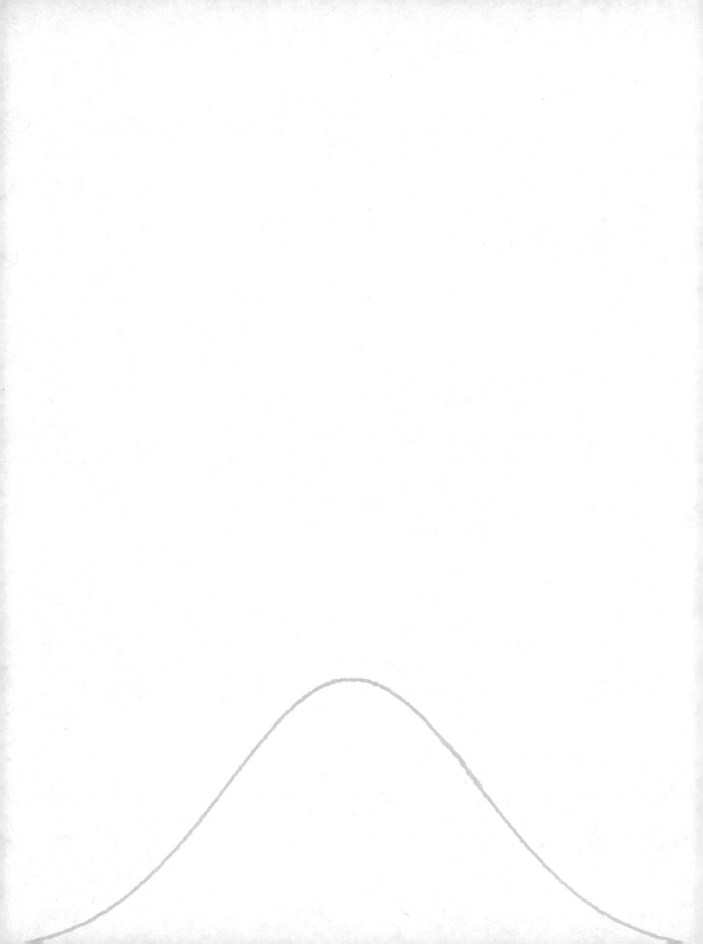

Nonparametric Tests

Many of the previously covered hypothesis tests require that the samples are drawn from populations that are normally distributed. If a set of sample data is extremely skewed, this is strong evidence that the population does not follow a normal distribution. In such a case we should not use these tests. In this chapter we introduce a series of tests that do not require the population of interest to be normally distributed.

In Section 11.1 we cover the sign test, which can be used to make inferences about a population median and two dependent samples. These tests are alternatives to tests about a single population mean (Section 7.2) and the paired difference test (Section 8.3). This test is the least efficient of the nonparametric tests, requiring a sample size of 100 to have the same results as these other tests had with sample sizes of 63.

In Section 11.2 we discuss the Wilcoxon signed-ranks test. This test is also an alternative to the paired difference test, and is much more efficient than the sign test. Section 11.3 deals with the Wilcoxon rank-sum test, which tests to see whether two samples are drawn from populations that have the same distribution. We use it to compare the median of one population to another as an alternative to the tests that compare two population means (Section 8.2).

In Section 11.4, we examine the Kruskal–Wallis test, which is a nonparametric alternative to ANOVA (Section 9.2). Finally, we conclude with a nonparametric alternative to correlation called rank correlation in Section 11.5.

Nonparametric tests involve easier calculations, and you will soon wonder why we do not abandon the previous tests for these. The reason is that when the proper conditions are met, the previous "parametric" tests are more likely to reject a false null hypothesis than the nonparametric tests. However, when the proper conditions are not met, nonparametric tests provide us with an arena to test a claim that we would not otherwise be able to test.

SECTION 11.1
Sign Test

The first nonparametric test that we shall explore is the **sign test.** We will use the sign test to test claims about a population median, in a situation similar to the test for a population mean (Section 7.2). We will also use the sign test for a test similar to the paired difference test (Section 8.3).

When will we need to test a claim about a population median? Often, claims about a population median are made for data that are skewed, such as family income or home sale prices. We also will test claims about a population median when the population does not follow a normal distribution. All of our tests for central tendency are based on the assumption that the populations we are sampling from follow a normal distribution. If we know that this is not the case, then we cannot use the t-tests developed earlier.

Before we fall in love with this new test, we should point out that it takes a sample of size 100 for the sign test to reach the same results that a t-test achieves with a sample size of 63. We say that the efficiency of the sign test is 0.63, which is very low compared to the nonparametric tests that follow in this chapter. The remaining tests all have efficiencies above 0.9.

Unless the sample indicates that the population most likely does not follow a normal distribution, we should apply the appropriate t-test, because it will increase the chances of rejecting a null hypothesis that is false.

Recall that the median is the value for which half of the values are above it and the other half are below it. When performing a hypothesis test for a median, we should therefore find that approximately half of the sample values are greater than the claimed median and half are less than it. We will record a positive sign (+) for each sample value above the claimed median, a negative sign (–) for all sample values that are below the median, and a zero (0) for all sample values that are equal to the median. We then disregard all 0s, and total the number of positive signs and the number of negative signs that we have. If the test is a two-tailed test, then the test statistic, x, is the smaller of these two totals. If the test is right-tailed, then the test statistic is the number of negative signs. For a left-tailed test, the test statistic is the number of positive signs. We let n represent the combined number of positive and negative signs.

To determine our decision rule, we will use the binomial probability distribution for n trials with a probability of success of $p = 0.5$. We use 0.5 because a sample value should be equally likely to be above the median as below it. For a single-tailed test, the critical value x^* is the lowest number of successes for which the probability of x^* successes or fewer is greater than the level of significance α. For a two-tailed test, we repeat the same after dividing the level of significance α by 2. We divide by 2 because in making the test statistic the smaller total of positive or negative signs, we are only interested in the "left tail" of the binomial probability distribution.

EXAMPLE
11.1

Find the decision rule for the following situations.

1. One-tailed, $n = 10$, $\alpha = 0.05$.

Here are the first few rows for the column for $p = 0.5$ from the binomial probability table for $n = 10$, with the cumulative probabilities next to the individual probabilities.

Number of Successes	Probability	Cumulative Probability
0	0.0010	0.0010
1	0.0098	0.0108
2	0.0439	0.0547
3	0.1172	0.1719
4	0.2051	0.3770

We see that the probability of 2 or fewer successes is 0.0547, which is greater than 0.05, so our critical value is 2.

Decision rule: Reject H_0 if $x < 2$.

2. One-tailed, $n = 18$, $\alpha = 0.01$.

Here are the first few rows of the column for $p = 0.5$ from the binomial probability table for $n = 18$, with the cumulative probabilities next to the individual probabilities. We are looking for the first number of successes that has a cumulative probability greater than 0.01.

Number of Successes	Probability	Cumulative Probability
0	0.0000	0.0000
1	0.0001	0.0001
2	0.0006	0.0007
3	0.0031	0.0038
4	0.0117	0.0155
5	0.0327	0.0482
6	0.0708	0.1190

The probability of 4 or fewer successes is 0.0155, so our critical value is 4.

Decision rule: Reject H_0 if $x < 4$.

3. Two-tailed, $n = 13$, $\alpha = 0.05$.

Here are the first few rows of the column for $p = 0.5$ from the binomial probability table for $n = 13$, with the cumulative probabilities next to the individual probabilities. We are looking for the first number of successes that has a cumulative probability greater than $0.05 \div 2 = 0.025$. We divide the level of significance by 2 for a two-tailed test.

Number of Successes	Probability	Cumulative Probability
0	0.0001	0.0001
1	0.0016	0.0017
2	0.0095	0.0112
3	0.0349	0.0461
4	0.0873	0.1334

The probability of 3 or fewer successes is 0.0461, so our critical value is 3.

Decision rule: Reject H_0 if $x < 3$. ∎

We are now ready to look at a complete example of this test. Note that we will use the symbol $\tilde{\mu}$ to represent the population median.

EXAMPLE 11.2

A math instructor has developed a test to determine a college student's math competency. He claims that the test is designed to produce a median score of 70. The test is given to 12 randomly selected college students, and here are their scores.

76	73	81	63	71	82
68	75	66	79	90	70

Test the instructor's claim that the median score for all students is 70 at the 0.05 level of significance.

We begin with the "calculations." We will write the appropriate sign, or 0, below each value.

76	73	81	63	71	82	68	75	66	79	90	70
+	+	+	−	+	+	−	+	−	+	+	0

We have 8 positive signs (8 values greater than the claimed median), 3 negative signs (3 values less than the claimed median), and one 0 (1 value equal to the claimed median). The total number of positive and negative signs is 11, so $n = 11$.

Step 1

Claim in words: The median score for all students is 70.
Claim: $\tilde{\mu} = 70$
Complement: $\tilde{\mu} \neq 70$
H_0: $\tilde{\mu} = 70$
H_A: $\tilde{\mu} \neq 70$

Step 2

Level of significance: $\alpha = 0.05$

Step 3

Test statistic: x = smaller number of positive or negative signs

Step 4

This is a two-tailed test, $\alpha = 0.05$, with $n = 11$. Here are the first few rows for the column for $p = 0.5$ from the binomial probability table for $n = 11$, with the cumulative probabilities next to the individual probabilities. We are looking for the first number of successes that has a cumulative probability greater than $0.05 \div 2 = 0.025$.

Number of Successes	Probability	Cumulative Probability
0	0.0005	0.0005
1	0.0054	0.0059
2	0.0269	0.0328
3	0.0806	0.1134

The probability of 2 or fewer successes is 0.0328, so our critical value is 2.

Decision rule: Reject H_0 if $x < 2$.

Step 5

Since there are only 3 negative signs and 8 positive signs, $x = 3$.

Decision: Fail to reject H_0.

Conclusion: There is not sufficient sample evidence to reject the claim that the median score for all students is 70.

p-value

To find the p-value for this test, we multiply the cumulative probability associated with $x = 3$ by 2.

$$p\text{-value} = 2 \cdot 0.1133$$
$$= 0.2266 \ \blacksquare$$

EXAMPLE 11.3   A college recruiter claims that the median annual income of a person who holds a bachelor's degree is over \$35,000. Here are the salaries of a random sample of 9 people with bachelor's degrees:

| \$32,485 | \$36,926 | \$40,857 | \$39,985 | \$45,510 |

| \$50,248 | \$39,637 | \$39,884 | \$33,392 |

At the 0.05 level of significance, test the recruiter's claim.

We begin by totaling the number of positive and negative signs.

\$32,485	\$36,926	\$40,857	\$39,985	\$45,510	\$50,248	\$39,637	\$39,884	\$33,392
–	+	+	+	+	+	+	+	–

We have 7 positive signs and 2 negative signs. The total number of positive and negative signs is 9, so $n = 9$.

Step 1

Claim in words: The median annual income of a person who holds a bachelor's degree is over \$35,000.

Claim: $\tilde{\mu} > 35,000$

Complement: $\tilde{\mu} \le 35,000$

H_0: $\tilde{\mu} \le 35,000$

H_A: $\tilde{\mu} > 35,000$

Step 2

Level of significance: $\alpha = 0.05$

Step 3

Test statistic: x = number of negative signs

Step 4

This is a one-tailed test, $\alpha = 0.05$, with $n = 9$. Here are the first few rows for the column for $p = 0.5$ from the binomial probability table for $n = 9$, with the cumulative probabilities next to the individual probabilities. We are looking for the first number of successes that has a cumulative probability greater than 0.05.

Number of Successes	Probability	Cumulative Probability
0	0.0020	0.0020
1	0.0176	0.0196
2	0.0703	0.0899
3	0.1641	0.2540

The probability of 2 or fewer successes is 0.0899, so our critical value is 2.

Decision rule: Reject H_0 if $x < 2$.

Step 5

Since there are 2 negative signs, $x = 2$.

Decision: Fail to reject H_0.

Conclusion: There is not sufficient sample evidence to support the claim that the median annual income of a person who holds a bachelor's degree is over $35,000.

p-value

The p-value for this test is the cumulative probability associated with $x = 2$, which is 0.0899. ∎

For sample sizes that are greater than 25, we will be able to use the normal approximation to the binomial distribution to calculate a test statistic for testing purposes. Here is the test statistic for such a situation.

$$z = \frac{(x + 0.5) - \dfrac{n}{2}}{\sqrt{n}/2}$$

 EXAMPLE 11.4 A city's chamber of commerce reports that the median home price is $98,000. A random sample of 40 recent home sales was taken. Of these homes, 27 houses sold for less than $98,000, and 13 sold for more than that. Test the chamber of commerce's claim at the 0.05 level of significance.

We have 13 positive signs and 27 negative signs, so $x = 13$. The total number of positive and negative signs is 40, so $n = 40$.

Step 1

Claim in words: The median home price is $98,000.
Claim: $\tilde{\mu} = 98{,}000$
Complement: $\tilde{\mu} \neq 98{,}000$
H_0: $\tilde{\mu} = 98{,}000$
H_A: $\tilde{\mu} \neq 98{,}000$

Step 2

Level of significance: $\alpha = 0.05$

Step 3

Test statistic: $z = \dfrac{(x + 0.5) - \dfrac{n}{2}}{\sqrt{n}\,/\,2}$

where x = smaller number of positive or negative signs.

Step 4

This is a two-tailed test and $\alpha = 0.05$

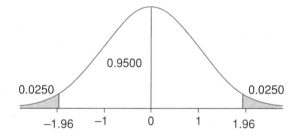

Decision rule: Reject H_0 if $z < -1.96$ or $z > 1.96$.

Step 5

For our calculations, $x = 13$ and $n = 40$.

$$z = \dfrac{(x + 0.5) - \dfrac{n}{2}}{\sqrt{n}\,/\,2}$$

$$= \dfrac{(13 + 0.5) - \dfrac{40}{2}}{\sqrt{40}\,/\,2}$$

$$= \dfrac{13.5 - 20}{\sqrt{40}\,/\,2}$$

$$= \dfrac{-6.5}{\sqrt{40}\,/\,2}$$

$$= -2.06$$

Decision: Reject H_0.

Conclusion: There is sufficient sample evidence to reject the claim that the median
  home price is $98,000.

p-value

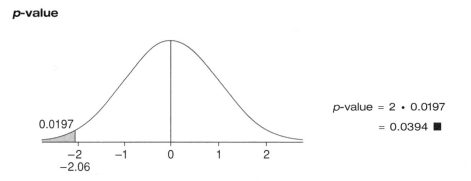

p-value = 2 • 0.0197

= 0.0394 ∎

The sign test can also be used to test the difference between paired samples, as did our paired difference test in Section 8.3. To come up with our positive and negative signs, we subtract the second sample values from their associated values from the first sample. Instead of recording the difference, however, we simply record the sign of the difference or 0 if there is no difference. Then we proceed with the sign test in the same fashion as the test for a population median, except that the claim will involve the population median difference $\tilde{\mu}_d$.

EXAMPLE 11.5

Many companies claim to raise students' SAT scores. When evaluating such a claim, we must consider something called the "testing effect." If a student has already taken the SAT exam once, there is a chance that the student's scores will improve the next time, due to familiarity with the test. Here are the scores of 12 randomly selected students on the SAT exam (combined math and English) on their first and second attempt at the exam. At the 0.05 level of significance, test the claim that scores improve from the first attempt to the second attempt.

Student	First Score	Second Score	Student	First Score	Second Score
1	680	650	7	710	740
2	1140	1140	8	830	870
3	910	910	9	1290	1270
4	1370	1420	10	1050	1130
5	1110	1110	11	1170	1160
6	1000	1090	12	1060	1110

We begin by determining in which order to subtract. Since the claim is that the scores improve, the second scores should be higher that the first scores. We will let d = second − first. If the claim is true, then these values should be positive in general, and the median of all population differences should be positive as well ($\tilde{\mu}_d > 0$). We will determine the number of positive and negative signs that we have before beginning.

Student	First Score	Second Score	d = Second − First
1	680	650	−
2	1140	1140	0
3	910	910	0

(continues)

(continued) Student	First Score	Second Score	d = Second − First
4	1370	1420	+
5	1110	1110	0
6	1000	1090	+
7	710	740	+
8	830	870	+
9	1290	1270	−
10	1050	1130	+
11	1170	1160	−
12	1060	1110	+

There are 6 positive signs, 3 negative signs, and 3 zeros, so $x = 3$ and $n = 9$.

Step 1

d = second − first
Claim in words: SAT scores improve the second time that students take the test.
Claim: $\tilde{\mu}_d > 0$
Complement: $\tilde{\mu}_d \leq 0$
H_0: $\tilde{\mu}_d \leq 0$
H_A: $\tilde{\mu}_d > 0$

Step 2

Level of significance: $\alpha = 0.05$

Step 3

Test statistic: x = number of negative signs

Step 4

This is a one-tailed test, $\alpha = 0.05$, with $n = 9$. Here are the first few rows for the column for $p = 0.5$ from the binomial probability table for $n = 9$, with the cumulative probabilities next to the individual probabilities. We are looking for the first number of successes that has a cumulative probability greater than 0.05.

Number of Successes	Probability	Cumulative Probability
0	0.0020	0.0020
1	0.0176	0.0196
2	0.0703	0.0899
3	0.1641	0.2540

The probability of 2 or fewer successes is 0.0898, so our critical value is 2.
Decision rule: Reject H_0 if $x < 2$.

Step 5

$x = 3$

Decision: Fail to reject H_0.

Conclusion: There is not sufficient sample evidence to support the claim that SAT scores improve the second time that students take the test.

p-value

The p-value for this test is the cumulative probability associated with $x = 3$, which is 0.2540. ■

MICROSOFT EXCEL   Critical Values for the Sign Test

We will now learn how to use Microsoft Excel to establish the decision rule for a sign test. We will be using Excel's built-in BINOMDIST function to construct a cumulative binomial probability distribution. This will be especially helpful if the number of nonzero signs (n) does not have its own table in the back of the book. At the very least, we will not have to add up the probabilities in an existing binomial table.

The BINOMDIST function requires four arguments to follow it in parentheses: number of successes, number of trials, probability of success, and a cumulative option. To construct a cumulative binomial probability distribution, the cumulative option should be *true*.

 EXAMPLE 11.6   Find the critical value for a sign test at the 0.05 level of significance, if there are 32 nonzero signs.

In a new Excel worksheet, type the numbers from 0 through 32 in cells **A1** through **A33.** These will represent the possible numbers of successes. In cell **B1,** type the following.

=BINOMDIST(A1,32,0.5,TRUE)

After you press the **Enter** key, you should see the value that is approximately 2.33×10^{-10}. We can convert that from scientific notation to a decimal number with four decimal places by clicking on cell **B1,** and selecting **Cells** from the **Format** menu. When a dialog box appears, click on the **Number** tab, and then click on **Number** under **Category.** Then, in the box labeled **Decimal Places,** type 4 and click on **OK.** This should change the value to 0.0000. Click on cell **B1** again, and select **Copy** from the **Edit** menu. Then highlight the cells from **B2** through **B33,** and select **Paste** from the **Edit** menu. This will complete the probability distribution. Here is the probability distribution, through 15 successes.

Number of Successes	Cumulative Probability
0	0.0000
1	0.0000
2	0.0000
3	0.0000
4	0.0000
5	0.0001
6	0.0003
7	0.0011
8	0.0035
9	0.0100
10	0.0251
11	0.0551
12	0.1077
13	0.1885
14	0.2983
15	0.4300

The first probability that is greater than 0.05 is associated with 11 successes, so our critical value is 11. The decision rule is "Reject H_0 if $x < 11$." ∎

EXERCISES 11.1

Use the standard procedure for all hypothesis tests.

1. A math instructor has written a new placement exam, and is concerned that the exam may be too long. The math department will use this new exam if the median time required to finish the exam for all students who take it is at most 50 minutes. The exam is given to 20 randomly selected students. Here are the times (in minutes) required to complete the exam for these students.

43	44	45	46	49	50	51	53	54	55
57	57	58	59	59	60	60	60	60	60

   Use these sample data to test the claim that the median length of time required to finish the exam for all students who take it is at most 50 minutes at the 0.05 level of significance.

2. An experimental algebra exam is given to 20 randomly selected college algebra students. Here are their scores.

68	94	50	60	92	14	49	68	41	67
76	71	32	15	54	54	65	92	58	68

   At the 0.05 level of significance, test the claim that the median score for all college algebra students on this exam is below 60.

3. Here are the number of trades reported on the New York Stock Exchange for eight randomly selected days in 1999.

713,908	698,287	733,686	746,847
583,053	854,294	606,022	690,561

   At the 0.05 level of significance, test the claim that the median number of trades is higher than 600,000.

4. Here are the winning mutual payouts for 20 randomly selected horse races at Santa Anita Park.

$3.40	$14.60	$3.40	$5.40	$44.40	$4.40	$8.00
$13.00	$4.20	$10.60	$14.80	$23.80	$5.20	$37.20
$23.80	$5.80	$4.80	$7.20	$7.00	$10.00	

   At the 0.05 level of significance, test the claim that the median winning mutual payout at Santa Anita Park is higher than $5.

5. Here are the weights of 11 randomly selected professional women's basketball players, in pounds.

195	148	168	142	170	165
170	165	180	172	132	

   At the 0.05 level of significance, test the claim that the median weight of professional women's basketball players is above 150 pounds.

6. These are the numbers of packages handled by a shipping office on 17 randomly selected days.

1103	1488	1713	1536	1037	1462
1625	1627	1080	1216	1639	1539
1545	907	1307	1387	1547	

   At the 0.01 level of significance, test the claim that the median number of packages handled per day is above 1100.

7. These are the total scores from 15 randomly selected NFL football games.

41	26	51	38	30	46	63	31
45	27	9	37	42	40	37	

At the 0.05 level of significance, test the claim that the median number of points scored in an NFL game is 45.

8. An algebra instructor has designed a final exam that he believes will produce a median score of 70. Here are the scores of 38 randomly selected algebra students on this exam.

32	41	43	44	46	46	47	48	49	50
52	54	54	56	57	58	58	60	60	61
65	66	66	67	67	68	68	68	68	71
72	76	83	87	92	92	94	100		

Test the instructor's claim at the 0.01 level of significance.

9. A football coach claims that consistent weight training over a period of time improves a person's strength. Here are the number of bench press repetitions performed before and after a 2-month training program for 18 randomly selected players.

Player	Before	After	Player	Before	After	Player	Before	After
1	4	11	7	2	2	13	0	3
2	16	21	8	2	4	14	9	16
3	14	19	9	11	19	15	4	7
4	4	6	10	15	17	16	4	10
5	12	14	11	2	9	17	0	1
6	10	16	12	11	10	18	6	9

Use the sign test to test the claim that the training program increases the number of repetitions a person can do at the 0.01 level of significance.

10. Here are the prices of a dose of 10 medications in Canada and the United States, all in U.S. dollars.

Drug	Canada	United States
Prilosec	1.47	3.31
Prozac	1.07	2.27
Lipitor	1.34	2.54
Prevacid	1.34	3.13
Epogen	21.44	23.40
Zocor	1.47	3.16
Zoloft	1.07	1.98
Zyprexa	3.39	5.27
Claritin	1.11	1.96
Paxil	1.13	2.22

(*Source:* USA Today.)

At the 0.05 level of significance, use the sign test to test the claim that prescription medications cost less in Canada than in the United States.

11. Some sports fans complain that referees show preferential treatment to the home team. Here are the number of free throws attempted by the home team and the away team in 13 randomly selected NBA games.

Home Team	28	27	34	26	29	30	34	27	37	16	46	48	29
Away Team	26	31	27	14	21	22	18	23	27	30	53	43	28

At the 0.05 level of significance, use the sign test to test the claim that the home team shoots more free throws than the away team.

12. A doctor believes that a sensible vegan diet and exercise program can be effective in lowering a person's serum cholesterol if it is high. He randomly selects seven patients that are willing to follow his program. Here are their cholesterol levels (mg/dL) before the program, and again 1 month after beginning the program.

Before	255	243	264	249	275	280	259
After	241	245	251	248	268	269	246

At the 0.05 level of significance, use the sign test to test the claim that the doctor's program lowers the cholesterol levels of people with high cholesterol.

13. Here are the midterm exam scores of 20 algebra students, along with their scores on the final exam.

Student	Midterm	Final	Student	Midterm	Final
1	65	85	11	38	56
2	50	77	12	92	95
3	75	90	13	59	60
4	70	84	14	66	87
5	68	61	15	61	83
6	58	70	16	71	77
7	49	76	17	68	79
8	92	78	18	82	74
9	68	80	19	85	94
10	93	92	20	67	80

At the 0.05 level of significance, use the sign test to test the claim that student's scores improve from the midterm to the final.

14. Here are the predicted high temperatures (°F) for 15 U.S. cities on August 16, 1999 that appeared in the *New York Times,* and the actual high temperatures for that day. The temperatures were predicted 2 days in advance.

City	Predicted High	Actual High
Atlantic City	80	84
Austin	99	103
Billings	78	77
Birmingham	92	90
Colorado Springs	86	77
Columbus	79	79
Fairbanks	65	57
Jackson	93	90
Minn.-St. Paul	81	79
Norfolk	85	83
Pittsburgh	81	74
St. Louis	86	80
Sioux Falls	87	85
Tampa	90	90
Wichita	95	97

At the 0.05 level of significance, use the sign test to test the claim that there is no difference between the predicted high temperatures and the actual high temperatures.

15. A group of 18 concertgoers was selected at random. Before the concert they were given a hearing test, and then given another one after the concert. (The volume

varied during the test, and the subject also had to state which ear the sound was in.) Here are the number of correctly identified sounds out of 10, both before and after the concert.

Before	After	Before	After
9	8	10	9
10	8	9	9
9	9	10	8
8	6	8	8
8	6	8	9
9	7	9	9
9	10	9	7
9	8	9	6
8	5	9	6

At the 0.05 level of significance, use the sign test to test the claim that a person's hearing is adversely affected by the noise of a concert.

16. For a random sample of 20 dairy cows, here are the number of pounds of milk produced during their first lactation (after their first calf) and their second lactation (after their second calf).

Cow	First Lactation	Second Lactation	Cow	First Lactation	Second Lactation
1	22,792	29,655	11	15,058	8,013
2	31,693	23,817	12	16,681	23,301
3	18,367	35,360	13	21,002	32,529
4	17,440	21,848	14	17,987	23,620
5	29,798	29,828	15	15,968	26,437
6	28,540	33,898	16	23,580	26,227
7	23,661	22,444	17	14,334	25,529
8	39,242	23,648	18	24,030	27,368
9	21,574	25,303	19	20,392	29,527
10	34,542	34,812	20	24,347	25,592

At the 0.05 level of significance, use the sign test to test the claim that cows produce more milk during their second lactation than during their first lactation.

17. Here are the results of 12 people who participated in a weight-loss program. The values represent the pounds lost (negative values) or gained (positive values) over a 3-month period.

–12   –10   20   4   5   –23   –12   –9   –16   1   –14   –3

At the 0.05 level of significance, use the sign test to test the claim that the diet is effective.

SECTION 11.2
Wilcoxon Signed-Rank Test

In the last section, we used the sign test to test a claim using two dependent samples. The sign test only uses the fact that the difference is positive or negative, but disregards the size of the differences. In the final example of the last section there were 6 positive differences and 3 negative differences, and we failed to reject the null hypothesis. Even if all of the positive differences had been large—say, 200 points—and if the 3 negative differences were small—say, only 10 points—we would still fail to reject the null hypothesis. Our intuition should tell us that such large positive differences and such small negative differences indicate there is a significant difference between the two populations. We say that the sign test is not very sensitive.

We now will examine the Wilcoxon signed-rank test, which does take into account the size of the differences. The Wilcoxon signed-rank test has an increased efficiency of 0.95, compared to the sign test's efficiency of 0.63. This test is based on the assumption that the two populations have distributions with similar shapes. We begin by finding the difference between each pair of values (sample 1 – sample 2). After discarding any differences that are 0, we rank the differences from lowest to highest, without considering the sign of the difference. The smallest difference would have a rank of 1, the next smallest difference's rank would be 2, and so on. If two or more differences are equal, we assign each the mean of the ranks in the tie. Next we total the ranks of the positive differences and the negative differences. It is a wise idea to employ technology to sort the differences from lowest to highest in order to assign the ranks, if that is at all possible.

The test statistic that we will use again depends on the number of nonzero differences (n) that we have. If there are 50 or fewer nonzero differences, then there is a table providing critical values for the test statistic W. For a two-tailed test, W is the smaller total of ranks. For a right-tailed test, W is the total of the negative ranks, whereas W is the total of the positive ranks for a left-tailed test. This table lists the largest value of W for which the null hypothesis can be rejected. If the number of nonzero differences is greater than 50, we use the following test statistic.

$$z = \frac{W - \dfrac{n(n+1)}{4}}{\sqrt{\dfrac{n(n+1)(2n+1)}{24}}}$$

We will begin by repeating the last example of Section 11.1, but using the Wilcoxon signed-rank test.

EXAMPLE 11.7	Many companies claim to raise students' SAT scores. When evaluating such a claim, we must consider something called the "testing effect." If a student has already taken the SAT exam once, there is a chance that the student's scores will improve the next time, due to familiarity with the test. Here are the scores of 12 randomly selected students on the SAT exam (combined math and English) on their first and second attempt at the exam. At the 0.05 level of significance, test the claim that scores improve from the first attempt to the second attempt.

Student	First Score	Second Score	Student	First Score	Second Score
1	680	650	7	710	740
2	1140	1140	8	830	870
3	910	910	9	1290	1270
4	1370	1420	10	1050	1130
5	1110	1110	11	1170	1160
6	1000	1090	12	1060	1110

We begin by determining in which order to subtract. Since the claim is that the scores improve, the second scores should be higher than the first scores. We will let d = second – first. If the claim is true, then these values should be positive in general, and the median of all population differences should be positive as well ($\tilde{\mu}_d > 0$). We will rank the differences before beginning.

Student	First Score	Second Score	d = Second – First	Rank
1	680	650	–30	3.5
2	1140	1140	0	
3	910	910	0	
4	1370	1420	50	6.5
5	1110	1110	0	
6	1000	1090	90	9
7	710	740	30	3.5
8	830	870	40	5
9	1290	1270	–20	2
10	1050	1130	80	8
11	1170	1160	–10	1
12	1060	1110	50	6.5

Note that there were two pairs with differences of 30 points. These two differences would have been assigned ranks 3 and 4, so we assign them both the mean of these two ranks, which is 3.5. In a similar fashion, the two differences of 50 points share a rank of 6.5.

The sum of the negative ranks is 6.5 (3.5 + 2 + 1). So W = 6.5, and n = 9 (9 nonzero differences).

Step 1

d = second – first
Claim in words: SAT scores improve the second time that students take the test.
Claim: $\tilde{\mu}_d > 0$
Complement: $\tilde{\mu}_d \leq 0$
H_0: $\tilde{\mu}_d \leq 0$
H_A: $\tilde{\mu}_d > 0$

Step 2

Level of significance: α = 0.05

Step 3

Test statistic: W (smaller total of the positive and negative ranks)

Step 4

This is a one-tailed test, α = 0.05, with n = 9. From the Wilcoxon signed-rank table, the critical value is 8.

Decision rule: Reject H_0 if $W \leq 8$.

Step 5

W = 6.5

Decision: Reject H_0.

Conclusion: There is sufficient sample evidence to support the claim that SAT scores improve the second time that students take the test.

p-value

To find the p-value, we read across the chart for n = 9. Our test statistic, W = 6.5, lies between 6 and 8 in that row, which tells us that the p-value is between 0.025 and 0.05. ∎

Note that the decision for this test is different from the decision based on the sign test. This shows that the sign test is not a very sensitive test. The Wilcoxon rank-sum test uses more information than the sign test, which makes it more dependable.

EXAMPLE 11.8

A doctor randomly selects nine of his patients for an experiment. At the end of an examination, he tells the patients that they need to lose a few pounds to improve their health. He gives them a sugar-based placebo, and tells them that it is an appetite suppressant that they are to take three times daily. Here are their weights, in pounds, at the examination and at a follow-up appointment 1 month later. At the 0.05 level of significance, test the claim that this "appetite suppressant" produces no difference in the weight of the patients.

Patient	Weight at Initial Exam	Weight at Follow-Up
A	209	201
B	249	245
C	185	179
D	207	204
E	227	231
F	174	182
G	157	162
H	196	197
I	177	177

We begin by finding the differences and ranking them. We will subtract the follow-up weight from the initial weight.

Patient	Weight at Initial Exam	Weight at Follow-Up	d	Rank
A	209	201	8	7.5
B	249	245	4	3.5
C	185	179	6	6
D	207	204	3	2
E	227	231	−4	3.5
F	174	182	−8	7.5
G	157	162	−5	5
H	196	197	−1	1
I	177	177	0	

The total of the positive ranks is 19, and the total of the negative ranks is 17. So, $W = 17$ and $n = 8$.

If there is no difference in the weights, as the claim suggests, then these differences should be close to 0, and the median of all population differences should equal 0 ($\tilde{\mu}_d = 0$).

Step 1

Claim in words: This "appetite suppressant" produces no difference in the weight of the patients.
Claim: $\tilde{\mu}_d = 0$
Complement: $\tilde{\mu}_d \neq 0$
H_0: $\tilde{\mu}_d = 0$
H_A: $\tilde{\mu}_d \neq 0$

Step 2

Level of significance: $\alpha = 0.05$

Step 3

Test statistic: W (smaller total of the positive and negative ranks)

Step 4

This is a two-tailed test, $\alpha = 0.05$, with $n = 8$. From the Wilcoxon signed-rank table, the critical value is 4.

Decision rule: Reject H_0 if $W \leq 4$.

Step 5

$W = 17$

Decision: Fail to reject H_0.

Conclusion: There is not sufficient sample evidence to reject the claim that this "appetite suppressant" produces no difference in the weight of the patients.

_p_-value

Since the value of W is greater than the greatest value in the row for $n = 8$, then the best we can say about the p-value is that it is greater than 0.10. ■

We finish with an example that has more than 50 nonzero differences.

EXAMPLE
11.9

Here are the serum cholesterol levels (mg/dL) of 60 randomly selected patients who had high cholesterol, before and after a program consisting of diet and exercise designed to lower cholesterol levels. At the 0.01 level of significance, test the claim that the program effectively lowers the serum cholesterol level of people who have high cholesterol.

Before	After	Before	After
241	217	251	237
260	255	264	263
263	261	266	267
262	259	238	211
260	255	253	241
262	259	261	257
261	257	261	257
250	235	261	257
251	237	251	237
260	255	254	243
268	271	273	281
274	283	286	307
247	229	252	239
270	275	267	269
266	267	254	243
254	243	254	243
256	247	255	245
250	235	274	283
258	251	267	269
265	265	261	257
259	253	257	249

(continues)

(continued)

Before	After	Before	After
257	249	266	267
273	281	255	245
255	245	263	261
269	273	260	255
245	225	253	241
273	281	270	275
266	267	255	245
266	267	255	245
270	275	252	239

We begin by calculating the differences, and ranking the absolute values of those differences.

Before	After	Difference	Rank	Before	After	Difference	Rank
241	217	−24	*58*	251	237	−14	*51*
260	255	−5	*23*	264	263	−1	*3.5*
263	261	−2	*8.5*	266	267	1	*3.5*
262	259	−3	*12*	238	211	−27	*59*
260	255	−5	*23*	253	241	−12	*46.5*
262	259	−3	*12*	261	257	−4	*16.5*
261	257	−4	*16.5*	261	257	−4	*16.5*
250	235	−15	*53.5*	261	257	−4	*16.5*
251	237	−14	*51*	251	237	−14	*51*
260	255	−5	*23*	254	243	−11	*43.5*
268	271	3	**12**	273	281	8	**31**
274	283	9	**35**	286	307	21	**57**
247	229	−18	*55*	252	239	−13	*48.5*
270	275	5	**23**	267	269	2	**8.5**
266	267	1	**3.5**	254	243	−11	*43.5*
254	243	−11	*43.5*	254	243	−11	*43.5*
256	247	−9	*35*	255	245	−10	*39*
250	235	−15	*53.5*	274	283	9	**35**
258	251	−7	*28*	267	269	2	**8.5**
265	265	0		261	257	−4	*16.5*
259	253	−6	*27*	257	249	−8	*31*
257	249	−8	*31*	266	267	1	**3.5**
273	281	8	**31**	255	245	−10	*39*
255	245	−10	*39*	263	261	−2	*8.5*
269	273	4	**16.5**	260	255	−5	*23*
245	225	−20	*56*	253	241	−12	*46.5*
273	281	8	**31**	270	275	5	**23**
266	267	1	**3.5**	255	245	−10	*39*
266	267	1	**3.5**	255	245	−10	*39*
270	275	5	**23**	252	239	−13	*48.5*

There was one patient whose cholesterol did not change, so we disregard that value. This leaves us $n = 59$ nonzero differences. The following table shows the ranked positions occupied by the different differences, so that we may see where the above ranks come from.

Difference	Rank Positions	Mean
1	1–6	3.5
2	7–10	8.5
3	11–13	12
4	14–19	16.5
5	20–26	23
6	27	27
7	28	28
8	29–33	31
9	34–36	35
10	37–41	39
11	42–45	43.5
12	46–47	46.5
13	48–49	48.5
14	50–52	51
15	53–54	53.5
18	55	55
20	56	56
21	57	57
24	58	58
27	59	59

Finally, note that the ranks of the positive differences are listed in **bold,** while the ranks of the negative differences are listed in *italics.* The total of the positive ranks is 352, and the total of the negative ranks is 1418. This tells us that we will be using 1418 for *W*.

If the program is effective in lowering serum cholesterol levels, as the claim suggests, then these differences (before – after) should be positive, and the median of all population differences should be less than 0 ($\tilde{\mu}_d < 0$).

Step 1

Claim in words: The program effectively lowers the serum cholesterol level of people who have high cholesterol.
Claim: $\tilde{\mu}_d < 0$
Complement: $\tilde{\mu}_d > 0$
H_0: $\tilde{\mu}_d > 0$
H_A: $\tilde{\mu}_d < 0$

Step 2

Level of significance: $\alpha = 0.01$

Step 3

Test statistic: $z = \dfrac{W - \dfrac{n(n+1)}{4}}{\sqrt{\dfrac{n(n+1)(2n+1)}{24}}}$

where *W* is the total of the positive ranks.

Step 4

This is a left-tailed test, with $\alpha = 0.01$.

Decision rule: Reject H_0 if $z < -2.33$.

Step 5

$W = 352$, $n = 59$

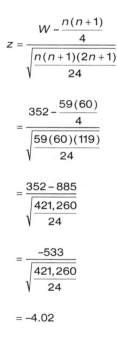

$$z = \frac{W - \dfrac{n(n+1)}{4}}{\sqrt{\dfrac{n(n+1)(2n+1)}{24}}}$$

$$= \frac{352 - \dfrac{59(60)}{4}}{\sqrt{\dfrac{59(60)(119)}{24}}}$$

$$= \frac{352 - 885}{\sqrt{\dfrac{421,260}{24}}}$$

$$= \frac{-533}{\sqrt{\dfrac{421,260}{24}}}$$

$$= -4.02$$

Decision: Reject H_0.

Conclusion: There is sufficient sample evidence to support the claim that this program effectively lowers the serum cholesterol level of people who have high cholesterol.

p-value

Since our value of z does not appear on our chart, we estimate the p-value by using $z = -3.50$. This gives us a p-value of 0.0002. We can show using a calculator or computer that the p-value is 0.000029. ■

EXERCISES 11.2

1. A football coach claims that consistent weight training over a period of time improves a person's strength. Here are the number of bench press repetitions performed before and after a 2-month training program for 18 randomly selected players.

Player	Before	After	Player	Before	After	Player	Before	After
1	4	11	7	2	2	13	0	3
2	16	21	8	2	4	14	9	16
3	14	19	9	11	19	15	4	7
4	4	6	10	15	17	16	4	10
5	12	14	11	2	9	17	0	1
6	10	16	12	11	10	18	6	9

Test the claim that the training program increases the number of repetitions a person can do at the 0.01 level of significance.

2. Here are the prices of a dose of 10 medications in Canada and the United States, all in U.S. dollars.

Drug	Canada	United States
Prilosec	1.47	3.31
Prozac	1.07	2.27
Lipitor	1.34	2.54
Prevacid	1.34	3.13
Epogen	21.44	23.40
Zocor	1.47	3.16
Zoloft	1.07	1.98
Zyprexa	3.39	5.27
Claritin	1.11	1.96
Paxil	1.13	2.22

(*Source:* USA Today.)

At the 0.05 level of significance, test the claim that prescription medications cost less in Canada than in the United States.

3. Some sports fans complain that referees show preferential treatment to the home team. Here are the number of free throws attempted by the home team and the away team in 13 randomly selected NBA games.

Home Team	28	27	34	26	29	30	34	27	37	16	46	48	29
Away Team	26	31	27	14	21	22	18	23	27	30	53	43	28

At the 0.05 level of significance, test the claim that the home team shoots more free throws than the away team.

4. A doctor believes that a sensible vegan diet and exercise program can be effective in lowering a person's serum cholesterol if it is high. He randomly selects seven patients that are willing to follow his program. Here are their cholesterol levels (mg/dL) before the program, and again 1 month after beginning the program.

Before	255	243	264	249	275	280	259
After	241	245	251	248	268	269	246

At the 0.05 level of significance, test the claim that the doctor's program lowers the cholesterol levels of people with high cholesterol.

5. Here are the midterm exam scores of 20 algebra students, along with their score on the final exam.

Student	Midterm	Final	Student	Midterm	Final
1	65	85	11	38	56
2	50	77	12	92	95
3	75	90	13	59	60
4	70	84	14	66	87
5	68	61	15	61	83
6	58	70	16	71	77
7	49	76	17	68	79
8	92	78	18	82	74
9	68	80	19	85	94
10	93	92	20	67	80

At the 0.05 level of significance, test the claim that student's scores improve from the midterm to the final.

6. Here are the predicted high temperatures (°F) for 15 U.S. cities on August 16, 1999 that appeared in the *New York Times,* and the actual high temperatures for that day. The temperatures were predicted 2 days in advance.

City	Predicted High	Actual High
Atlantic City	80	84
Austin	99	103
Billings	78	77
Birmingham	92	90
Colorado Springs	86	77
Columbus	79	79
Fairbanks	65	57
Jackson	93	90
Minn.-St. Paul	81	79
Norfolk	85	83
Pittsburgh	81	74
St. Louis	86	80
Sioux Falls	87	85
Tampa	90	90
Wichita	95	97

At the 0.05 level of significance, test the claim that there is no difference between the predicted high temperatures and the actual high temperatures.

7. A group of 18 concertgoers was selected at random. Before the concert they were given a hearing test, and then given another one after the concert. (The volume varied during the test, and the subject also had to state which ear the sound was in.) Here are the number of correctly identified sounds out of 10, both before and after the concert.

Before	After	Before	After
9	8	10	9
10	8	9	9

(continues)

(continued)

Before	After	Before	After
9	9	10	8
8	6	8	8
8	6	8	9
9	7	9	9
9	10	9	7
9	8	9	6
8	5	9	6

At the 0.05 level of significance, test the claim that a person's hearing is adversely affected by the noise of a concert.

8. For a random sample of 20 dairy cows, here are the number of pounds of milk produced during their first lactation (after their first calf) and their second lactation (after their second calf).

Cow	First Lactation	Second Lactation	Cow	First Lactation	Second Lactation
1	22,792	29,655	11	15,058	8,013
2	31,693	23,817	12	16,681	23,301
3	18,367	35,360	13	21,002	32,529
4	17,440	21,848	14	17,987	23,620
5	29,798	29,828	15	15,968	26,437
6	28,540	33,898	16	23,580	26,227
7	23,661	22,444	17	14,334	25,529
8	39,242	23,648	18	24,030	27,368
9	21,574	25,303	19	20,392	29,527
10	34,542	34,812	20	24,347	25,592

At the 0.05 level of significance, test the claim that cows produce more milk during their second lactation than during their first lactation.

9. Here are the results of 12 people who participated in a weight-loss program. The values represent the pounds lost (negative values) or gained (positive values) over a 3-month period.

−12	−10	20	4	5	−23	−12	−9	−16	1	−14	−3

At the 0.05 level of significance, test the claim that the diet is effective.

10. An instructor claims that test scores drop from the Unit 4 test to the Unit 5 test. Here are the scores of 15 randomly selected students on the Unit 4 test and the Unit 5 test. At the 0.05 level of significance, test the instructor's claim.

Student	1	2	3	4	5	6	7	8
Unit 4	67	96	91	100	100	96	81	93
Unit 5	41	92	95	96	100	82	89	96

Student	9	10	11	12	13	14	15	16
Unit 4	65	89	90	88	100	100	87	84
Unit 5	69	74	87	72	93	86	94	80

11. Eight faculty members at a community college decided to get a flu shot for the first time. Here are the number of absences due to illness each person had during the flu season the year before the shot and also for the flu season following the shot.

Before Shot	1	4	4	4	6	3	3	2
After Shot	0	3	1	2	3	2	3	2

At the 0.05 level of significance, test the claim that the flu shot is effective in lowering the number of absences due to illness.

12. Here are the prices of a dose of eight medications in Canada and Great Britain, all in U.S. dollars.

Drug	Canada	Great Britain
Prilosec	1.47	1.67
Prozac	1.07	1.08
Lipitor	1.34	1.67
Prevacid	1.34	0.82
Zocor	1.47	1.73
Zoloft	1.07	0.95
Claritin	1.11	0.41
Paxil	1.13	1.70

(Source: USA Today.)

At the 0.01 level of significance, test the claim that prices of medications are cheaper in Canada than in Great Britain.

SECTION 11.3
Wilcoxon Rank-Sum Test

In this section, we examine a nonparametric test for comparing the central tendency of two populations using independent samples. This test is called the **Wilcoxon rank-sum test.** The only requirement that we shall impose is that each sample has a size of at least 10. To use a t-test for two independent samples, we assume the populations that the samples were drawn from are normally distributed, but that is not required for the Wilcoxon rank-sum test. This nonparametric test has an efficiency of 0.95.

We will be testing to see whether the two samples come from the same distribution. We begin the calculations by combining the two samples, and ranking the individual values from lowest to highest as we did in the last section. If one sample has a majority of the higher ranks, this suggests that it comes from a distribution centered to the right of the other distribution. If the two samples are equally dispersed throughout the high and low ranks, this suggests that the two samples come from distributions with the same central tendency. We then total the ranks for the first sample, and denote this by R. Here is the test statistic that we will use.

$$z = \frac{R - \mu_R}{\sigma_R}$$

where $\mu_R = \dfrac{n_1(n_1 + n_2 + 1)}{2}$ and $\sigma_R = \sqrt{\dfrac{n_1 n_2(n_1 + n_2 + 1)}{12}}$.

EXAMPLE 11.10

A business student wondered whether the stock market would be busier on Mondays or Fridays. She took a random sample of 23 Mondays and 16 Fridays and found the volume (number of shares sold) for the New York Stock Exchange. Here are the results, rounded to the nearest million.

Monday

731	542	559	592	548	629	597	692
542	543	530	564	560	620	564	714
610	690	610	592	774	689	531	

Friday

| 689 | 669 | 613 | 569 | 579 | 635 | 622 | 557 |
| 558 | 759 | 725 | 736 | 637 | 785 | 683 | 602 |

At the 0.05 level of significance, test the claim that the median volume on Mondays is the same as the median volume on Fridays.

Step 1

Population 1: Mondays; population 2: Fridays
Claim in words: The median volume on Mondays is the same as the median volume on Fridays.
Claim: $\tilde{\mu}_1 = \tilde{\mu}_2$
Complement: $\tilde{\mu}_1 \neq \tilde{\mu}_2$
H_0: $\tilde{\mu}_1 = \tilde{\mu}_2$
H_A: $\tilde{\mu}_1 \neq \tilde{\mu}_2$

Step 2

Level of significance: $\alpha = 0.05$

Step 3

Test statistic: $z = \dfrac{R - \mu_R}{\sigma_R}$, where $\mu_R = \dfrac{n_1(n_1 + n_2 + 1)}{2}$ and $\sigma_R = \sqrt{\dfrac{n_1 n_2(n_1 + n_2 + 1)}{12}}$

Step 4

Two-tailed test, $\alpha = 0.05$

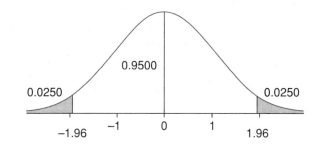

Decision rule: Reject H_0 if $z < -1.96$ or if $z > 1.96$.

Step 5

We begin the calculations by combining the two samples, sorting them in ascending order, and then ranking these values. Here is a table showing the results.

Volume	Day	Rank	Volume	Day	Rank
530	Monday	1	613	Friday	21
531	Monday	2	620	Monday	22
542	Monday	3.5	622	Friday	23
542	Monday	3.5	629	Monday	24
543	Monday	5	635	Friday	25
548	Monday	6	637	Friday	26
557	Friday	7	669	Friday	27
558	Friday	8	683	Friday	28
559	Monday	9	689	Monday	29.5
560	Monday	10	689	Friday	29.5
564	Monday	11.5	690	Monday	31
564	Monday	11.5	692	Monday	32
569	Friday	13	714	Monday	33
579	Friday	14	725	Friday	34
592	Monday	15.5	731	Monday	35
592	Monday	15.5	736	Friday	36
597	Monday	17	759	Friday	37
602	Friday	18	774	Monday	38
610	Monday	19.5	785	Friday	39
610	Monday	19.5			

We next total the ranks for sample 1, Monday. This total is 394.5, so $R = 394.5$.

$$\mu_R = \frac{n_1(n_1 + n_2 + 1)}{2} \qquad \sigma_R = \sqrt{\frac{n_1 n_2 (n_1 + n_2 + 1)}{12}}$$

$$= \frac{23(23 + 16 + 1)}{2} \qquad = \sqrt{\frac{23 \cdot 16 \cdot (23 + 16 + 1)}{12}}$$

$$= \frac{23(40)}{2} \qquad = \sqrt{\frac{23 \cdot 16 \cdot 40}{12}}$$

$$= 460 \qquad = 35.02$$

$$z = \frac{R - \mu_R}{\sigma_R}$$

$$= \frac{394.5 - 460}{35.02}$$

$$= -1.87$$

Decision: Fail to reject H_0.

Conclusion: There is not sufficient sample evidence to reject the claim that the median volume on Mondays is the same as the median volume on Fridays.

p-value

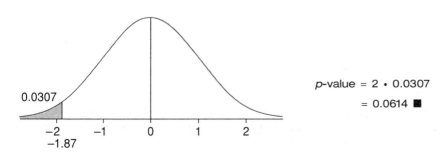

$$p\text{-value} = 2 \cdot 0.0307$$
$$= 0.0614 \ ∎$$

EXAMPLE 11.11

A golfer is considering buying a new driver. He hits 12 balls with his old driver, and then 15 balls with the new driver. Here are the distances, in yards.

Old

238	261	263	236	269	268
270	250	262	242	251	251

New

254	266	275	279	261	268	280	269
285	254	276	272	283	262	267	

At the 0.05 level of significance, test the claim that the mean driving distance with his old driver is less than the mean driving distance with the new driver.

Step 1

Population 1: old driver; population 2: new driver
Claim in words: The median driving distance with his old driver is less than the median driving distance with the new driver.
Claim: $\tilde{\mu}_1 < \tilde{\mu}_2$
Complement: $\tilde{\mu}_1 \geq \tilde{\mu}_2$
H_0: $\tilde{\mu}_1 \geq \tilde{\mu}_2$
H_A: $\tilde{\mu}_1 < \tilde{\mu}_2$

Step 2

Level of significance: $\alpha = 0.05$

Step 3

Test statistic: $z = \dfrac{R - \mu_R}{\sigma_R}$, where $\mu_R = \dfrac{n_1(n_1 + n_2 + 1)}{2}$ and $\sigma_R = \sqrt{\dfrac{n_1 n_2(n_1 + n_2 + 1)}{12}}$

Step 4

Left-tailed test, $\alpha = 0.05$

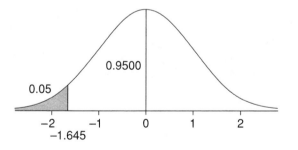

Decision rule: Reject H_0 if $z < -1.645$.

Step 5

We begin the calculations by combining the two samples, sorting them in ascending order, and then ranking these values. Here is a table showing the results.

Distance	Driver	Rank	Distance	Driver	Rank
236	Old	1	267	New	15
238	Old	2	268	Old	16.5
242	Old	3	268	New	16.5
250	Old	4	269	Old	18.5
251	Old	5.5	269	New	18.5
251	Old	5.5	270	Old	20
254	New	7.5	272	New	21
254	New	7.5	275	New	22
261	Old	9.5	276	New	23
261	New	9.5	279	New	24
262	Old	11.5	280	New	25
262	New	11.5	283	New	26
263	Old	13	285	New	27
266	New	14			

Next we total the ranks for sample 1, the old driver. This total is 110, so $R = 110$.

$$\mu_R = \frac{n_1(n_1 + n_2 + 1)}{2} \qquad \sigma_R = \sqrt{\frac{n_1 n_2 (n_1 + n_2 + 1)}{12}}$$

$$= \frac{12(12 + 15 + 1)}{2} \qquad = \sqrt{\frac{12 \cdot 15 \cdot (12 + 15 + 1)}{12}}$$

$$= \frac{12(28)}{2} \qquad = \sqrt{\frac{12 \cdot 15 \cdot 28}{12}}$$

$$= 168 \qquad = 20.49$$

$$z = \frac{R - \mu_R}{\sigma_R}$$

$$= \frac{110 - 168}{20.49}$$

$$= -2.83$$

Decision: Reject H_0.

Conclusion: There is sufficient sample evidence to support the claim that the median driving distance with his old driver is less than the median driving distance with the new driver.

p-value

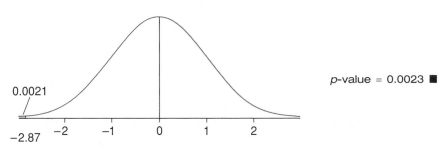

p-value $= 0.0023$ ∎

0.0021

−2.87

EXAMPLE 11.12

Here are the lengths, in minutes, of 12 randomly selected American League baseball games and 10 randomly selected National League baseball games.

American League (AL)

166	168	182	219	235	224
202	171	159	153	169	171

National League (NL)

160	177	171	198	159	160
146	188	149	151		

At the 0.05 level of significance, test the claim that the median length of American League games is greater than the median length of National League games.

Step 1

Population 1: American League games; population 2: National League games
Claim in words: The median length of American League games is greater than the median length of National League games.
Claim: $\tilde{\mu}_1 > \tilde{\mu}_2$
Complement: $\tilde{\mu}_1 \leq \tilde{\mu}_2$
H_0: $\tilde{\mu}_1 \leq \tilde{\mu}_2$
H_A: $\tilde{\mu}_1 > \tilde{\mu}_2$

Step 2

Level of significance: $\alpha = 0.05$

Step 3

Test statistic: $z = \dfrac{R - \mu_R}{\sigma_R}$, where $\mu_R = \dfrac{n_1(n_1 + n_2 + 1)}{2}$ and $\sigma_R = \sqrt{\dfrac{n_1 n_2(n_1 + n_2 + 1)}{12}}$

Step 4

Right-tailed test, $\alpha = 0.05$

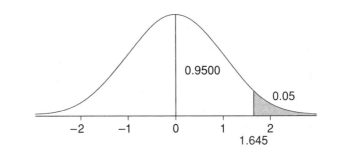

Decision rule: Reject H_0 if $z > 1.645$.

Step 5

We begin the calculations by combining the two samples, sorting them in ascending order, and then ranking these values. Here is a table showing the results.

Length	League	Rank	Length	League	Rank
146	NL	1	159	AL	5.5
149	NL	2	159	NL	5.5
151	NL	3	160	NL	7.5
153	AL	4	160	NL	7.5

(continues)

(continued)

Length	League	Rank	Length	League	Rank
166	AL	9	182	AL	16
168	AL	10	188	NL	17
169	AL	11	198	NL	18
171	AL	13	202	AL	19
171	AL	13	219	AL	20
171	NL	13	224	AL	21
177	NL	15	235	AL	22

Next we total the ranks for sample 1, the American League. This total is 163.5, so $R = 163.5$.

$$\mu_R = \frac{n_1(n_1 + n_2 + 1)}{2} \qquad \sigma_R = \sqrt{\frac{n_1 n_2(n_1 + n_2 + 1)}{12}}$$

$$= \frac{12(12 + 10 + 1)}{2} \qquad = \sqrt{\frac{12 \cdot 10 \cdot (12 + 10 + 1)}{12}}$$

$$= \frac{12(23)}{2} \qquad = \sqrt{\frac{12 \cdot 10 \cdot 23}{12}}$$

$$= 138 \qquad = 15.17$$

$$z = \frac{R - \mu_R}{\sigma_R}$$

$$= \frac{163.5 - 138}{15.17}$$

$$= 1.68$$

Decision: Reject H_0.

Conclusion: There is sufficient sample evidence to support the claim that the median length of American League games is greater than the median length of National League games.

p-value

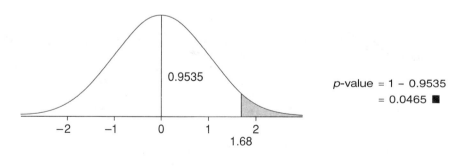

$$p\text{-value} = 1 - 0.9535$$
$$= 0.0465 \ \blacksquare$$

EXERCISES 11.3

Use the standard procedure for all hypothesis tests.

1. An instructor gives one version of a test to his 7:30 A.M. statistics class and another version to his 6 P.M. class. Here are their scores.

Morning

85	73	83	100	86	70	98	80
97	66	41	72	83	89	57	
64	76	91	77	68	81	63	

Evening

74	80	92	73	82	96	82	84
76	98	94	63	94	91	46	58
54	37	86	79	55	73	64	

At the 0.05 level of significance, test the claim that the two versions were of equal difficulty; in other words, the two exams produce the same median score.

2. A high school instructor is curious to see the effect that an open-notes policy would have on tests. He allows one of his classes to use their notes on their test, but has the other class take the test without them. Here are the scores.

With Notes

86	95	97	98	53	84	91	64
97	97	97	84	64	94	73	

Without Notes

70	92	97	50	81	97	84	61
98	98	58	23	69	84	91	78

At the 0.05 level, test the claim that the use of notes produces a higher median test score.

3. Here are the serum cholesterol levels, in mg/dL, of 12 men and 26 women ages 60–69.

Men

176	196	189	184	230	169	173
159	180	284	221	196		

Women

168	239	213	197	170	226	200
205	237	219	246	247	268	273
213	151	188	196	273	202	236
214	295	221	255	260		

At the 0.01 level of significance, test the claim that men and women ages 60–69 have the same median serum cholesterol level.

4. A female college student majoring in math has heard many times that males have higher SAT math scores than females. She believes that this is not true, but instead that the two groups have the same median SAT math scores. She randomly samples 20 females and 25 males at her school. Here are their scores.

Females

320	420	370	420	590	680	600
520	500	360	530	530	340	590
520	640	620	670	520	450	

Males

370	600	600	540	450	610	600
550	570	580	690	640	420	370
690	760	490	440	490	360	630
560	330	410	660			

Use the data to test the claim that females and males have the same median SAT math score at the 0.05 level of significance.

5. A mathematics instructor holds a supplementary review session the day before each exam. Fourteen of the 24 students in his algebra class attended his review session before the exam. Here are the scores of the students.

Attended

97	94	92	85	84	84	77
75	72	70	62	84	82	77

Did Not Attend

73	65	55	50	48
47	30	28	28	28

At the 0.01 level of significance, test the claim that students who attend review sessions have a higher median score than students who do not attend.

6. Here are the winning times of 50 one-mile races at Santa Anita Park. Twenty-one of the races were restricted to fillies and mares (female horses), while male horses ran in the other 29.

Females

97.4	100.0	99.0	99.0	99.0	95.8	97.2	95.6
98.8	97.0	97.2	98.2	96.8	96.2	96.2	96.0
99.2	100.0	100.0	99.0	95.6			

Males

96.4	97.0	95.6	97.2	96.8	98.4	96.4	94.8
98.2	97.4	96.6	96.4	97.2	97.6	97.8	96.8
96.6	95.4	98.2	95.6	98.8	97.8	97.6	97.4
95.6	97.0	100.0	95.6	99.8			

At the 0.05 level of significance, test the claim that female racehorses and male racehorses have the same median winning time in 1-mile races.

7. A mathematics instructor has changed his instruction style to include more group activities, both inside and outside of class. He wants to know whether the results of such an instruction style have any effect on students' performance on a standard department final exam. Here are the final exam scores of 30 students from the semester before the change in styles, and 32 students from the semester after the change. Scores are out of a maximum of 50 points. At the 0.05 level of significance, test the claim that the median score for both styles is the same.

Before

34	21	36	28	46	31	45	40
48	42	27	41	39	29	37	37
29	46	34	38	43	46	35	34
32	45	27	40	47	41		

After

35	39	50	25	25	36	34	31
45	35	37	32	36	31	44	38
34	23	31	36	21	30	44	45
30	39	39	42	28	47	40	27

8. The PSAT mathematics scores of 50 high school juniors and 35 high school sophomores were selected at random.

Juniors

24	36	72	48	28	50	44	46	38	72
57	69	48	64	39	48	59	51	59	48
42	37	52	72	47	55	38	58	40	60
60	59	38	42	49	40	31	61	42	47
56	37	62	59	59	44	69	41	60	33

Sophomores

29	52	42	68	38	60	58	49	45	55
30	60	52	37	42	49	43	36	38	52
58	61	31	61	46	58	49	36	59	49
48	47	37	43	31					

At the 0.01 level of significance, test the claim that the median score for juniors on this test is greater than the mean score of sophomores on this test.

9. PSAT verbal scores were obtained for a random sample of 40 high school juniors and PSAT mathematics scores were obtained for a different sample of 40 high school juniors.

Verbal

28	41	49	41	40	41	48	47	48	37
56	41	36	49	40	47	44	72	42	50
64	47	56	54	58	71	38	63	40	50
42	53	68	43	38	39	30	70	46	60

Mathematics

34	52	46	56	29	51	49	41	66	42
50	65	50	48	48	45	48	44	67	65
45	44	52	54	46	30	32	58	39	49
52	72	46	59	29	39	35	53	51	41

At the 0.01 level of significance, test the claim that the median score for the two exams is the same.

10. It is a widely held belief that males have better mathematical skills than verbal skills. Here are the SAT math scores for 36 randomly selected male students, and the SAT verbal scores for 30 randomly selected male students.

Math

370	670	460	530	690	640	740	540	610
470	610	660	410	450	720	740	430	480
460	520	580	580	560	510	320	640	650
580	540	580	440	310	260	420	320	410

Verbal

320	280	750	450	520	460	630	580	490
570	450	490	590	530	320	380	520	440
480	610	660	740	580	330	480	590	330
580	500	480						

At the 0.05 level of significance, test the claim that the median SAT math score for males is greater than the median SAT verbal score for males.

11. It is a widely held belief that females have better verbal skills than mathematical skills. Here are the SAT math scores for 42 randomly selected female students, and the SAT verbal scores for 33 randomly selected female students.

Math

320	490	470	280	390	560	520	760	480
460	470	490	460	520	530	580	370	490
550	600	310	480	460	340	400	330	700
570	420	570	750	730	470	640	530	470
310	560	570	330	570	500			

Verbal

280	250	510	680	570	360	480	540	600
350	590	390	580	540	510	550	320	510
290	550	550	350	450	530	630	630	630
490	570	690	620	500	380			

At the 0.05 level of significance, test the claim that the median SAT math score for females is less than the median SAT verbal score for females.

12. A professional bowler claims that his game is better suited to synthetic lanes than to natural lanes. Here are the scores of 42 randomly selected games on synthetic lanes, and the scores of 42 randomly selected games on natural lanes.

Synthetic

256	269	279	245	290	214	210	235	238
213	279	236	218	222	247	188	249	214
196	222	237	177	246	227	279	267	236
237	244	217	258	264	266	279	238	257
238	224	226	267	256	228			

Natural

248	221	246	255	289	244	223	222	255
192	224	203	204	179	244	234	225	218
201	219	214	233	204	212	217	207	203
244	163	215	182	192	203	198	210	225
207	236	224	264	228	164			

At the 0.05 level of significance, test the claim that the bowler's median score on synthetic lanes is higher than his median score on natural lanes.

13. A math instructor is not sure that collecting homework from his class on a daily basis is beneficial to his students. He conducts an experiment by collecting homework from one algebra class of 37 students that he is teaching, but not collecting homework from another algebra class of 32 students. Here are the test scores for both classes.

Collected Homework

53	88	68	64	70	67	57	88	83	79
78	55	84	78	78	62	82	76	54	74
85	76	74	71	84	87	75	73	63	64
73	48	82	87	78	78	75			

Did Not Collect Homework

60	79	65	74	76	62	71	72	76	77
62	73	66	80	74	74	81	68	61	68
76	76	78	70	79	88	68	82	63	68
76	87								

At the 0.05 level of significance, test the claim that collecting homework produces the same median score as not collecting homework.

SECTION 11.4
Kruskal–Wallis Test

In this section we will learn a nonparametric alternative to ANOVA, called the **Kruskal–Wallis test**. Recall from Section 9.2 that to use ANOVA the populations sampled from had to be normally distributed with equal variances. This new test drops the assumption that the populations are normally distributed. We will only require that the samples are each of size 5 or greater. The efficiency of this test is 0.95, which again means that a sample size of 100 for the Kruskal–Wallis test is equivalent to a sample size of 95 for ANOVA.

As with ANOVA, we will let k represent the number of samples or populations, and let N represent the total of all the sample sizes. As in the last section, we will combine all of the samples, and rank all of the values from smallest to largest. We then total the ranks of each sample. The total of the first sample's ranks is denoted R_1, the total of the second sample's ranks is denoted R_2, and so on. The size of the first sample is denoted n_1, the size of the second sample is denoted n_2, and so on.

Once each of the samples has had its ranks totaled, we can calculate the test statistic H using the following formula.

$$H = \frac{12}{N(N+1)}\left(\frac{R_1^2}{n_1} + \frac{R_2^2}{n_2} + \cdots + \frac{R_k^2}{n_k}\right) - 3(N+1)$$

If there is a disparity between the individual sample's rank totals, this will produce a larger value of H. Thus, larger values of H will suggest that the rank sums are not all equal and that the distributions of the populations in question are not all the same. This means that our tests about H will be right-tailed tests. The distribution of H can be approximated by a chi-square distribution with $k - 1$ degrees of freedom.

EXAMPLE 11.13

Football players who play defense can be broken into three categories: linemen, linebackers, and defensive backs. The position requirements essentially dictate the size of the players. Linemen must be large and powerful to get by the offensive line, defensive backs must be quick and nimble to cover fleet wide receivers, and linebackers must be a combination of both. Here are the weights of 28 randomly selected football players—10 linemen, 7 linebackers, and 11 defensive backs. At the 0.05 level of significance, test the claim that the median weight for all three positions is the same.

Linemen

275	310	260	295	302		
260	268	280	305	276		

Linebackers

220	234	241	223	250	229	220

Defensive Backs

200	195	215	195	197	190
200	191	185	196	168	

There are three samples here, so $k = 3$. Also, $N = 28$, which is the total of the three sample sizes. We will let linemen be population 1, linebackers be population 2, and defensive backs be population 3. Other choices for the three populations would work as well.

Step 1

Population 1: linemen; population 2: linebackers; population 3: defensive backs
Claim in words: The median weight for all three positions is the same.
Claim: $\tilde{\mu}_1 = \tilde{\mu}_2 = \tilde{\mu}_3$
Complement: At least one of the medians is different from the others.
H_0: $\tilde{\mu}_1 = \tilde{\mu}_2 = \tilde{\mu}_3$
H_A: At least one of the medians is different from the others.

Step 2

Level of significance: $\alpha = 0.05$

Step 3

Test statistic: $H = \dfrac{12}{N(N+1)} \left(\dfrac{R_1^2}{n_1} + \dfrac{R_2^2}{n_2} + \dfrac{R_3^2}{n_3} \right) - 3(N+1)$

Step 4

There are $3 - 1 = 2$ degrees of freedom, and $\alpha = 0.05$. Using the χ^2 table, we see that the critical value is 5.991.

Decision rule: Reject H_0 if $H > 5.991$.

Step 5

Here are all of the values ranked, with the position of the player listed.

Weight	Position	Rank
275	Lineman	22
310	Lineman	28
260	Lineman	19.5
295	Lineman	25
302	Lineman	26
260	Lineman	19.5
268	Lineman	21
280	Lineman	24
305	Lineman	27
276	Lineman	23
220	Linebacker	12.5
234	Linebacker	16
241	Linebacker	17
223	Linebacker	14
250	Linebacker	18
229	Linebacker	15
220	Linebacker	12.5
200	Defensive back	9.5
195	Defensive back	5.5
215	Defensive back	11
195	Defensive back	5.5
197	Defensive back	8
190	Defensive back	3
200	Defensive back	9.5
191	Defensive back	4
185	Defensive back	2
196	Defensive back	7
168	Defensive back	1

The total of the ranks for sample 1 (linemen) is 235, the total of the ranks for sample 2 (linebackers) is 105, and the total of the ranks for sample 3 (defensive backs) is 66.

Now we can calculate the test statistic.

$$H = \frac{12}{N(N+1)} \left(\frac{R_1^2}{n_1} + \frac{R_2^2}{n_2} + \frac{R_3^2}{n_3} \right) - 3(N+1)$$

$$= \frac{12}{28(29)} \left(\frac{235^2}{10} + \frac{105^2}{7} + \frac{66^2}{11} \right) - 3(29)$$

$$= \frac{12}{812} \left(\frac{55{,}225}{10} + \frac{11{,}025}{7} + \frac{4356}{11} \right) - 87$$

$$= 23.741$$

Decision: Reject H_0.

Conclusion: There is sufficient sample evidence to reject the claim that the median weight for all three positions is the same.

p-value

From our chi-square table, we know that the p-value is less than 0.01. A calculator or a computer tells us that the p-value is approximately 0.000007, which is a very small p-value. ■

Look at the rankings for the previous example, and note that the 11 lowest ranks were all defensive backs. The next 7 lowest ranks all belonged to linebackers, and the highest 10 ranks all belonged to linemen. We cannot expect to see such a division often, and it should be no surprise that we were able to reject the null hypothesis in this example.

 EXAMPLE 11.14

Here are the waiting times, in minutes, at the airport check-in counters for four different airlines for randomly selected fliers.

Airline A	Airline B	Airline C	Airline D
3	19	8	7
11	13	11	16
19	11	17	15
11	13	7	9
7	21	12	20
10			

At the 0.05 level of significance, test the claim that the median waiting times at the checkout counters of the four airlines are equal.

There are four samples here, so $k = 4$. Also, $N = 21$, which is the total of the four sample sizes. We will let airline A be population 1, airline B be population 2, airline C be population 3, and airline D be population 4.

Step 1

Population 1: airline A; population 2: airline B; population 3: airline C; population 4: airline D

Claim in words: The median waiting times at the checkout counters of the four airlines are equal.

Claim: $\tilde{\mu}_1 = \tilde{\mu}_2 = \tilde{\mu}_3 = \tilde{\mu}_4$

Complement: At least one of the medians is different from the others.

H_0: $\tilde{\mu}_1 = \tilde{\mu}_2 = \tilde{\mu}_3 = \tilde{\mu}_4$

H_A: At least one of the medians is different from the others.

Step 2

Level of significance: $\alpha = 0.05$

Step 3

Test statistic: $H = \dfrac{12}{N(N+1)}\left(\dfrac{R_1^2}{n_1} + \dfrac{R_2^2}{n_2} + \dfrac{R_3^2}{n_3} + \dfrac{R_4^2}{n_4}\right) - 3(N+1)$

Step 4

There are $4 - 1 = 3$ degrees of freedom, and $\alpha = 0.05$. Using the χ^2 table, we see that the critical value is 7.815.

Decision rule: Reject H_0 if $H > 7.815$.

Step 5

Here are all of the waiting times ranked, with the airline listed.

Waiting Time	Airline	Rank
3	A	1
11	A	9.5
19	A	18.5
11	A	9.5
7	A	3
10	A	7
19	B	18.5
13	B	13.5
21	B	9.5
11	B	13.5
13	B	21
8	C	5
11	C	9.5
17	C	17
7	C	3
12	C	12
7	D	3
16	D	16
15	D	15
9	D	6
20	D	20

Here are the totals of the ranks for each sample.

Airline	A	B	C	D
Total of Ranks	48.5	76	46.5	60

Now we can calculate the test statistic.

$$H = \frac{12}{N(N+1)}\left(\frac{R_1^2}{n_1} + \frac{R_2^2}{n_2} + \frac{R_3^2}{n_3} + \frac{R_4^2}{n_4}\right) - 3(N+1)$$

$$= \frac{12}{21(22)}\left(\frac{48.5^2}{6} + \frac{76^2}{5} + \frac{46.5^2}{5} + \frac{60^2}{5}\right) - 3(22)$$

$$= \frac{12}{462}\left(\frac{2352.5}{6} + \frac{5776}{5} + \frac{2162.5}{5} + \frac{3600}{5}\right) - 66$$

$$= 4.122$$

Decision: Fail to reject H_0.

Conclusion: There is not sufficient sample evidence to reject the claim that the median waiting times at the checkout counters of the four airlines are equal.

p-value

From our chi-square table, we know that the *p*-value is greater than 0.10. Technology tells us that the *p*-value is approximately 0.2486. ∎

We complete the section by looking at one more example.

A random sample of 30 major league baseball players was divided into three groups: pitchers, infielders/catchers, and outfielders. Here are the salaries of those 30 players, broken down by position.

Pitchers	Infield/Catcher	Outfield
567,666	242,500	6,000,000
205,000	6,000,000	2,500,000
1,750,000	725,000	175,000
197,500	200,000	2,200,000
1,000,000	4,000,000	318,750
775,000	2,550,000	2,000,000
180,000	4,000,000	4,200,000
275,000	175,000	5,275,000
1,233,334	2,150,000	
	240,000	
	500,000	
	3,000,000	
	300,000	

At the 0.05 level of significance, test the claim that the median salary is the same for all three positions.

There are three samples here, so $k = 3$. Also, $N = 30$, which is the total of the three sample sizes. We will let pitchers be population 1, infielders/catchers be population 2, and outfielders be population 3.

Step 1

Population 1: pitchers; population 2: infielders/catchers; population 3: outfielders
Claim in words: The median salary for all three positions is the same.
Claim: $\tilde{\mu}_1 = \tilde{\mu}_2 = \tilde{\mu}_3$
Complement: At least one of the medians is different from the others.
H_0: $\tilde{\mu}_1 = \tilde{\mu}_2 = \tilde{\mu}_3$
H_A: At least one of the medians is different from the others.

Step 2

Level of significance: $\alpha = 0.05$

Step 3

Test statistic: $H = \dfrac{12}{N(N+1)} \left(\dfrac{R_1^2}{n_1} + \dfrac{R_2^2}{n_2} + \dfrac{R_3^2}{n_3} \right) - 3(N+1)$

Step 4

There are $3 - 1 = 2$ degrees of freedom, and $\alpha = 0.05$. Using the χ^2 table, we see that the critical value is 5.991.

Decision rule: Reject H_0 if $H > 5.991$.

Step 5

Here are all of the values ranked, with the position of the player listed.

Salary	Position	Rank
567,666	Pitcher	13
205,000	Pitcher	6
1,750,000	Pitcher	18
197,500	Pitcher	4
1,000,000	Pitcher	16
775,000	Pitcher	15
180,000	Pitcher	3
275,000	Pitcher	9
1,233,334	Pitcher	17
242,500	Infielder/catcher	8
6,000,000	Infielder/catcher	29.5
725,000	Infielder/catcher	14
200,000	Infielder/catcher	5
4,000,000	Infielder/catcher	25.5
2,550,000	Infielder/catcher	23
4,000,000	Infielder/catcher	25.5
175,000	Infielder/catcher	1.5
2,150,000	Infielder/catcher	20
240,000	Infielder/catcher	7
500,000	Infielder/catcher	12
3,000,000	Infielder/catcher	24
300,000	Infielder/catcher	10
6,000,000	Outfielder	29.5
2,500,000	Outfielder	22
175,000	Outfielder	1.5
2,200,000	Outfielder	21
318,750	Outfielder	11
2,000,000	Outfielder	19
4,200,000	Outfielder	27
5,275,000	Outfielder	28

Here are the totals of the ranks for each sample.

Position	Pitcher	Infielder/Catcher	Outfielder
Total of Ranks	101	205	159

Now we can calculate the test statistic.

$$H = \frac{12}{N(N + 1)}\left(\frac{R_1^2}{n_1} + \frac{R_2^2}{n_2} + \frac{R_3^2}{n_3}\right) - 3(N + 1)$$

$$= \frac{12}{30(31)}\left(\frac{101^2}{9} + \frac{205^2}{13} + \frac{159^2}{8}\right) - 3(31)$$

$$= \frac{12}{930}\left(\frac{10,201}{9} + \frac{42,025}{13} + \frac{25,281}{8}\right) - 93$$

$$= 4.113$$

Decision: Fail to reject H_0.

Conclusion: There is not sufficient sample evidence to reject the claim that the median salary for all three positions is the same.

p-value

From our chi-square table, we know that the p-value is greater than 0.10. A calculator or computer tells us that the p-value is approximately 0.1279. ∎

Note that we would also fail to reject the null hypothesis in the previous example using ANOVA.

MICROSOFT EXCEL   Kruskal–Wallis p-Values

Microsoft Excel can help us to calculate the p-value for the Kruskal–Wallis test. We will use Excel's built-in function CHIDIST, which requires two arguments in parentheses: the value of the test statistic and the number of degrees of freedom. In the last example, there were 2 degrees of freedom and the test statistic calculated to be 4.113. To find the p-value for this test, type the following in any empty cell.

    =CHIDIST(4.113,2)

Excel should return the value 0.1279.

TI-83   Kruskal–Wallis p-Values

The TI-83 calculator can be used to calculate the test statistic as well. We will be using the χ^2cdf function. This function can be used to give us the area to the left of our calculated test statistic, and then we can subtract that area from 1 to find the area of the right tail. In the last example, there were 2 degrees of freedom and the test statistic calculated to be 4.113. To find the p-value for this test, press (2nd)(VARS) to access the DISTR menu. Select option 7: χ^2cdf and press (ENTER). When you are taken to the main screen, enter 0 (,) 4.113 (,) 2 ( )). This will give us the area under the chi-square curve between 0 and 4.113 if there are 2 degrees of freedom. The result should be 0.8721. We then take the complement of this to find the p-value, which is 0.1279.

EXERCISES 11.4

Use the standard procedure for all hypothesis tests.

1. A random sample of 25 beer drinkers was asked how many servings of beer they had in the past week. Here are the responses, broken down by age group.

| | Age Groups | |
21–29	30–39	40 and above
5	4	2
7	5	4
5	5	4
6	5	3
6	5	4
6	5	3
5	6	
6	8	
6	5	
	5	

At the 0.05 level of significance, test the claim that the median number of servings is the same for the three age groups (21–29, 30–39, 40 and above).

2. A random sample of 20 adult readers was asked to report how many minutes they read per day. Here are their responses, broken down into four age groups: 18–25, 26–45, 46–65, and over 65.

	Age Groups		
18–25	26–45	46–65	Over 65
10	15	30	65
20	25	60	75
20	40	40	80
15	20	45	75
25	30	40	80

At the 0.01 level of significance, test the claim that the median reading times for the four age groups are equal.

3. Here are the GPAs of five sorority members, six fraternity members, and eight students that do not belong to either.

Sorority	Fraternity	Neither
2.85	2.92	3.55
2.93	3.31	3.22
3.06	3.05	3.15
3.04	3.17	3.32
2.89	3.25	3.06
	3.18	3.01
		3.03
		3.30

At the 0.05 level of significance, test the claim that the median GPAs of the three populations are equal.

4. Hole selections for professional golf courses are changed each day of a PGA tournament. Are any of the days set up to be more difficult or easier? Here are the scores of 10 golfers on the four days of a tournament.

	Round					Round			
Golfer	1	2	3	4	Golfer	1	2	3	4
1	63	65	68	65	6	71	68	70	71
2	70	69	67	69	7	66	67	73	75
3	67	71	69	70	8	71	68	71	73
4	67	68	72	72	9	72	68	74	71
5	72	66	71	71	10	68	72	71	81

At the 0.05 level of significance, test the claim that the median scores produced by the four different rounds are equal.

5. Hole selections for professional golf courses are changed each day of an LPGA tournament. Are any of the days set up to be more difficult or easier? Here are the scores of 12 golfers on the four days of a tournament.

	Round					Round			
Golfer	1	2	3	4	Golfer	1	2	3	4
1	65	75	70	71	7	73	81	76	72
2	74	72	76	71	8	71	78	78	77
3	70	78	79	69	9	73	82	76	75
4	70	81	76	70	10	79	87	75	71
5	71	80	76	72	11	81	84	79	71
6	73	80	75	72	12	73	88	83	78

At the 0.01 level of significance, test the claim that the median scores produced by the four different rounds are equal.

6. Here are the prices of a dose of eight medications in the United States, Canada, Great Britain, Australia, and Mexico, all in U.S. dollars.

Drug	United States	Canada	Great Britain	Australia	Mexico
Prilosec	3.31	1.47	1.67	1.29	0.99
Prozac	2.27	1.07	1.08	0.82	0.79
Lipitor	2.54	1.34	1.67	1.32	3.60
Prevacid	3.13	1.34	0.82	0.83	1.18
Zocor	3.16	1.47	1.73	1.75	3.66
Zoloft	1.98	1.07	0.95	0.84	1.96
Claritin	1.96	1.11	0.41	0.48	0.92
Paxil	2.22	1.13	1.70	0.82	1.83

(Source: USA Today.)

At the 0.01 level of significance, test the claim that the median price of medications in these five countries is equal.

7. An onion farmer has three different locations, each planted by a different method. The farmer randomly selects eight 3-foot beds, and counts the number of plants in each bed. Here are the totals.

Farm A	Farm B	Farm C
57	55	52
66	59	49
62	55	55
57	65	54
68	66	70
52	52	59
61	52	67
54	71	64

At the 0.05 level of significance, test the claim that the three farms have the same median number of plants per 3-foot bed.

8. Here are the number of contacts that 30 people had with a physician during the past 12 months, summarized by family income.

	Family Income		
Under $10,000	$10,000–$19,999	$20,000–$34,999	$35,000 or more
5	4	3	7
9	8	7	7
10	8	6	5
9	8	6	4
9	7	4	4
8	7	3	6
	8	4	5
		6	
		5	
		4	

At the 0.01 level of significance, test the claim that the median number of physician contacts are the same for all four income levels.

9. A cotton farmer is investigating three fertilizers, trying to find the one that will produce the greatest yield of cotton. She randomly assigns a fertilizer to each 1-acre parcel of her 21-acre field in such a way that each fertilizer is used on 7 acres. Here are the yields for each field, in pounds.

Fertilizer A	Fertilizer B	Fertilizer C
888	705	788
686	607	820
1000	837	920
1213	1240	1003
1197	861	938
1307	1300	840
500	1105	1165

At the 0.05 level of significance, test the claim that the three fertilizers produce equal cotton yields.

10. At a community college, students who place into a prealgebra course take a math competency exam at the end of the course. (The passing score is 39 out of 55.) There are three types of courses that a student could enroll in. One course meets 4 days per week for 17 weeks, the second course meets 5 days per week for 17 weeks, and the third is an intensive short-term class (3 hours/day, 5 days/week, 4 weeks). Here are the competency exam results for 14 students who met 4 days per week, 9 students who met 5 days per week, and 11 students who took the short-term class.

4 Days/Week	5 Days/Week	Short-Term
26	22	29
42	44	40
37	41	41
45	38	41
42	44	40
41	39	47
39	37	45
47	42	47
46	35	49
44		43
36		39
46		
44		
41		

At the 0.01 level of significance, test the claim that the three different courses produce the same median scores on the math competency exam.

SECTION 11.5
Rank Correlation

In this section we will learn about a nonparametric alternative to the correlation coefficient that we examined in Section 10.1, called **Spearman's rank correlation coefficient.** We will also learn how to use this coefficient to test whether there is a relation between two variables.

Recall that when we made inferences about the population correlation, for any x-value the y-values had to be normally distributed, and vice versa. Such an assumption is not necessary here. Another advantage to rank correlation is that it can detect relations that are not linear.

To calculate Spearman's rank correlation coefficient, r_s, we begin by ranking the x-values from lowest to highest, and then we do the same for the y-values. Next we subtract the rank of each y-value from its corresponding x-value. These differences are denoted by d. Spearman's rank correlation coefficient can then be found using the following formula.

$$r_s = 1 - \frac{6\Sigma d^2}{n(n^2 - 1)}$$

The value of r_s will always be between -1 and 1, and can be interpreted in the same fashion as the correlation coefficient r. As a matter of fact, if there are no ties in the rankings, r_s and r will be equal, and they will differ slightly if there are ties. If we lack availability to technology, using rank correlation is a faster and easier way to examine the relationship between two variables.

We begin with an example.

 EXAMPLE 11.16   Here are the number of hours that 10 students spent studying for a final exam, and their scores on that exam. Calculate Spearman's rank correlation coefficient for these two variables.

Hours	7	8	4	9	13	5	9	6	16	3
Score	70	76	57	77	91	66	82	64	96	50

We begin by ranking the *x*-values (hours studied) and also ranking the *y*-values (score on exam).

Hours	Rank	Score	Rank
7	5	70	5
8	6	76	6
4	2	57	2
9	7.5	77	7
13	9	91	9
5	3	66	4
9	7.5	82	8
6	4	64	3
16	10	96	10
3	1	50	1

Next, we calculate the differences between the ranks, and square the differences.

Rank (hours)	Rank (score)	d (hours – score)	d^2
5	5	0	0
6	6	0	0
2	2	0	0
7.5	7	0.5	0.25
9	9	0	0
3	4	−1	1
7.5	8	−0.5	0.25
4	3	1	1
10	10	0	0
1	1	0	0

$$\Sigma d^2 = 2.5$$

We can now calculate r_s.

$$r_s = 1 - \frac{6\Sigma d^2}{n(n^2 - 1)}$$

$$= 1 - \frac{6(2.5)}{10(10^2 - 1)}$$

$$= 1 - \frac{15}{10(99)}$$

$$= 1 - \frac{15}{990}$$

$$= 0.9848$$

Our result, $r_s = 0.9848$, indicates that there is strong correlation between the two variables. ■

Before proceeding to inferences about correlation, we will work through one more example of calculating Spearman's rank correlation coefficient.

EXAMPLE 11.17

A tutorial lab on campus offers free tutoring for any student on campus. Here are the GPAs for 12 randomly selected students, and the number of tutoring appointments that those students missed. Calculate r_s, Spearman's rank correlation coefficient.

GPA	2.66	2.05	2.07	2.62	1.30	3.00
Missed Appointments	3	1	2	0	7	0

GPA	3.25	2.58	2.36	2.81	3.11	2.56
Missed Appointments	2	0	3	1	1	2

We begin by ranking the *x*-values (GPA) and also ranking the *y*-values (number of missed appointments).

GPA	Rank	Missed Appointments	Rank
2.66	8	3	10.5
2.05	2	1	5
2.07	3	2	8
2.62	7	0	2
1.30	1	7	12
3.00	10	0	2
3.25	12	2	8
2.58	6	0	2
2.36	4	3	10.5
2.81	9	1	5
3.11	11	1	5
2.56	5	2	8

Next, we calculate the differences between the ranks, and square the differences.

Rank (GPA)	Rank (missed appointments)	d (GPA—missed appointments)	d²
8	10.5	−2.5	6.25
2	5	−3	9
3	8	−5	25
7	2	5	25
1	12	−11	121
10	2	8	64
12	8	4	16
6	2	4	16
4	10.5	−6.5	42.25
9	5	4	16
11	5	6	36
5	8	−3	9

$$\Sigma d^2 = 385.5$$

We can now calculate r_s.

$$r_s = 1 - \frac{6\Sigma d^2}{n(n^2 - 1)}$$

$$= 1 - \frac{6(385.5)}{12(12^2 - 1)}$$

$$= 1 - \frac{2313}{12(143)}$$

$$= 1 - \frac{2313}{1716}$$

$$= -0.3479$$

Our result, $r_s = -0.3479$, indicates that there is weak negative correlation between the two variables. ∎

Spearman's rank correlation coefficient is the test statistic that we use when we try to test for significant correlation between two variables. The efficiency of this test is 0.91. We will use Table H in the back of the text to find the critical value for the test. In the table, look down the left column until you find the correct value for *n*. Then scan across the row until you reach the correct level of significance. We read the critical values in the same fashion that we read critical values for Student's *t*-distribution. Let's look at an example.

EXAMPLE
11.18

A tutorial lab on campus offers free tutoring for any student on campus. Here are the GPAs for 12 randomly selected students, and the number of tutoring appointments that those students missed. Spearman's rank correlation coefficient, r_s, was calculated to be −0.3479 in the previous example. At the 0.05 level of significance, test the claim that there is significant correlation between the two variables.

GPA	2.66	2.05	2.07	2.62	1.30	3.00
Missed Appointments	3	1	2	0	7	0
GPA	3.25	2.58	2.36	2.81	3.11	2.56
Missed Appointments	2	0	3	1	1	2

Step 1

Claim in words: There is a significant correlation between the GPA of a student and the number of tutoring appointments missed. (In other words, the rank correlation coefficient for the population ρ_s is not equal to 0.)

Claim: $\rho_s \neq 0$

Complement: $\rho_s = 0$

H_0: $\rho_s = 0$

H_A: $\rho_s \neq 0$

Step 2

Level of significance: $\alpha = 0.05$

Step 3

Test statistic: r_s

Step 4

Two-tailed, $\alpha = 0.05$, $n = 12$

Decision rule: Reject H_0 if $r_s < -0.591$ or $r_s > 0.591$.

Step 5

$r_s = -0.3479$

Decision: Fail to reject H_0.

Conclusion: There is not sufficient sample evidence to support the claim that there is a significant correlation between the GPA of a student and the number of tutoring appointments missed.

p-value

The best we can say is that the p-value is greater than 0.10, since the value of the test statistic is less than the critical value associated with a 0.10 level of significance. ∎

Here are the number of hits and runs during 11 randomly selected National League baseball games. At the 0.05 level of significance, use the rank correlation coefficient to test the claim that there is a significant correlation between the number of hits in a game and the number of runs scored in a game.

Hits	19	20	14	12	25	27	16	17	23	19	20
Runs	11	11	6	1	15	19	9	11	12	13	6

Here are the rankings for hits and runs.

Hits	*Rank*	*Runs*	*Rank*
19	5.5	11	6
20	7.5	11	6
14	2	6	2.5
12	1	1	1
25	10	15	10
27	11	19	11
16	3	9	4

(continues)

(continued)

Hits	Rank	Runs	Rank
17	4	11	6
23	9	12	8
19	5.5	13	9
20	7.5	6	2.5

Next, we calculate the differences between the ranks, and square the differences.

Rank (hits)	Rank (runs)	d (hits – runs)	d^2
5.5	6	-0.5	0.25
7.5	6	1.5	2.25
2	2.5	-0.5	0.25
1	1	0	0
10	10	0	0
11	11	0	0
3	4	1	1
4	6	2	4
9	8	1	1
5.5	9	3.5	12.25
7.5	2.5	5	25

$$\Sigma d^2 = 46$$

We can now calculate r_s.

$$r_s = 1 - \frac{6\Sigma d^2}{n(n^2 - 1)}$$

$$= 1 - \frac{6(46)}{11(11^2 - 1)}$$

$$= 1 - \frac{276}{11(120)}$$

$$= 1 - \frac{276}{1320}$$

$$= 0.7909$$

Step 1

Claim in words: There is a significant correlation between the number of hits in a National League baseball game and the number of runs in a game. (In other words, the rank correlation coefficient for the population ρ_s is not equal to 0.)

Claim: $\rho_s \neq 0$

Complement: $\rho_s = 0$

H_0: $\rho_s = 0$

H_A: $\rho_s \neq 0$

Step 2

Level of significance: $\alpha = 0.05$

Step 3

Test statistic: r_s

Step 4

Two-tailed, $\alpha = 0.05$, $n = 11$

Decision rule: Reject H_0 if $r_s < -0.618$ or $r_s > 0.618$.

Step 5

$r_s = 0.7909$

Decision: Reject H_0.

Conclusion: There is sufficient sample evidence to support the claim that there is a significant correlation between the number of hits in a National League baseball game and the number of runs in a game.

p-value

The best we can say is that the p-value is between 0.01 and 0.02, since those are the critical values that it fits between. ■

EXERCISES 11.5

1. Here are the shoe sizes of 10 randomly selected men, along with their heights in inches. Calculate Spearman's rank correlation coefficient r_s.

Shoe Size	4	7	7.5	9	9.5	10.5	9	10	11	11.5
Height (in.)	62	64	66	68	68	69	69	70	71	72

2. Here are the waist and hip measurements, in inches, of twelve 4-year-old dance students. Calculate Spearman's rank correlation coefficient r_s.

Waist	24	26	22	23	23	25	24	27	23	23	25	31
Hips	29	31	28	28	28	31	30	34	28	28	28	37

(Source: Dancer's Edge dance studio.)

3. Here are the midterm exam scores of 20 algebra students, along with their scores on the final exam. Calculate Spearman's rank correlation coefficient r_s.

Student	Midterm	Final	Student	Midterm	Final
1	65	85	11	38	56
2	50	77	12	92	95
3	75	90	13	59	60
4	70	84	14	66	87
5	68	61	15	61	83
6	58	70	16	71	77
7	49	76	17	68	79
8	92	78	18	82	74
9	68	80	19	85	94
10	93	92	20	67	80

4. Here are the scores of 16 randomly selected statistics students on a test on Unit 3 and on the final exam. Calculate Spearman's rank correlation coefficient r_s.

Student	Unit 3	Final	Student	Unit 3	Final
1	68	59	3	90	75
2	93	91	4	97	100

(continues)

(continued)

Student	Unit 3	Final	Student	Unit 3	Final
5	97	98	11	90	84
6	48	59	12	54	78
7	89	89	13	78	95
8	96	97	14	86	94
9	84	81	15	56	86
10	74	85	16	79	57

5. Here are the gross values, in millions of dollars, of milk produced and cattle raised in Tulare County, California for the ten years from 1989 through 1998. Calculate Spearman's rank correlation coefficient r_s.

Year	Milk ($ millions)	Cattle ($ millions)
1989	287	177
1990	363	214
1991	413	212
1992	411	237
1993	455	238
1994	477	223
1995	547	223
1996	569	229
1997	712	252
1998	718	271

(Source: Visalia Times Delta.*)*

6. Wilt Chamberlain played 14 NBA seasons in Philadelphia, San Francisco, and Los Angeles. Here are his point totals for those 14 seasons.

Season	Team	Points	Rebounds
1959–60	Philadelphia	2707	1941
1960–61	Philadelphia	3033	2149
1961–62	Philadelphia	4029	2052
1962–63	San Francisco	3586	1946
1963–64	San Francisco	2948	1787
1964–65	San Francisco	2534	1673
1965–66	Philadelphia	2649	1943
1966–67	Philadelphia	1956	1957
1967–68	Philadelphia	1992	1952
1968–69	Los Angeles	1664	1712
1969–70	Los Angeles	328	221
1970–71	Los Angeles	1696	1493
1971–72	Los Angeles	1213	1572
1972–73	Los Angeles	1084	1526

(a) Calculate Spearman's rank correlation coefficient ρ_s for this set of data. (Since the data represent a population, this is ρ_s.)

(b) Note that Wilt's totals for the 1969–70 season are significantly different from those for the rest of his career. (He missed most of the season with an injury.) Calculate Spearman's rank correlation coefficient without including data from the 1969–70 season. Does this radically change the coefficient that was calculated in part (a)?

7. It seems that there should be a strong relation between one season's NBA ticket prices and the previous season's prices. Here are the average ticket prices for NBA arenas for the 1998–1999 and 1999–2000 seasons. Calculate Spearman's rank correlation coefficient ρ_s. (Since the data represent a population, this is ρ_s, not r_s.)

Team	1998–1999	1999–2000
New York	79.34	86.82
L.A. Lakers	51.11	81.89
Seattle	63.47	64.60
Houston	58.18	62.63
Washington	61.40	59.65
New Jersey	49.24	59.22
Utah	43.47	54.60
Chicago	53.17	52.84
Portland	52.28	52.28
Boston	49.79	49.79
Indiana	43.36	48.39
Golden State	36.79	48.10
Miami	36.55	46.57
Atlanta	36.79	45.75
Phoenix	48.84	45.39
Sacramento	34.11	44.68
Philadelphia	41.96	44.26
Orlando	44.46	44.18
L.A. Clippers	31.75	43.89
Toronto	26.17	42.76
Dallas	34.84	40.76
Detroit	33.32	40.04
Cleveland	39.75	39.75
Minnesota	38.61	39.08
San Antonio	38.01	38.92
Denver	30.53	38.34
Vancouver	31.90	34.71
Charlotte	28.12	32.04
Milwaukee	29.06	30.83

(Source: Team Marketing Report, USA Today.)

8. Barry Sanders was an NFL running back with the Detroit Lions for ten NFL seasons. Here are his yearly statistics for the 1989 through 1998 seasons. Included are the number of rushing attempts, number of rushing touchdowns, and the number of rushing yards and receiving yards. Calculate Spearman's rank correlation coefficient ρ_s between the number of rushing attempts and the number of rushing yards.

Year	Attempts	Rush TD	Rushing Yards	Receiving Yards
1989	280	14	1470	282
1990	255	13	1304	480
1991	342	16	1548	307
1992	312	9	1352	225
1993	243	3	1115	205
1994	331	7	1883	283
1995	314	11	1500	398
1996	307	11	1553	147
1997	335	11	2053	305
1998	343	4	1491	289

9. Jerry Rice began his NFL career as a wide receiver with the San Francisco Forty-Niners in 1985. Here are the number of receptions, receiving yards, and touchdown receptions for the 1985–1996 seasons. Calculate Spearman's rank correlation coefficient ρ_s between the number of receptions and receiving yards.

Year	Receptions	Yards	Touchdowns
1985	49	927	3
1986	86	1570	15
1987	65	1078	22
1988	64	1306	9
1989	82	1483	17
1990	100	1502	13
1991	80	1206	14
1992	84	1201	10
1993	98	1503	15
1994	112	1499	13
1995	122	1848	15
1996	108	1254	8

10. Here are the median family incomes for 16 California cities, and the median sales price for homes in those cities. Calculate Spearman's rank correlation coefficient r_s. Then, at the 0.05 level of significance, test the claim that there is a significant correlation between these two variables.

City	Median Income ($ thousands)	Median Sales Price ($ thousands)
Bakersfield	38.7	94
Riverside	47.2	133
Modesto	43.1	128
Visalia	34.3	98
Sacramento	51.9	158
Redding	37.5	113
Fresno	37.2	109
Merced	36.9	122
Stockton	44.3	154
Ventura	65.3	235
Los Angeles	51.3	189
Santa Barbara	52.1	210
San Diego	52.5	208
San Luis Obispo	48.0	192
San Jose	82.6	355
San Francisco	72.4	407

(Source: National Association of Home Builders, Fresno Bee.)

11. For a random sample of 20 dairy cows, here are the number of pounds of milk produced during their first lactation (after their first calf) and their second lactation (after their second calf). Is there a relation between these two variables? Calculate Spearman's rank correlation coefficient r_s for the data. Then, at the 0.05 level of significance, test the claim that there is a significant correlation between the amount of milk produced during the first lactation and the second lactation.

Cow	First Lactation	Second Lactation	Cow	First Lactation	Second Lactation
1	22,792	29,655	8	39,242	23,648
2	31,693	23,817	9	21,574	25,303
3	18,367	35,360	10	34,542	34,812
4	17,440	21,848	11	15,058	8,013
5	29,798	29,828	12	16,681	23,301
6	28,540	33,898	13	21,002	32,529
7	23,661	22,444	14	17,987	23,620

(continues)

(continued)

Cow	First Lactation	Second Lactation	Cow	First Lactation	Second Lactation
15	15,968	26,437	18	24,030	27,368
16	23,580	26,227	19	20,392	29,527
17	14,339	25,529	20	24,347	25,592

12. If a company has a quick Web server, does that mean that it is reliable as well? Here is a listing of 10 major electronic commerce Web sites, the average length of time for the site to come up on a user's computer, and the percentage of time the site is available. Calculate Spearman's rank correlation coefficient r_s. Then, at the 0.05 level of significance, test the claim that there is a significant correlation between these two variables.

Site	Seconds	Availability (%)
Amazon.com	17.66	90.1
Barnesandnoble.com	15.85	94.2
CDnow	15.52	89.2
eBay	17.25	78.4
eToys.com	20.11	84.0
Gateway	19.01	86.1
Landsend.com	13.49	94.2
Macys.com	31.42	93.0
Wal Mart Online	22.24	92.0
Wine.com	20.58	93.6

(Source: Keynote Systems.)

13. Many politicians and citizen groups complain about the cost of prescription medications in the United States. Here are the prices of a dose of 10 medications in Canada and the United States, all in U.S. dollars.

Drug	Canada	United States
Prilosec	1.47	3.31
Prozac	1.07	2.27
Lipitor	1.34	2.54
Prevacid	1.34	3.13
Epogen	21.44	23.40
Zocor	1.47	3.16
Zoloft	1.07	1.98
Zyprexa	3.39	5.27
Claritin	1.11	1.96
Paxil	1.13	2.22

(Source: USA Today.)

(a) Calculate Spearman's rank correlation coefficient r_s. Then, at the 0.01 level of significance, test the claim that there is a significant correlation between these two variables.

(b) Note that the costs associated with Epogen are far removed from the rest of the costs. Calculate Spearman's rank correlation coefficient r_s without including the data for Epogen. Does this radically change the coefficient that was calculated in part (a)?

14. Here are the number of hours that 10 students spent working the week before a final exam, and their scores on that exam. Calculate Spearman's rank correlation coefficient r_s.

Hours	20	15	0	30	20	15	0	30	0	0
Score	87	71	89	60	86	89	91	65	94	92

15. Here are the heights (in inches) and weights (in pounds) of 9 randomly selected major league baseball players. Calculate Spearman's rank correlation coefficient r_s.

Height	73	69	72	70	72	66	72	72	74
Weight	201	170	180	200	190	175	205	185	186

16. Here are the heights and circumferences, both in feet, of 12 giant sequoia trees. At the 0.05 level of significance, test the claim that there is a significant correlation between the height and circumference of giant sequoia trees.

Tree	Height (ft.)	Circumference (ft.)
1	274.9	102.6
2	246.1	101.1
3	267.4	107.6
4	240.9	93.0
5	255.8	98.3
6	243.0	109.0
7	257.5	85.3
8	268.8	113.0
9	223.8	94.8
10	270.3	104.2
11	247.8	91.3
12	254.7	88.3

Overview

The hypothesis tests introduced in this chapter are used when the assumptions of a previously introduced hypothesis test are not met. The sign test is used as an alternative to the one-mean hypothesis test that uses the t-distribution when the sample appears to have been taken from a population that is not normally distributed. The sign test can also be used as an alternative to the paired difference test, but the Wilcoxon signed-rank test is a preferable alternative. The assumptions for a paired difference test require that both of the dependent samples are drawn from populations that are normally distributed.

An alternative to the two-mean hypothesis test that uses the t-distribution is the Wilcoxon rank-sum test. To use the t-distribution for a two-mean test requires that both independent samples are drawn from populations that are normally distributed. A similar requirement is needed for ANOVA, with the added requirement that the variances of the k populations are all equal. When these conditions are not met, the nonparametric alternative is the Kruskal–Wallis test.

Finally, rank correlation was introduced as an alternative to linear correlation that requires no assumption of normality.

Review Exercises

1. Here are the 1999 salaries of 40 randomly selected CEOs, in millions of dollars. (These salaries do not include changes in paper value of stocks owned prior to 1999.) At the 0.05 level of significance, test the claim that the median is $2.0 million.

1.2	2.8	2.6	3.2	2.1	6.7	4.5	1.7	8.7	9.0
14.5	1.5	2.9	2.7	4.2	6.0	4.2	2.5	7.9	1.2
7.7	14.8	1.8	2.2	2.3	2.7	2.1	8.3	0.6	15.5
1.8	4.9	1.0	3.8	4.3	3.7	4.2	3.8	1.4	0.8

   (Source: USA Today.)

2. While on the campaign trail, a politician arguing for improved medical benefits states that senior citizens have more than 10 "physician contacts" each year. Here are the number of physician contacts last year for 18 randomly selected senior citizens.

   | 2 | 13 | 16 | 14 | 12 | 19 | 13 | 17 | 17 |
   | 13 | 8 | 18 | 17 | 14 | 12 | 12 | 18 | 8 |

   At the 0.05 level of significance, test the claim that the median number of physician contacts per year is above 10 for senior citizens.

3. Here are the reading proficiency scores of 12 randomly selected 9-year-olds both before and after the implementation of a new reading program. Use the sign test to test the claim that the reading program is effective at the 0.05 level of significance.

Before	After
177	173
200	198
264	270
170	176
186	197
169	183
276	286

   (continues)

(continued)

Before	After
207	219
121	126
175	191
185	189
212	217

4. At PGA golf tournaments two drives are measured for each player for statistical purposes. When selecting which holes to measure, the PGA selects one hole that is "with" the wind and a second that is "against" the wind. Here are the drives of 15 randomly selected golfers from a PGA tournament, measured in yards. Test the claim that drives with the wind go farther than drives against the wind at the 0.05 level of significance.

With the Wind	Against the Wind
305	290
255	269
249	275
293	274
291	275
303	280
291	264
287	248
288	257
282	283
282	284
284	272
282	275
284	266
287	265

5. Here are the reading proficiency scores of 25 randomly selected 17-year-olds. Sixteen of the students reported spending less than 2 hours on homework each day, while 9 reported spending more than 2 hours on homework each day.

Less Than 2 Hours

213	343	201	286	335	244	276	290
200	267	315	260	279	211	288	339

More Than 2 Hours

236	366	224	309	358	267	299	313
223							

At the 0.01 level of significance, test the claim that students who do less than 2 hours of homework each day have lower reading proficiency scores than students who spend more than 2 hours each day on homework.

6. The owner of a flower shop chain is looking to buy a fleet of delivery vans. She is concerned about the mileage of the vans, and wants to compare four different models of vans. She obtains the following chart that shows the average mileage (miles per gallon) for various different vans of the four different models under test conditions.

Model A	Model B	Model C	Model D
11.4	13.1	9.3	17.7
8.3	9.9	13.5	12.1
13.1	15.0	12.5	14.3
16.3	8.5	12.8	12.2

(continues)

(continued)

Model A	Model B	Model C	Model D
16.0	8.9	12.4	11.5
14.7	9.4	12.3	11.9
11.6	8.4	10.5	16.4
	9.3	14.5	9.3
	11.3	15.1	14.7
	12.2	11.5	
	10.8		

Use the Kruskal–Wallis test to test the claim that the four different models have the same mean mileage at the 0.05 level of significance.

7. Is there a relationship between the amount of cholesterol in food and the number of calories? Here is the nutritional analysis for 12 randomly selected food items.

Cholesterol (mg)	Calories
31	368
43	388
90	640
25	290
15	510
65	560
95	530
30	410
0	222
70	398
30	340
110	640

Calculate Spearman's rank correlation coefficient r_s. Then, at the 0.05 level of significance, test the claim that there is a significant correlation between these two variables.

Research Project

This project calls for you to carry out a statistical project from start to finish. Begin by selecting a claim to test. (You may use any of the hypothesis tests that have been covered in this text.) Hopefully you will be able to come up with a claim based on your major, your hobbies, or your interests. If not, perhaps the examples or exercises in this text will provide inspiration. Take a sample to gather data, and use these data to test your claim.

Begin your paper with an introduction. Describe how you came up with your claim, and provide background information that will be useful to the reader. Explain how you arrived at the plan to gather your data. Which sampling technique did you employ? Was your sampling truly random? Were there any biases? How would you improve upon your sampling if you were to test the claim again?

The introduction should be followed by a conclusion with a summary of your results, including important sample statistics, confidence intervals for each sample when appropriate, your conclusion about the claim, and the p-value for the test. In your conclusion, address whether the test turned out the way you thought it would. Did you come up with any new ideas during the process? Were there any ways that you could expand the project? Be sure to include anything that you learned throughout the project.

After the body of the paper, you should have appendices of useful information. These should include the raw data and relevant sample statistics, appropriate graphs and charts (pie charts, histograms, etc.), confidence intervals for each sample if appropriate, and the formal hypothesis test. The following is an example of a research project.

RESEARCH PROJECT

RESEARCH PROJECT

Sample Research Project

Every time there is a presidential election it seems that the issue of the death penalty is involved in the campaign. I am very interested in this year's election, and often wonder how the candidates will fare in my city. I decided to test the claim that more than half of the people in my city are in favor of the death penalty. If this claim is true, the candidate who is in favor of the death penalty should get more votes in my city than the candidate who opposes it.

I decided to get my data through telephone interviews. As this is a strongly charged issue, I felt uncomfortable asking people in person because I didn't want to get involved in a debate. I just wanted to gather information. Also, I thought that some people might change their answers in person to give the answers that they thought I wanted to hear, and perhaps the anonymity of a phone survey would produce more valid responses.

There are twelve 3-digit prefixes in my city, and I decided to ask 10 people from each prefix for a total of 120 people. I began with the first prefix and randomly generated a 4-digit number to call using Microsoft Excel. This sampling technique is random sampling. I repeated this process until I had 10 responses for that prefix, and then moved on to the next prefix. I was amazed at how many calls I had to make until I could get 10 responses. If a number belonged to a business I politely apologized and hung up. Often someone at a residence would hang up before I could even ask my question. Some people did not speak English and could not understand me. The biggest problem was the number of residences for which I got no answer or reached an answering machine. Not counting the number of times I reached a business, I made 272 calls to get 120 responses, meaning that I was unsuccessful on nearly 56% of my attempts. If the people that I was not able to get a response from were mostly in favor or mostly opposed then my sample is biased. If I had enough time, I would continue to call the numbers I selected until I received a response.

Of the 120 responses, 77 people (64.2%) said that they were in favor of the death penalty. A 95% confidence interval for the proportion of people in my city that are in favor of the death penalty extends from 0.5559 to 0.7275 (55.59% to 72.75%). This proportion was significantly higher than 50% ($\alpha = 0.05$), supporting my claim with a p-value of 0.0010.

I originally thought that the percentage would be closer to 50%. It will be interesting to see whether the percentage of votes that the candidate who supports the death penalty receives is close to 64.2%. About halfway through the sampling, several ideas came to me. I could test a claim about the percentage of nonresponses that a phone survey would get. I also began to think about several tests of independence, examining whether or not a person supports the death penalty is independent of age or the candidate that a voter supports. I also thought about a two-proportion test, as it seemed that a higher percentage of men supported the death penalty than did women.

I did learn that the process of gathering a sample is time consuming, and I can appreciate the work done by survey takers much more than I ever had before. The amazing thing is that I feel that my survey had a lot of flaws even though I put in so much time and effort. Gathering data in such a way that it eliminates bias must be a cross between art and science.

In conclusion, it was nice to apply something I learned in class to an area of interest to me.

Appendix A

In this project, 120 people were surveyed: 77 supported the death penalty and 43 opposed it.

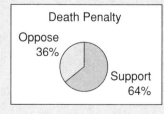

RESEARCH PROJECT

Confidence Interval for the Proportion of People Who Support the Death Penalty

$$E = 1.96 \sqrt{\frac{\frac{77}{120} \cdot \frac{43}{120}}{120}}$$

$$= 0.0858$$

$$\frac{77}{120} - 0.0858 \leq \pi \leq \frac{77}{120} - 0.0858$$

$$0.5559 \leq \pi \leq 0.7775$$

Hypothesis Test

Claim: More than half of the people in my city are in favor of the death penalty.

Claim in symbols: $\pi > 0.5$

Complement: $\pi \leq 0.5$

H_0: $\pi \leq 0.5$

H_A: $\pi > 0.5$

Level of significance: $\alpha = 0.05$

Test statistic: $z = \dfrac{p - \pi}{\sqrt{\dfrac{\pi(1-\pi)}{n}}}$

Decision Rule: Reject H_0 if $z > 1.645$.

Calculation:

$$z = \frac{\frac{77}{120} - 0.5}{\sqrt{\frac{0.5 \cdot 0.5}{120}}}$$

$$= 3.10$$

Reject H_0.

There is sufficient sample evidence to support the claim that more than half of the people in my city are in favor of the death penalty.

p-value: 0.0010

Appendix Tables

TABLE A ■ Binomial Probabilities

$n = 1$						p					
x	0.05	0.1	0.2	0.3	0.4	0.5	0.6	0.7	0.8	0.9	0.95
0	0.9500	0.9000	0.8000	0.7000	0.6000	0.5000	0.4000	0.3000	0.2000	0.1000	0.0500
1	0.0500	0.1000	0.2000	0.3000	0.4000	0.5000	0.6000	0.7000	0.8000	0.9000	0.9500

$n = 2$						p					
x	0.05	0.1	0.2	0.3	0.4	0.5	0.6	0.7	0.8	0.9	0.95
0	0.9025	0.8100	0.6400	0.4900	0.3600	0.2500	0.1600	0.0900	0.0400	0.0100	0.0025
1	0.0950	0.1800	0.3200	0.4200	0.4800	0.5000	0.4800	0.4200	0.3200	0.1800	0.0950
2	0.0025	0.0100	0.0400	0.0900	0.1600	0.2500	0.3600	0.4900	0.6400	0.8100	0.9025

$n = 3$						p					
x	0.05	0.1	0.2	0.3	0.4	0.5	0.6	0.7	0.8	0.9	0.95
0	0.8574	0.7290	0.5120	0.3430	0.2160	0.1250	0.0640	0.0270	0.0080	0.0010	0.0001
1	0.1354	0.2430	0.3840	0.4410	0.4320	0.3750	0.2880	0.1890	0.0960	0.0270	0.0071
2	0.0071	0.0270	0.0960	0.1890	0.2880	0.3750	0.4320	0.4410	0.3840	0.2430	0.1354
3	0.0001	0.0010	0.0080	0.0270	0.0640	0.1250	0.2160	0.3430	0.5120	0.7290	0.8574

$n = 4$						p					
x	0.05	0.1	0.2	0.3	0.4	0.5	0.6	0.7	0.8	0.9	0.95
0	0.8145	0.6561	0.4096	0.2401	0.1296	0.0625	0.0256	0.0081	0.0016	0.0001	0
1	0.1715	0.2916	0.4096	0.4116	0.3456	0.2500	0.1536	0.0756	0.0256	0.0036	0.0005
2	0.0135	0.0486	0.1536	0.2646	0.3456	0.3750	0.3456	0.2646	0.1536	0.0486	0.0135
3	0.0005	0.0036	0.0256	0.0756	0.1536	0.2500	0.3456	0.4116	0.4096	0.2916	0.1715
4	0	0.0001	0.0016	0.0081	0.0256	0.0625	0.1296	0.2401	0.4096	0.6561	0.8145

$n = 5$						p					
x	0.05	0.1	0.2	0.3	0.4	0.5	0.6	0.7	0.8	0.9	0.95
0	0.7738	0.5905	0.3277	0.1681	0.0778	0.0313	0.0102	0.0024	0.0003	0	0
1	0.2036	0.3281	0.4096	0.3602	0.2592	0.1563	0.0768	0.0284	0.0064	0.0005	0
2	0.0214	0.0729	0.2048	0.3087	0.3456	0.3125	0.2304	0.1323	0.0512	0.0081	0.0011
3	0.0011	0.0081	0.0512	0.1323	0.2304	0.3125	0.3456	0.3087	0.2048	0.0729	0.0214
4	0	0.0005	0.0064	0.0284	0.0768	0.1563	0.2592	0.3602	0.4096	0.3281	0.2036
5	0	0	0.0003	0.0024	0.0102	0.0313	0.0778	0.1681	0.3277	0.5905	0.7738

TABLE A ■ Binomial Probabilities—*(continued)*

n = 6

x	0.05	0.1	0.2	0.3	0.4	0.5	0.6	0.7	0.8	0.9	0.95
0	0.7351	0.5314	0.2621	0.1176	0.0467	0.0156	0.0041	0.0007	0.0001	0	0
1	0.2321	0.3543	0.3932	0.3025	0.1866	0.0938	0.0369	0.0102	0.0015	0.0001	0
2	0.0305	0.0984	0.2458	0.3241	0.3110	0.2344	0.1382	0.0595	0.0154	0.0012	0.0001
3	0.0021	0.0146	0.0819	0.1852	0.2765	0.3125	0.2765	0.1852	0.0819	0.0146	0.0021
4	0.0001	0.0012	0.0154	0.0595	0.1382	0.2344	0.3110	0.3241	0.2458	0.0984	0.0305
5	0	0.0001	0.0015	0.0102	0.0369	0.0938	0.1866	0.3025	0.3932	0.3543	0.2321
6	0	0	0.0001	0.0007	0.0041	0.0156	0.0467	0.1176	0.2621	0.5314	0.7351

n = 7

x	0.05	0.1	0.2	0.3	0.4	0.5	0.6	0.7	0.8	0.9	0.95
0	0.6983	0.4783	0.2097	0.0824	0.0280	0.0078	0.0016	0.0002	0	0	0
1	0.2573	0.3720	0.3670	0.2471	0.1306	0.0547	0.0172	0.0036	0.0004	0	0
2	0.0406	0.1240	0.2753	0.3177	0.2613	0.1641	0.0774	0.0250	0.0043	0.0002	0
3	0.0036	0.0230	0.1147	0.2269	0.2903	0.2734	0.1935	0.0972	0.0287	0.0026	0.0002
4	0.0002	0.0026	0.0287	0.0972	0.1935	0.2734	0.2903	0.2269	0.1147	0.0230	0.0036
5	0	0.0002	0.0043	0.0250	0.0774	0.1641	0.2613	0.3177	0.2753	0.1240	0.0406
6	0	0	0.0004	0.0036	0.0172	0.0547	0.1306	0.2471	0.3670	0.3720	0.2573
7	0	0	0	0.0002	0.0016	0.0078	0.0280	0.0824	0.2097	0.4783	0.6983

n = 8

x	0.05	0.1	0.2	0.3	0.4	0.5	0.6	0.7	0.8	0.9	0.95
0	0.6634	0.4305	0.1678	0.0576	0.0168	0.0039	0.0007	0.0001	0	0	0
1	0.2793	0.3826	0.3355	0.1977	0.0896	0.0313	0.0079	0.0012	0.0001	0	0
2	0.0515	0.1488	0.2936	0.2965	0.2090	0.1094	0.0413	0.0100	0.0011	0	0
3	0.0054	0.0331	0.1468	0.2541	0.2787	0.2188	0.1239	0.0467	0.0092	0.0004	0
4	0.0004	0.0046	0.0459	0.1361	0.2322	0.2734	0.2322	0.1361	0.0459	0.0046	0.0004
5	0	0.0004	0.0092	0.0467	0.1239	0.2188	0.2787	0.2541	0.1468	0.0331	0.0054
6	0	0	0.0011	0.0100	0.0413	0.1094	0.2090	0.2965	0.2936	0.1488	0.0515
7	0	0	0.0001	0.0012	0.0079	0.0313	0.0896	0.1977	0.3355	0.3826	0.2793
8	0	0	0	0.0001	0.0007	0.0039	0.0168	0.0576	0.1678	0.4305	0.6634

n = 9

x	0.05	0.1	0.2	0.3	0.4	0.5	0.6	0.7	0.8	0.9	0.95
0	0.6302	0.3874	0.1342	0.0404	0.0101	0.0020	0.0003	0	0	0	0
1	0.2985	0.3874	0.3020	0.1556	0.0605	0.0176	0.0035	0.0004	0	0	0
2	0.0629	0.1722	0.3020	0.2668	0.1612	0.0703	0.0212	0.0039	0.0003	0	0
3	0.0077	0.0446	0.1762	0.2668	0.2508	0.1641	0.0743	0.0210	0.0028	0.0001	0
4	0.0006	0.0074	0.0661	0.1715	0.2508	0.2461	0.1672	0.0735	0.0165	0.0008	0
5	0	0.0008	0.0165	0.0735	0.1672	0.2461	0.2508	0.1715	0.0661	0.0074	0.0006
6	0	0.0001	0.0028	0.0210	0.0743	0.1641	0.2508	0.2668	0.1762	0.0446	0.0077
7	0	0	0.0003	0.0039	0.0212	0.0703	0.1612	0.2668	0.3020	0.1722	0.0629
8	0	0	0	0.0004	0.0035	0.0176	0.0605	0.1556	0.3020	0.3874	0.2985
9	0	0	0	0	0.0003	0.0020	0.0101	0.0404	0.1342	0.3874	0.6302

TABLE A ■ Binomial Probabilities—*(continued)*

n = 10						*p*					
x	*0.05*	*0.1*	*0.2*	*0.3*	*0.4*	*0.5*	*0.6*	*0.7*	*0.8*	*0.9*	*0.95*
0	0.5987	0.3487	0.1074	0.0282	0.0060	0.0010	0.0001	0	0	0	0
1	0.3151	0.3874	0.2684	0.1211	0.0403	0.0098	0.0016	0.0001	0	0	0
2	0.0746	0.1937	0.3020	0.2335	0.1209	0.0439	0.0106	0.0014	0.0001	0	0
3	0.0105	0.0574	0.2013	0.2668	0.2150	0.1172	0.0425	0.0090	0.0008	0	0
4	0.0010	0.0112	0.0881	0.2001	0.2508	0.2051	0.1115	0.0368	0.0055	0.0001	0
5	0.0001	0.0015	0.0264	0.1029	0.2007	0.2461	0.2007	0.1029	0.0264	0.0015	0.0001
6	0	0.0001	0.0055	0.0368	0.1115	0.2051	0.2508	0.2001	0.0881	0.0112	0.0010
7	0	0	0.0008	0.0090	0.0425	0.1172	0.2150	0.2668	0.2013	0.0574	0.0105
8	0	0	0.0001	0.0014	0.0106	0.0439	0.1209	0.2335	0.3020	0.1937	0.0746
9	0	0	0	0.0001	0.0016	0.0098	0.0403	0.1211	0.2684	0.3874	0.3151
10	0	0	0	0	0.0001	0.0010	0.0060	0.0282	0.1074	0.3487	0.5987

n = 11						*p*					
x	*0.05*	*0.1*	*0.2*	*0.3*	*0.4*	*0.5*	*0.6*	*0.7*	*0.8*	*0.9*	*0.95*
0	0.5688	0.3138	0.0859	0.0198	0.0036	0.0005	0	0	0	0	0
1	0.3293	0.3835	0.2362	0.0932	0.0266	0.0054	0.0007	0	0	0	0
2	0.0867	0.2131	0.2953	0.1998	0.0887	0.0269	0.0052	0.0005	0	0	0
3	0.0137	0.0710	0.2215	0.2568	0.1774	0.0806	0.0234	0.0037	0.0002	0	0
4	0.0014	0.0158	0.1107	0.2201	0.2365	0.1611	0.0701	0.0173	0.0017	0	0
5	0.0001	0.0025	0.0388	0.1321	0.2207	0.2256	0.1471	0.0566	0.0097	0.0003	0
6	0	0.0003	0.0097	0.0566	0.1471	0.2256	0.2207	0.1321	0.0388	0.0025	0.0001
7	0	0	0.0017	0.0173	0.0701	0.1611	0.2365	0.2201	0.1107	0.0158	0.0014
8	0	0	0.0002	0.0037	0.0234	0.0806	0.1774	0.2568	0.2215	0.0710	0.0137
9	0	0	0	0.0005	0.0052	0.0269	0.0887	0.1998	0.2953	0.2131	0.0867
10	0	0	0	0	0.0007	0.0054	0.0266	0.0932	0.2362	0.3835	0.3293
11	0	0	0	0	0	0.0005	0.0036	0.0198	0.0859	0.3138	0.5688

n = 12						*p*					
x	*0.05*	*0.1*	*0.2*	*0.3*	*0.4*	*0.5*	*0.6*	*0.7*	*0.8*	*0.9*	*0.95*
0	0.5404	0.2824	0.0687	0.0138	0.0022	0.0002	0	0	0	0	0
1	0.3413	0.3766	0.2062	0.0712	0.0174	0.0029	0.0003	0	0	0	0
2	0.0988	0.2301	0.2835	0.1678	0.0639	0.0161	0.0025	0.0002	0	0	0
3	0.0173	0.0852	0.2362	0.2397	0.1419	0.0537	0.0125	0.0015	0.0001	0	0
4	0.0021	0.0213	0.1329	0.2311	0.2128	0.1208	0.0420	0.0078	0.0005	0	0
5	0.0002	0.0038	0.0532	0.1585	0.2270	0.1934	0.1009	0.0291	0.0033	0	0
6	0	0.0005	0.0155	0.0792	0.1766	0.2256	0.1766	0.0792	0.0155	0.0005	0
7	0	0	0.0033	0.0291	0.1009	0.1934	0.2270	0.1585	0.0532	0.0038	0.0002
8	0	0	0.0005	0.0078	0.0420	0.1208	0.2128	0.2311	0.1329	0.0213	0.0021
9	0	0	0.0001	0.0015	0.0125	0.0537	0.1419	0.2397	0.2362	0.0852	0.0173
10	0	0	0	0.0002	0.0025	0.0161	0.0639	0.1678	0.2835	0.2301	0.0988
11	0	0	0	0	0.0003	0.0029	0.0174	0.0712	0.2062	0.3766	0.3413
12	0	0	0	0	0	0.0002	0.0022	0.0138	0.0687	0.2824	0.5404

TABLE A ■ Binomial Probabilities—*(continued)*

n = 13   p

x	0.05	0.1	0.2	0.3	0.4	0.5	0.6	0.7	0.8	0.9	0.95
0	0.5133	0.2542	0.0550	0.0097	0.0013	0.0001	0	0	0	0	0
1	0.3512	0.3672	0.1787	0.0540	0.0113	0.0016	0.0001	0	0	0	0
2	0.1109	0.2448	0.2680	0.1388	0.0453	0.0095	0.0012	0.0001	0	0	0
3	0.0214	0.0997	0.2457	0.2181	0.1107	0.0349	0.0065	0.0006	0	0	0
4	0.0028	0.0277	0.1535	0.2337	0.1845	0.0873	0.0243	0.0034	0.0001	0	0
5	0.0003	0.0055	0.0691	0.1803	0.2214	0.1571	0.0656	0.0142	0.0011	0	0
6	0	0.0008	0.0230	0.1030	0.1968	0.2095	0.1312	0.0442	0.0058	0.0001	0
7	0	0.0001	0.0058	0.0442	0.1312	0.2095	0.1968	0.1030	0.0230	0.0008	0
8	0	0	0.0011	0.0142	0.0656	0.1571	0.2214	0.1803	0.0691	0.0055	0.0003
9	0	0	0.0001	0.0034	0.0243	0.0873	0.1845	0.2337	0.1535	0.0277	0.0028
10	0	0	0	0.0006	0.0065	0.0349	0.1107	0.2181	0.2457	0.0997	0.0214
11	0	0	0	0.0001	0.0012	0.0095	0.0453	0.1388	0.2680	0.2448	0.1109
12	0	0	0	0	0.0001	0.0016	0.0113	0.0540	0.1787	0.3672	0.3512
13	0	0	0	0	0	0.0001	0.0013	0.0097	0.0550	0.2542	0.5133

n = 14   p

x	0.05	0.1	0.2	0.3	0.4	0.5	0.6	0.7	0.8	0.9	0.95
0	0.4877	0.2288	0.0440	0.0068	0.0008	0.0001	0	0	0	0	0
1	0.3593	0.3559	0.1539	0.0407	0.0073	0.0009	0.0001	0	0	0	0
2	0.1229	0.2570	0.2501	0.1134	0.0317	0.0056	0.0005	0	0	0	0
3	0.0259	0.1142	0.2501	0.1943	0.0845	0.0222	0.0033	0.0002	0	0	0
4	0.0037	0.0349	0.1720	0.2290	0.1549	0.0611	0.0136	0.0014	0	0	0
5	0.0004	0.0078	0.0860	0.1963	0.2066	0.1222	0.0408	0.0066	0.0003	0	0
6	0	0.0013	0.0322	0.1262	0.2066	0.1833	0.0918	0.0232	0.0020	0	0
7	0	0.0002	0.0092	0.0618	0.1574	0.2095	0.1574	0.0618	0.0092	0.0002	0
8	0	0	0.0020	0.0232	0.0918	0.1833	0.2066	0.1262	0.0322	0.0013	0
9	0	0	0.0003	0.0066	0.0408	0.1222	0.2066	0.1963	0.0860	0.0078	0.0004
10	0	0	0	0.0014	0.0136	0.0611	0.1549	0.2290	0.1720	0.0349	0.0037
11	0	0	0	0.0002	0.0033	0.0222	0.0845	0.1943	0.2501	0.1142	0.0259
12	0	0	0	0	0.0005	0.0056	0.0317	0.1134	0.2501	0.2570	0.1229
13	0	0	0	0	0.0001	0.0009	0.0073	0.0407	0.1539	0.3559	0.3593
14	0	0	0	0	0	0.0001	0.0008	0.0068	0.0440	0.2288	0.4877

n = 15   p

x	0.05	0.1	0.2	0.3	0.4	0.5	0.6	0.7	0.8	0.9	0.95
0	0.4633	0.2059	0.0352	0.0047	0.0005	0	0	0	0	0	0
1	0.3658	0.3432	0.1319	0.0305	0.0047	0.0005	0	0	0	0	0
2	0.1348	0.2669	0.2309	0.0916	0.0219	0.0032	0.0003	0	0	0	0
3	0.0307	0.1285	0.2501	0.1700	0.0634	0.0139	0.0016	0.0001	0	0	0
4	0.0049	0.0428	0.1876	0.2186	0.1268	0.0417	0.0074	0.0006	0	0	0
5	0.0006	0.0105	0.1032	0.2061	0.1859	0.0916	0.0245	0.0030	0.0001	0	0
6	0	0.0019	0.0430	0.1472	0.2066	0.1527	0.0612	0.0116	0.0007	0	0
7	0	0.0003	0.0138	0.0811	0.1771	0.1964	0.1181	0.0348	0.0035	0	0
8	0	0	0.0035	0.0348	0.1181	0.1964	0.1771	0.0811	0.0138	0.0003	0
9	0	0	0.0007	0.0116	0.0612	0.1527	0.2066	0.1472	0.0430	0.0019	0
10	0	0	0.0001	0.0030	0.0245	0.0916	0.1859	0.2061	0.1032	0.0105	0.0006
11	0	0	0	0.0006	0.0074	0.0417	0.1268	0.2186	0.1876	0.0428	0.0049
12	0	0	0	0.0001	0.0016	0.0139	0.0634	0.1700	0.2501	0.1285	0.0307
13	0	0	0	0	0.0003	0.0032	0.0219	0.0916	0.2309	0.2669	0.1348
14	0	0	0	0	0	0.0005	0.0047	0.0305	0.1319	0.3432	0.3658
15	0	0	0	0	0	0	0.0005	0.0047	0.0352	0.2059	0.4633

TABLE A ■ Binomial Probabilities—(continued)

n = 16						p					
x	0.05	0.1	0.2	0.3	0.4	0.5	0.6	0.7	0.8	0.9	0.95
0	0.4401	0.1853	0.0281	0.0033	0.0003	0	0	0	0	0	0
1	0.3706	0.3294	0.1126	0.0228	0.0030	0.0002	0	0	0	0	0
2	0.1463	0.2745	0.2111	0.0732	0.0150	0.0018	0.0001	0	0	0	0
3	0.0359	0.1423	0.2463	0.1465	0.0468	0.0085	0.0008	0	0	0	0
4	0.0061	0.0514	0.2001	0.2040	0.1014	0.0278	0.0040	0.0002	0	0	0
5	0.0008	0.0137	0.1201	0.2099	0.1623	0.0667	0.0142	0.0013	0	0	0
6	0.0001	0.0028	0.0550	0.1649	0.1983	0.1222	0.0392	0.0056	0.0002	0	0
7	0	0.0004	0.0197	0.1010	0.1889	0.1746	0.0840	0.0185	0.0012	0	0
8	0	0.0001	0.0055	0.0487	0.1417	0.1964	0.1417	0.0487	0.0055	0.0001	0
9	0	0	0.0012	0.0185	0.0840	0.1746	0.1889	0.1010	0.0197	0.0004	0
10	0	0	0.0002	0.0056	0.0392	0.1222	0.1983	0.1649	0.0550	0.0028	0.0001
11	0	0	0	0.0013	0.0142	0.0667	0.1623	0.2099	0.1201	0.0137	0.0008
12	0	0	0	0.0002	0.0040	0.0278	0.1014	0.2040	0.2001	0.0514	0.0061
13	0	0	0	0	0.0008	0.0085	0.0468	0.1465	0.2463	0.1423	0.0359
14	0	0	0	0	0.0001	0.0018	0.0150	0.0732	0.2111	0.2745	0.1463
15	0	0	0	0	0	0.0002	0.0030	0.0228	0.1126	0.3294	0.3706
16	0	0	0	0	0	0	0.0003	0.0033	0.0281	0.1853	0.4401

n = 17						p					
x	0.05	0.1	0.2	0.3	0.4	0.5	0.6	0.7	0.8	0.9	0.95
0	0.4181	0.1668	0.0225	0.0023	0.0002	0	0	0	0	0	0
1	0.3741	0.3150	0.0957	0.0169	0.0019	0.0001	0	0	0	0	0
2	0.1575	0.2800	0.1914	0.0581	0.0102	0.0010	0.0001	0	0	0	0
3	0.0415	0.1556	0.2393	0.1245	0.0341	0.0052	0.0004	0	0	0	0
4	0.0076	0.0605	0.2093	0.1868	0.0796	0.0182	0.0021	0.0001	0	0	0
5	0.0010	0.0175	0.1361	0.2081	0.1379	0.0472	0.0081	0.0006	0	0	0
6	0.0001	0.0039	0.0680	0.1784	0.1839	0.0944	0.0242	0.0026	0.0001	0	0
7	0	0.0007	0.0267	0.1201	0.1927	0.1484	0.0571	0.0095	0.0004	0	0
8	0	0.0001	0.0084	0.0644	0.1606	0.1855	0.1070	0.0276	0.0021	0	0
9	0	0	0.0021	0.0276	0.1070	0.1855	0.1606	0.0644	0.0084	0.0001	0
10	0	0	0.0004	0.0095	0.0571	0.1484	0.1927	0.1201	0.0267	0.0007	0
11	0	0	0.0001	0.0026	0.0242	0.0944	0.1839	0.1784	0.0680	0.0039	0.0001
12	0	0	0	0.0006	0.0081	0.0472	0.1379	0.2081	0.1361	0.0175	0.0010
13	0	0	0	0.0001	0.0021	0.0182	0.0796	0.1868	0.2093	0.0605	0.0076
14	0	0	0	0	0.0004	0.0052	0.0341	0.1245	0.2393	0.1556	0.0415
15	0	0	0	0	0.0001	0.0010	0.0102	0.0581	0.1914	0.2800	0.1575
16	0	0	0	0	0	0.0001	0.0019	0.0169	0.0957	0.3150	0.3741
17	0	0	0	0	0	0	0.0002	0.0023	0.0225	0.1668	0.4181

TABLE A ■ Binomial Probabilities—*(continued)*

n = 18   **p**

x	0.05	0.1	0.2	0.3	0.4	0.5	0.6	0.7	0.8	0.9	0.95
0	0.3972	0.1501	0.0180	0.0016	0.0001	0	0	0	0	0	0
1	0.3763	0.3002	0.0811	0.0126	0.0012	0.0001	0	0	0	0	0
2	0.1683	0.2835	0.1723	0.0458	0.0069	0.0006	0	0	0	0	0
3	0.0473	0.1680	0.2297	0.1046	0.0246	0.0031	0.0002	0	0	0	0
4	0.0093	0.0700	0.2153	0.1681	0.0614	0.0117	0.0011	0	0	0	0
5	0.0014	0.0218	0.1507	0.2017	0.1146	0.0327	0.0045	0.0002	0	0	0
6	0.0002	0.0052	0.0816	0.1873	0.1655	0.0708	0.0145	0.0012	0	0	0
7	0	0.0010	0.0350	0.1376	0.1892	0.1214	0.0374	0.0046	0.0001	0	0
8	0	0.0002	0.0120	0.0811	0.1734	0.1669	0.0771	0.0149	0.0008	0	0
9	0	0	0.0033	0.0386	0.1284	0.1855	0.1284	0.0386	0.0033	0	0
10	0	0	0.0008	0.0149	0.0771	0.1669	0.1734	0.0811	0.0120	0.0002	0
11	0	0	0.0001	0.0046	0.0374	0.1214	0.1892	0.1376	0.0350	0.0010	0
12	0	0	0	0.0012	0.0145	0.0708	0.1655	0.1873	0.0816	0.0052	0.0002
13	0	0	0	0.0002	0.0045	0.0327	0.1146	0.2017	0.1507	0.0218	0.0014
14	0	0	0	0	0.0011	0.0117	0.0614	0.1681	0.2153	0.0700	0.0093
15	0	0	0	0	0.0002	0.0031	0.0246	0.1046	0.2297	0.1680	0.0473
16	0	0	0	0	0	0.0006	0.0069	0.0458	0.1723	0.2835	0.1683
17	0	0	0	0	0	0.0001	0.0012	0.0126	0.0811	0.3002	0.3763
18	0	0	0	0	0	0	0.0001	0.0016	0.0180	0.1501	0.3972

n = 19   **p**

x	0.05	0.1	0.2	0.3	0.4	0.5	0.6	0.7	0.8	0.9	0.95
0	0.3774	0.1351	0.0144	0.0011	0.0001	0	0	0	0	0	0
1	0.3774	0.2852	0.0685	0.0093	0.0008	0	0	0	0	0	0
2	0.1787	0.2852	0.1540	0.0358	0.0046	0.0003	0	0	0	0	0
3	0.0533	0.1796	0.2182	0.0869	0.0175	0.0018	0.0001	0	0	0	0
4	0.0112	0.0798	0.2182	0.1491	0.0467	0.0074	0.0005	0	0	0	0
5	0.0018	0.0266	0.1636	0.1916	0.0933	0.0222	0.0024	0.0001	0	0	0
6	0.0002	0.0069	0.0955	0.1916	0.1451	0.0518	0.0085	0.0005	0	0	0
7	0	0.0014	0.0443	0.1525	0.1797	0.0961	0.0237	0.0022	0	0	0
8	0	0.0002	0.0166	0.0981	0.1797	0.1442	0.0532	0.0077	0.0003	0	0
9	0	0	0.0051	0.0514	0.1464	0.1762	0.0976	0.0220	0.0013	0	0
10	0	0	0.0013	0.0220	0.0976	0.1762	0.1464	0.0514	0.0051	0	0
11	0	0	0.0003	0.0077	0.0532	0.1442	0.1797	0.0981	0.0166	0.0002	0
12	0	0	0	0.0022	0.0237	0.0961	0.1797	0.1525	0.0443	0.0014	0
13	0	0	0	0.0005	0.0085	0.0518	0.1451	0.1916	0.0955	0.0069	0.0002
14	0	0	0	0.0001	0.0024	0.0222	0.0933	0.1916	0.1636	0.0266	0.0018
15	0	0	0	0	0.0005	0.0074	0.0467	0.1491	0.2182	0.0798	0.0112
16	0	0	0	0	0.0001	0.0018	0.0175	0.0869	0.2182	0.1796	0.0533
17	0	0	0	0	0	0.0003	0.0046	0.0358	0.1540	0.2852	0.1787
18	0	0	0	0	0	0	0.0008	0.0093	0.0685	0.2852	0.3774
19	0	0	0	0	0	0	0.0001	0.0011	0.0144	0.1351	0.3774

TABLE A ■ Binomial Probabilities—*(continued)*

n = 20　　　　　　　　　　　　　　　　　　　　　　　　*p*

x	0.05	0.1	0.2	0.3	0.4	0.5	0.6	0.7	0.8	0.9	0.95
0	0.3585	0.1216	0.0115	0.0008	0	0	0	0	0	0	0
1	0.3774	0.2702	0.0576	0.0068	0.0005	0	0	0	0	0	0
2	0.1887	0.2852	0.1369	0.0278	0.0031	0.0002	0	0	0	0	0
3	0.0596	0.1901	0.2054	0.0716	0.0123	0.0011	0	0	0	0	0
4	0.0133	0.0898	0.2182	0.1304	0.0350	0.0046	0.0003	0	0	0	0
5	0.0022	0.0319	0.1746	0.1789	0.0746	0.0148	0.0013	0	0	0	0
6	0.0003	0.0089	0.1091	0.1916	0.1244	0.0370	0.0049	0.0002	0	0	0
7	0	0.0020	0.0545	0.1643	0.1659	0.0739	0.0146	0.0010	0	0	0
8	0	0.0004	0.0222	0.1144	0.1797	0.1201	0.0355	0.0039	0.0001	0	0
9	0	0.0001	0.0074	0.0654	0.1597	0.1602	0.0710	0.0120	0.0005	0	0
10	0	0	0.0020	0.0308	0.1171	0.1762	0.1171	0.0308	0.0020	0	0
11	0	0	0.0005	0.0120	0.0710	0.1602	0.1597	0.0654	0.0074	0.0001	0
12	0	0	0.0001	0.0039	0.0355	0.1201	0.1797	0.1144	0.0222	0.0004	0
13	0	0	0	0.0010	0.0146	0.0739	0.1659	0.1643	0.0545	0.0020	0
14	0	0	0	0.0002	0.0049	0.0370	0.1244	0.1916	0.1091	0.0089	0.0003
15	0	0	0	0	0.0013	0.0148	0.0746	0.1789	0.1746	0.0319	0.0022
16	0	0	0	0	0.0003	0.0046	0.0350	0.1304	0.2182	0.0898	0.0133
17	0	0	0	0	0	0.0011	0.0123	0.0716	0.2054	0.1901	0.0596
18	0	0	0	0	0	0.0002	0.0031	0.0278	0.1369	0.2852	0.1887
19	0	0	0	0	0	0	0.0005	0.0068	0.0576	0.2702	0.3774
20	0	0	0	0	0	0	0	0.0008	0.0115	0.1216	0.3585

n = 25　　　　　　　　　　　　　　　　　　　　　　　　*p*

x	0.05	0.1	0.2	0.3	0.4	0.5	0.6	0.7	0.8	0.9	0.95
0	0.2774	0.0718	0.0038	0.0001	0	0	0	0	0	0	0
1	0.3650	0.1994	0.0236	0.0014	0	0	0	0	0	0	0
2	0.2305	0.2659	0.0708	0.0074	0.0004	0	0	0	0	0	0
3	0.0930	0.2265	0.1358	0.0243	0.0019	0.0001	0	0	0	0	0
4	0.0269	0.1384	0.1867	0.0572	0.0071	0.0004	0	0	0	0	0
5	0.0060	0.0646	0.1960	0.1030	0.0199	0.0016	0	0	0	0	0
6	0.0010	0.0239	0.1633	0.1472	0.0442	0.0053	0.0002	0	0	0	0
7	0.0001	0.0072	0.1108	0.1712	0.0800	0.0143	0.0009	0	0	0	0
8	0	0.0018	0.0623	0.1651	0.1200	0.0322	0.0031	0.0001	0	0	0
9	0	0.0004	0.0294	0.1336	0.1511	0.0609	0.0088	0.0004	0	0	0
10	0	0.0001	0.0118	0.0916	0.1612	0.0974	0.0212	0.0013	0	0	0
11	0	0	0.0040	0.0536	0.1465	0.1328	0.0434	0.0042	0.0001	0	0
12	0	0	0.0012	0.0268	0.1140	0.1550	0.0760	0.0115	0.0003	0	0
13	0	0	0.0003	0.0115	0.0760	0.1550	0.1140	0.0268	0.0012	0	0
14	0	0	0.0001	0.0042	0.0434	0.1328	0.1465	0.0536	0.0040	0	0
15	0	0	0	0.0013	0.0212	0.0974	0.1612	0.0916	0.0118	0.0001	0
16	0	0	0	0.0004	0.0088	0.0609	0.1511	0.1336	0.0294	0.0004	0
17	0	0	0	0.0001	0.0031	0.0322	0.1200	0.1651	0.0623	0.0018	0
18	0	0	0	0	0.0009	0.0143	0.0800	0.1712	0.1108	0.0072	0.0001
19	0	0	0	0	0.0002	0.0053	0.0442	0.1472	0.1633	0.0239	0.0010
20	0	0	0	0	0	0.0016	0.0199	0.1030	0.1960	0.0646	0.0060
21	0	0	0	0	0	0.0004	0.0071	0.0572	0.1867	0.1384	0.0269
22	0	0	0	0	0	0.0001	0.0019	0.0243	0.1358	0.2265	0.0930
23	0	0	0	0	0	0	0.0004	0.0074	0.0708	0.2659	0.2305
24	0	0	0	0	0	0	0	0.0014	0.0236	0.1994	0.3650
25	0	0	0	0	0	0	0	0.0001	0.0038	0.0718	0.2774

TABLE B ■ Poisson Distribution: Probability of *x* Successes

					λ				
x	*0.1*	*0.2*	*0.3*	*0.4*	*0.5*	*0.6*	*0.7*	*0.8*	*0.9*
0	0.9048	0.8187	0.7408	0.6703	0.6065	0.5488	0.4966	0.4493	0.4066
1	0.0905	0.1637	0.2222	0.2681	0.3033	0.3293	0.3476	0.3595	0.3659
2	0.0045	0.0164	0.0333	0.0536	0.0758	0.0988	0.1217	0.1438	0.1647
3	0.0002	0.0011	0.0033	0.0072	0.0126	0.0198	0.0284	0.0383	0.0494
4	0.0000	0.0001	0.0003	0.0007	0.0016	0.0030	0.0050	0.0077	0.0111
5	0.0000	0.0000	0.0000	0.0001	0.0002	0.0004	0.0007	0.0012	0.0020
6	0.0000	0.0000	0.0000	0.0000	0.0000	0.0000	0.0001	0.0002	0.0003

					λ				
x	*1*	*2*	*3*	*4*	*5*	*6*	*7*	*8*	*9*
0	0.3679	0.1353	0.0498	0.0183	0.0067	0.0025	0.0009	0.0003	0.0001
1	0.3679	0.2707	0.1494	0.0733	0.0337	0.0149	0.0064	0.0027	0.0011
2	0.1839	0.2707	0.2240	0.1465	0.0842	0.0446	0.0223	0.0107	0.0050
3	0.0613	0.1804	0.2240	0.1954	0.1404	0.0892	0.0521	0.0286	0.0150
4	0.0153	0.0902	0.1680	0.1954	0.1755	0.1339	0.0912	0.0573	0.0337
5	0.0031	0.0361	0.1008	0.1563	0.1755	0.1606	0.1277	0.0916	0.0607
6	0.0005	0.0120	0.0504	0.1042	0.1462	0.1606	0.1490	0.1221	0.0911
7	0.0001	0.0034	0.0216	0.0595	0.1044	0.1377	0.1490	0.1396	0.1171
8	0.0000	0.0009	0.0081	0.0298	0.0653	0.1033	0.1304	0.1396	0.1318
9	0.0000	0.0002	0.0027	0.0132	0.0363	0.0688	0.1014	0.1241	0.1318
10	0.0000	0.0000	0.0008	0.0053	0.0181	0.0413	0.0710	0.0993	0.1186
11	0.0000	0.0000	0.0002	0.0019	0.0082	0.0225	0.0452	0.0722	0.0970
12	0.0000	0.0000	0.0001	0.0006	0.0034	0.0113	0.0263	0.0481	0.0728
13	0.0000	0.0000	0.0000	0.0002	0.0013	0.0052	0.0142	0.0296	0.0504
14	0.0000	0.0000	0.0000	0.0001	0.0005	0.0022	0.0071	0.0169	0.0324
15	0.0000	0.0000	0.0000	0.0000	0.0002	0.0009	0.0033	0.0090	0.0194
16	0.0000	0.0000	0.0000	0.0000	0.0000	0.0003	0.0014	0.0045	0.0109
17	0.0000	0.0000	0.0000	0.0000	0.0000	0.0001	0.0006	0.0021	0.0058
18	0.0000	0.0000	0.0000	0.0000	0.0000	0.0000	0.0002	0.0009	0.0029
19	0.0000	0.0000	0.0000	0.0000	0.0000	0.0000	0.0001	0.0004	0.0014
20	0.0000	0.0000	0.0000	0.0000	0.0000	0.0000	0.0000	0.0002	0.0006
21	0.0000	0.0000	0.0000	0.0000	0.0000	0.0000	0.0000	0.0001	0.0003
22	0.0000	0.0000	0.0000	0.0000	0.0000	0.0000	0.0000	0.0000	0.0001

TABLE C ■ Standard Normal Distribution (Area to the Left of z)

Negative z-Values

	0.00	0.01	0.02	0.03	0.04	0.05	0.06	0.07	0.08	0.09
0.0	0.5000	0.4960	0.4920	0.4880	0.4840	0.4801	0.4761	0.4721	0.4681	0.4641
−0.1	0.4602	0.4562	0.4522	0.4483	0.4443	0.4404	0.4364	0.4325	0.4286	0.4247
−0.2	0.4207	0.4168	0.4129	0.4090	0.4052	0.4013	0.3974	0.3936	0.3897	0.3859
−0.3	0.3821	0.3783	0.3745	0.3707	0.3669	0.3632	0.3594	0.3557	0.3520	0.3483
−0.4	0.3446	0.3409	0.3372	0.3336	0.3300	0.3264	0.3228	0.3192	0.3156	0.3121
−0.5	0.3085	0.3050	0.3015	0.2981	0.2946	0.2912	0.2877	0.2843	0.2810	0.2776
−0.6	0.2743	0.2709	0.2676	0.2643	0.2611	0.2578	0.2546	0.2514	0.2483	0.2451
−0.7	0.2420	0.2389	0.2358	0.2327	0.2296	0.2266	0.2236	0.2206	0.2177	0.2148
−0.8	0.2119	0.2090	0.2061	0.2033	0.2005	0.1977	0.1949	0.1922	0.1894	0.1867
−0.9	0.1841	0.1814	0.1788	0.1762	0.1736	0.1711	0.1685	0.1660	0.1635	0.1611
−1.0	0.1587	0.1562	0.1539	0.1515	0.1492	0.1469	0.1446	0.1423	0.1401	0.1379
−1.1	0.1357	0.1335	0.1314	0.1292	0.1271	0.1251	0.1230	0.1210	0.1190	0.1170
−1.2	0.1151	0.1131	0.1112	0.1093	0.1075	0.1056	0.1038	0.1020	0.1003	0.0985
−1.3	0.0968	0.0951	0.0934	0.0918	0.0901	0.0885	0.0869	0.0853	0.0838	0.0823
−1.4	0.0808	0.0793	0.0778	0.0764	0.0749	0.0735	0.0721	0.0708	0.0694	0.0681
−1.5	0.0668	0.0655	0.0643	0.0630	0.0618	0.0606	0.0594	0.0582	0.0571	0.0559
−1.6	0.0548	0.0537	0.0526	0.0516	0.0505	0.0495	0.0485	0.0475	0.0465	0.0455
−1.7	0.0446	0.0436	0.0427	0.0418	0.0409	0.0401	0.0392	0.0384	0.0375	0.0367
−1.8	0.0359	0.0351	0.0344	0.0336	0.0329	0.0322	0.0314	0.0307	0.0301	0.0294
−1.9	0.0287	0.0281	0.0274	0.0268	0.0262	0.0256	0.0250	0.0244	0.0239	0.0233
−2.0	0.0228	0.0222	0.0217	0.0212	0.0207	0.0202	0.0197	0.0192	0.0188	0.0183
−2.1	0.0179	0.0174	0.0170	0.0166	0.0162	0.0158	0.0154	0.0150	0.0146	0.0143
−2.2	0.0139	0.0136	0.0132	0.0129	0.0125	0.0122	0.0119	0.0116	0.0113	0.0110
−2.3	0.0107	0.0104	0.0102	0.0099	0.0096	0.0094	0.0091	0.0089	0.0087	0.0084
−2.4	0.0082	0.0080	0.0078	0.0075	0.0073	0.0071	0.0069	0.0068	0.0066	0.0064
−2.5	0.0062	0.0060	0.0059	0.0057	0.0055	0.0054	0.0052	0.0051	0.0049	0.0048
−2.6	0.0047	0.0045	0.0044	0.0043	0.0041	0.0040	0.0039	0.0038	0.0037	0.0036
−2.7	0.0035	0.0034	0.0033	0.0032	0.0031	0.0030	0.0029	0.0028	0.0027	0.0026
−2.8	0.0026	0.0025	0.0024	0.0023	0.0023	0.0022	0.0021	0.0021	0.0020	0.0019
−2.9	0.0019	0.0018	0.0018	0.0017	0.0016	0.0016	0.0015	0.0015	0.0014	0.0014
−3.0	0.0013	0.0013	0.0013	0.0012	0.0012	0.0011	0.0011	0.0011	0.0010	0.0010
−3.1	0.0010	0.0009	0.0009	0.0009	0.0008	0.0008	0.0008	0.0008	0.0007	0.0007
−3.2	0.0007	0.0007	0.0006	0.0006	0.0006	0.0006	0.0006	0.0005	0.0005	0.0005
−3.3	0.0005	0.0005	0.0005	0.0004	0.0004	0.0004	0.0004	0.0004	0.0004	0.0003
−3.4	0.0003	0.0003	0.0003	0.0003	0.0003	0.0003	0.0003	0.0003	0.0003	0.0002
−3.5	0.0002									

TABLE C ■ Standard Normal Distribution (Area to the Left of *z*)—*(continued)*

Positive *z*-Values

	0.00	0.01	0.02	0.03	0.04	0.05	0.06	0.07	0.08	0.09
0.0	0.5000	0.5040	0.5080	0.5120	0.5160	0.5199	0.5239	0.5279	0.5319	0.5359
0.1	0.5398	0.5438	0.5478	0.5517	0.5557	0.5596	0.5636	0.5675	0.5714	0.5753
0.2	0.5793	0.5832	0.5871	0.5910	0.5948	0.5987	0.6026	0.6064	0.6103	0.6141
0.3	0.6179	0.6217	0.6255	0.6293	0.6331	0.6368	0.6406	0.6443	0.6480	0.6517
0.4	0.6554	0.6591	0.6628	0.6664	0.6700	0.6736	0.6772	0.6808	0.6844	0.6879
0.5	0.6915	0.6950	0.6985	0.7019	0.7054	0.7088	0.7123	0.7157	0.7190	0.7224
0.6	0.7257	0.7291	0.7324	0.7357	0.7389	0.7422	0.7454	0.7486	0.7517	0.7549
0.7	0.7580	0.7611	0.7642	0.7673	0.7704	0.7734	0.7764	0.7794	0.7823	0.7852
0.8	0.7881	0.7910	0.7939	0.7967	0.7995	0.8023	0.8051	0.8078	0.8106	0.8133
0.9	0.8159	0.8186	0.8212	0.8238	0.8264	0.8289	0.8315	0.8340	0.8365	0.8389
1.0	0.8413	0.8438	0.8461	0.8485	0.8508	0.8531	0.8554	0.8577	0.8599	0.8621
1.1	0.8643	0.8665	0.8686	0.8708	0.8729	0.8749	0.8770	0.8790	0.8810	0.8830
1.2	0.8849	0.8869	0.8888	0.8907	0.8925	0.8944	0.8962	0.8980	0.8997	0.9015
1.3	0.9032	0.9049	0.9066	0.9082	0.9099	0.9115	0.9131	0.9147	0.9162	0.9177
1.4	0.9192	0.9207	0.9222	0.9236	0.9251	0.9265	0.9279	0.9292	0.9306	0.9319
1.5	0.9332	0.9345	0.9357	0.9370	0.9382	0.9394	0.9406	0.9418	0.9429	0.9441
1.6	0.9452	0.9463	0.9474	0.9484	0.9495	0.9505	0.9515	0.9525	0.9535	0.9545
1.7	0.9554	0.9564	0.9573	0.9582	0.9591	0.9599	0.9608	0.9616	0.9625	0.9633
1.8	0.9641	0.9649	0.9656	0.9664	0.9671	0.9678	0.9686	0.9693	0.9699	0.9706
1.9	0.9713	0.9719	0.9726	0.9732	0.9738	0.9744	0.9750	0.9756	0.9761	0.9767
2.0	0.9772	0.9778	0.9783	0.9788	0.9793	0.9798	0.9803	0.9808	0.9812	0.9817
2.1	0.9821	0.9826	0.9830	0.9834	0.9838	0.9842	0.9846	0.9850	0.9854	0.9857
2.2	0.9861	0.9864	0.9868	0.9871	0.9875	0.9878	0.9881	0.9884	0.9887	0.9890
2.3	0.9893	0.9896	0.9898	0.9901	0.9904	0.9906	0.9909	0.9911	0.9913	0.9916
2.4	0.9918	0.9920	0.9922	0.9925	0.9927	0.9929	0.9931	0.9932	0.9934	0.9936
2.5	0.9938	0.9940	0.9941	0.9943	0.9945	0.9946	0.9948	0.9949	0.9951	0.9952
2.6	0.9953	0.9955	0.9956	0.9957	0.9959	0.9960	0.9961	0.9962	0.9963	0.9964
2.7	0.9965	0.9966	0.9967	0.9968	0.9969	0.9970	0.9971	0.9972	0.9973	0.9974
2.8	0.9974	0.9975	0.9976	0.9977	0.9977	0.9978	0.9979	0.9979	0.9980	0.9981
2.9	0.9981	0.9982	0.9982	0.9983	0.9984	0.9984	0.9985	0.9985	0.9986	0.9986
3.0	0.9987	0.9987	0.9987	0.9988	0.9988	0.9989	0.9989	0.9989	0.9990	0.9990
3.1	0.9990	0.9991	0.9991	0.9991	0.9992	0.9992	0.9992	0.9992	0.9993	0.9993
3.2	0.9993	0.9993	0.9994	0.9994	0.9994	0.9994	0.9994	0.9995	0.9995	0.9995
3.3	0.9995	0.9995	0.9995	0.9996	0.9996	0.9996	0.9996	0.9996	0.9996	0.9997
3.4	0.9997	0.9997	0.9997	0.9997	0.9997	0.9997	0.9997	0.9997	0.9997	0.9998
3.5	0.9998									

TABLE D ■ Student's *t*-Distribution

Critical value for which the area under the curve in the right tail is α

	One-Tailed					
	0.005	0.01	0.025	0.05	0.1	0.25
	Two-Tailed					
d.f.	0.01	0.02	0.05	0.1	0.2	0.5
1	63.656	31.821	12.706	6.314	3.078	1.000
2	9.925	6.965	4.303	2.920	1.886	0.816
3	5.841	4.541	3.182	2.353	1.638	0.765
4	4.604	3.747	2.776	2.132	1.533	0.741
5	4.032	3.365	2.571	2.015	1.476	0.727
6	3.707	3.143	2.447	1.943	1.440	0.718
7	3.499	2.998	2.365	1.895	1.415	0.711
8	3.355	2.896	2.306	1.860	1.397	0.706
9	3.250	2.821	2.262	1.833	1.383	0.703
10	3.169	2.764	2.228	1.812	1.372	0.700
11	3.106	2.718	2.201	1.796	1.363	0.697
12	3.055	2.681	2.179	1.782	1.356	0.695
13	3.012	2.650	2.160	1.771	1.350	0.694
14	2.977	2.624	2.145	1.761	1.345	0.692
15	2.947	2.602	2.131	1.753	1.341	0.691
16	2.921	2.583	2.120	1.746	1.337	0.690
17	2.898	2.567	2.110	1.740	1.333	0.689
18	2.878	2.552	2.101	1.734	1.330	0.688
19	2.861	2.539	2.093	1.729	1.328	0.688
20	2.845	2.528	2.086	1.725	1.325	0.687
21	2.831	2.518	2.080	1.721	1.323	0.686
22	2.819	2.508	2.074	1.717	1.321	0.686
23	2.807	2.500	2.069	1.714	1.319	0.685
24	2.797	2.492	2.064	1.711	1.318	0.685
25	2.787	2.485	2.060	1.708	1.316	0.684
26	2.779	2.479	2.056	1.706	1.315	0.684
27	2.771	2.473	2.052	1.703	1.314	0.684
28	2.763	2.467	2.048	1.701	1.313	0.683
29	2.756	2.462	2.045	1.699	1.311	0.683
30	2.750	2.457	2.042	1.697	1.310	0.683
35	2.724	2.438	2.030	1.690	1.306	0.682
40	2.704	2.423	2.021	1.684	1.303	0.681
50	2.678	2.403	2.009	1.676	1.299	0.679
60	2.660	2.390	2.000	1.671	1.296	0.679
100	2.626	2.364	1.984	1.660	1.290	0.677
∞	2.576	2.326	1.960	1.645	1.282	0.674

TABLE E ■ Right-Tailed Critical Values for the *F*-Distribution (0.01 Level of Significance)

							Degrees of Freedom—Numerator												
	1	*2*	*3*	*4*	*5*	*6*	*7*	*8*	*9*	*10*	*12*	*15*	*20*	*24*	*30*	*40*	*60*	*120*	∞
1	4052	4999	5404	5624	5764	5859	5928	5981	6022	6056	6107	6157	6209	6234	6260	6286	6313	6340	6366
2	98.5	99.0	99.2	99.3	99.3	99.3	99.4	99.4	99.4	99.4	99.4	99.4	99.5	99.5	99.5	99.5	99.5	99.5	99.5
3	34.1	30.8	29.5	28.7	28.2	27.9	27.7	27.5	27.3	27.2	27.1	26.9	26.7	26.6	26.5	26.4	26.3	26.2	26.1
4	21.2	18.0	16.7	16.0	15.5	15.2	15.0	14.8	14.7	14.6	14.4	14.2	14.0	13.9	13.8	13.8	13.7	13.6	13.5
5	16.3	13.3	12.1	11.4	11.0	10.7	10.5	10.3	10.2	10.1	9.89	9.72	9.55	9.47	9.38	9.29	9.20	9.11	9.02
6	13.8	10.9	9.78	9.15	8.75	8.47	8.26	8.10	7.98	7.87	7.72	7.56	7.40	7.31	7.23	7.14	7.06	6.97	6.88
7	12.3	9.55	8.45	7.85	7.46	7.19	6.99	6.84	6.72	6.62	6.47	6.31	6.16	6.07	5.99	5.91	5.82	5.74	5.65
8	11.3	8.65	7.59	7.01	6.63	6.37	6.18	6.03	5.91	5.81	5.67	5.52	5.36	5.28	5.20	5.12	5.03	4.95	4.86
9	10.6	8.02	6.99	6.42	6.06	5.80	5.61	5.47	5.35	5.26	5.11	4.96	4.81	4.73	4.65	4.57	4.48	4.40	4.31
10	10.0	7.56	6.55	5.99	5.64	5.39	5.20	5.06	4.94	4.85	4.71	4.56	4.41	4.33	4.25	4.17	4.08	4.00	3.91
11	9.65	7.21	6.22	5.67	5.32	5.07	4.89	4.74	4.63	4.54	4.40	4.25	4.10	4.02	3.94	3.86	3.78	3.69	3.60
12	9.33	6.93	5.95	5.41	5.06	4.82	4.64	4.50	4.39	4.30	4.16	4.01	3.86	3.78	3.70	3.62	3.54	3.45	3.36
13	9.07	6.70	5.74	5.21	4.86	4.62	4.44	4.30	4.19	4.10	3.96	3.82	3.66	3.59	3.51	3.43	3.34	3.25	3.17
14	8.86	6.51	5.56	5.04	4.69	4.46	4.28	4.14	4.03	3.94	3.80	3.66	3.51	3.43	3.35	3.27	3.18	3.09	3.00
15	8.68	6.36	5.42	4.89	4.56	4.32	4.14	4.00	3.89	3.80	3.67	3.52	3.37	3.29	3.21	3.13	3.05	2.96	2.87
16	8.53	6.23	5.29	4.77	4.44	4.20	4.03	3.89	3.78	3.69	3.55	3.41	3.26	3.18	3.10	3.02	2.93	2.84	2.75
17	8.40	6.11	5.19	4.67	4.34	4.10	3.93	3.79	3.68	3.59	3.46	3.31	3.16	3.08	3.00	2.92	2.83	2.75	2.65
18	8.29	6.01	5.09	4.58	4.25	4.01	3.84	3.71	3.60	3.51	3.37	3.23	3.08	3.00	2.92	2.84	2.75	2.66	2.57
19	8.18	5.93	5.01	4.50	4.17	3.94	3.77	3.63	3.52	3.43	3.30	3.15	3.00	2.92	2.84	2.76	2.67	2.58	2.49
20	8.10	5.85	4.94	4.43	4.10	3.87	3.70	3.56	3.46	3.37	3.23	3.09	2.94	2.86	2.78	2.69	2.61	2.52	2.42
21	8.02	5.78	4.87	4.37	4.04	3.81	3.64	3.51	3.40	3.31	3.17	3.03	2.88	2.80	2.72	2.64	2.55	2.46	2.36
22	7.95	5.72	4.82	4.31	3.99	3.76	3.59	3.45	3.35	3.26	3.12	2.98	2.83	2.75	2.67	2.58	2.50	2.40	2.31
23	7.88	5.66	4.76	4.26	3.94	3.71	3.54	3.41	3.30	3.21	3.07	2.93	2.78	2.70	2.62	2.54	2.45	2.35	2.26
24	7.82	5.61	4.72	4.22	3.90	3.67	3.50	3.36	3.26	3.17	3.03	2.89	2.74	2.66	2.58	2.49	2.40	2.31	2.21
25	7.77	5.57	4.68	4.18	3.85	3.63	3.46	3.32	3.22	3.13	2.99	2.85	2.70	2.62	2.54	2.45	2.36	2.27	2.17
30	7.56	5.39	4.51	4.02	3.70	3.47	3.30	3.17	3.07	2.98	2.84	2.70	2.55	2.47	2.39	2.30	2.21	2.11	2.01
40	7.31	5.18	4.31	3.83	3.51	3.29	3.12	2.99	2.89	2.80	2.66	2.52	2.37	2.29	2.20	2.11	2.02	1.92	1.80
60	7.08	4.98	4.13	3.65	3.34	3.12	2.95	2.82	2.72	2.63	2.50	2.35	2.20	2.12	2.03	1.94	1.84	1.73	1.60
120	6.85	4.79	3.95	3.48	3.17	2.96	2.79	2.66	2.56	2.47	2.34	2.19	2.03	1.95	1.86	1.76	1.66	1.53	1.38
∞	6.63	4.61	3.78	3.32	3.02	2.80	2.64	2.51	2.41	2.32	2.18	2.04	1.88	1.79	1.70	1.59	1.47	1.32	1.00

Degrees of Freedom—Denominator

TABLE E ■ Right-Tailed Critical Values for the *F*-Distribution (0.05 Level of Significance)

<table>
<thead>
<tr><th rowspan="2"></th><th colspan="19">Degrees of Freedom—Numerator</th></tr>
<tr><th>1</th><th>2</th><th>3</th><th>4</th><th>5</th><th>6</th><th>7</th><th>8</th><th>9</th><th>10</th><th>12</th><th>15</th><th>20</th><th>24</th><th>30</th><th>40</th><th>60</th><th>120</th><th>∞</th></tr>
</thead>
<tbody>
<tr><td>1</td><td>161</td><td>200</td><td>216</td><td>225</td><td>230</td><td>234</td><td>237</td><td>239</td><td>241</td><td>242</td><td>244</td><td>246</td><td>248</td><td>249</td><td>250</td><td>251</td><td>252</td><td>253</td><td>254</td></tr>
<tr><td>2</td><td>18.5</td><td>19.0</td><td>19.2</td><td>19.3</td><td>19.3</td><td>19.3</td><td>19.4</td><td>19.4</td><td>19.4</td><td>19.4</td><td>19.4</td><td>19.4</td><td>19.5</td><td>19.5</td><td>19.5</td><td>19.5</td><td>19.5</td><td>19.5</td><td>19.5</td></tr>
<tr><td>3</td><td>10.1</td><td>9.55</td><td>9.28</td><td>9.12</td><td>9.01</td><td>8.94</td><td>8.89</td><td>8.85</td><td>8.81</td><td>8.79</td><td>8.74</td><td>8.70</td><td>8.66</td><td>8.64</td><td>8.62</td><td>8.59</td><td>8.57</td><td>8.55</td><td>8.53</td></tr>
<tr><td>4</td><td>7.71</td><td>6.94</td><td>6.59</td><td>6.39</td><td>6.26</td><td>6.16</td><td>6.09</td><td>6.04</td><td>6.00</td><td>5.96</td><td>5.91</td><td>5.86</td><td>5.80</td><td>5.77</td><td>5.75</td><td>5.72</td><td>5.69</td><td>5.66</td><td>5.63</td></tr>
<tr><td>5</td><td>6.61</td><td>5.79</td><td>5.41</td><td>5.19</td><td>5.05</td><td>4.95</td><td>4.88</td><td>4.82</td><td>4.77</td><td>4.74</td><td>4.68</td><td>4.62</td><td>4.56</td><td>4.53</td><td>4.50</td><td>4.46</td><td>4.43</td><td>4.40</td><td>4.37</td></tr>
<tr><td>6</td><td>5.99</td><td>5.14</td><td>4.76</td><td>4.53</td><td>4.39</td><td>4.28</td><td>4.21</td><td>4.15</td><td>4.10</td><td>4.06</td><td>4.00</td><td>3.94</td><td>3.87</td><td>3.84</td><td>3.81</td><td>3.77</td><td>3.74</td><td>3.70</td><td>3.67</td></tr>
<tr><td>7</td><td>5.59</td><td>4.74</td><td>4.35</td><td>4.12</td><td>3.97</td><td>3.87</td><td>3.79</td><td>3.73</td><td>3.68</td><td>3.64</td><td>3.57</td><td>3.51</td><td>3.44</td><td>3.41</td><td>3.38</td><td>3.34</td><td>3.30</td><td>3.27</td><td>3.23</td></tr>
<tr><td>8</td><td>5.32</td><td>4.46</td><td>4.07</td><td>3.84</td><td>3.69</td><td>3.58</td><td>3.50</td><td>3.44</td><td>3.39</td><td>3.35</td><td>3.28</td><td>3.22</td><td>3.15</td><td>3.12</td><td>3.08</td><td>3.04</td><td>3.01</td><td>2.97</td><td>2.93</td></tr>
<tr><td>9</td><td>5.12</td><td>4.26</td><td>3.86</td><td>3.63</td><td>3.48</td><td>3.37</td><td>3.29</td><td>3.23</td><td>3.18</td><td>3.14</td><td>3.07</td><td>3.01</td><td>2.94</td><td>2.90</td><td>2.86</td><td>2.83</td><td>2.79</td><td>2.75</td><td>2.71</td></tr>
<tr><td>10</td><td>4.96</td><td>4.10</td><td>3.71</td><td>3.48</td><td>3.33</td><td>3.22</td><td>3.14</td><td>3.07</td><td>3.02</td><td>2.98</td><td>2.91</td><td>2.85</td><td>2.77</td><td>2.74</td><td>2.70</td><td>2.66</td><td>2.62</td><td>2.58</td><td>2.54</td></tr>
<tr><td>11</td><td>4.84</td><td>3.98</td><td>3.59</td><td>3.36</td><td>3.20</td><td>3.09</td><td>3.01</td><td>2.95</td><td>2.90</td><td>2.85</td><td>2.79</td><td>2.72</td><td>2.65</td><td>2.61</td><td>2.57</td><td>2.53</td><td>2.49</td><td>2.45</td><td>2.40</td></tr>
<tr><td>12</td><td>4.75</td><td>3.89</td><td>3.49</td><td>3.26</td><td>3.11</td><td>3.00</td><td>2.91</td><td>2.85</td><td>2.80</td><td>2.75</td><td>2.69</td><td>2.62</td><td>2.54</td><td>2.51</td><td>2.47</td><td>2.43</td><td>2.38</td><td>2.34</td><td>2.30</td></tr>
<tr><td>13</td><td>4.67</td><td>3.81</td><td>3.41</td><td>3.18</td><td>3.03</td><td>2.92</td><td>2.83</td><td>2.77</td><td>2.71</td><td>2.67</td><td>2.60</td><td>2.53</td><td>2.46</td><td>2.42</td><td>2.38</td><td>2.34</td><td>2.30</td><td>2.25</td><td>2.21</td></tr>
<tr><td>14</td><td>4.60</td><td>3.74</td><td>3.34</td><td>3.11</td><td>2.96</td><td>2.85</td><td>2.76</td><td>2.70</td><td>2.65</td><td>2.60</td><td>2.53</td><td>2.46</td><td>2.39</td><td>2.35</td><td>2.31</td><td>2.27</td><td>2.22</td><td>2.18</td><td>2.13</td></tr>
<tr><td>15</td><td>4.54</td><td>3.68</td><td>3.29</td><td>3.06</td><td>2.90</td><td>2.79</td><td>2.71</td><td>2.64</td><td>2.59</td><td>2.54</td><td>2.48</td><td>2.40</td><td>2.33</td><td>2.29</td><td>2.25</td><td>2.20</td><td>2.16</td><td>2.11</td><td>2.07</td></tr>
<tr><td>16</td><td>4.49</td><td>3.63</td><td>3.24</td><td>3.01</td><td>2.85</td><td>2.74</td><td>2.66</td><td>2.59</td><td>2.54</td><td>2.49</td><td>2.42</td><td>2.35</td><td>2.28</td><td>2.24</td><td>2.19</td><td>2.15</td><td>2.11</td><td>2.06</td><td>2.01</td></tr>
<tr><td>17</td><td>4.45</td><td>3.59</td><td>3.20</td><td>2.96</td><td>2.81</td><td>2.70</td><td>2.61</td><td>2.55</td><td>2.49</td><td>2.45</td><td>2.38</td><td>2.31</td><td>2.23</td><td>2.19</td><td>2.15</td><td>2.10</td><td>2.06</td><td>2.01</td><td>1.96</td></tr>
<tr><td>18</td><td>4.41</td><td>3.55</td><td>3.16</td><td>2.93</td><td>2.77</td><td>2.66</td><td>2.58</td><td>2.51</td><td>2.46</td><td>2.41</td><td>2.34</td><td>2.27</td><td>2.19</td><td>2.15</td><td>2.11</td><td>2.06</td><td>2.02</td><td>1.97</td><td>1.92</td></tr>
<tr><td>19</td><td>4.38</td><td>3.52</td><td>3.13</td><td>2.90</td><td>2.74</td><td>2.63</td><td>2.54</td><td>2.48</td><td>2.42</td><td>2.38</td><td>2.31</td><td>2.23</td><td>2.16</td><td>2.11</td><td>2.07</td><td>2.03</td><td>1.98</td><td>1.93</td><td>1.88</td></tr>
<tr><td>20</td><td>4.35</td><td>3.49</td><td>3.10</td><td>2.87</td><td>2.71</td><td>2.60</td><td>2.51</td><td>2.45</td><td>2.39</td><td>2.35</td><td>2.28</td><td>2.20</td><td>2.12</td><td>2.08</td><td>2.04</td><td>1.99</td><td>1.95</td><td>1.90</td><td>1.84</td></tr>
<tr><td>21</td><td>4.32</td><td>3.47</td><td>3.07</td><td>2.84</td><td>2.68</td><td>2.57</td><td>2.49</td><td>2.42</td><td>2.37</td><td>2.32</td><td>2.25</td><td>2.18</td><td>2.10</td><td>2.05</td><td>2.01</td><td>1.96</td><td>1.92</td><td>1.87</td><td>1.81</td></tr>
<tr><td>22</td><td>4.30</td><td>3.44</td><td>3.05</td><td>2.82</td><td>2.66</td><td>2.55</td><td>2.46</td><td>2.40</td><td>2.34</td><td>2.30</td><td>2.23</td><td>2.15</td><td>2.07</td><td>2.03</td><td>1.98</td><td>1.94</td><td>1.89</td><td>1.84</td><td>1.78</td></tr>
<tr><td>23</td><td>4.28</td><td>3.42</td><td>3.03</td><td>2.80</td><td>2.64</td><td>2.53</td><td>2.44</td><td>2.37</td><td>2.32</td><td>2.27</td><td>2.20</td><td>2.13</td><td>2.05</td><td>2.01</td><td>1.96</td><td>1.91</td><td>1.86</td><td>1.81</td><td>1.76</td></tr>
<tr><td>24</td><td>4.26</td><td>3.40</td><td>3.01</td><td>2.78</td><td>2.62</td><td>2.51</td><td>2.42</td><td>2.36</td><td>2.30</td><td>2.25</td><td>2.18</td><td>2.11</td><td>2.03</td><td>1.98</td><td>1.94</td><td>1.89</td><td>1.84</td><td>1.79</td><td>1.73</td></tr>
<tr><td>25</td><td>4.24</td><td>3.39</td><td>2.99</td><td>2.76</td><td>2.60</td><td>2.49</td><td>2.40</td><td>2.34</td><td>2.28</td><td>2.24</td><td>2.16</td><td>2.09</td><td>2.01</td><td>1.96</td><td>1.92</td><td>1.87</td><td>1.82</td><td>1.77</td><td>1.71</td></tr>
<tr><td>30</td><td>4.17</td><td>3.32</td><td>2.92</td><td>2.69</td><td>2.53</td><td>2.42</td><td>2.33</td><td>2.27</td><td>2.21</td><td>2.16</td><td>2.09</td><td>2.01</td><td>1.93</td><td>1.89</td><td>1.84</td><td>1.79</td><td>1.74</td><td>1.68</td><td>1.62</td></tr>
<tr><td>40</td><td>4.08</td><td>3.23</td><td>2.84</td><td>2.61</td><td>2.45</td><td>2.34</td><td>2.25</td><td>2.18</td><td>2.12</td><td>2.08</td><td>2.00</td><td>1.92</td><td>1.84</td><td>1.79</td><td>1.74</td><td>1.69</td><td>1.64</td><td>1.58</td><td>1.51</td></tr>
<tr><td>60</td><td>4.00</td><td>3.15</td><td>2.76</td><td>2.53</td><td>2.37</td><td>2.25</td><td>2.17</td><td>2.10</td><td>2.04</td><td>1.99</td><td>1.92</td><td>1.84</td><td>1.75</td><td>1.70</td><td>1.65</td><td>1.59</td><td>1.53</td><td>1.47</td><td>1.39</td></tr>
<tr><td>120</td><td>3.92</td><td>3.07</td><td>2.68</td><td>2.45</td><td>2.29</td><td>2.18</td><td>2.09</td><td>2.02</td><td>1.96</td><td>1.91</td><td>1.83</td><td>1.75</td><td>1.66</td><td>1.61</td><td>1.55</td><td>1.50</td><td>1.43</td><td>1.35</td><td>1.25</td></tr>
<tr><td>∞</td><td>3.84</td><td>3.00</td><td>2.60</td><td>2.37</td><td>2.21</td><td>2.10</td><td>2.01</td><td>1.94</td><td>1.88</td><td>1.83</td><td>1.75</td><td>1.67</td><td>1.57</td><td>1.52</td><td>1.46</td><td>1.39</td><td>1.32</td><td>1.22</td><td>1.00</td></tr>
</tbody>
</table>

Degrees of Freedom—Denominator

TABLE F ■ Right-Tailed Critical Values for the Chi-Square (χ^2) Distribution

d.f.	0.10	0.05	0.02	0.01
			α	
1	2.706	3.841	5.412	6.635
2	4.605	5.991	7.824	9.210
3	6.251	7.815	9.837	11.345
4	7.779	9.488	11.668	13.277
5	9.236	11.070	13.388	15.086
6	10.645	12.592	15.033	16.812
7	12.017	14.067	16.622	18.475
8	13.362	15.507	18.168	20.090
9	14.684	16.919	19.679	21.666
10	15.987	18.307	21.161	23.209
11	17.275	19.675	22.618	24.725
12	18.549	21.026	24.054	26.217
13	19.812	22.362	25.471	27.688
14	21.064	23.685	26.873	29.141
15	22.307	24.996	28.259	30.578
16	23.542	26.296	29.633	32.000
17	24.769	27.587	30.995	33.409
18	25.989	28.869	32.346	34.805
19	27.204	30.144	33.687	36.191
20	28.412	31.410	35.020	37.566
21	29.615	32.671	36.343	38.932
22	30.813	33.924	37.659	40.289
23	32.007	35.172	38.968	41.638
24	33.196	36.415	40.270	42.980
25	34.382	37.652	41.566	44.314
26	35.563	38.885	42.856	45.642
27	36.741	40.113	44.140	46.963
28	37.916	41.337	45.419	48.278
29	39.087	42.557	46.693	49.588
30	40.256	43.773	47.962	50.892
40	51.805	55.758	60.436	63.691
50	63.167	67.505	72.613	76.154
60	74.397	79.082	84.580	88.379
70	85.527	90.531	96.387	100.425
80	96.578	101.879	108.069	112.329
90	107.565	113.145	119.648	124.116
100	118.498	124.342	131.142	135.807

TABLE G ■ Critical Values for the Wilcoxon Signed-Rank Test

Reject the null hypothesis for values of W that are less than or equal to the critical value listed in the table.

	One-Tailed α			
	0.005	0.01	0.025	0.05
	Two-Tailed α			
	0.01	0.02	0.05	0.10
n				
5				1
6			1	2
7		0	2	4
8	0	2	4	6
9	2	3	6	8
10	3	5	8	11
11	5	7	11	14
12	7	10	14	17
13	10	13	17	21
14	13	16	21	26
15	16	20	25	30
16	19	24	30	36
17	23	28	35	41
18	28	33	40	47
19	32	38	46	54
20	37	43	52	60
21	43	49	59	68
22	49	56	66	75
23	55	62	73	83
24	61	69	81	92
25	68	77	90	101
26	76	85	98	110
27	84	93	107	120
28	92	102	117	130
29	100	111	127	141
30	109	120	137	152
31	118	130	148	163
32	128	141	159	175
33	138	151	171	188
34	149	162	183	201
35	160	174	195	214
36	171	186	208	228
37	183	198	222	242
38	195	211	235	256
39	208	224	250	271
40	221	238	264	287
41	234	252	279	303
42	248	267	295	319
43	262	281	311	336
44	277	297	327	353
45	292	313	344	371
46	307	329	361	389
47	323	345	379	408
48	339	362	397	427
49	356	380	415	446
50	373	398	434	466

TABLE H ■ Critical Values for Spearman's Rank Correlation Coefficient

	One-Tailed α			
	0.005	*0.01*	*0.025*	*0.05*
	Two-Tailed α			
n	*0.01*	*0.02*	*0.05*	*0.1*
5				0.900
6		0.943	0.886	0.829
7		0.893	0.786	0.714
8	0.881	0.833	0.738	0.643
9	0.833	0.783	0.700	0.600
10	0.794	0.745	0.648	0.564
11	0.818	0.736	0.618	0.536
12	0.780	0.703	0.591	0.497
13	0.745	0.673	0.566	0.475
14	0.716	0.646	0.545	0.457
15	0.689	0.623	0.525	0.441
16	0.666	0.601	0.507	0.425
17	0.645	0.582	0.490	0.412
18	0.625	0.564	0.476	0.399
19	0.608	0.549	0.462	0.388
20	0.591	0.534	0.450	0.377
21	0.576	0.521	0.438	0.368
22	0.562	0.508	0.428	0.359
23	0.549	0.496	0.418	0.351
24	0.537	0.485	0.409	0.343
25	0.526	0.475	0.400	0.336
26	0.515	0.465	0.392	0.329
27	0.505	0.456	0.385	0.323
28	0.496	0.448	0.377	0.317
29	0.487	0.440	0.370	0.311
30	0.478	0.432	0.364	0.305

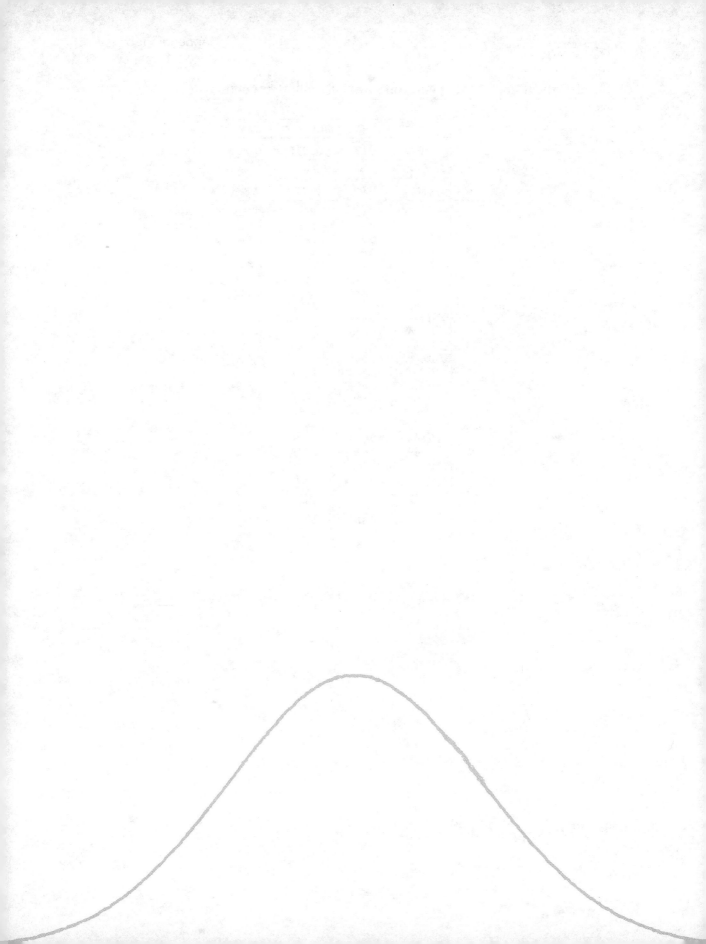

Answers to Odd-Numbered Exercises

SECTION 1.1

1. Descriptive; fact
3. Inferential; predicting 6 years away
5. Inferential; not presented
7. During the three seasons, the AL never had more than 141 complete games.
9. Descriptive; fact
11. Inferential; beyond the data
13. Inferential; prediction
15. Inferential; prediction
17. Nominal level
19. Interval level; continuous
21. Ordinal level
23. Ratio level; discrete
25. Interval level; continuous
27. Ratio level; continuous
29. Interval level; discrete
31. Ordinal level
33. Interval level; continuous
35. Ratio level; discrete
37. Cluster sampling; may not be representative
39. Cluster sampling; may not be representative
41. Random sampling; self-reporting/ placebo effect
43. Randomly select certain floors of the dorms and ask each person.
45. Divide the state into groups based on gender, party affiliation, income, race, work status.

SECTION 1.2

1.

Stem	Leaf
14	0 3 4
15	3 8
16	7 7
17	1 5 5 6 6 7
18	0 2 7
19	2 2 2 9
20	1 3
21	
22	1 7
23	0

3.

Stem	Leaf
47	9
48	2 7
49	3 4 4 5 6 7 8 8 9
50	3 4 7 7 8
51	0 1 2 4 6
52	0 4 5 5 7
53	4 6
54	2 5 6 7 9
55	7 9
56	1 1 3 3 7 8 9
57	0 2 8
58	4 5 6
59	4 4

5.

Stem	Leaf
5	69 83
6	05 49 61 82 83 84 91 92 99
7	19 32 39 45 89 93
8	36 59 60 64

7.

Stem	Leaf
13	1 4
14	1 7
15	3 3 3 6 9
16	5 6 7
17	2 6
18	1 3 8
19	6
20	
21	0
22	
23	
24	0

9. Missing stems

11.

Republicans	Frequency
140–160	5
160–180	8
180–200	7
200–220	2
220–240	3

13.

SAT	Frequency
460–480	1
480–500	10
500–520	10
520–540	7
540–560	7
560–580	10
580–600	5

15.

Age	Frequency	Relative Frequency
40–45	2	0.05
45–50	6	0.14
50–55	12	0.29
55–60	12	0.29
60–65	7	0.17
65–70	3	0.07

17.

Earthquakes	Frequency	Relative Frequency
5–10	10	0.101
10–15	19	0.192
15–20	19	0.192
20–25	18	0.182
25–30	13	0.131
30–35	5	0.051
35–40	4	0.040
40–45	1	0.010

19.

Price($)	Frequency	Less-Than	Greater-Than
0–5	2	2	50
5–10	21	23	48
10–15	12	35	27
15–20	4	39	15
20–25	8	47	11
25–30	3	50	3

21.

Income	Frequency	Less-Than	Greater-Than
$25k–$30k	3	3	51
$30k–$35k	13	16	48
$35k–$40k	18	34	35
$40k–$45k	10	44	17
$45k–$50k	7	51	7

23.

Number of Cigarettes	Frequency
1–6	95
6–11	161
11–16	25
16–21	100
21–31	13
31–41	6

25.

SAT Verbal Score	Frequency
200–300	6
300–400	26
400–500	62
500–600	64
600–700	32
700–800	10

SECTION 2.1

1. 44.8
3. 73.9
5. Q_1: 44; Med.: 47.5; Q_3: 55
7. Q_1: 41; Med.: 60; Q_3: 85
9. 66
11. 78.5
13. Mode: 66.7, 67.3; Mid.: 69.05
15. $\bar{x} = 53.25$; Med.: 57
17. Med.: 77.5; $\bar{x} = 77.9$
19. $\bar{x} = 39.3$; Mode: 38
21. $\bar{x} = 719$; Med.: 695.5; Q_1: 682; Q_3: 789; Mid.: 716.5; Mode: 684
23. $\bar{x} = 175.3$; Med.: 171; Q_1: 157.5; Q_3: 193.5; Mid.: 181.5; Mode: 154, 194
25. $\bar{x} = 54.9$; Med.: 55; Q_1: 51; Q_3: 58; Mid.: 55.5; Mode: 51
27. $\bar{x} = 45.7$; Med.: 46.5; Q_1: 37.5; Q_3: 52; Mid.: 47; Mode: 57
29. $\bar{x} = 505$; Med.: 505; Q_1: 415; Q_3: 585; Mid.: 540; Mode: 590
31. $\bar{x} = 432.5$; Med.: 424.5; Q_1: 374; Q_3: 495; Mid.: 416; Mode: None
33. $\bar{x} = 12.3$; Med.: 6; Q_1: 3.5; Q_3: 7.8; Mid.: 50.75; Mode: 6, 6.5
35. (a) 262.8   (b) 323.3, 180, 263.3, 240.7
    (c) No, use weighted mean.
37. 1, 2, 3, 4, 5, 30
39. 50, 60, 70, 80, 90, 100
41. 8, 9, 10, 20, 30, 40
43. 1, 2, 3, 4, 5, 6
45. (a) $\bar{x} = 2244.2$; Med.: 2263
    (b) $\bar{x} = 2391.6$; Med.: 2534
47. 1.7
49. 76.7
51. 68
53. 43.8
55. (a)   1023.25
    (b)

Score	Frequency
550–750	3
750–950	10
950–1150	17
1150–1350	9
1350–1550	1

    (c) 1025
57. (a) AL: 776.3; NL: 733.4
    (b) AL: 790; NL: 733.5
    (c) Perhaps more earned runs in AL

SECTION 2.2

1. 71
3. 39
5. 11
7. 44
13. 2.32
15. 7.875

17. Var: 9.3; S.D.: 3.05
19. Var: 123.44; S.D.: 11.11
21. Var: 103.27; S.D.: 10.16
23. Var: 61.01; S.D.: 7.81
25. (a) 208   (b) 97   (d) 52.59   (e) 66.09
    (f) 4368
27. (a) 83   (b) 34.5   (d) 19.47   (e) 23.41
    (f) 547.91
29. (a) 73   (b) 31.5   (d) 17.08   (e) 20.65
    (f) 426.58
31. (a) 10,445   (b) 3533   (d) 2496.58
    (e) 3300.66   (f) 10,894,355.79
33. (a) $38.20   (b) $7.10   (d) $7.09   (e) $10.10
    (f) 101.91
35. (a) 17,650   (b) 3820.5   (d) 2468.46
    (e) 3329.85   (f) 11,087,094.36
37. 0.95, positive, moderate
39. −0.83, negative, moderate
41. 0.81, positive, moderate
43. Outliers are high-priced homes.
45. Negatively skewed
47. Positively skewed (stretched to "high end")
49. Negatively skewed
51.

CT	ME	MA	NH	RI	VT
−0.06	−0.28	2.18	−0.42	−0.88	−0.52

53. 1.23, −0.54, 0.77, −0.86, 0.06, −0.73, 0.16, 0.86, −1.91, 0.95
55. At least 88.9%
57. At least 55.6%
59. At least 67.3%
61. At least 12.7%
67. (a) Range: 294; I.Q. range: 60.5; S.D.: 75.24
    (b) Range: 212; I.Q. range: 48; S.D.: 54.77
    (c) The range and standard deviation are greatly affected, whereas the interquartile range is affected in a minor way.

SECTION 3.1

1. (a) 1   (b) 4   (c) 6   (d) 4   (e) 1
3. (a) 5   (b) 8   (c) 26   (d) 24
5. 105
7. 310
9. 210
11. (a) 1024   (b) 1   (c) 243
13. 90
15. 560
17. 59,280
19. 220
21. 18,009,460
23. 1287
25. 220
27. 310,080

SECTION 3.2

1. 1/6
3. 1/2
5. 1/12
7. 2/9
9. 1/2
11. 1/2
13. 1/13
15. 1/3
17. 1/5
19. 1
21. 0
23. 1
25. 2/13
27. 1/2
29. 0.55
31. 0.15
33. 1
35. 1
37. 12/13
39. 1/2
41. 0.75
43. 0.65
45. 4/13
47. 13/36
49. 0.95

SECTION 3.3

1. 3/5
3. 11/20
5. 9/20
7. 11/100
9. 7/10
11. 173/200
13. 3/4
15. 1/4
17. 9/11
19. 22/75
21. 12/25
23. 51/100
25. 1/3
27. 1/2
29. 49/52
31. 3/11
33. 1/9
35. 1/3
37. 10/49
39. 1/4
41. 0.25
43. 0.625
45. 0.36

SECTION 3.4

1. Not independent
3. Independent
5. 1/8
7. 1/8
9. 1/36
11. 0.0001889568
13. 0.2649
15. 1/17
17. 1/221
19. 13/204
21. 13/102
23. 11/850
25. 13/850
27. 1/16; there are more diamonds in the deck for the second card.
29. 1/8
31. 77/13455

SECTION 4.1

1. (a) 0.3750
   (b) 0.6875
   (c) 0.9375
   (d) $\mu = 2$; We should expect two heads.

3. (a) 0.4 (b) 0.35 (c) $\mu = 1.1$
   The child should expect to sell 1.1 candy bars.

5.

x	P(x)
2	1/36
3	2/36
4	3/36
5	4/36
6	5/36
7	6/36
8	5/36
9	4/36
10	3/36
11	2/36
12	1/36

(a) $\frac{26}{36} = \frac{13}{18}$ (b) $\frac{6}{36} = \frac{1}{6}$ (c) $\frac{18}{36} = \frac{1}{2}$ (d) $\mu = 7$
We should expect a sum of 7.

7. Since the probabilities have to add up to equal 1, the missing probability is 0.13. $\mu = 4.5$

9. $\mu = 1.439$
   $\sigma = 1.0854$

11. $\mu = 1.400$
    $\sigma = 1.0586$

13. $\mu = 2.100$
    $\sigma = 1.4500$

15. $\mu = 0.700$
    $\sigma = 0.8371$

SECTION 4.2

1. 0.1460		13. 0.9804	
3. 0.1954		15. 0.6562	
5. 0.5139		17. 0.1876	
7. 0.9806		19. 0.0012	
9. 0.4487		21. 0.7623	
11. 0.8723		23. 0.9520	

SECTION 4.3

1.

Number of Hits	Probability
0	0.3430
1	0.3670
2	0.1964
3	0.0700
4	0.0187
5	0.0040
6	0.0007
7	0.0001

3.

Number of Fires	Probability
0	0.7408
1	0.2222
2	0.0333
3	0.0033
4	0.0003

5. 0.1633
7. 0.5372

9. (a) 0.0498 (b) 0.6160 (c) 0.0840
11. (a) 0.0916 (b) 0.0996 (c) 0.1912 (d) 0.6247
13. 0.1527
15. 0.0037
17. 0.1848
19. 0.1157
21. 0.0839

23.

Number of Accidents	Probability	Expected Frequency	Actual Frequency
0	0.2231	8.924	6
1	0.3347	13.388	11
2	0.2510	10.040	12
3	0.1255	5.020	6
4	0.0471	1.884	2
5	0.0141	0.564	2
6	0.0035	0.140	1
7	0.0008	0.032	0
8	0.0001	0.004	0

The number of accidents may follow a Poisson distribution with a mean of 1.5 accidents per day, as each expected frequency is within 3 of the actual frequency.

25. 1.34
27. 2.68

SECTION 5.1

1. 0.8944	29. 0.0994
3. 0.0154	31. 0.9115
5. 0.8577	33. 0.0143
7. 0.0099	35. 0.3821
9. 0.6664	37. 0.4531
11. 0.4641	39. 0.0139
13. 0.0819	41. 0
15. 0.1090	43. 0.1762
17. 0.9544	45. 0.1430
19. 0	47. 0.5000
21. 0.1423	49. 0.0668
23. 0.6406	51. 0.0336
25. 0.2611	53. 0.0097
27. 0.0002	

SECTION 5.2

1. 0.1003   3. 0.3612
5. 72.91%. The difference (2.91%) may be due to the fact that these scores are *approximately* normally distributed.

7. 0.6664	17. 0.7823
9. 0.0968	19. 16.85%
11. 0.5398	21. 0.0516
13. 0.0129	23. 0.3563
15. 0.7193	

25. If the random variable x is discrete, then we must make a continuity correction.
27. $\mu = 250$, $\sigma = 11.18$, 0.0409
29. $\mu = 250$, $\sigma = 11.18$, 0.0143

31. $\mu = 250$, $\sigma = 11.18$, 0.3110
33. $\mu = 250$, $\sigma = 11.18$, 0.3446
35. $\mu = 250$, $\sigma = 11.18$, 0.3758
37. $\mu = 180$, $\sigma = 8.49$, 0.0436
39. $\mu = 180$, $\sigma = 8.49$, 0.0019
41. $\mu = 120$, $\sigma = 8.49$, 0.1314
43. $\mu = 16.32$, $\sigma = 2.83$
    The probability is less than 0.0002, or 1 in 5000.
45. $\mu = 11$, $\sigma = 2.97$, 0.0643
47. $\mu = 43.2$, $\sigma = 5.88$, 0.0495

SECTION 5.3

1. –2.33
3. –1.28
5. 1.88
7. 1.04
9. –0.25
11. Between –1.28 and 1.28
13. Between –1.81 and 1.81
15. –0.67
17. 65.052 inches
19. 71.912 inches
21. Between 66.088 inches and 71.912 inches
23. 74.264 inches
25. 30.328 inches
27. 28.996 inches
29. The lowest score that will be sent a letter is 61 points.
31. The highest score to be invited is 39 points.
33. 103.75 points (104 or above)
35. The lowest possible IQ is 131 points.
37. 211.6 points (212 or above)
39. 271.6 points (272 or above)
41. 722.56 points (730 or above)
43. Between 417.92 points and 606.08 points (420 through 600)

SECTION 6.1

1. (a) 3.5 (b) 1.71 (c) 3.5 (d) 1.21

(e)

	1	2	3	4	5	6
1	1.0	1.5	2.0	2.5	3.0	3.5
2	1.5	2.0	2.5	3.0	3.5	4.0
3	2.0	2.5	3.0	3.5	4.0	4.5
4	2.5	3.0	3.5	4.0	4.5	5.0
5	3.0	3.5	4.0	4.5	5.0	5.5
6	3.5	4.0	4.5	5.0	5.5	6.0

(f) $\mu_{\bar{x}} = 3.5$, $\sigma_{\bar{x}} = 1.21$
    Yes, they are the same.
3. (a) 69.0 (b) 0.23
5. (a) 512 (b) 17.71
7. 0.9998
9. 0.0002
11. 0.0002
13. 0.0002
15. 0.9912

17. (a) 0.0104
    (b) The chances of obtaining a sample that is at least this extreme is so small; there is a strong possibility that the sample is not representative of the population.
    (c) Answers vary.
19. $n = 49 : 0.1762$
    $n = 121 : 0.0708$
    $n = 400 : 0.0038$

SECTION 6.2

1. $25.4 - 1.92 \leq \mu \leq 25.4 + 1.92$
    $23.48 \leq \mu \leq 27.32$
3. (a) $3.2 - 0.78 \leq \mu \leq 3.2 + 0.78$
        $2.42 \leq \mu \leq 3.98$
5. $28.3 - 0.79 \leq \mu \leq 28.3 + 0.79$
    $27.51 \leq \mu \leq 29.09$
7. $515 - 20.0 \leq \mu \leq 515 + 20.0$
    $495.0 \leq \mu \leq 535.0$
9. (a) $11.9 - 3.81 \leq \mu \leq 11.9 + 3.81$
        $8.09 \leq \mu \leq 15.71$
    (b) Yes, 15 miles is contained in the interval.
11. (a) $22.4 - 0.75 \leq \mu \leq 22.4 + 0.75$
        $21.65 \leq \mu \leq 23.15$
    (b) No, 25 years is higher than the upper limit of the interval.
13. $5235 - 440.03 \leq \mu \leq 5235 + 440.03$
    $\$4794.97 \leq \mu \leq \$5675.03$
15. $64.5 - 1.35 \leq \mu \leq 64.5 + 1.35$
    $63.15 \leq \mu \leq 65.85$
17. $8.9 - 0.54 \leq \mu \leq 8.9 + 0.54$
    $8.36 \leq \mu \leq 9.44$
19. $527 - 51.75 \leq \mu \leq 527 + 51.75$
    $\$475.25 \leq \mu \leq \$578.75$
21. She needs to sample at least 217 students.
23. A sample of at least 115 students is needed.
25. A sample of at least 61 students is needed.
27. An increase in sample size generally decreases the margin of error.

SECTION 6.3

1. $17.9 - 2.16 \leq \mu \leq 17.9 + 2.16$
    $15.74 \leq \mu \leq 20.06$
3. $39.7 - 8.53 \leq \mu \leq 39.7 + 8.53$
    $31.17 \leq \mu \leq 48.23$
5. $\bar{x} = 35.0$, $s = 11.10$
    $35.0 - 18.27 \leq \mu \leq 35.0 + 18.27$
    $16.73 \leq \mu \leq 53.27$
7. (a) $7.1 - 0.88 \leq \mu \leq 7.1 + 0.88$
        $6.22 \leq \mu \leq 7.98$
    (b) Yes it is consistent, since all of the values in the interval are less than 10.
9. (a) $63.7 - 2.87 \leq \mu \leq 63.7 + 2.87$
        $60.83 \leq \mu \leq 66.57$
    (b) Yes it is consistent, since 65 inches is contained in the interval.

11. $\bar{x} = 703,332.25$, $s = 84,381.04$
$703,332.25 - 70,555.52 \leq \mu \leq 703,332.25$
$+ 70,555.52$
$632,776.73 \leq \mu \leq 773,887.77$

13. $\bar{x} = \$128.4$ thousand, $s = \$67.78$ thousand
$128.4 - 28.63 \leq \mu \leq 128.4 + 28.63$
$\$99.77$ thousand $\leq \mu \leq \$157.03$ thousand

15. (a) $11.3 - 1.32 \leq \mu \leq 11.3 + 1.32$
$9.98 \leq \mu \leq 12.62$

17. $120.9 - 3.62 \leq \mu \leq 120.9 + 3.62$
$117.28 \leq \mu \leq 124.52$

19. $74 - 13.01 \leq \mu \leq 74 + 13.01$
$60.99 \leq \mu \leq 87.01$

SECTION 6.4

1. $0.55 - 0.0488 \leq \pi \leq 0.55 + 0.0488$
$0.5012 \leq \pi \leq 0.5988$

3. $0.11 - 0.0570 \leq \pi \leq 0.11 + 0.0570$
$0.0530 \leq \pi \leq 0.1670$

5. $0.252 - 0.0452 \leq \pi \leq 0.252 + 0.0452$
$0.2068 \leq \pi \leq 0.2972$

7. (a) $0.66 - 0.1313 \leq \pi \leq 0.66 + 0.1313$
$0.5287 \leq \pi \leq 0.7913$
(b) Perhaps

9. (a) $0.46 - 0.1048 \leq \pi \leq 0.46 + 0.1048$
$0.3552 \leq \pi \leq 0.5648$
(b) Yes

11. $\frac{553}{1005} - 0.0308 \leq \pi \leq \frac{553}{1005} + 0.0308$
$0.5194 \leq \pi \leq 0.5810$

13. $\frac{40}{75} - 0.1483 \leq \pi \leq \frac{40}{75} + 0.1483$
$0.3850 \leq \pi \leq 0.6816$

15. $0.09 - 0.0221 \leq \pi \leq 0.09 + 0.0221$
$0.0679 \leq \pi \leq 0.1121$

17. $\frac{89}{124} - 0.0665 \leq \pi \leq \frac{89}{124} + 0.0665$
$0.6512 \leq \pi \leq 0.7842$

19. $\frac{79}{139} - 0.0823 \leq \pi \leq \frac{79}{139} + 0.0823$
$0.4860 \leq \pi \leq 0.6506$

21. $0.2 - 0.0576 \leq \pi \leq 0.2 + 0.0576$
$0.1424 \leq \pi \leq 0.2576$

23. $\frac{20}{340} - 0.0210 \leq \pi \leq \frac{20}{340} + 0.0210$
$0.0378 \leq \pi \leq 0.0798$

25. $\frac{74}{619} - 0.0336 \leq \pi \leq \frac{74}{619} + 0.0336$
$0.0859 \leq \pi \leq 0.1531$

27. (a) $0.42 - 0.0612 \leq \pi \leq 0.42 + 0.0612$
$0.3588 \leq \pi \leq 0.4812$
(b) Yes.

29. $\frac{802}{4155} - 0.0158 \leq \pi \leq \frac{802}{4155} + 0.0158$
$0.1772 \leq \pi \leq 0.2088$

31. At least 385 high school students

33. At least 335 college students

35. At least 871 high school athletes

SECTION 7.1

1. We write the claim using μ for the population mean, comparing it to the claimed mean using the appropriate symbol: $<, \leq, >, \geq, =,$ or $\neq$.

3. To find the null hypothesis, look for the statement that contains equality (out of the claim and its complement).

5. The level of significance is the probability of making a Type I error, rejecting a true null hypothesis.

7. By examining the alternate hypothesis. If the alternate hypothesis involves the symbol $<$, then the test is left-tailed. If the alternate hypothesis involves the symbol $>$, then the test is right-tailed. If the alternate hypothesis involves the symbol $\neq$, then the test is two-tailed.

9. We place α in the left tail, and find the z-value associated with it. We reject the null hypothesis if the value of the test statistic is less than the z-value.

11. The null hypothesis is a statement, assumed to be true, that is established for testing purposes. However, when testing a claim, we must finish by making a conclusion about the claim.

13. (a) The mean height of adult males is greater than 69 inches.
(b) Claim: $\mu > 69$
(c) Complement: $\mu \leq 69$
(d) H_0: $\mu \leq 69$
(e) H_A: $\mu > 69$
(f) Level of significance: $\alpha = 0.05$
(g) Test statistic: $z = \dfrac{\bar{x} - \mu}{s/\sqrt{n}}$
(h) This is a right-tailed test.
(i) Decision rule: Reject H_0 if $z > 1.645$.
(j) $z = \dfrac{70.4 - 69}{2.44/\sqrt{45}} = 3.85$
(k) Reject H_0.
(l) There is sufficient sample evidence to support the claim.

15. H_0: $\mu = 70$
H_A: $\mu \neq 70$
Reject H_0 if $z < -1.96$ or if $z > 1.96$.
$z = -0.62$
Fail to reject H_0. There is not sufficient sample evidence to reject the claim.
p-value: 0.5352

17. H_0: $\mu \geq 120$
H_A: $\mu < 120$
Reject H_0 if $z < -2.33$.
$z = -6.11$
Reject H_0. There is sufficient sample evidence to support the claim.
p-value: 0.0002

19. (a) H_0: $\mu \leq 5000$
H_A: $\mu > 5000$
Reject H_0 if $z > 1.645$.
$z = 1.05$
Fail to reject H_0. There is not sufficient sample evidence to support the claim.
p-value: 0.1469

21. (a) H_0: $\mu \leq 25$
H_A: $\mu > 25$

Reject H_0 if $z > 1.645$.
$z = 1.67$
Reject H_0. There is sufficient sample evidence to support the claim.
p-value: 0.0475

23. H_0: $\mu = 70.8$
H_A: $\mu \neq 70.8$
Reject H_0 if $z < -2.575$ or if $z > 2.575$.
$\bar{x} = 70.47$, $s = 0.988$
$z = -2.65$
Reject H_0. There is sufficient sample evidence to reject the claim.
p-value: 0.0080

25. H_0: $\mu \geq 8.5$
H_A: $\mu < 8.5$
Reject H_0 if $z < -1.645$.
$z = -2.21$
Reject H_0. There is sufficient sample evidence to support the claim.
p-value: 0.0136

27. H_0: $\mu \geq 20$
H_A: $\mu < 20$
Reject H_0 if $z < -2.33$.
$z = -3.33$
Reject H_0. There is sufficient sample evidence to support the claim.
p-value: 0.0008

29. H_0: $\mu \leq 40$
H_A: $\mu > 40$
Reject H_0 if $z > 1.645$.
$z = 2.20$
Reject H_0. There is sufficient sample evidence to support the claim.
p-value: 0.0139

31. H_0: $\mu \leq 5$
H_A: $\mu > 5$
Reject H_0 if $z > 1.645$.
$z = 3.71$
Reject H_0. There is sufficient sample evidence to support the claim.
p-value: 0.0002

33. H_0: $\mu = 5$
H_A: $\mu \neq 5$
Reject H_0 if $z < -2.575$ or if $z > 2.575$.
$z = 2.05$
Fail to reject H_0. There is not sufficient sample evidence to reject the claim.
p-value: 0.0404

35. H_0: $\mu = 25$
H_A: $\mu \neq 25$
Reject H_0 if $z < -1.96$ or if $z > 1.96$.
$z = 1.85$
Fail to reject H_0. There is not sufficient sample evidence to reject the claim.
p-value: 0.0644

37. H_0: $\mu = 270$
H_A: $\mu \neq 270$
Reject H_0 if $z < -1.96$ or if $z > 1.96$.
$\bar{x} = 272.9$, $s = 17.01$
$z = 1.35$
Fail to reject H_0. There is not sufficient sample evidence to reject the claim.
p-value: 0.1770

39. There is sufficient sample evidence to support the claim that the mean is greater than 17.1.

41. There is not sufficient sample evidence to support the claim that the mean is not 25.

SECTION 7.2

1. H_0: $\mu \geq 50000$
H_A: $\mu < 50000$
d.f. = 24
Reject H_0 if $t < -1.711$.
$t = -2.377$
Reject H_0. There is sufficient sample evidence to support the claim.
p-value: Between 0.01 and 0.025

3. H_0: $\mu = 2$
H_A: $\mu \neq 2$
d.f. = 26
Reject H_0 if $t < -2.056$ or if $t > 2.056$.
$t = 1.237$
Fail to reject H_0. There is not sufficient sample evidence to reject the claim.
p-value: Between 0.2 and 0.5

5. (a) H_0: $\mu \geq 19$
H_A: $\mu < 19$
d.f. = 16
Reject H_0 if $t < -1.746$.
$t = -1.269$
Fail to reject H_0. There is not sufficient sample evidence to support the claim.
p-value: Between 0.1 and 0.25

7. H_0: $\mu \geq 10$
H_A: $\mu < 10$
d.f. = 14
Reject H_0 if $t < -1.761$.
$t = -5.923$
Reject H_0. There is sufficient sample evidence to support the claim.
p-value: Less than 0.005

9. H_0: $\mu \geq 15$
H_A: $\mu < 15$
d.f. = 23
Reject H_0 if $t < -1.714$.
$t = -1.443$
Fail to reject H_0. There is not sufficient sample evidence to support the claim.
p-value: Between 0.05 and 0.1

11. H_0: $\mu \leq 72$
H_A: $\mu > 72$
d.f. = 10

Reject H_0 if $t > 1.812$.
$\bar{x} = 72.6$, $s = 3.59$
$t = 0.554$
Fail to reject H_0. There is not sufficient sample evidence to support the claim.
p-value: Greater than 0.25

13. H_0: $\mu \leq 1200$
H_A: $\mu > 1200$
$d.f. = 16$
Reject H_0 if $t > 2.584$.
$\bar{x} = 1397.5$, $s = 244.60$
$t = 3.329$
Reject H_0. There is sufficient sample evidence to support the claim.
p-value: Less than 0.005

15. H_0: $\mu = 34$
H_A: $\mu \neq 34$
$d.f. = 14$
Reject H_0 if $t < -2.145$ or if $t > 2.145$.
$\bar{x} = 37.5$, $s = 12.40$
$t = 1.093$
Fail to reject H_0. There is not sufficient sample evidence to reject the claim.
p-value: Between 0.2 and 0.5

17. H_0: $\mu \leq 90$
H_A: $\mu > 90$
$d.f. = 19$
Reject H_0 if $t > 1.729$.
$t = 1.704$
Fail to reject H_0. There is not sufficient sample evidence to support the claim.
p-value: Between 0.05 and 0.1

19. H_0: $\mu \leq 290$
H_A: $\mu > 290$
$d.f. = 7$
Reject H_0 if $t > 1.895$.
$\bar{x} = 299.25$, $s = 8.58$
$t = 3.049$
Reject H_0. There is sufficient sample evidence to support the claim.
p-value: Between 0.005 and 0.01

21. H_0: $\mu = 100$
H_A: $\mu \neq 100$
$d.f. = 15$
Reject H_0 if $t < -2.131$ or if $t > 2.131$.
$\bar{x} = 95.1$, $s = 17.38$
$t = -1.128$
Fail to reject H_0. There is not sufficient sample evidence to reject the claim.
p-value: Between 0.2 and 0.5

SECTION 7.3

1. H_0: $\pi \leq 0.5$
H_A: $\pi > 0.5$
Reject H_0 if $z > 2.33$.
$z = 1.69$

Fail to reject H_0. There is not sufficient sample evidence to support the claim.
p-value: 0.0455

3. H_0: $\pi \geq 0.7$
H_A: $\pi < 0.7$
Reject H_0 if $z < -1.645$.
$z = -0.55$
Fail to reject H_0. There is not sufficient sample evidence to support the claim.
p-value: 0.2912

5. H_0: $\pi = 0.525$
H_A: $\pi \neq 0.525$
Reject H_0 if $z < -1.96$ or if $z > 1.96$.
$z = 1.10$
Fail to reject H_0. There is not sufficient sample evidence to reject the claim.
p-value: 0.2714

7. H_0: $\pi \leq 0.5$
H_A: $\pi > 0.5$
Reject H_0 if $z > 1.645$.
$z = 2.97$
Reject H_0. There is sufficient sample evidence to support the claim.
p-value: 0.0015

9. H_0: $\pi = 0.6$
H_A: $\pi \neq 0.6$
Reject H_0 if $z < -1.96$ or if $z > 1.96$.
$z = 1.46$
Fail to reject H_0. There is not sufficient sample evidence to reject the claim.
p-value: 0.1442

11. (a) H_0: $\pi \leq 0.5$
H_A: $\pi > 0.5$
Reject H_0 if $z > 1.645$.
$z = 2.02$
Reject H_0. There is sufficient sample evidence to support the claim.
p-value: 0.0217

13. H_0: $\pi \leq \frac{2}{3}$
H_A: $\pi > \frac{2}{3}$
Reject H_0 if $z > 1.645$.
$z = 4.86$
Reject H_0. There is sufficient sample evidence to support the claim.
p-value: 0.0002

15. H_0: $\pi \geq 0.5$
H_A: $\pi < 0.5$
Reject H_0 if $z < -1.645$.
$z = -4.71$
Reject H_0. There is sufficient sample evidence to support the claim.
p-value: 0.0002

17. H_0: $\pi = 0.6$
H_A: $\pi \neq 0.6$
Reject H_0 if $z < -1.96$ or if $z > 1.96$.
$z = 2.75$

Reject H_0. There is sufficient sample evidence to reject the claim.

p-value: 0.0060

19. H_0: $\pi \leq 0.2$
    H_A: $\pi > 0.2$
    Reject H_0 if $z > 1.645$.
    $z = 2.00$
    Reject H_0. There is sufficient sample evidence to support the claim.
    p-value: 0.0228

21. H_0: $\pi \leq 0.01$
    H_A: $\pi > 0.01$
    Reject H_0 if $z > 1.645$.
    $z = 1.53$
    Fail to reject H_0. There is not sufficient sample evidence to support the claim.
    p-value: 0.0630

23. H_0: $\pi \leq 0.6$
    H_A: $\pi > 0.6$
    Reject H_0 if $z > 1.645$.
    $z = 0.92$
    Fail to reject H_0. There is not sufficient sample evidence to support the claim.
    p-value: 0.1788

25. H_0: $\pi \geq 0.5$
    H_A: $\pi < 0.5$
    Reject H_0 if $z < -1.645$.
    $z = -3.06$
    Reject H_0. There is sufficient sample evidence to support the claim.
    p-value: 0.0011

27. H_0: $\pi \leq 0.5$
    H_A: $\pi > 0.5$
    Reject H_0 if $z > 1.645$.
    $z = 2.03$
    Reject H_0. There is sufficient sample evidence to support the claim.
    p-value: 0.0212

29. H_0: $\pi \leq 0.25$
    H_A: $\pi > 0.25$
    Reject H_0 if $z > 1.645$.
    $z = 2.12$
    Reject H_0. There is sufficient sample evidence to support the claim.
    p-value: 0.0170

31. H_0: $\pi \geq 0.25$
    H_A: $\pi < 0.25$
    Reject H_0 if $z < -2.33$.
    $z = -2.64$
    Reject H_0. There is sufficient sample evidence to support the claim.
    p-value: 0.0041

33. H_0: $\pi = 0.05$
    H_A: $\pi \neq 0.05$
    Reject H_0 if $z < -1.96$ or if $z > 1.96$.
    $z = 0.54$

Fail to reject H_0. There is not sufficient sample evidence to reject the claim.

p-value: 0.5892

35. H_0: $\pi = 0.05$
    H_A: $\pi \neq 0.05$
    Reject H_0 if $z < -2.575$ or if $z > 2.575$.
    $z = -1.22$
    Fail to reject H_0. There is not sufficient sample evidence to reject the claim.
    p-value: 0.2224

37. H_0: $\pi - 0.1$
    H_A: $\pi \neq 0.1$
    Reject H_0 if $z < -1.96$ or if $z > 1.96$.
    $z = 0.66$
    Fail to reject H_0. There is not sufficient sample evidence to reject the claim.
    p-value: 0.5092

39. H_0: $\pi = 0.1$
    H_A: $\pi \neq 0.1$
    Reject H_0 if $z < -1.96$ or if $z > 1.96$
    $z = -3.38$
    Reject H_0. There is sufficient sample evidence to reject the claim.
    p-value: 0.0008

SECTION 8.1

1. (a) Population 1: Female students
       H_0: $\mu_1 \leq \mu_2$
       H_A: $\mu_1 > \mu_2$
       Reject H_0 if $z > 1.645$.
       $z = 1.53$
       Fail to reject H_0. There is not sufficient sample evidence to support the claim.
       p-value: 0.0630
   (b) Students may not give their real GPA.

3. Population 1: Male students
   H_0: $\mu_1 = \mu_2$
   H_A: $\mu_1 \neq \mu_2$
   Reject H_0 if $z < -1.96$ or $z > 1.96$.
   $z = -0.15$
   Fail to reject H_0. There is not sufficient sample evidence to reject the claim.
   p-value: 0.8808

5. Population 1: Spring students
   H_0: $\mu_1 = \mu_2$
   H_A: $\mu_1 \neq \mu_2$
   Reject H_0 if $z < -2.575$ or $z > 2.575$.
   $z = -2.14$
   Fail to reject H_0. There is not sufficient sample evidence to support the claim.
   p-value: 0.0324

7. Population 1: Juniors
   H_0: $\mu_1 \leq \mu_2$
   H_A: $\mu_1 > \mu_2$
   Reject H_0 if $z > 2.33$.
   $\bar{x}_1 = 49.8$, $s_1 = 11.96$

$\bar{x}_2 = 47.1$, $s_2 = 10.36$
$z = 1.11$
Fail to reject H_0. There is not sufficient sample evidence to support the claim.
p-value: 0.1335

9. Population 1: Math
H_0: $\mu_1 \leq \mu_2$
H_A: $\mu_1 > \mu_2$
Reject H_0 if $z > 1.645$.
$\bar{x}_1 = 525$, $s_1 = 127.04$
$\bar{x}_2 = 504.3$, $s_2 = 119.36$
$z = 0.68$
Fail to reject H_0. There is not sufficient sample evidence to support the claim.
p-value: 0.2483

11. (a) Population 1: Wait for takeoff
H_0: $\mu_1 \geq \mu_2$
H_A: $\mu_1 < \mu_2$
Reject H_0 if $z < -1.645$.
$z = -2.63$
Reject H_0. There is sufficient sample evidence to support the claim.
p-value: 0.0043

13. Population 1: Synthetic lanes
H_0: $\mu_1 \leq \mu_2$
H_A: $\mu_1 > \mu_2$
Reject H_0 if $z > 1.645$.
$\bar{x}_1 = 240.2$, $s_1 = 25.93$
$\bar{x}_2 = 218.8$, $s_2 = 25.63$
$z = 3.80$
Reject H_0. There is sufficient sample evidence to support the claim.
p-value: 0.0002

15. Population 1: American League
H_0: $\mu_1 \leq \mu_2$
H_A: $\mu_1 > \mu_2$
Reject H_0 if $z > 2.33$.
$\bar{x}_1 = 10.2$, $s_1 = 4.98$
$\bar{x}_2 = 9.7$, $s_2 = 4.94$
$z = 0.58$
Fail to reject H_0. There is not sufficient sample evidence to support the claim.
p-value: 0.2810

17. Population 1: American League
H_0: $\mu_1 \leq \mu_2$
H_A: $\mu_1 > \mu_2$
Reject H_0 if $z > 1.645$.
$\bar{x}_1 = 176.8$, $s_1 = 23.34$
$\bar{x}_2 = 170.9$, $s_2 = 25.05$
$z = 1.41$
Fail to reject H_0. There is not sufficient sample evidence to support the claim.
p-value: 0.0793

19. Population 1: Women married in 1970
H_0: $\mu_1 \geq \mu_2$
H_A: $\mu_1 < \mu_2$
Reject H_0 if $z < -1.645$.

$z = -8.87$
Reject H_0. There is sufficient sample evidence to support the claim.
p-value: 0.0002

21. Population 1: Households with cable TV
H_0: $\mu_1 \leq \mu_2$
H_A: $\mu_1 > \mu_2$
Reject H_0 if $z > 1.645$.
$z = 5.52$
Reject H_0. There is sufficient sample evidence to support the claim.
p-value: 0.0002

23. (a) Population 1: 2- to 5-year-olds
$-35.2 - 30.68 \leq \mu \leq -35.2 + 30.68$
$-65.88 \leq \mu \leq -4.52$
(b) Population 1: 2- to 5-year-olds
H_0: $\mu_1 = \mu_2$
H_A: $\mu_1 \neq \mu_2$
Reject H_0 if $z < -1.96$ or $z > 1.96$.
$z = -2.25$
Reject H_0: There is sufficient sample evidence to support the claim.
p-value: 0.0122
(c) Yes, since the value 0 is not contained in the interval, the two means are most likely not equal and therefore the correct decision is to reject H_0.

SECTION 8.2

1. Population 1: Morning
H_0: $\mu_1 = \mu_2$
H_A: $\mu_1 \neq \mu_2$
$d.f. = 21$
Reject H_0 if $t < -2.080$ or $t > 2.080$.
$\bar{x}_1 = 77.3$, $s_1 = 14.32$
$\bar{x}_2 = 75.3$, $s_2 = 16.88$
$t = 0.429$
Fail to reject H_0. There is not sufficient sample evidence to reject the claim.
p-value: Greater than 0.5

3. Population 1: Day
H_0: $\mu_1 \geq \mu_2$
H_A: $\mu_1 < \mu_2$
$d.f. = 24$
Reject H_0 if $t < -2.492$.
$t = -3.831$
Reject H_0. There is sufficient sample evidence to support the claim.
p-value: Less than 0.005

5. Population 1: Male
H_0: $\mu_1 \leq \mu_2$
H_A: $\mu_1 > \mu_2$
$d.f. = 10$
Reject H_0 if $t > 1.812$.
$t = 2.041$
Reject H_0. There is sufficient sample evidence to support the claim.
p-value: Between 0.025 and 0.05

7. Population 1: Female
   H_0: $\mu_1 \leq \mu_2$
   H_A: $\mu_1 > \mu_2$
   $d.f. = 18$
   Reject H_0 if $t > 1.734$.
   $t = 1.523$
   Fail to reject H_0. There is not sufficient sample evidence to support the claim.
   p-value: Between 0.05 and 0.1

9. Population 1: Men
   H_0: $\mu_1 = \mu_2$
   H_A: $\mu_1 \neq \mu_2$
   $d.f. = 11$
   Reject H_0 if $t < -3.106$ or $t > 3.106$.
   $\bar{x}_1 = 196.4$, $s_1 = 34.38$
   $\bar{x}_2 = 223.5$, $s_2 = 35.54$
   $t = -2.235$
   Fail to reject H_0. There is not sufficient sample evidence to reject the claim.
   p-value: Between 0.02 and 0.05

11. Population 1: System 1
   H_0: $\mu_1 = \mu_2$
   H_A: $\mu_1 \neq \mu_2$
   $d.f. = 19$
   Reject H_0 if $t < -2.093$ or $t > 2.093$.
   $t = -2.790$
   Reject H_0. There is sufficient sample evidence to reject the claim.
   p-value: Between 0.01 and 0.02

13. Population 1: Large luxury cars
   H_0: $\mu_1 \leq \mu_2$
   H_A: $\mu_1 > \mu_2$
   $d.f. = 3$
   Reject H_0 if $t > 2.353$.
   $\bar{x}_1 = \$1009.25$, $s_1 = \$595.26$
   $\bar{x}_2 = \$684.50$, $s_2 = \$293.81$
   $t = 0.978$
   Fail to reject H_0. There is not sufficient sample evidence to support the claim.
   p-value: Between 0.1 and 0.25

15. (a) Population 1: Laparoscopic appendectomy
   H_0: $\mu_1 = \mu_2$
   H_A: $\mu_1 \neq \mu_2$
   $d.f. = 17$
   Reject H_0 if $t < -2.110$ or $t > 2.110$.
   $t = 0.815$
   Fail to reject H_0. There is not sufficient sample evidence to reject the claim.
   p-value: Between 0.2 and 0.5

   (b) Population 1: Laparoscopic appendectomy
   H_0: $\mu_1 = \mu_2$
   H_A: $\mu_1 \neq \mu_2$
   $d.f. = 17$
   Reject H_0 if $t < -2.110$ or $t > 2.110$.
   $t = -1.130$
   Fail to reject H_0. There is not sufficient sample evidence to reject the claim.
   p-value: Between 0.2 and 0.5

(c) Population 1: Laparoscopic appendectomy
   H_0: $\mu_1 \leq \mu_2$
   H_A: $\mu_1 > \mu_2$
   $d.f. = 17$
   Reject H_0 if $t > 1.740$.
   $t = 32.546$
   Reject H_0. There is sufficient sample evidence to support the claim.
   p-value: Less than 0.005

17. Population 1: Women with Turner's syndrome
   H_0: $\mu_1 \geq \mu_2$
   H_A: $\mu_1 < \mu_2$
   $d.f. = 24$
   Reject H_0 if $t < -2.492$.
   $t = -21.909$
   Reject H_0. There is sufficient sample evidence to support the claim.
   p-value: Less than 0.005

19. Population 1: Students majoring in mathematics or statistics
   H_0: $\mu_1 \geq \mu_2$
   H_A: $\mu_1 < \mu_2$
   $d.f. = 13$
   Decision rule: Reject H_0 if $t < -1.771$.
   $t = -2.504$
   Reject H_0. There is sufficient sample evidence to support the claim.
   p-value: Between 0.01 and 0.025

21. Population 1: Women
   H_0: $\mu_1 = \mu_2$
   H_A: $\mu_1 \neq \mu_2$
   $d.f. = 25$
   Decision rule: Reject H_0 if $t < -2.060$ or $t > 2.060$.
   $t = 0.327$
   Fail to reject H_0. There is not sufficient sample evidence to reject the claim.
   p-value: Greater than 0.5

23. Population 1: Children's cereals
   H_0: $\mu_1 \leq \mu_2$
   H_A: $\mu_1 > \mu_2$
   $d.f. = 9$
   Decision rule: Reject H_0 if $t < 1.833$.
   $t = 1.431$
   Fail to reject H_0. There is not sufficient sample evidence to support the claim.
   p-value: Between 0.05 and 0.1

25. (a) Population 1: Sandy soil
   $d.f. = 5$
   $(941.4 - 870.3) - 236.06 \leq (\mu_1 - \mu_2) \leq (941.4 - 870.3) + 236.06$
   $-164.96 \leq (\mu_1 - \mu_2) \leq 307.16$

   (b) Population 1: Sandy soil
   H_0: $\mu_1 = \mu_2$
   H_A: $\mu_1 \neq \mu_2$
   Decision rule: Reject H_0 if $t < -2.571$ or $t > 2.571$.
   $t = 0.774$
   Fail to reject H_0. There is not sufficient sample evidence to reject the claim.
   p-value: Between 0.2 and 0.5

SECTION 8.3

1. (a) d = United States − Canada
   H_0: $\mu_d \leq 0$
   H_A: $\mu_d > 0$
   $d.f.$ = 9
   Reject H_0 if $t > 1.833$.
   $\bar{d}$ = \$1.44, s_d = \$0.431
   t = 10.565
   Reject H_0. There is sufficient sample evidence to support the claim.
   p-value: Less than 0.005
   (b) The samples are dependent because the prices listed in the two samples are for the same medication.

3. (a) d = Home − Away
   H_0: $\mu_d \leq 0$
   H_A: $\mu_d > 0$
   $d.f.$ = 12
   Reject H_0 if $t > 1.782$.
   $\bar{d}$ = 3.7, s_d = 8.20
   t = 1.627
   Fail to reject H_0. There is not sufficient sample evidence to support the claim.
   p-value: Between 0.05 and 0.1
   (b) These samples are paired because the home and away totals are taken from the same game.

5. (a) d = Second Lactation − First Lactation
   H_0: $\mu_d \leq 0$
   H_A: $\mu_d > 0$
   $d.f.$ = 19
   Reject H_0 if $t > 1.729$.
   $\bar{d}$ = 3386.4, s_d = 7509.45
   t = 2.017
   Reject H_0. There is sufficient sample evidence to support the claim.
   p-value: Between 0.025 and 0.05
   (b) The samples are dependent because the totals come from the same cow.

7. d = Before Flu Shot − After Flu Shot
   H_0: $\mu_d \leq 0$
   H_A: $\mu_d > 0$
   $d.f.$ = 7
   Reject H_0 if $t > 1.895$.
   $\bar{d}$ = 1.4, s_d = 1.19
   t = 3.328
   Reject H_0. There is sufficient sample evidence to support the claim.
   p-value: Between 0.005 and 0.01

9. d = After − Before
   H_0: $\mu_d \geq 0$
   H_A: $\mu_d < 0$
   $d.f.$ = 11
   Reject H_0 if $t < -1.796$.
   $\bar{d}$ = −5.8, s_d = 11.69
   t = −1.719
   Fail to reject H_0. There is not sufficient sample evidence to support the claim.
   p-value: Between 0.05 and 0.1

11. d = Web Site − Over the Phone
    H_0: $\mu_d = 0$
    H_A: $\mu_d \neq 0$
    $d.f.$ = 7
    Reject H_0 if $t < -3.499$ or $t > 3.499$.
    $\bar{d}$ = −0.85, s_d = 5.33
    t = −0.451
    Fail to reject H_0. There is not sufficient sample evidence to reject the claim.
    p-value: Greater than 0.5

13. (a) d = Final Exam − Midterm
    $d.f.$ = 19
    $\bar{d}$ = 10.1, s_d = 11.52
    $10.1 - 5.391 \leq \mu_d \leq 10.1 + 5.391$
    $4.709 \leq \mu_d \leq 15.491$
    (b) d = Final Exam − Midterm
    H_0: $\mu_d \leq 0$
    H_A: $\mu_d > 0$
    $d.f.$ = 19
    Reject H_0 if $t > 2.093$.
    $\bar{d}$ = 10.1, s_d = 11.52
    t = 3.921
    Reject H_0. There is sufficient sample evidence to support the claim.
    p-value: Less than 0.005

15. (a) d = After Program − Before Program
    $d.f.$ = 19
    $\bar{d}$ = 35, s_d = 42.36
    $35 - 19.825 \leq \mu_d \leq 35 + 19.825$
    $15.175 \leq \mu_d \leq 54.825$
    (b) d = After Program − Before Program
    H_0: $\mu_d \leq 0$
    H_A: $\mu_d > 0$
    $d.f.$ = 19
    Reject H_0 if $t > 1.729$.
    $\bar{d}$ = 35, s_d = 42.36
    t = 3.695
    Reject H_0. There is sufficient sample evidence to support the claim.
    p-value: Less than 0.005

SECTION 8.4

1. Population 1: Health-care professionals
   H_0: $\pi_1 \geq \pi_2$
   H_A: $\pi_1 < \pi_2$
   Reject H_0 if $z < -2.33$.
   $\bar{p} = \frac{21}{125}$ $z = -1.95$
   Fail to reject H_0. There is not sufficient sample evidence to support the claim.
   p-value: 0.0256

3. Population 1: Males
   H_0: $\pi_1 \leq \pi_2$
   H_A: $\pi_1 > \pi_2$
   Reject H_0 if $z > 1.645$.
   $\bar{p} = \frac{60}{250}$ $z = 1.68$
   Reject H_0. There is sufficient sample evidence to support the claim.
   p-value: 0.0465

5. Population 1: Drivers 35 years old or younger
H_0: $\pi_1 \leq \pi_2$
H_A: $\pi_1 > \pi_2$
Reject H_0 if $z > 1.645$.
$\bar{p} = \frac{118}{208}$ $z = 1.58$
Fail to reject H_0. There is not sufficient sample evidence to support the claim.
p-value: 0.0566

7. Population 1: Home
H_0: $\pi_1 \leq \pi_2$
H_A: $\pi_1 > \pi_2$
Reject H_0 if $z > 2.33$.
$\bar{p} = \frac{573}{775}$ $z = 1.50$
Fail to reject H_0. There is not sufficient sample evidence to support the claim.
p-value: 0.0668

9. Population 1: 18 to 24 years old
H_0: $\pi_1 = \pi_2$
H_A: $\pi_1 \neq \pi_2$
Reject H_0 if $z < -1.96$ or $z > 1.96$.
$\bar{p} = \frac{29}{232}$ $z = 0.63$
Fail to reject H_0. There is not sufficient sample evidence to reject the claim.
p-value: 0.5286

11. Population 1: Drivers (Label running red lights as hazardous)
H_0: $\pi_1 \leq \pi_2$
H_A: $\pi_1 > \pi_2$
Reject H_0 if $z > 2.33$.
$\bar{p} = \frac{968}{1120}$ $z = 21.57$
Reject H_0. There is sufficient sample evidence to support the claim.
p-value: 0.0002

13. Population 1: Italians
H_0: $\pi_1 \leq \pi_2$
H_A: $\pi_1 > \pi_2$
Reject H_0 if $z > 1.645$.
$\bar{p} = \frac{159}{515}$ $z = 2.89$
Reject H_0.
There is sufficient sample evidence to support the claim.
p-value: 0.0019

15. Population 1: Jury trials
H_0: $\pi_1 \leq \pi_2$
H_A: $\pi_1 > \pi_2$
Reject H_0 if $z > 1.645$.
$\bar{p} = \frac{61}{140}$ $z = 1.67$
Reject H_0. There is sufficient sample evidence to support the claim.
p-value: 0.0475

17. Population 1: American men
H_0: $\pi_1 = \pi_2$
H_A: $\pi_1 \neq \pi_2$
Reject H_0 if $z < -1.96$ or $z > 1.96$.
$\bar{p} = \frac{111}{391}$ $z = -0.03$
Fail to reject H_0. There is not sufficient sample evidence to support the claim.
p-value: 0.9760

19. Population 1: Men
H_0: $\pi_1 = \pi_2$
H_A: $\pi_1 \neq \pi_2$
Reject H_0 if $z < -1.96$ or $z > 1.96$.
$\bar{p} = \frac{222}{320}$ $z = -1.88$
Fail to reject H_0. There is not sufficient sample evidence to reject the claim.
p-value: 0.0602

21. Population 1: Senior-level bosses
H_0: $\pi_1 \geq \pi_2$
II_A: $\pi_1 < \pi_2$
Reject H_0 if $z < -1.645$.
$\bar{p} = \frac{232}{557}$ $z = -2.17$
Reject H_0. There is sufficient sample evidence to support the claim.
p-value: 0.0150

23. Population 1: Men ages 30–34
H_0: $\pi_1 \leq \pi_2$
H_A: $\pi_1 > \pi_2$
Reject H_0 if $z > 1.645$.
$\bar{p} = \frac{52}{200}$ $z = 1.09$
Fail to reject H_0. There is not sufficient sample evidence to support the claim.
p-value: 0.1379

25. Population 1: High school males
H_0: $\pi_1 \leq \pi_2$
H_A: $\pi_1 > \pi_2$
Reject H_0 if $z > 2.33$.
$\bar{p} = \frac{854}{2227}$ $z = 4.68$
Reject H_0. There is sufficient sample evidence to support the claim.
p-value: 0.0002

27. Population 1: Washington, D.C. students
H_0: $\pi_1 \leq \pi_2$
H_A: $\pi_1 > \pi_2$
Reject H_0 if $z > 1.645$.
$\bar{p} = \frac{57}{374}$ $z = 2.73$
Reject H_0. There is sufficient sample evidence to support the claim.
p-value: 0.0032

29. Population 1: Male high school students
H_0: $\pi_1 = \pi_2$
H_A: $\pi_1 \neq \pi_2$
Reject H_0 if $z < -1.96$ or $z > 1.96$.
$\bar{p} = \frac{241}{2891}$ $z = -5.05$
Reject H_0. There is sufficient sample evidence to reject the claim.
p-value: 0.0004

31. Population 1: Male high school students
H_0: $\pi_1 \geq \pi_2$
H_A: $\pi_1 < \pi_2$
Reject H_0 if $z < -1.645$.
$\bar{p} = \frac{260}{379}$ $z = -1.98$
Reject H_0. There is sufficient sample evidence to support the claim.
p-value: 0.0239

33. Part (a) $\bar{p} = \frac{31}{90}$
$-0.3876 \leq \pi_1 - \pi_2 \leq 0.0076$

Part (b) Population 1: Female inner-city high school students

H_0: $\pi_1 = \pi_2$

H_A: $\pi_1 \neq \pi_2$

Reject H_0 if $z < -1.96$ or $z > 1.96$.

$z = -1.88$

Fail to reject H_0. There is not sufficient sample evidence to reject the claim.

p-value: 0.0602

Part (c) Since the interval contains 0, the two means could be equal. Therefore, we are unable to reject the claim that they are the same.

35. Part (a) $\bar{p} = \frac{1267}{25,216}$

$0.0562 \leq \pi_1 - \pi_2 \leq 0.2188$

Part (b) Population 1: Low thyroid level group

H_0: $\pi_1 = \pi_2$

H_A: $\pi_1 \neq \pi_2$

Reject H_0 if $z < -2.575$ or $z > 2.575$.

$z = 4.36$

Reject H_0. There is sufficient sample evidence to reject the claim.

p-value: 0.0002

SECTION 9.1

1. Reject H_0 if $F > 3.36$.

3. Reject H_0 if $F > 2.05$.

5. Reject H_0 if $F > 3.15$.

7. Reject H_0 if $F > 2.05$.

9. Reject H_0 if $F > 2.52$.

11. Population 1: Males

H_0: $\sigma_1^2 = \sigma_2^2$

H_A: $\sigma_1^2 \neq \sigma_2^2$

$d.f._{num} = 87$, $d.f._{den} = 83$

Reject H_0 if $F > 1.84$.

$F = 1.14$

Fail to reject H_0. There is not sufficient sample evidence to reject the claim.

13. Population 1: Men

H_0: $\sigma_1^2 = \sigma_2^2$

H_A: $\sigma_1^2 \neq \sigma_2^2$

$d.f._{num} = 102$, $d.f._{den} = 103$

Reject H_0 if $F > 1.53$.

$F = 1.26$

Fail to reject H_0. There is not sufficient sample evidence to reject the claim.

15. Population 1: 6–12 years old

H_0: $\sigma_1^2 \leq \sigma_2^2$

H_A: $\sigma_1^2 > \sigma_2^2$

$d.f._{num} = 119$, $d.f._{den} = 44$

Reject H_0 if $F > 1.64$.

$F = 1.70$

Reject H_0. There is sufficient sample evidence to support the claim.

17. Population 1: Unit 2

H_0: $\sigma_1^2 \leq \sigma_2^2$

H_A: $\sigma_1^2 > \sigma_2^2$

$d.f._{num} = 37$, $d.f._{den} = 20$

Reject H_0 if $F > 2.04$.

$F = 2.25$

Reject H_0. There is sufficient sample evidence to support the claim.

19. Population 1: General population

H_0: $\sigma_1^2 \leq \sigma_2^2$

H_A: $\sigma_1^2 > \sigma_2^2$

$d.f._{num} = 154$, $d.f._{den} = 24$

Reject H_0 if $F > 2.31$.

$F = 3.62$

Reject H_0. There is sufficient sample evidence to support the claim.

21. Population 1: Qualifying

H_0: $\sigma_1^2 = \sigma_2^2$

H_A: $\sigma_1^2 \neq \sigma_2^2$

$d.f._{num} = 35$, $d.f._{den} = 47$

Reject H_0 if $F > 1.74$.

$s_1 = 28.56$, $s_2 = 25.95$

$F = 1.21$

Fail to reject H_0. There is not sufficient sample evidence to reject the claim.

23. Population 1: Verbal

H_0: $\sigma_1^2 = \sigma_2^2$

H_A: $\sigma_1^2 \neq \sigma_2^2$

$d.f._{num} = 39$, $d.f._{den} = 39$

Reject H_0 if $F > 2.39$.

$s_1 = 10.97$, $s_2 = 10.48$

$F = 1.10$

Fail to reject H_0. There is not sufficient sample evidence to reject the claim.

25. Population 1: Collected

H_0: $\sigma_1^2 \leq \sigma_2^2$

H_A: $\sigma_1^2 > \sigma_2^2$

$d.f._{num} = 36$, $d.f._{den} = 31$

Reject H_0 if $F > 1.84$.

$s_1 = 10.67$, $s_2 = 7.31$

$F = 2.13$

Reject H_0. There is sufficient sample evidence to support the claim.

27. Population 1: San Francisco

H_0: $\sigma_1^2 \leq \sigma_2^2$

H_A: $\sigma_1^2 > \sigma_2^2$

$d.f._{num} = 38$, $d.f._{den} = 47$

Reject H_0 if $F > 1.74$.

$s_1 = \$0.048$, $s_2 = \$0.038$

$F = 1.60$

Fail to reject H_0. There is not sufficient sample evidence to support the claim.

SECTION 9.2

1. H_0: $\mu_1 = \mu_2 = \mu_3 = \mu_4$

H_A: At least one of the means is different from the others.

$d.f._{num} = 3$, $d.f._{den} = 15$

Reject H_0 if $F > 3.29$.

$F = 0.62$
Fail to reject H_0. There is not sufficient sample evidence to reject the claim.

3. H_0: $\mu_1 = \mu_2 = \mu_3 = \mu_4$
H_A: At least one of the means is different from the others.
$d.f._{num} = 3$, $d.f._{den} = 12$
Reject H_0 if $F > 5.95$.
$F = 27.48$
Reject H_0. There is sufficient sample evidence to reject the claim.

5. H_0: $\mu_1 = \mu_2 = \mu_3$
H_A: At least one of the means is different from the others.
$d.f._{num} = 2$, $d.f._{den} = 132$
Reject H_0 if $F > 3.07$.
$\bar{x} = 931.8$
$SST = 154{,}642.05$ $SSE = 4{,}873{,}383.51$
$F = 2.09$
Fail to reject H_0. There is not sufficient sample evidence to reject the claim.

7. H_0: $\mu_1 = \mu_2 = \mu_3 = \mu_4 = \mu_5 = \mu_6 = \mu_7$
H_A: At least one of the means is different from the others.
$d.f._{num} = 6$, $d.f._{den} = 343$
Reject H_0 if $F > 2.96$.
$\bar{x} = 6.96$
$SST = 3618.86$ $SSE = 1376.64$
$F = 150.28$
Reject H_0. There is sufficient sample evidence to reject the claim.

9. H_0: $\mu_1 = \mu_2 = \mu_3$
H_A: At least one of the means is different from the others.
$d.f._{num} = 2$, $d.f._{den} = 68$
Reject H_0 if $F > 4.98$.
$F = 0.59$
Fail to reject H_0. There is not sufficient sample evidence to reject the claim.

11. H_0: $\mu_1 = \mu_2 = \mu_3$
H_A: At least one of the means is different from the others.
$d.f._{num} = 2$, $d.f._{den} = 177$
Reject H_0 if $F > 4.79$.
$F = 6.97$
Reject H_0. There is sufficient sample evidence to reject the claim.

13. H_0: $\mu_1 = \mu_2 = \mu_3 = \mu_4$
H_A: At least one of the means is different from the others.
$d.f._{num} = 3$, $d.f._{den} = 292$
Reject H_0 if $F > 3.95$.
$F = 27.55$
Reject H_0. There is sufficient sample evidence to reject the claim.

15. H_0: $\mu_1 = \mu_2 = \mu_3$
H_A: At least one of the means is different from the others.

$d.f._{num} = 2$, $d.f._{den} = 27$
Reject H_0 if $F > 3.39$.
$F = 3.35$
Fail to reject H_0. There is not sufficient sample evidence to reject the claim.

17. H_0: $\mu_1 = \mu_2 = \mu_3 = \mu_4 = \mu_5$
H_A: At least one of the means is different from the others.
$d.f._{num} = 4$, $d.f._{den} = 35$
Reject H_0 if $F > 4.02$.
$F = 7.53$
Reject H_0. There is sufficient sample evidence to reject the claim.

19. H_0: $\mu_1 = \mu_2 = \mu_3$
H_A: At least one of the means is different from the others.
$d.f._{num} = 2$, $d.f._{den} = 412$
Reject H_0 if $F > 3.07$.
$\bar{x} = 365.9$
$SST = 120{,}540.15$ $SSE = 975{,}814.6$
$F = 25.45$
Reject H_0. There is sufficient sample evidence to reject the claim.

21. H_0: $\mu_1 = \mu_2 = \mu_3$
H_A: At least one of the means is different from the others.
$d.f._{num} = 2$, $d.f._{den} = 78$
Reject H_0 if $F > 3.15$.
$\bar{x} = 5.0$
$SST = 1066.5$ $SSE = 755.56$
$F = 55.05$
Reject H_0. There is sufficient sample evidence to reject the claim.

SECTION 9.3

1. H_0: $\pi_G = \pi_R = \pi_Y = \pi_O = 0.25$
H_A: At least one of the proportions is different than claimed.
$d.f. = 3$
Reject H_0 if $\chi^2 > 7.815$.
$\chi^2 = 2.480$
Fail to reject H_0. There is not sufficient sample evidence to reject the claim.

3. H_0: $\pi_R = \pi_B = 0.474$, $\pi_G = 0.052$
H_A: At least one of the proportions is different than claimed.
$d.f. = 2$
Reject H_0 if $\chi^2 > 5.991$.
$\chi^2 = 3.764$
Fail to reject H_0. There is not sufficient sample evidence to reject the claim.

5. H_0: $\pi_A = 0.28$, $\pi_B = 0.34$, $\pi_C = 0.38$
H_A: At least one of the proportions is different than claimed.
$d.f. = 2$
Reject H_0 if $\chi^2 > 5.991$.
$\chi^2 = 3.534$

Fail to reject H_0. There is not sufficient sample evidence to reject the claim.

7. H_0: $\pi_2 = 1/36$, $\pi_3 = 2/36$, $\pi_4 = 3/36$, $\pi_5 = 4/36$, $\pi_6 = 5/36$, $\pi_7 = 6/36$, $\pi_8 = 5/36$, $\pi_9 = 4/36$, $\pi_{10} = 3/36$, $\pi_{11} = 2/36$, $\pi_{12} = 1/36$
H_A: At least one of the proportions is different than claimed.
$d.f. = 10$
Reject H_0 if $\chi^2 > 23.209$.
$\chi^2 = 11.826$
Fail to reject H_0. There is not sufficient sample evidence to reject the claim.

9. H_0: $\pi_O = 0.45$, $\pi_A = 0.40$, $\pi_B = 0.11$, $\pi_{AB} = 0.04$
H_A: At least one of the proportions is different than claimed.
$d.f. = 3$
Reject H_0 if $\chi^2 > 7.815$.
$\chi^2 = 12.043$
Reject H_0. There is sufficient sample evidence to reject the claim.

11. H_0: $\pi_{SUN} = \pi_{MON} = \pi_{TUE} = \pi_{WED} = \pi_{THU} = \pi_{FRI} = \pi_{SAT} = 1/7$
H_A: At least one of the proportions is different than claimed.
$d.f. = 6$
Reject H_0 if $\chi^2 > 16.812$.
$\chi^2 = 331.847$
Reject H_0. Conclusion: There is sufficient sample evidence to reject the claim.

13. H_0: $\pi_{WINTER} = \pi_{SPRING} = \pi_{SUMMER} = \pi_{FALL} = 0.25$
H_A: At least one of the proportions is different than claimed.
$d.f. = 3$
Reject H_0 if $\chi^2 > 7.815$.
$\chi^2 = 1.559$
Fail to reject H_0. There is not sufficient sample evidence to reject the claim.

15. H_0: $\pi_{NoH.S.Graduation} = 0.245$, $\pi_{Diploma} = 0.489$, $\pi_{Associate'sOrHigher} = 0.266$
H_A: At least one of the proportions is different than claimed.
$d.f. = 2$
Reject H_0 if $\chi^2 > 5.991$.
$\chi^2 = 9.297$
Reject H_0. There is sufficient sample evidence to support the claim.

17. H_0: $\pi_{Excellent} = 0.14$, $\pi_{Good} = 0.38$, $\pi_{Fair} = 0.37$, $\pi_{Poor} = 0.11$
H_A: At least one of the proportions is different than claimed.
$d.f. = 3$
Reject H_0 if $\chi^2 > 7.815$.
$\chi^2 = 27.918$
Reject H_0. There is sufficient sample evidence to reject the claim.

19. H_0: $\pi_{JanFebMar} = \pi_{AprMayJun} = \pi_{JulAugSep} = \pi_{OctNovDec} = 0.25$

H_A: At least one of the proportions is different than claimed.
$d.f. = 3$
Reject H_0 if $\chi^2 > 7.815$.
$\chi^2 = 18.587$
Reject H_0. There is sufficient sample evidence to reject the claim.

21. H_0: The data come from a population that follows a normal distribution with $\mu = 5.65$ and $\sigma = 2.90$.
($\pi_1 = 0.1401$, $\pi_2 = 0.2045$, $\pi_3 = 0.2695$, $\pi_4 = 0.2224$, $\pi_5 = 0.1635$)
H_A: The data do not come from a population that follows a normal distribution with $\mu = 5.65$ and $\sigma = 2.90$.
(At least one of the proportions is different than claimed.)
$d.f. = 2$
Reject H_0 if $\chi^2 > 5.991$.
$\chi^2 = 3.360$
Fail to reject H_0. There is not sufficient sample evidence to reject the claim.

23. H_0: The data come from a population that follows a normal distribution with $\mu = 1.67$ and $\sigma = 1.94$.
($\partial_1 = 0.2743$, $\partial_2 = 0.1898$, $\partial_3 = 0.2023$, $\partial_4 = 0.1600$, $\partial_5 = 0.1736$)
H_A: The data do not come from a population that follows a normal distribution with $\mu = 1.67$ and $\sigma = 1.94$.
(At least one of the proportions is different than claimed.)
$d.f. = 2$
Reject H_0 if $\chi^2 > 5.991$.
$\chi^2 = 6.125$
Reject H_0. There is sufficient sample evidence to reject the claim.

SECTION 9.4

1. H_0: Race is independent of health insurance status.
H_A: They are dependent.
$d.f. = 3$
Reject H_0 if $\chi^2 > 11.345$.
$\chi^2 = 10.602$
Fail to reject H_0. There is not sufficient sample evidence to reject claim.

3. H_0: Voting status and age are independent.
H_A: They are dependent.
$d.f. = 3$
Reject H_0 if $\chi^2 > 11.345$.
$\chi^2 = 54.530$
Reject H_0. There is sufficient sample evidence to reject claim.

5. H_0: Blood type and Rh factor are independent.
H_A: They are dependent.
$d.f. = 3$

Reject H_0 if $\chi^2 > 7.815$
$\chi^2 = 10.660$
Reject H_0. There is sufficient sample evidence to reject claim.

7. H_0: The proportion of women who suffer a spinal fracture is the same for all three medication treatments.
H_A: They are different.
d.f. = 2
Reject H_0 if $\chi^2 > 5.991$
$\chi^2 = 19.408$
Reject H_0. There is sufficient sample evidence to reject claim.

9. H_0: Proportion of drivers who run red lights is the same for all 5 age groups.
H_A: They are different.
d.f. = 4
Reject H_0 if $\chi^2 > 13.277$
$\chi^2 = 28.748$
Reject H_0. There is sufficient sample evidence to reject claim.

11. H_0: The proportional distribution is the same for both age groups.
H_A: They are different.
d.f. = 3
Reject H_0 if $\chi^2 > 7.815$
$\chi^2 = 8.642$
Reject H_0. There is sufficient sample evidence to reject claim.

13. H_0: Educational attainment and smoking status are independent.
H_A: They are dependent.
d.f. = 3
Reject H_0 if $\chi^2 > 11.345$
$\chi^2 = 12.755$
Reject H_0. There is sufficient sample evidence to reject claim.

15. H_0: A person's response to the question is independent of the person's gender.
H_A: They are dependent.
d.f. = 4
Reject H_0 if $\chi^2 > 9.488$
$\chi^2 = 0.251$
Fail to reject H_0. There is not sufficient sample evidence to reject claim.

17. H_0: Whether or not a female high school student has considered suicide is independent of their grade.
H_A: They are dependent.
d.f. = 3
Reject H_0 if $\chi^2 > 7.815$
$\chi^2 = 16.103$
Reject H_0. There is sufficient sample evidence to reject claim.

19. H_0: A person's response is independent of his or her race.
H_A: They are dependent.
d.f. = 6

Reject H_0 if $\chi^2 > 12.592$
$\chi^2 = 17.259$
Reject H_0. There is sufficient sample evidence to reject claim.

SECTION 10.1

1.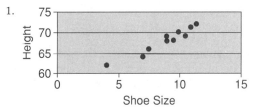

The correlation appears to be positive and strong.

3.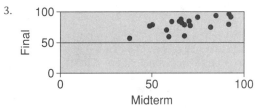

The correlation appears to be positive and moderate.

5. $r = 0.967$

7. $r = 0.633$

9. $r = 0.849$ $r^2 = 0.721$
72.1% of the variation in cattle values is explained by the variation in milk production values.

11. (a) $\rho = 0.753$

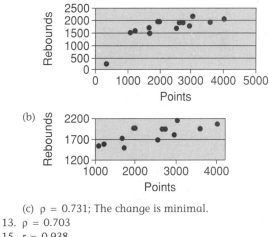

(c) $\rho = 0.731$; The change is minimal.

13. $\rho = 0.703$

15. $r = 0.938$
$H_0: \rho = 0$
$H_A: \rho \neq 0$
d.f. = 14
Reject H_0 if $t > 2.145$ or if $t < -2.145$.
$t = 9.374$
Reject H_0. There is sufficient sample evidence to support the claim.
p-value: Less than 0.01

17. $r = 0.132$
$H_0: \rho = 0$
$H_A: \rho \neq 0$
$d.f. = 8$
Reject H_0 if $t > 2.306$ or if $t < -2.306$.
$t = 0.377$
Fail to reject H_0. There is not sufficient sample
evidence to support the claim.
p-value: Greater than 0.5

19. (a) $r = 0.326$
(b) $r = 0.558$
(c) The number of shots would be a better predictor
of goals given up, because the correlation coefficient is higher.

21. $r = 0.970$

23. (a) $r = -0.884$
(b) Yes, because there is a strong negative
correlation.

25. $r = -0.940$
This makes sense, because many students take only
one of these two exams. As the percentage that take
one of the exams increases, the percentage that
takes the other exam decreases.

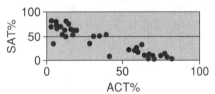

27. (a) $r = 0.804$
(b) $H_0: \rho = 0$
$H_A: \rho \neq 0$
$d.f. = 23$
Reject H_0 if $t > 2.807$ or if $t < -2.807$.
$t = 6.484$
Reject H_0. There is sufficient sample evidence to
support the claim.
p-value: Less than 0.01
(c) $r = 0.715$
(d) $H_0: \rho = 0$
$H_A: \rho \neq 0$
$d.f. = 98$
Reject H_0 if $t > 2.660$ or if $t < -2.660$.
$t = 10.124$
Reject H_0. There is sufficient sample evidence to
support the claim.
p-value: Less than 0.01
(e) The larger sample size makes the sample results
more outstanding.

SECTION 10.2

1. (a) $\hat{y} = 55.928 + 1.3452x$
(b) 66.6896 inches
(c) 63.9992 inches
(d) 85.5224 inches

(e) It is inappropriate to use this equation because it
is based on shoe sizes that are much lower than
size 22.

3. (a) $\hat{y} = 46.73 + 0.4672x$
(b) 84.106
(c) 72.426
(d) 49.807

5. (a) $\hat{y} = 153.03 + 0.1506x$
(b) \$243.39 million

7. (a) $\hat{y} = 921.39 + 0.3509x$
(b) $\hat{y} = 1408.8 + 0.1733x$
(c) The y-intercept increased by nearly 500 points,
while the slope was approximately cut in half.

9. (a) $\hat{y} = -125.56 + 6.1788x$
(b) 277.91564 or \$277,915.64
The residual is approximately 32.9, meaning
that the predicted value was 32.9 above the
actual value.
(c) 321.78512 or \$321,785.12
The residual is approximately -85.2, meaning
that the predicted value was 85.2 below the
actual value.

11. (a) $\hat{y} = 0.8682 + 0.0014x$ (b) 0.8857 or 88.57%
(c) The slope, 0.0014, means that for each additional second that it takes for the Web site to
appear, the predicted value of reliability
increases by 0.0014 or 0.14%.

13. (a) $\hat{y} = -10.637 + 3.2007x$
(b) 8.5672 or approximately \$8.60
(c) The slope, 3.2007, means that for each additional horse in the race, the predicted winning
payout increases by 3.2007 or approximately
\$3.20.

15. (a) $\hat{y} = 17217 + 6.1702x$ (b) 61,383.2916 beds
(c) Nevada has more complaints listed than beds.
Another state in this situation is New Mexico.
(d) 41,996.5232 beds
(e) Ohio has significantly more beds than complaints. This is the situation for Illinois and
New York as well.

17. $\hat{y} = 34.92 + 0.9337x$

19. (a) $\hat{y} = 1145.4 - 2.165x$
(b) The predicted value for 49% is 1039.315 points.
Her students are scoring approximately 28
points below the predicted score.

21. (a) $\hat{y} = 62.613 + 0.4714x$
(b) The equation supports the idea. The positive
slope indicates that as spending increases, the
number of wins increases.

23. (a) $\hat{y} = 112.73 + 3.2177x$
(b) The slope, 3.2177, indicates that each hit in a
game extends the length of the game by 3.2177
minutes.
(c) $\hat{y} = 145.41 + 2.6319x$
(d) The slope, 2.6319, indicates that each run scored
in a game extends the length of the game by
2.6319 minutes.

SECTION 10.3

1. (a) $\hat{y} = 55.928 + 1.3452x$
    (b) $s_e = 0.840$
3. (a) $\hat{y} = 46.73 + 0.4672x$
    (b) $s_e = 8.655$
5. (a) $\hat{y} = 153.03 + 0.1506x$
    (b) $s_e = 14.122$
    (c) $H_0: \beta = 0$
    $H_A: \beta \neq 0$
    $d.f. = 8$
    Reject H_0 if $t > 2.306$ or if $t < -2.306$.
    $t = 4.542$
    Reject H_0. There is sufficient sample evidence to support the claim.
    p-value: Less than 0.01
7. (a) $\hat{y} = -125.56 + 6.1788x$ (b) $s_e = 32.012$
    (c) $H_0: \beta = 0$
    $H_A: \beta \neq 0$
    $d.f. = 14$
    Reject H_0 if $t > 2.145$ or if $t < -2.145$.
    $t = 10.160$
    Reject H_0. There is sufficient sample evidence to support the claim.
    p-value: Less than 0.01
    (d) $d.f. = 14$
    $6.1788 - 1.3044 \leq \beta \leq 6.1788 + 1.3044$
    $\qquad 4.8744 \leq \beta \leq 7.4832$
    We are 95% confident that the population regression slope is somewhere between 4.8744 and 7.4832.
9. (a) $\hat{y} = 0.8682 + 0.0014x$ (b) $s_e = 0.055$
    (c) $H_0: \beta = 0$
    $H_A: \beta \neq 0$
    $d.f. = 8$
    Reject H_0 if $t > 2.306$ or if $t < -2.306$.
    $t = 0.382$
    Fail to reject H_0. There is not sufficient sample evidence to support the claim.
    p-value: Greater than 0.5
    (d) $d.f. = 8$
    $0.0014 - 0.0085 \leq \beta \leq 0.0014 + 0.0085$
    $\qquad -0.0071 \leq \beta \leq 0.0099$
    We are 95% confident that the population regression slope is somewhere between -0.0071 and 0.0099.
11. (a) $\hat{y} = -10.637 + 3.2007x$ (b) 8.5672
    (c) $s_e = 17.441$ $d.f. = 17$
    $8.5672 - 38.1862 \leq y \leq 8.5672 + 38.1862$
    $\qquad -29.619 \leq y \leq 46.7534$
13. (a) $\hat{y} = 17217 + 6.1702x$ (b) 61,383.2916 beds
    (c) $s_e = 29,283.1472$ $d.f. = 49$
    $61,383.2916 - 60,608.5595 \leq y \leq 61,383.2916$
    $+ 60,608.5595$
    $\qquad 774.7321 \leq y \leq 121,991.8511$
    The interval does contain the actual number of beds (4250).

(d) 41,996.5232 beds
(e) $s_e = 29,283.1472$ $d.f. = 49$
$41,996.5232 - 59,803.3383 \leq y \leq 41,996.5232$
$+ 59,803.3383$
$\qquad -17,806.8151 \leq y \leq 101,799.8615$
The interval does not contain the actual number of beds (128,181).

15. (a) $\hat{y} = 112.73 + 3.2177x$ (b) $s_e = 19.608$
    (c) $H_0: \beta = 0$
    $H_A: \beta \neq 0$
    $d.f. = 69$
    Reject H_0 if $t > 2.000$ or if $t < -2.000$.
    $t = 6.729$
    Reject H_0. There is sufficient sample evidence to support the claim.
    p-value: Less than 0.1
    (d) $d.f. = 69$
    $3.2177 - 0.9563 \leq \beta \leq 3.2177 + 0.9563$
    $\qquad 2.2614 \leq \beta \leq 4.1740$
    (e) 12 Hits: 151.3424 minutes
    $d.f. = 69$
    $151.3424 - 39.9198 \leq y \leq 151.3424 + 39.9198$
    $\qquad 111.4226 \leq y \leq 191.2622$
17. (a) $\hat{y} = 125.79 + 2.6533x$ (b) $s_e = 17.564$
    (c) $H_0: \beta = 0$
    $H_A: \beta \neq 0$
    $d.f. = 60$
    Reject H_0 if $t > 2.000$ or if $t < -2.000$.
    $t = 6.907$
    Reject H_0. There is sufficient sample evidence to support the claim.
    p-value: Less than 0.1
    (d) $d.f. = 60$
    $2.6533 - 0.7683 \leq \beta \leq 2.6533 + 0.7683$
    $\qquad 1.8850 \leq \beta \leq 3.4216$
    (e) 12 Hits: 157.6296 minutes
    $d.f. = 60$
    $157.6296 - 35.8397 \leq y \leq 157.6296 + 35.8397$
    $\qquad 121.7899 \leq y \leq 193.4693$

SECTION 11.1

1. $H_0: \tilde{\mu} \leq 50$
    $H_A: \tilde{\mu} > 50$
    Reject H_0 if $x < 6$.
    $x = 5$
    Reject H_0. There is sufficient sample evidence to reject the claim.
    p-value: 0.0317
3. $H_0: \tilde{\mu} \leq 600,000$
    $H_A: \tilde{\mu} > 600,000$
    Reject H_0 if $x < 2$.
    $x = 1$
    Reject H_0. There is sufficient sample evidence to support the claim.
    p-value: 0.0352

5. H_0: $\tilde{\mu} \leq 150$
   H_A: $\tilde{\mu} > 150$
   Reject H_0 if $x < 3$.
   $x = 3$
   Fail to reject H_0. There is not sufficient sample evidence to support the claim.
   p-value: 0.1134

7. H_0: $\tilde{\mu} = 45$
   H_A: $\tilde{\mu} \neq 45$
   Reject H_0 if $x < 3$.
   $x = 3$
   Fail to reject H_0. There is not sufficient sample evidence to reject the claim.
   p-value: 0.0576

9. d = After − Before
   H_0: $\tilde{\mu}_d \leq 0$
   H_A: $\tilde{\mu}_d > 0$
   Reject H_0 if $x < 4$.
   $x = 1$
   Reject H_0. There is sufficient sample evidence to support the claim.
   p-value: 0.0001

11. d = Home − Away
    H_0: $\tilde{\mu}_d \leq 0$
    H_A: $\tilde{\mu}_d > 0$
    Reject H_0 if $x < 4$.
    $x = 3$
    Reject H_0. There is sufficient sample evidence to support the claim.
    p-value: 0.0461

13. d = Final − Midterm
    H_0: $\tilde{\mu}_d \leq 0$
    H_A: $\tilde{\mu}_d > 0$
    Reject H_0 if $x < 6$.
    $x = 4$
    Reject H_0. There is sufficient sample evidence to support the claim.
    p-value: 0.0059

15. d = Before − After
    H_0: $\tilde{\mu}_d \leq 0$
    H_A: $\tilde{\mu}_d > 0$
    Reject H_0 if $x < 4$.
    $x = 2$
    Reject H_0. There is sufficient sample evidence to support the claim.
    p-value: 0.0066

17. d = Before − After
    H_0: $\tilde{\mu}_d \geq 0$
    H_A: $\tilde{\mu}_d < 0$
    Reject H_0 if $x < 3$.
    $x = 4$
    Fail to reject H_0. There is not sufficient sample evidence to support the claim.
    p-value: 0.1937

SECTION 11.2

1. d = After − Before
   H_0: $\tilde{\mu}_d \leq 0$
   H_A: $\tilde{\mu}_d > 0$
   Reject H_0 if $W \leq 28$.
   $W = 1.5$
   Reject H_0. There is sufficient sample evidence to support the claim.
   p-value < 0.005

3. d = Home − Away
   H_0: $\tilde{\mu}_d \leq 0$
   H_A: $\tilde{\mu}_d > 0$
   Reject H_0 if $W \leq 21$.
   $W = 22$
   Fail to reject H_0. There is not sufficient sample evidence to support the claim.
   p-value > 0.05

5. d = Final − Midterm
   H_0: $\tilde{\mu}_d \leq 0$
   H_A: $\tilde{\mu}_d > 0$
   Reject H_0 if $W \leq 60$.
   $W = 25$
   Reject H_0. There is sufficient sample evidence to support the claim.
   p-value < 0.005

7. d = Before − After
   H_0: $\tilde{\mu}_d \leq 0$
   H_A: $\tilde{\mu}_d > 0$
   Reject H_0 if $W \leq 30$.
   $W = 6$
   Reject H_0. There is sufficient sample evidence to support the claim.
   p-value < 0.005

9. d = Before − After
   H_0: $\tilde{\mu}_d \geq 0$
   H_A: $\tilde{\mu}_d < 0$
   Reject H_0 if $W \leq 17$.
   $W = 19$
   Fail to reject H_0. There is not sufficient sample evidence to support the claim.
   p-value > 0.05

11. d = Before − After
    H_0: $\tilde{\mu}_d \geq 0$
    H_A: $\tilde{\mu}_d < 0$
    Reject H_0 if $W \leq 2$.
    $W = 0$
    Reject H_0. There is sufficient sample evidence to support the claim.
    p-value < 0.025

SECTION 11.3

1. #1: Morning test
   H_0: $\tilde{\mu}_1 = \tilde{\mu}_2$
   H_A: $\tilde{\mu}_1 \neq \tilde{\mu}_2$
   Reject H_0 if $z < -1.96$ or if $z > 1.96$.

$R = 517.5$
$\mu_R = 506$
$\sigma_R = 44.04$
$z = 0.26$
Fail to reject H_0. There is not sufficient sample evidence to reject the claim.
p-value $= 0.7948$

3. #1: Men
   $H_0: \tilde{\mu}_1 = \tilde{\mu}_2$
   $H_A: \tilde{\mu}_1 \neq \tilde{\mu}_2$
   Reject H_0 if $z < -2.575$ or if $z > 2.575$.
   $R = 159.5$
   $\mu_R = 234$
   $\sigma_R = 31.84$
   $z = -2.34$
   Fail to reject H_0. There is not sufficient sample evidence to reject the claim.
   p-value $= 0.0192$

5. #1: Students who attend
   $H_0: \tilde{\mu}_1 \leq \tilde{\mu}_2$
   $H_A: \tilde{\mu}_1 > \tilde{\mu}_2$
   Reject H_0 if $z > 2.33$.
   $R = 241$
   $\mu_R = 175$
   $\sigma_R = 17.08$
   $z = 3.86$
   Reject H_0. There is sufficient sample evidence to support the claim.
   p-value $= 0.0002$

7. #1: Style before the change
   $H_0: \tilde{\mu}_1 = \tilde{\mu}_2$
   $H_A: \tilde{\mu}_1 \neq \tilde{\mu}_2$
   Reject H_0 if $z < -1.96$ or if $z > 1.96$.
   $R = 1026.5$
   $\mu_R = 945$
   $\sigma_R = 70.99$
   $z = 1.15$
   Fail to reject H_0. There is not sufficient sample evidence to reject the claim.
   p-value $= 0.2502$

9. #1: Verbal
   $H_0: \tilde{\mu}_1 = \tilde{\mu}_2$
   $H_A: \tilde{\mu}_1 \neq \tilde{\mu}_2$
   Reject H_0 if $z < -2.575$ or if $z > 2.575$.
   $R = 1581$
   $\mu_R = 1620$
   $\sigma_R = 103.92$
   $z = -0.38$
   Fail to reject H_0. There is not sufficient sample evidence to reject the claim.
   p-value $= 0.7040$

11. #1: Math
    $H_0: \tilde{\mu}_1 \geq \tilde{\mu}_2$
    $H_A: \tilde{\mu}_1 < \tilde{\mu}_2$
    Reject H_0 if $z < -1.645$.
    $R = 1533$
    $\mu_R = 1596$

$\sigma_R = 93.69$
$z = -0.67$
Fail to reject H_0. There is not sufficient sample evidence to support the claim.
p-value $= 0.2514$

13. #1: Collected homework
    $H_0: \tilde{\mu}_1 = \tilde{\mu}_2$
    $H_A: \tilde{\mu}_1 \neq \tilde{\mu}_2$
    Reject H_0 if $z < -1.96$ or if $z > 1.96$.
    $R = 1351$
    $\mu_R = 1295$
    $\sigma_R = 79.85$
    $z = 0.70$
    Fail to reject H_0. There is not sufficient sample evidence to reject the claim.
    p-value $= 0.4840$

SECTION 11.4

1. #1: 21–29; #2: 30–39; #3: 40 and above
   $H_0: \tilde{\mu}_1 = \tilde{\mu}_2 = \tilde{\mu}_3$
   $H_A:$ At least one of the medians is different from the others.
   Reject H_0 if $H > 5.991$.
   $H = 14.142$
   Reject H_0. There is sufficient sample evidence to reject the claim.
   p-value: Less than 0.01

3. #1: Sorority; #2: Fraternity; #3: Neither
   $H_0: \tilde{\mu}_1 = \tilde{\mu}_2 = \tilde{\mu}_3$
   $H_A:$ At least one of the medians is different from the others.
   Reject H_0 if $H > 5.991$.
   $H = 6.123$
   Reject H_0. There is sufficient sample evidence to reject the claim.
   p-value: Between 0.02 and 0.05

5. #1: Round 1; #2: Round 2; #3: Round 3; #4: Round 4
   $H_0: \tilde{\mu}_1 = \tilde{\mu}_2 = \tilde{\mu}_3 = \tilde{\mu}_4$
   $H_A:$ At least one of the medians is different from the others.
   Reject H_0 if $H > 11.345$.
   $H = 21.253$
   Reject H_0. There is sufficient sample evidence to reject the claim.
   p-value: Less than 0.01

7. #1: Farm A; #2: Farm B; #3: Farm C
   $H_0: \tilde{\mu}_1 = \tilde{\mu}_2 = \tilde{\mu}_3$
   $H_A:$ At least one of the medians is different from the others.
   Reject H_0 if $H > 5.991$.
   $H = 0.151$
   Fail to reject H_0. There is not sufficient sample evidence to reject the claim.
   p-value: Greater than 0.1

9. #1: Fertilizer A; #2: Fertilizer B; #3: Fertilizer C
   $H_0: \tilde{\mu}_1 = \tilde{\mu}_2 = \tilde{\mu}_3$

H_A: At least one of the medians is different from the others.

Reject H_0 if $H > 5.991$.

$H = 0.230$

Fail to reject H_0. There is not sufficient sample evidence to reject the claim.

p-value: Greater than 0.1

SECTION 11.5

1. $r_s = 0.9424$
3. $r_s = 0.5771$
5. $r_s = 0.7424$
7. $\rho_s = 0.8415$
9. $\rho_s = 0.7063$
11. $r_s = 0.3278$
   H_0: $\rho_s = 0$
   H_A: $\rho_s \neq 0$
   Reject H_0 if $r_s < -0.450$ or $r_s > 0.450$.
   $r_s = 0.3278$
   Fail to reject H_0. There is not sufficient sample evidence to support the claim that there is a significant correlation.
   p-value: Greater than 0.1
13. (a) $r_s = 0.9242$
   H_0: $\rho_s = 0$
   H_A: $\rho_s \neq 0$
   Reject H_0 if $r_s < -0.794$ or $r_s > 0.794$.
   $r_s = 0.9242$
   Reject H_0. There is sufficient sample evidence to support the claim that there is a significant correlation.
   p-value: Less than 0.01
   (b) $r_s = 0.8958$
   No, the change is not drastic.
15. $r_s = 0.5417$

Answers to Unit Review Exercises

UNIT 1

1. (a) Descriptive   (b) Inferential   (c) Inferential
   (d) Inferential
2. (a) This would be a sample if we were interested in the mean amount spent on textbooks by all students at your school.
   (b) This would be a population if we were only interested in the mean amount spent on textbooks by all students in your classes.
3. (a) Numerical, discrete   (b) Categorical, ordinal
   (c) Categorical, nominal   (d) Numerical, continuous
4. (a) Obtain a list of all of the students at your school, and randomly generate numbers to decide which students to include in the sample. Ask them how many hours they study per week.
   (b) Obtain a list of all of the students at your school, and randomly select a number for which

student in the list to start with. Then select every 25th student (for example) from there. Ask them how many hours they study per week.
   (c) Select your students at a public location on campus. Ask them how many hours they study per week.
   (d) One way would be to break down your campus by gender and year in class, and then select students in such a way that your sample reflects the percentages at your school. Ask them how many hours they study per week.
   (e) Randomly select classes to survey, and then ask each student in those classes. Ask them how many hours they study per week.
   (f) Students may not tell the truth, or they may not accurately remember.

5. a.

Stem	Leaf
14	0 4
15	0 2 7 8
16	0 3 5 6 6
17	7 8
18	1
19	5
20	
21	0 1

(b)

Hits	Frequency
140–155	4
155–170	7
170–185	3
185–200	1
200–215	2

(c)

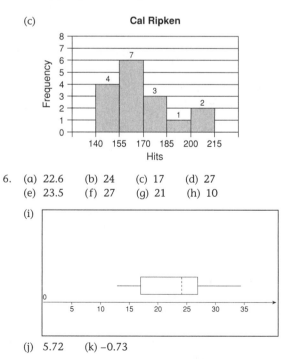

6. (a) 22.6   (b) 24   (c) 17   (d) 27
   (e) 23.5   (f) 27   (g) 21   (h) 10
   (i)
   (j) 5.72   (k) −0.73

7. (a) 70.09   (b) 115.64   (c) 13,372.61

(d)

Year	1990	1991	1992
Value	387	93	359
z	0.11	−2.43	−0.13
Year	1993	1994	1995
Value	356	375	472
z	−0.16	0.01	0.84
Year	1996	1997	1998
Value	396	454	478
z	0.19	0.69	0.90

8. $z = 1.58$; at least 60%

9. $x_g = 49.5$; $s_g = 10.10$

UNIT 2

1. No heads: 1; one head: 3; two heads: 3; three heads: 1
2. 2640
3. 538,200
4. (a) 65,536   (b) 4   (c) 6561
5. 324,632
6. (a) 32,760   (b) 1365   (c) 5460
7. 0.8
8. (a) 0.5   (b) 0.99   (c) 0.474552
9. 0.68
10. (a) 7/20   (b) 7/11   (c) 4/7   (d) 3/4
    (e) 2/3
11. 0.4
12. No
13. (a) 1/4   (b) 25/102
14. 1/11
15. (a) 4,394,000,000   (b) 3,450,000,000
16. 24
17. (a) 12,650   (b) 303,600   (c) 151,800
18. (a) 0.677   (b) 0.034
19. (a) 1/17   (b) 1/16
20. (a) 1/99   (b) 98/99
21. (a) 25/38   (b) 7/19   (c) 3/38   (d) 1/2
    (e) 15/31

UNIT 3

1. 0.9502	11. 0.9251
2. 0.2119	12. 0.3300
3. 0.7580	13. 66.648 inches
4. 0.2340	14. 0.0840
5. 0.8665	15. 0.9991
6. 31.336 inches	16. 0.9162
7. 0.1336	17. 0.0668
8. 0.1719	18. 0.4886
9. 0.4232	19. 0.2197
10. 0.9827	20. 0.0001

UNIT 4

1. $E = \$352.15$
   $\$253.85 \leq \mu \leq \$958.15$

2. $E = 0.6350$
   $2.0120 \leq \mu \leq 3.2820$
3. $H_0: \partial \leq 0.5$ $H_A: \partial > 0.5$
   Reject H_0 if $z > 1.645$.
   $z = 0.82$
   Fail to reject H_0. Fail to support the claim.
   p-value: 0.2061
4. $E = 0.0740$
   $0.3248 \leq \mu \leq 0.4728$
5. At least 385 people
6. $H_0: \mu \leq 10$ $H_A: \mu > 10$
   Reject H_0 if $t > 1.703$.
   $t = 2.993$
   Reject H_0. Support the claim.
   p-value: Less than 0.005
7. At least 865 people
8. $H_0: \mu \leq 15$ $H_A: \mu > 15$
   Reject H_0 if $z > 1.645$.
   $z = 2.25$
   Reject H_0. Support the claim.
   p-value: 0.0122

UNIT 5

1. Population 1: Elementary school children that attend private schools
   $H_0: \mu_1 \leq \mu_2$ $H_A: \mu_1 > \mu_2$
   Reject H_0 if $z > 2.33$.
   $z = 2.45$
   Reject H_0. Support the claim.
   p-value: 0.0071
2. H_0: A person's view of the economy is independent of his or her political affiliation.
   H_A: They are dependent.
   Reject H_0 if $\chi^2 > 5.991$.
   $\chi^2 = 28.799$
   Reject H_0. Reject the claim.
3. Population 1: Students who pass the prealgebra course
   $H_0: \mu_1 \leq \mu_2$ $H_A: \mu_1 > \mu_2$
   Reject H_0 if $z > 2.33$
   $z = 2.58$
   Reject H_0. Support the claim.
   p-value: 0.0049
4. Population 1: Professors at public four-year universities
   $H_0: \sigma_1^2 = \sigma_2^2$ $H_A: \sigma_1^2 \neq \sigma_2^2$
   Reject H_0 if $F > 1.53$.
   $F = 1.08$
   Fail to reject H_0. Fail to reject the claim.
5. $H_0: \mu_0 = 0.0625, \mu_1 = 0.25, \mu_2 = 0.375,$
   $\mu_3 = 0.25, \mu_4 = 0.0625$
   H_A: At least one proportion is different than claimed.
   Reject H_0 if $\chi^2 > 9.488$
   $\chi^2 = 24.747$
   Reject H_0. Reject the claim.

6. Population 1: Response times for first company
H_0: $\mu_1 = \mu_2$ H_A: $\mu_1 \neq \mu_2$
Reject H_0 if $t < -3.106$ or $t > 3.106$.
$t = -2.033$
Fail to reject H_0. Fail to support the claim.
p-value: Between 0.05 and 0.1

7. H_0: $\sigma_1^2 = \sigma_2^2$ H_A: $\sigma_1^2 \neq \sigma_2^2$
Reject H_0 if $F > 2.92$.
$F = 2.80$
Fail to reject H_0. Fail to reject the claim.

8. d = Swimsuit – Bodysuit
H_0: $\mu_d \leq 0$ H_A: $\mu_d > 0$
Reject H_0 if $t > 1.833$.
$t = 0.163$
Fail to reject H_0. Fail to support the claim.
p-value: Greater than 0.25

UNIT 6

1. (a)

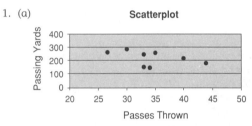

Scatterplot

(b) For these data the comment does not seem to
be accurate. As the number of passes increases,
the amount of passing yards appears to be
decreasing.
(c) -0.4980
(d) H_0: $\rho = 0$ H_A: $\rho \neq 0$
Reject H_0 if $t < -2.447$ or $t > 2.447$.
$t = -1.407$
Fail to reject H_0. Fail to support the claim.
p-value: Between 0.2 and 0.5

2. (a) $\hat{y} = 207.1776 + 11.1926x$
(b) 341.4888 calories

3. (a) $s_e = 53.843$
(b) H_0: $\beta = 0$ H_A: $\beta \neq 0$
Reject H_0 if $t < -2.201$ or $t > 2.201$.
$t = 8.066$
Reject H_0. Support the claim.
p-value: Less than 0.01
(c) $E = 3.0541$
$8.1385 \leq \beta \leq 14.2467$
(d) $E = 123.8892$
$251.1774 \leq y \leq 498.9558$

UNIT 7

1. H_0: $\tilde{\mu} = 2.0$ H_A: $\tilde{\mu} \neq 2.0$
Reject H_0 if $x < 13$.
$x = 10$
Reject H_0. Reject the claim.
p-value: 0.0022

2. H_0: $\tilde{\mu} \leq 10$ H_A: $\tilde{\mu} > 10$
Reject H_0 if $x < 6$.
$x = 3$
Reject H_0. Reject the claim.
p-value: 0.0038

3. d = After – Before
H_0: $\tilde{\mu}_d \leq 0$ H_A: $\tilde{\mu}_d > 0$
Reject H_0 if $x < 3$.
$x = 2$
Reject H_0. Support the claim.
p-value: 0.0193

4. #1: Drives with the wind
H_0: $\tilde{\mu}_1 \leq \tilde{\mu}_2$ H_A: $\tilde{\mu}_1 > \tilde{\mu}_2$
Reject H_0 if $z > 1.645$.
$R = 302$
$\mu_R = 232.5$
$\sigma_R = 24.11$
$z = 3.30$
Reject H_0. Support the claim.
p-value = 0.0005

5. #1: Students who do less than two hours of
homework each day
H_0: $\tilde{\mu}_1 \geq \tilde{\mu}_2$ H_A: $\tilde{\mu}_1 < \tilde{\mu}_2$
Reject H_0 if $z < -2.33$.
$R = 193.5$
$\mu_R = 208$
$\sigma_R = 17.66$
$z = -0.82$
Fail to reject H_0. Fail to support the claim.
p-value = 0.2061

6. #1: Model A; #2: Model B; #3: Model C;
#4: Model D
H_0: $\tilde{\mu}_1 = \tilde{\mu}_2 = \tilde{\mu}_3 = \tilde{\mu}_4$
H_A: At least one of the medians is different from
the others.
Reject H_0 if $H > 7.815$.
$H = 6.710$
Fail to reject the claim.
p-value: Between 0.05 and 0.10

7. $r_s = 0.7448$
H_0: $\rho_s = 0$ H_A: $\rho_s \neq 0$
Reject H_0 if $r_s < -0.591$ or $r_s > 0.591$.
$r_s = 0.7448$
Reject H_0. Support the claim.
p-value: Between 0.1 and 0.2

Index